Antibody Engineering

METHODS IN MOLECULAR BIOLOGY™

John M. Walker, Series Editor

270. **Parasite Genomics Protocols,** edited by *Stuart N. Isaacs, 2004*
269. **Vaccina Virus and Poxvirology:** *Methods and Protocols,* edited by *Stuart N. Isaacs, 2004*
268. **Public Health Microbiology:** *Methods and Protocols,* edited by *John F. T. Spencer and Alicia L. Ragout de Spencer, 2004*
267. **Recombinant Gene Expression:** *Reviews and Protocols, Second Edition,* edited by *Paulina Balbas and Argelia Johnson, 2004*
266. **Genomics, Proteomics, and Clinical Bacteriology**: *Methods and Reviews*, edited by Neil Woodford and Alan Johnson, 2004
265. **RNA Interference, Editing, and Modification:** *Methods and Protocols,* edited by *Jonatha M. Gott, 2004*
264. **Protein Arrays:** *Methods and Protocols,* edited by *Eric Fung, 2004*
263. **Flow Cytometry, Second Edition**, edited by *Teresa S. Hawley and Robert G. Hawley, 2004*
262. **Genetic Recombination Protocols**, edited by *Alan S. Waldman, 2004*
261. **Protein–Protein Interactions:** *Methods and Applications,* edited by *Haian Fu, 2004*
260. **Mobile Genetic Elements:** *Protocols and Genomic Applications,* edited by *Wolfgang J. Miller and Pierre Capy, 2004*
259. **Receptor Signal Transduction Protocols, Second Edition,** edited by *Gary B. Willars and R. A. John Challiss, 2004*
258. **Gene Expression Profiling:** *Methods and Protocols,* edited by *Richard A. Shimkets, 2004*
257. **mRNA Processing and Metabolism:** *Methods and Protocols,* edited by *Daniel R. Schoenberg, 2004*
256. **Bacterial Artifical Chromosomes,** *Volume 2: Functional Studies,* edited by *Shaying Zhao and Marvin Stodolsky, 2004*
255. **Bacterial Artifical Chromosomes,** *Volume 1: Library Construction, Physical Mapping, and Sequencing,* edited by *Shaying Zhao and Marvin Stodolsky, 2004*
254. **Germ Cell Protocols,** *Volume 2: Molecular Embryo Analysis, Live Imaging, Transgenesis, and Cloning,* edited by *Heide Schatten, 2004*
253. **Germ Cell Protocols,** *Volume 1: Sperm and Oocyte Analysis,* edited by *Heide Schatten, 2004*
252. **Ribozymes and siRNA Protocols, Second Edition,** edited by *Mouldy Sioud, 2004*
251. **HPLC of Peptides and Proteins:** *Methods and Protocols,* edited by *Marie-Isabel Aguilar, 2004*
250. **MAP Kinase Signaling Protocols,** edited by *Rony Seger, 2004*
249. **Cytokine Protocols,** edited by *Marc De Ley, 2004*
248. **Antibody Engineering:** *Methods and Protocols,* edited by *Benny K. C. Lo, 2004*
247. ***Drosophila* Cytogenetics Protocols,** edited by *Daryl S. Henderson, 2004*
246. **Gene Delivery to Mammalian Cells:** *Volume 2: Viral Gene Transfer Techniques,* edited by *William C. Heiser, 2004*
245. **Gene Delivery to Mammalian Cells:** *Volume 1: Nonviral Gene Transfer Techniques,* edited by *William C. Heiser, 2004*
244. **Protein Purification Protocols, Second Edition,** edited by *Paul Cutler, 2004*
243. **Chiral Separations:** *Methods and Protocols,* edited by *Gerald Gübitz and Martin G. Schmid, 2004*
242. **Atomic Force Microscopy:** *Biomedical Methods and Applications,* edited by *Pier Carlo Braga and Davide Ricci, 2004*
241. **Cell Cycle Checkpoint Control Protocols,** edited by *Howard B. Lieberman, 2004*
240. **Mammalian Artificial Chromosomes:** *Methods and Protocols,* edited by *Vittorio Sgaramella and Sandro Eridani, 2003*
239. **Cell Migration in Inflammation and Immunity:** *Methods and Protocols,* edited by *Daniele D'Ambrosio and Francesco Sinigaglia, 2003*
238. **Biopolymer Methods in Tissue Engineering,** edited by *Anthony P. Hollander and Paul V. Hatton, 2003*
237. **G Protein Signaling:** *Methods and Protocols,* edited by *Alan V. Smrcka, 2003*
236. **Plant Functional Genomics**: *Methods and Protocols,* edited by *Erich Grotewold, 2003*
235. ***E. coli* Plasmid Vectors**: *Methods and Applications,* edited by *Nicola Casali and Andrew Preston, 2003*
234. **p53 Protocols**, edited by *Sumitra Deb and Swati Palit Deb, 2003*
233. **Protein Kinase C Protocols**, edited by *Alexandra C. Newton, 2003*
232. **Protein Misfolding and Disease:** *Principles and Protocols,* edited by *Peter Bross and Niels Gregersen, 2003*
231. **Directed Evolution Library Creation**: *Methods and Protocols,* edited by *Frances H. Arnold and George Georgiou, 2003*
230. **Directed Enzyme Evolution**: *Screening and Selection Methods,* edited by *Frances H. Arnold and George Georgiou, 2003*
229. **Lentivirus Gene Engineering Protocols**, edited by *Maurizio Federico, 2003*
228. **Membrane Protein Protocols**: *Expression, Purification, and Characterization,* edited by *Barry S. Selinsky, 2003*

METHODS IN MOLECULAR BIOLOGY™

Antibody Engineering

Methods and Protocols

Edited by

Benny K. C. Lo

MRC Laboratory of Molecular Biology
University of Cambridge, Cambridge, UK

Humana Press **Totowa, New Jersey**

999 Riverview Drive, Suite 208
Totowa, New Jersey 07512

www.humanapress.com

This publication is printed on acid-free paper. ∞
ANSI Z39.48-1984 (American Standards Institute) Permanence of Paper for Printed Library Materials.

Cover design by Patricia F. Cleary.

Cover illustration: Figure 11C, Chapter 5, S. E. Harding et al.

For additional copies, pricing for bulk purchases, and/or information about other Humana titles, contact Humana at the above address or at any of the following numbers: Tel.: 973-256-1699; Fax: 973-256-8341; E-mail: humana@humanapr.com; or visit our Website: www.humanapress.com

Printed in the United States of America. 10 9 8 7 6 5 4 3 2 1

Library of Congress Cataloging in Publication Data
Antibody engineering : methods and protocols / edited by Benny K.C. Lo.
p. ; cm. — (Methods in molecular biology ; 248)
Includes bibliographical references and index.
ISBN 1-58829-092-1 (alk. paper) eISBN 1-59259-666-5
1. Immunotechnology—Laboratory manuals.
[DNLM: 1. Antibodies—therapeutic use—Laboratory Manuals. 2. Antibodies—genetics—Laboratory Manuals. 3. Protein Engineering—methods—Laboratory Manuals. QW 25 A6297 2004] I. Lo, Benny K. C. II. Series : Methods in molecular biology (Totowa, N.J.) ; v. 248.
TP248.65.I49A58 2004
616.07'98—dc22 2003020281

Preface

The exquisite binding specificity of antibodies has made them valuable tools from the laboratory to the clinic. Since the description of the murine hybridoma technology by Köhler and Milstein in 1975, a phenomenal number of monoclonal antibodies have been generated against a diverse array of targets. Some of these have become indispensable reagents in biomedical research, while others were developed for novel therapeutic applications. The attractiveness of antibodies in this regard is obvious—high target specificity, adaptability to a wide range of disease states, and the potential ability to direct the host's immune system for a therapeutic response. The initial excitement in finding Paul Ehrlich's "magic bullet," however, was met with widespread disappointment when it was demonstrated that murine antibodies frequently elicit the human anti-murine antibody (HAMA) response, thus rendering them ineffective and potentially unsafe in humans. Despite this setback, advances in recombinant DNA techniques over the last 15–20 years have empowered the engineering of recombinant antibodies with desired characteristics, including properties to avoid HAMA. The ability to produce bulk quantities of recombinant proteins from bacterial fermentation also fueled the design of numerous creative antibody constructs. To date, the United States Food and Drug Administration has approved more than 10 recombinant antibodies for human use, and hundreds more are in the development pipeline. The recent explosion in genomic and proteomic information appears ready to deliver many more disease targets amenable to antibody-based therapy. Without doubt, the continued use of antibodies in the 21st century is ensured by virtue of their powerful recognition properties and, as now demonstrated, by their successful partnership with protein engineering.

Antibody Engineering: Methods and Protocols presents cutting-edge techniques in antibody engineering research. In Part I, popular resources for antibody sequence analysis are described, together with in-depth discussions on antibody structural modeling. A directory summarizing useful websites relevant to antibody engineering is also included. Part II presents protocols for antibody lead generation from the cloning of immunoglobulin genes to the selection and generation of human recombinant antibodies by molecular display technologies and transgenic animals. For well-characterized murine antibodies with clinical potential, humanization by CDR grafting offers a proven solution to minimizing HAMA, while sparing the additional efforts in generating a completely new human antibody entity. Procedures are also described on reformatting antibody leads into monovalent, multivalent, and bispecific binding fragments for a wide range of in vivo applications. Part III focuses on the expression and optimization of antibody leads. Traditional antibody expression systems such as bacterial and mammalian cell culture are described, followed by more recent developments in insect cell cultures and transgenic plants. The use of plants is particularly important as it provides the scope for the mass production of antibodies at a fraction of the cost compared to conventional systems, hence making therapeutic antibodies more economical. Besides lead expression, chapters are also devoted to the in vitro affinity maturation of recombinant antibodies using phage display and a rational approach in the design of minimally immunogenic antibodies. Finally, Part IV details state-of-the-art technologies for the characterization of antigen-binding affinity and specificity. Some novel applications of recombinant antibodies in radioimmunotargeting, cancer immunotherapy, drug abuse, and the emerging field of proteomics are also presented. Although *Antibody Engineering: Methods and Protocols* cannot cover every facet of antibody engineering research, it is hoped that these chapters will provide the antibody engineer with the fundamental techniques upon which further imaginative technologies can be developed.

I am indebted to all the authors for their expert contributions. In addition, I thank my colleagues at the Laboratory of Molecular Biology for discussions and support, John Walker for editorial guidance, and Humana Press for publishing this book.

Benny K. C. Lo

Contents

Contributors

RICHARD BEERS • *Laboratory of Molecular Biology, National Cancer Institute, National Institutes of Health, Bethesda, MD*
TAPAN K. BERA • *Laboratory of Molecular Biology, National Cancer Institute, National Institutes of Health, Bethesda, MD*
Diane A. Blake • *Department of Ophthalmology, Tulane University School of Medicine, New Orleans, LA*
ROBERT C. BLAKE II • *Department of Basic Pharmaceutical Sciences, Xavier University, New Orleans, LA, and Tulane/Xavier Center for Bioenvironmental Research, New Orleans, LA*
ANDREW R. M. BRADBURY • *Biosciences Division, Los Alamos National Laboratory, Los Alamos, NM, and International School for Advanced Studies (SISSA), Trieste, Italy*
KEVIN BRADY • *School of Chemistry, University of Nottingham, Nottingham, United Kingdom*
BEATRIZ CARRASCO • *Departamento de Química Física, Facultad de Química, Universidad de Murcia, Murcia, Spain*
KEITH A. CHARLTON • *Haptogen Ltd., Polwarth Building, Foresterhill, Aberdeen, United Kingdom*
LESLIE CHASTEEN • *Biosciences Division, Los Alamos National Laboratory, Los Alamos, NM*
YU WAI CHEN • *Medical Research Council Centre for Protein Engineering, Medical Research Council Centre, Hills Road, Cambridge, United Kingdom*
KERRY A. CHESTER • *Cancer Research UK Targeting and Imaging Group, Department of Oncology, Royal Free and University College Medical School, London, United Kingdom*
NEIL COOLEY • *Discerna Ltd, Babraham Hall, Babraham, Cambridge, United Kingdom*
PAUL CHRISTOU • *Fraunhofer Institute for Molecular Biology and Applied Ecology, IME, Schmallenberg, Germany*
C. GEOFFREY DAVIS • *Abgenix Inc., Fremont, CA*
JOSE GARCIA DE LA TORRE • *Departamento de Química Física, Facultad de Química, Universidad de Murcia, Murcia, Spain*
ROBERTO DE PASCALIS • *Laboratory of Tumor Immunology and Biology, Center for Cancer Research, National Cancer Institute, National Institutes of Health, Bethesda, MD*
PALOMA DE PRADA • *Division of Clinical Pharmacology and Experimental Therapeutics, Department of Medicine, Columbia University, New York, NY*

XIAO FENG • *Abgenix Inc., Fremont, CA*
RAINER FISCHER • *Institute for Molecular Biology, RWTH Aachen, Aachen, Germany, and Fraunhofer Institute for Molecular Biology and Applied Ecology, IME, Schmallenberg, Germany*
IKUO FUJII • *Biomolecular Engineering Research Institute, Suita, Osaka, Japan*
NOREEN R. GONZALES • *Laboratory of Tumor Immunology and Biology, Center for Cancer Research, National Cancer Institute, National Institutes of Health, Bethesda, MD*
C. ROSAMUND L. GRAVES • *Division of Breast Surgery, University of Nottingham, City Hospital, Nottingham, United Kingdom*
JÜRGEN GRÜNBERG • *Center for Radiopharmaceutical Science ETH-PSI-USZ, Paul Scherrer Institute, Villigen, Switzerland*
MARY C. GUTTIERI • *Virology Division, United States Army Medical Research Institute of Infectious Diseases, Fort Detrick, Frederick, MD*
MARY HAAK-FRENDSCHO • *Abgenix, Inc., Fremont, CA*
STEPHEN E. HARDING • *National Centre for Macromolecular Dynamics, University of Nottingham, School of Biosciences, Loughborough, United Kingdom*
MINGYUE HE • *Discerna Ltd, Babraham Hall, Babraham, Cambridge, United Kingdom*
ALEXANDRA HUHALOV • *Cancer Research UK Targeting and Imaging Group, Department of Oncology, Royal Free and University College Medical School, London, United Kingdom*
ALISON JACKSON • *Discerna Ltd, Babraham Hall, Babraham, Cambridge, United Kingdom*
XIAO-CHI JIA • *Abgenix Inc., Fremont, CA*
GEORGE JOHNSON • *Department of Biomedical Engineering, Northwestern University, Evanston, IL*
ROBERT KARLSSON • *Biacore AB, Uppsala, Sweden*
SYED V. S. KASHMIRI • *Laboratory of Tumor Immunology and Biology, Center for Cancer Research, National Cancer Institute, National Institutes of Health, Bethesda, MD*
ROLAND E. KONTERMANN • *Vectron Therapeutics AG, Marburg, Germany*
TINA KORN • *Institut für Molekularbiologie und Tumorforschung, Marburg, Germany*
DONALD W. LANDRY • *Division of Clinical Pharmacology and Experimental Therapeutics, Department of Medicine, Columbia University, New York, NY*
ANITA LARSSON • *Biacore AB, Uppsala, Sweden*
MARIE-PAULE LEFRANC • *IMGT, Laboratoire d'ImmunoGénétique Moléculaire, LIGM, Université Montpellier II, UPR CNRS 1142, Institut de Génétique Humaine, IGH, Montpellier, France*
MIFANG LIANG • *Institute of Virology, Chinese Academy of Preventive Medicine, Beijing, People's Republic of China*
BENNY K. C. LO • *Division of Protein and Nucleic Acid Chemistry, Medical Research Council Laboratory of Molecular Biology, University of Cambridge, Cambridge, United Kingdom (Contact e-mail: bkclo@yahoo.co.uk)*

Emma Longman • *National Centre for Macromolecular Dynamics, School of Biosciences, University of Nottingham, United Kingdom*
James D. Marks • *Departments of Anesthesia and Pharmaceutical Chemistry, University of California, San Francisco, San Francisco General Hospital, San Francisco, CA*
Roberto Marzari • *University of Triste, Triste, Italy*
Sotiris Missailidis • *Department of Chemistry, The Open University, Milton Keynes, United Kingdom*
Andrea Murray • *School of Biomedical Sciences, University of Nottingham Medical School, Queen's Medical Centre, Nottingham, United Kingdom*
Ilse Novak-Hofer • *Center for Radiopharmaceutical Science ETH-PSI-USZ, Paul Scherrer Institute, Villigen, Switzerland*
Alvaro Ortega • *Departamento de Química Física, Facultad de Química, Universidad de Murcia, Murcia, Spain*
Milan Ovecka • *Biosciences Division, Los Alamos National Laboratory, Los Alamos, NM*
Ira Pastan • *Laboratory of Molecular Biology, National Cancer Institute, National Institutes of Health, Bethesda, MD*
Peter Pavlik • *Biosciences Division, Los Alamos National Laboratory, Los Alamos, NM*
Alan C. Perkins • *Department of Medical Physics, University of Nottingham, Queens Medical Center, Nottingham, United Kingdom*
Anthony R. Rees • *Syntem, Parc Scientifique Georges Besse, Nimes, France, and Centre for Protein Analysis and Design, University of Bath, Swindon, United Kingdom*
Ulrich Reineke • *Jerini AG, Berlin, Germany*
Daniele Sblattero • *University of Trieste, Trieste, Italy*
Roger Schibli • *Center for Radiopharmaceutical Science ETH-PSI-USZ, Paul Scherrer Institute, Villigen, Switzerland*
Stefan Schillberg • *Fraunhofer Institute for Molecular Biology and Applied Ecology, IME, Schmallenberg, Germany*
P. August Schubiger • *Center for Radiopharmaceutical Science ETH-PSI-USZ, Paul Scherrer Institute, Villigen, Switzerland*
Robert Siegel • *Pacific Northwest National Lab, Richland, WA*
Daniel I. R. Spencer • *Cancer Research UK Targeting and Imaging Group, Department of Oncology, Royal Free and University College Medical School, Royal Free Campus, Rowland Hill Street, London, United Kingdom*
Eva Stoger • *Institute for Molecular Biology, RWTH Aachen, Aachen, Germany*
Michael J. Taussig • *Technology Research Group, Babraham Institute, Cambridge, United Kingdom*
Richard M. Twyman • *Department of Biology, University of York, York, United Kingdom*
Nileena Velappan • *Biosciences Division, Los Alamos National Laboratory, Los Alamos, NM*

VITTORIO VERZILLO • *Biosciences Division, Los Alamos National Laboratory, Los Alamos, NM*

TINA VÖLKEL • *Vectron Therapeutics AG, Marburg, Germany*

ROBERT WAIBEL • *Center for Radiopharmaceutical Science ETH-PSI-USZ, Paul Scherrer Institute, Villigen, Switzerland*

NICHOLAS WHITELEGG • *Centre for Protein Analysis and Design, University of Bath, Swindon, United Kingdom*

ANNA M. WU • *Crump Institute for Molecular Imaging, Department of Molecular and Medical Pharmacology, David Geffen School of Medicine at UCLA, Los Angeles, CA*

TAI TE WU • *Department of Biomedical Engineering, Northwestern University, Evanston, IL*

PAUL J. YAZAKI • *Beckman Research Institute of the City of Hope National Medical Center, Duarte, CA*

KURT ZIMMERMANN, *Center for Radiopharmaceutical Science ETH-PSI-USZ, Paul Scherrer Institute, Villigen, Switzerland*

I

Antibody Sequence and Structure

1

Internet Resources for the Antibody Engineer

Benny K. C. Lo and Yu Wai Chen

1. Introduction

The Internet contains a wealth of information and tools that are relevant to various aspects of antibody engineering. Here, we present a collection of useful websites and software that is specific to antibody structure analysis and engineering, as well as for general protein analysis. Although this survey is by no means complete, it represents a good starting point. This list is accurate at the time of writing (August 2003).

2. List of Websites

2.1. Antibody-Specific Sites

2.1.1. The Kabat Database (G. Johnson and T. T. Wu, 2002; http://www.kabatdatabase.com)

Created by E. A. Kabat and T. T. Wu in 1966, the Kabat database publishes aligned sequences of antibodies, T-cell receptors, major histocompatibility complex (MHC) class I and II molecules, and other proteins of immunological interest. A searchable interface is provided by the SeqhuntII tool, and a range of utilities is available for sequence alignment, sequence subgroup classification, and the generation of variability plots (*see* Chapter 2 for more details).

2.1.2. KabatMan (A. C. R. Martin, 2002; http://www.bioinf.org.uk/abs/simkab.html)

This is a web interface to make simple queries to the Kabat sequence database. For more complex cases, queries should be sent directly in the KabatMan SQL-like query language.

From: *Methods in Molecular Biology, Vol. 248: Antibody Engineering: Methods and Protocols*
Edited by: B. K. C. Lo © Humana Press Inc., Totowa, NJ

2.1.3. IMGT, the International ImMunoGeneTics Information System® (M. -P. Lefranc, 2002; http://imgt.cines.fr)

IMGT is an integrated information system that specializes in antibodies, T-cell receptors, and MHC molecules of all vertebrate species. It provides a common portal to standardized data that include nucleotide and protein sequences, oligonucleotide primers, gene maps, genetic polymorphisms, specificities, and two-dimensional (2D) and three-dimensional (3D) structures. IMGT includes three sequence databases *(IMGT/LIGM-DB, IMGT/MHC-DB, IMGT/PRIMER-DB),* one genome database *(IMGT/GENE-DB),* one 3D structure database *(IMGT/3Dstructure-DB),* and a range of web resources ("*IMGT* Marie-Paule page") and interactive tools (*see* Chapter 3 for more details).

2.1.4. V-BASE (I. M. Tomlinson, 2002; http://www.mrc-cpe.cam.ac.uk/vbase)

V-BASE is a comprehensive directory of all human antibody germline variable region sequences compiled from more than one thousand published sequences. It includes a version of the alignment software DNAPLOT (developed by Hans-Helmar Althaus and Werner Müller) that allows the assignment of rearranged antibody V genes to their closest germline gene segments.

2.1.5. Antibodies—Structure and Sequence (A. C. R. Martin, 2002; http://www.bioinf.org.uk/abs)

This page summarizes useful information on antibody structure and sequence. It provides a query interface to the Kabat antibody sequence data, general information on antibodies, crystal structures, and links to other antibody-related information. It also distributes an automated summary of all antibody structures deposited in the Protein Databank (PDB). Of particular interest is a thorough description and comparison of the various numbering schemes for antibody variable regions.

2.1.6. AAAAA—AHo's Amazing Atlas of Antibody Anatomy (A. Honegger, 2001; http://www.unizh.ch/~antibody)

This resource includes tools for structural analysis, modeling, and engineering. It adopts a unifying scheme for comprehensive structural alignment of antibody and T-cell-receptor sequences, and includes Excel macros for antibody analysis and graphical representation.

2.1.7. WAM—Web Antibody Modeling (N. Whitelegg and A. R. Rees, 2001; http://antibody.bath.ac.uk)

Hosted by the Centre for Protein Analysis and Design at the University of Bath, United Kingdom.

Based on the AbM package (formerly marketed by Oxford Molecular) to construct 3D models of antibody Fv sequences using a combination of established theoretical methods, this site also includes the latest antibody structural information. It is free for academic use (*see* Chapter 4 for more details).

2.1.8. Mike's Immunoglobulin Structure/Function Page (M. R. Clark, 2001; http://www.path.cam.ac.uk/~mrc7/mikeimages.html)

These pages provide educational materials on immunoglobulin structure and function, and are illustrated by many color images, models, and animations. Additional information is available on antibody humanization and Mike Clark's Therapeutic Antibody Human Homology Project, which aims to correlate clinical efficacy and anti-immunoglobulin responses with variable region sequences of therapeutic antibodies.

2.1.9. The Antibody Resource Page (The Antibody Resource Page, 2000; http://www.antibodyresource.com)

This site describes itself as the "complete guide to antibody research and suppliers." Links to amino acid sequencing tools, nucleotide antibody sequencing tools, and hybridoma/cell-culture databases are provided. It also includes information on commercial suppliers, which is particularly useful for searching multiple suppliers for antibodies to your antigen of interest.

2.1.10. The Recombinant Antibody Pages (S. Dübel, 2000; http://www.mgen.uni-heidelberg.de/SD/SDscFvSite.html)

This is a large collection of links and information on recombinant antibody technology and general immunology that provides links to companies that exploit antibody technology.

2.1.11. Humanization bY Design (J. Saldanha, 2000; http://people.cryst.bbk.ac.uk/~ubcg07s)

This resource provides an overview on antibody humanization technology (*see* Chapter 7). The most useful feature is a searchable database (by sequence and text) of more than 40 published humanized antibodies including information on design issues, framework choice, framework back-mutations, and binding affinity of the humanized constructs.

2.2. Primary Structure Analysis

2.2.1. ExPASy Molecular Biology Server (ExPASy, 2002; http://www.expasy.org)

This all-in-one portal provides links to many other protein sequence and structure analysis sites, and includes the following sections: Databases, Tools and Software, Education, Documentation, and Links. Of these, the proteomic tools and databases are the most useful.

2.3. Three-Dimensional Structure Analysis and Graphics

2.3.1. O (A. Jones, 2002; http://xray.bmc.uu.se/~alwyn/o_related.html; note that the "official WWW server for O": the O Files, http://www.imsb.au.dk/~mok/o, is now officially outdated).

Love it or, hate it, O is still *the* indispensable graphics tool for structure rebuilding and analysis among protein crystallographers. However, the learning curve is very steep.

2.3.2. Rasmol (Rasmol Home Page, 2000; http://www.umass.edu/microbio/rasmol/index2.htm)

For ease of use, there is no replacement for Roger Sayle's free program. This is a simple molecular graphics viewer that has an easy-to-use graphical interface. A newer version known as the Protein Explorer is gradually taking over (Eric Martz, 2002; http://molvis.sdsc.edu/protexpl/frntdoor.htm).

2.3.3. PyMOL (DeLano Scientific, 2002; http://pymol.sourceforge.net)

This is a relatively new development with the ambition to be *the* complete program to replace all other molecular graphics programs. It offers plenty of graphical features, such as an electron-density map and surface representations, includes an internal ray-tracer, and can produce publication-quality images.

2.3.4. WebLab ViewerLite (MSI, now Accelrys, 1999; http://molsim.vei.co.uk/weblab)

Another molecular graphics program with a graphical user interface, this resource offers good rendering output. Development of this program has come to a halt. ViewerLite is free, but the extended-version ViewerPro is commercial.

2.3.5. DeepView (Swiss-Pdb Viewer) (N. Guex and T. Schwede, 2002; http://ca.expasy.org/spdbv)

Swiss-PdbViewer is also a user-friendly graphics program that allows several proteins to be compared for structural alignments. It also offers many tools for structure analysis. Moreover, Swiss-PdbViewer is tightly linked to Swiss-Model, an automated homology modeling server (*see* **Subheading 2.5.1.**).

2.3.6. GRASP (Graphical Representation and Analysis of Structural Properties) (A. Nicholls; http://trantor.bioc.columbia.edu/grasp)

This is a highly original graphics program for the calculation and visualization of molecular properties. It is mostly used for analyzing electrostatic potentials and surface complementarities. Although it has a graphical user interface, this program is not easy to use. Both academic and industrial users must buy a license. It is only available on the Silicon Graphics platform.

2.3.7. Uppsala Software Factory (G. J. Kleywegt, 2002; http://xray.bmc.uu.se/~gerard/manuals)

Gerard Kleywegt's huge collection of programs for structure analysis and structure data handling offers many utilities and macros that can enhance the power of the graphics program O (*see* **Subheading 2.3.1.**).

2.4. Structural Analysis Databases

2.4.1. The Protein Data Bank (Research Collaboratory for Structural Bioinformatics, 2002; http://www.rcsb.org/pdb)

This is the single worldwide repository for the processing and distribution of 3D biological macromolecular structure data.

2.4.2. SCOP (Structural Classification of Proteins) (The SCOP authors, 2002; http://scop.mrc-lmb.cam.ac.uk/scop)

Originally developed by A. Murzin, S. Brenner, T. Hubbard, and C. Chothia, the SCOP database (hosted by the Medical Research Council Centre, Cambridge, UK) provides a detailed and comprehensive description of the structural and evolutionary relationships between all proteins with a known structure.

2.4.3. FSSP (Fold classification based on structure-structure alignment of proteins) (L. Holm, 1995; http://www.ebi.ac.uk/dali/fssp)

Developed by L. Holm and C. Sander, the FSSP database is based on exhaustive all-against-all 3D structure comparison of protein structures in the Protein Data Bank.

2.5. Homology Modeling and Docking

2.5.1. Swiss-Model (T. Schwede , M. C. Peitsch and N. Guex, 2002; http://www.expasy.org/swissmod)

This is a fully automated protein structure homology-modeling server, accessible via the ExPASy web server, or from the molecular graphics program DeepView (Swiss Pdb-Viewer; *see* **Subheading 2.3.5.**).

2.5.2. Modeller (A. Sali group, 2002; http://www.salilab.org/modeller/modeller.html)

Modeller is designed for homology or comparative modeling of protein 3D structures from a structure-based sequence alignment. This program, which has proven to be very popular among protein chemists, is a Unix-based program that is free for academic use.

2.5.3. CNS (Crystallography and NMR System) (Yale University, 2000; http://cns.csb.yale.edu)

This is a very popular structure refinement package for structural scientists that includes many tools for structure analysis. For modeling purposes, it offers effective energy minimization protocols, including conventional energy minimization and simulated annealing. The commercial version, CNX, is marketed by Accelrys (http://www.accelrys.com/products/cnx).

2.5.4. CCP4 (Collaborative Computational Project, Number 4) Suite (CCP4, 2002; http://www.ccp4.ac.uk)

Another very popular suite of programs among X-ray crystallographers, this suite consists of state-of-the-art utility programs covering all stages of protein crystallography. Among these, Refmac5 is a refinement program that offers structure idealization after homology model building.

2.5.5. XtalView (Scripps XtalView WWW Page, 2002; http://www.scripps.edu/pub/dem-web)

XtalView is another highly regarded complete package for X-ray crystallography developed by D. McRee et al. at the Scripps Research Institute. It features a graphical user interface, and is relatively easy to use. It is very well-documented, and is accompanied by a textbook. Although it is free for academic use, commercial users must contact dem@scripps.edu.

2.5.6. Dock (Kuntz group, 1997; http://dock.compbio.ucsf.edu)

This program, developed at the University of California, San Francisco, evaluates the chemical and geometric complementarity between a ligand and a receptor-binding site, and searches for favorable interacting orientations.

2.5.7. AutoDock (G. M. Morris, 2002; http://www.scripps.edu/pub/olson-web/doc/autodock/)

AutoDock is a suite of automated docking tools developed at the Scripps Research Institute, La Jolla, CA, that enables users to predict how small ligands bind to a receptor of known structure.

2.5.8. ICM-Dock (MolSoft, 2002; http://www.molsoft.com/products/modules/dock.htm)

ICM (Internal Coordinate Mechanics) uses an efficient and general global optimization method for structure design, simulation, and analysis. Within the

ICM-Main bundle, there is a module ICM-Dock that claims success in predicting protein-protein interactions and protein-ligand docking. **Note:** this is a commercial product.

2.6. Miscellaneous

2.6.1. Delphion (Delphion, Inc.; 2002; http://www.delphion.com)

This is an excellent gateway to information on granted U.S. and worldwide patents and patent applications. It requires mandatory registration and payment for selected services.

2

The Kabat Database and a Bioinformatics Example

George Johnson and Tai Te Wu

1. Introduction

In 1969, Elvin A. Kabat of Columbia University College of Physicians and Surgeons and Tai Te Wu of Cornell University Medical College began to collect and align amino acid sequences of human and mouse Bence Jones proteins and immunoglobulin (Ig) light chains. This was the beginning of the *Kabat Database.* They used a simple mathematical formula to calculate the various amino acid substitutions at each position and predict the precise locations of segments of the light-chain variable region that would form the antibody-combining site from a variability plot *(1)*. The *Kabat Database* is one of the oldest biological sequence databases, and for many years was the only sequence database with alignment information.

The *Kabat Database* was available in book form free to the scientific community starting in 1976 *(2)*, with an updated second edition released in 1979 *(3)*, third edition in 1983 *(4)*, fourth edition in 1987 *(5),* and fifth printed edition in 1991 *(6)*. Because of the inclusion of amino acid as well as nucleotide sequences of antibodies, T-cell receptors for antigens (TCR), major histocompatibility complex (MHC) class I and II molecules, and other related proteins of immunological interest, it became impossible to provide printed versions after 1991. In that same year, George Johnson of Northwestern University created a website to electronically distribute the database located temporarily at:

http://kabatdatabase.com

During the following decade, the *Kabat Database* had grown more than five times. Thanks to the generous financial support from the National Institutes of Health, access to this website had been free for both academic and commercial use.

With the completion of the human genome project as well as several other genome projects, scientific emphasis has gradually shifted from determining

From: *Methods in Molecular Biology, Vol. 248: Antibody Engineering: Methods and Protocols*
Edited by: B. K. C. Lo © Humana Press Inc., Totowa, NJ

more sequences to analyzing the information content of the existing sequence data. With regard to the *Kabat Database*, the collection and alignment of amino acid and nucleotide sequences of proteins of immunological interest has been progressing side-by-side with the ability to determine structure and function information from these sequences, from its very start.

1.1. Historical Analysis and Use

After the pioneering work of Hilschmann and Craig *(7)* on the sequencing of three human Bence Jones proteins, many research groups joined the effort of determining Ig light chain amino acid sequences. By 1970, there were 77 published complete or partial Ig light chain sequences: 24 human κ-I, 4 human κ-II, 17 human κ-III, 10 human λ-I, 2 human λ-II, 6 human λ-III, 5 human λ-IV, 2 human λ-V, 2 mouse κ-I, and 5 mouse κ-II proteins *(1)*. The invariant Cys residues were aligned at positions 23 and 88, the invariant Trp residue positioned at 35, and the two invariant Gly residues at positions 99 and 101. To align the variable region of kappa and lambda light chains, single-residue gaps were placed at positions 10 and 106A. Longer gaps were introduced between positions 27 and 28 (27A, 27B, 27C, 27D, 27E, and 27F) and between 97 and 98 (97A and 97B), which was later changed to between 95 and 96 (95A, 95B, 95C, 95D, 95E and 95F). A similar alignment technique with a different numbering system was introduced for the Ig heavy-chain variable regions *(8)*. The invariant Cys residues were located at positions 22 and 92, the Trp residue at position 36, and the two invariant Gly residues at positions 104 and 106.

The most important discovery to come from alignment of the Ig heavy- and light-chain sequences was the location of segments forming the antibody-combining site, known as the complementarity (initially called hypervariable)-determining regions (CDRs). Since different antibodies bind different antigens, numerous amino acid substitutions occur in these segments, leading to large, calculated variability values. The first variability plot of the 77 complete and partial amino acid sequences of human and mouse light chains showed three distinct peaks of variability, located between positions 24 to 34, 50 to 56, and 89 to 97 *(1)*. Three similar peaks were discovered in heavy chains at positions 31 to 35, 50 to 65, and 95 to 102. These six short segments were hypothesized to form the antigen-binding site and were designated as CDRL1, CDRL2, CDRL3 for light chains, and CDRH1, CDRH2, and CDRH3 for heavy chains, respectively.

Initial Ig three-dimensional (3D) X-ray diffraction experiments suggested that the six binding-site segments were indeed physically located on one side of the Ig macromolecule. Final verification of this theoretical prediction came after the development of hybridoma technology *(9)*. An anti-lysozyme monoclonal antibody F_{ab} fragment was co-crystallized with lysozyme *(10)*, and the combined 3D structure was determined by X-ray diffraction analysis. Several amino acid residues in each of the six CDRs of the antibody were found to be in direct contact with the antigen. As theoretically predicted, antibody specificity thus resided exclusively in the CDRs. During the past decade, designer antibodies have been constructed genetically by selecting these CDRs for their affinity for the target antigen.

By comparing the amino acid sequences of the CDRs as well the stretches of sequence that connect them, known as framework regions (FR), Kabat and Wu hypothesized that the Ig variable regions were assembled from short genetic segments ***(11,12)***. This hypothesis was verified experimentally by Bernard et al. ***(13)*** with the discovery of the J-minigenes, reminiscent of the switch peptide proposed by Milstein ***(14)***. The D-minigenes were soon identified as another component of the heavy-chain variable region ***(15,16)***. In addition, the idea of gene conversion ***(17)*** was proposed as a possible mechanism of antibody diversification, and appears to play a central role in chickens ***(18)***, and to a varying extent in humans, rabbits, and sheep.

For precisely aligned amino acid sequences of Ig heavy-chain variable regions, CDRH3 is defined as the segment from position 95 to position 102, with possible insertions between positions 100 and 101. The CDRH3-binding loop is the result of the joining of the V-genes, D-minigenes, and J-minigenes. This intriguing process has been studied extensively ***(19,20)***, and suggests the CDRH3 plays a unique role in conferring fine specificity to antibodies ***(21,22)***. Indeed, a particular amino acid sequence of CDRH3 is almost always associated with one unique antibody specificity. The CDRH3 sequences within the *Kabat Database* have further been analyzed by their length distributions ***(23)***, for which the length distributions of 2,500 complete and distinct CDRH3s of human, mouse, and other species were found to be more-or-less in agreement with the Poisson distribution. Interestingly, the longest mouse CDRH3 had a length of 19 amino acid residues, and that of human had 32 residues, and only one of them was shared by both species ***(24)***, suggesting that CDRH3 may be species-specific.

Because of the subtle differences between the variable regions of the Ig light and heavy chains, their alignment position numberings are independent. For example, in light chains, the first invariant Cys is located at position 23 and CDRL1 is from position 24 to 34—e.g., immediately after the Cys residue. However, in heavy chains, the invariant Cys is located at position 22 and CDRH1 is from position 31 to 35—e.g., eight amino residues after that Cys. Because of this important difference, the Kabat numbering systems are separate for Ig light and heavy chains. Attempts to combine these two numbering systems into one in other databases have resulted in the presence of many gaps and confusions. Similarly, variable regions of TCR alpha, beta, gamma, and

Table 1
FRs and CDRs of Antibody and TCR Variable Regions

FR or CDR	V_L	V_H	V_α	V_β	V_γ	V_δ
FR1	1–23	1–22	1–22	1–23	1–21	1–22
CDR1	24–34	31–35B	23–33	24–33	22–34	23–34A
FR2	35–49	36–49	34–47	34–49	35–49	35–49
CDR2	50–56	50–65	48–56	50–56	50–59	50–57
FR3	57–88	66–91	57–92	57–94	60–95	58–89
CDR3	89–97	95–102	93–105	95–107	96–107	90–105
FR4	98–107	103–113	106–116	108–116A	108–116C	106–116

delta chains are aligned using different numbering systems. The alignments are summarized in **Table 1,** with the locations of CDRs indicated.

1.2. Current Analysis and Use

There are approx 25,000 unique yearly logins to the website of the *Kabat Database* by immunologists and other researchers around the world. The website is designed to be simple to use by those who are familiar with computers and those who are not. A description of the tools currently available is shown in **Table 2.** We encourage researchers who use the database to share their suggestions for improving the access and searching tools.

A common but extremely important question asked by researchers is whether a new sequence of protein of immunological interest has been determined before and stored in the database. Without asking this simple question, one may encounter the following situation: a heavy-chain V-gene from goldfish was sequenced *(25)* and found to be nearly identical to some of the human V-genes. Subsequently, the authors suggested that it might be of human origin, possibly because of the extremely sensitive amplification method used in the study and minute contamination of the sample by human tissue.

Another common use of the database is to confirm the reading frame of an immunologically related nucleotide sequence. Comparing short segments of sequence with stored database sequences can easily identify inadvertent omission of a nucleotide in the sequencing gel. Of course, if the missing nucleotide is real, this can suggest the presence of a pseudogene. Researchers also use the website to calculate variability for groupings of similar sequences of interest. For example, the variability plots of the variable regions of the Ig heavy and light chains of human anti-DNA antibodies are shown in **Figs. 1** and **2.** These two plots seem to indicate that CDRH3 may contribute most to the binding of DNA.

In many instances, investigators would like to identify the germline gene that is closest to their gene of interest, as well as the classification of that par-

Table 2
Listing of Tools Available on the Kabat Database Website

Tool	Description
Seqhunt II	The *SeqhuntII* tool is a collection of searching programs for retrieving sequence entries and performing pattern matches, with allowable mismatches, on the nucleotide and amino acid sequence data. The majority of fields in the database are searchable—for example, a sequence's journal citation. Matching entries may be viewed as HTML files or downloaded and printed. Pattern matching results show the matching database sequence aligned with the target pattern, with differences highlighted.
Align-A-Sequence	The Align-A-Sequence tool attempts to programmatically align different types of user-entered sequences. Currently kappa and lambda Ig light-chain variable regions may be aligned using the program.
Subgrouping	The Subgrouping tool takes a user-entered sequence of either Ig heavy, kappa, or lambda light-chain variable region and attempts to assign it a subgroup designation based on those described in the 1991 edition of the database. In many cases the assignment is ambiguous because of a sequence's similarity to more than one subgroup.
Find Your Families	The Find Your Family tool attempts to assign a "family" designation to a user-entered sequence. The user-entered target sequence is compared to previously assembled groupings of sequences, based on sequence homology. Please note that the assigned family number is arbitrary, since the groupings usually change as new data is added to the database.
Current Counts	Current amino acid, nucleotide, and entry counts may be made for various groupings of sequences.
Variability	Variability calculations may be made over a user-specified collection of sequences. The distributions used to calculate the variability are also available for viewing and printing. Variability plots can be customized for scale, axis labels, and title, or downloaded for printing.

ticular gene to a specific family or subgroup. *SEQHUNT* ***(26)*** can pinpoint the sequence available in the database with the least number of amino acid or nucleotide differences.

The previous examples represent most of the current uses of the *Kabat Database* by immunologists and other scientists. However, many more detailed

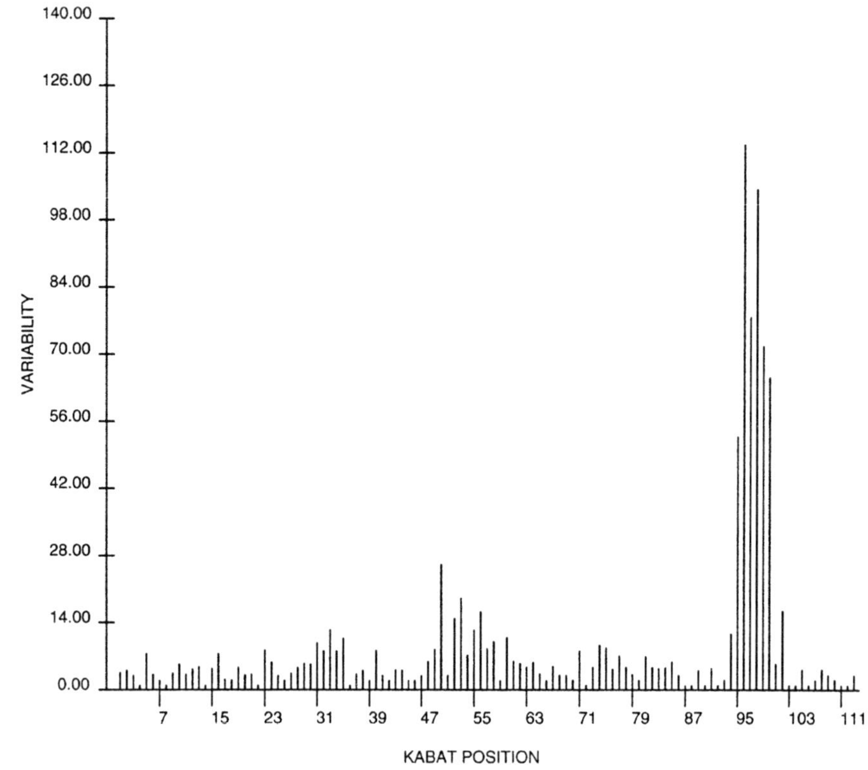

Fig. 1. Variability plot for human anti-DNA heavy-chain variable region.

analyses are possible from the data stored in the *Kabat Database,* as shown in **Table 3.**

In the following section, a current bioinformatics example is illustrated, using the uniquely aligned data contained in the *Kabat Database.*

2. Kabat Database Bioinformatics Example: HIV gp120 V3-loop and Human CDRH3 Amino Acid Sequences

The human immunodeficiency virus (HIV) has intrigued the scientific community for several decades. It is a retrovirus with two copies of RNA as its genetic material. Upon infecting humans, HIV uses its reverse-transcriptase molecules to convert its RNA into DNA, which are in turn transported into the nucleus and incorporated into the host chromosomes of CD4+ T cells. Although the infected individual produces antibodies against the initial viral strain, not all viruses can be eliminated because of the integration of its genetic material into the host cells. Gradually, the viral-coat proteins change in sequence, rendering the host's antibodies less effective. Eventually, acquired

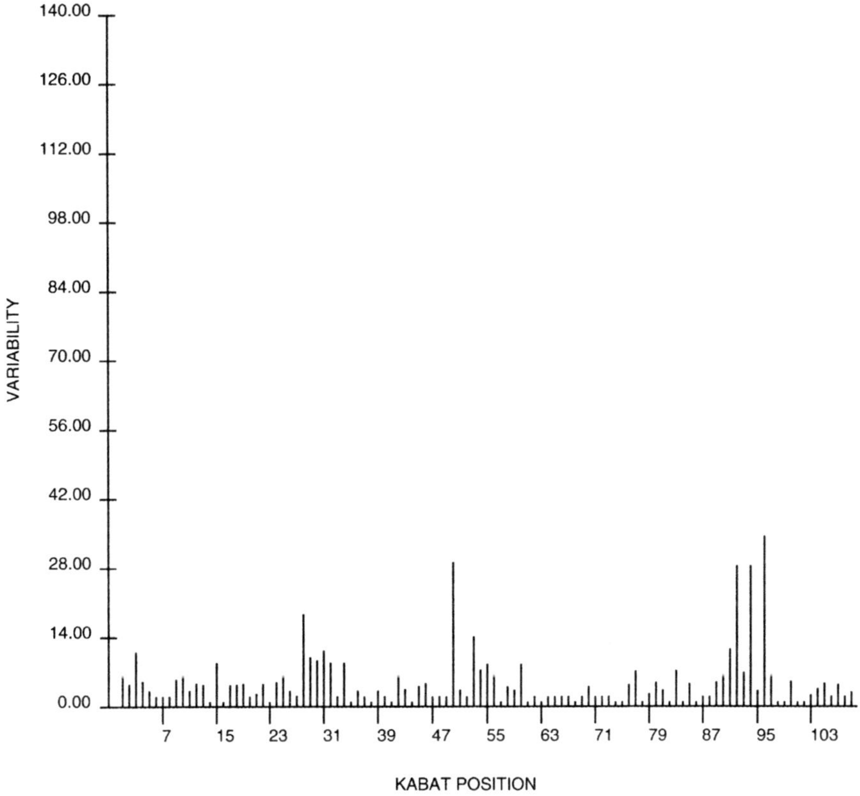

Fig. 2. Variability plot for human anti-DNA kappa light-chain variable region.

immunodeficiency syndrome (AIDS) develops with a latent period of approx 10 ± 3 yr. Because of this, HIV is classified as a lentivirus or slow virus.

Several specific drugs have been synthesized during recent years to treat HIV infection and AIDS. They include reverse-transcriptase inhibitors, protease inhibitors, and fusion inhibitors. However, these drugs have serious side effects, and most are very expensive, making the cost of treatment prohibitive in countries with a large percentage of HIV-positive patients. For years, the ideal solution has been to develop an inexpensive vaccine. Unfortunately, because of the rapid changes of its envelope coat proteins, especially gp120, HIV strains cannot be singled out as candidates for vaccine. Many research laboratories around the world have undertaken the task of sequencing gp120, and these sequences have been stored on two websites:

http://ncbi.nlm.nih.gov and http://www.lanl.gov

Figure 3 shows a variability plot for the 302 nearly complete sequences of HIV-1 stored at the latter site. For comparison, a variability plot of 138

Table 3
Partial Listing of Bioinformatics Studies Performed Using the Kabat Database

Subject	Summary
Binding Site Prediction	The CDRs of Ig heavy and light chains were predicted from variability calculations made over the sequence alignments ***(1,8)***.
Antibody Humanization	It is possible to identify the most similar framework regions between the mouse antibody and all existing human antibodies stored in the database ***(30)***.
Gene Count Estimation	From the existing sequences, it is possible to estimate the total number of human and mouse V-genes for antibody light and heavy chains, as well as TCR alpha and beta chains ***(31,32)***.
MHC Class I gene assortment	The known sequences of human MHC class I sequences suggest that their a1 and a2 regions can be assorted ***(33)***.
TCR CDR3 length distribution	The lengths of CDR3s in antibodies and TCRs have distinct features ***(34,35)***. In the case of TCR alpha and beta chains, their CDR3 lengths follow a narrow and random distribution. That may be a result of the relatively fixed size and shape of the processed peptide in the groove of MHC class I or II molecules. On the other hand, although the TCR gamma chain CDR3 lengths are similarly distributed, those of TCR delta chains exhibit a bimodal distribution ***(35)***. TCR delta chains with shorter CDR3s may be MHC-restricted, although those with longer CDR3s MHC-unrestricted.
Antibody and TCR evolution	Possible mechanisms of antibody and TCR evolution can also be investigated by comparing aligned sequences from different species ***(36,37)***.
Designer Antibodies	More specific/potent antibodies may be designed using the preferred CDR lengths calculated from database sequences against the same antigen ***(34)***.
Autoimmunity	Similarities between non-self antigens such as influenza virus and Ig autoantibodies have been found. Certain antigens may help initially trigger autoimmunity, and certain antibody clones may help to stimulate the autoimmune response ***(36)***.

aligned human influenza virus A hemagglutinin amino acid sequences is shown in **Fig. 4.**

Based on various studies, the V3-loop has been singled out for vaccine development. Although the V3-loop has the least amount of variation among

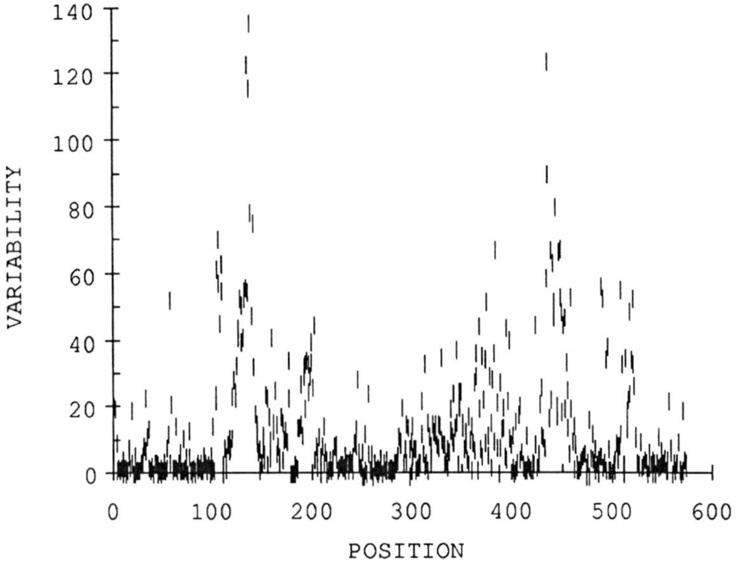

Fig. 3. Variability plot for HIV-1 gp120.

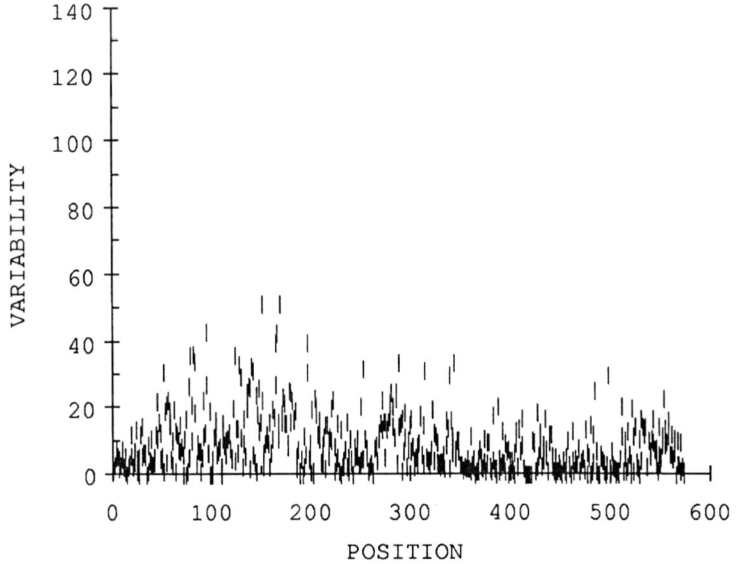

Fig. 4. Variability plot for influenza virus A hemagglutinin.

the five V-loops, there are still many different sequences from various strains of HIV. How these different sequences are related to the pathogenesis and progression of HIV infection is unclear. Longitudinal analysis of sequences of the V3-loop as the disease progresses is of vital importance in understanding the changes that occur during infection, so that an effective vaccine can be developed. Unfortunately, there is only one published report for a 10-yr sequence analysis, and in that case, the authors were unable to describe how the V3-loop amino acid sequences are related to disease progression *(27)*.

When HIV infects a person, its gp120 is a foreign protein and the patient produces antibodies toward this foreign antigen. However, once the HIV gene is integrated into the host chromosome, as in various human endogenous retroviruses, the gp120 becomes a self-protein. This transition from foreign to self usually cannot occur instantaneously, but as it occurs the host will have increasing difficulty producing effective antibodies. Indeed, initial antibodies from patients who are infected with HIV are usually ineffective in binding HIV at later stages of the disease.

The V3-loop has been described as being located on the surface of gp120. One way for the gp120 to become less antigenic would be for the virus to replace portions of the exposed V3-loop with segments of the host chromosome. Although any human protein could serve this purpose, we investigate the possibility that human CDRH3 regions are being used. CDRH3 is particularly attractive, because they can assume many possible configurations and they are on the surface of normal human proteins.

To locate matches between the V3-loop and CDRH3, the *Kabat Database* is uniquely useful. BLAST *(http://www.ncbi.nlm.nih.gov)* has recently allowed matches of short amino acid sequences, and eMOTIF *(http://emotif.stanford.edu/emotif/)* can be used to search for various length sequences. However, both programs use sequence databases containing large numbers of HIV-1 sequences and relatively few antibody heavy-chain variable region sequences. A search for short V3-loop sequences at these two websites usually results in a listing of other V3-loop sequences, and few, if any, CDRH3 sequences. By using the *SEQHUNTII* program, we picked the human heavy-chain variable regions and searched for all penta-peptides in the sequences of V3-loops determined in the 10-yr longitudinal study. The result of matching is listed in **Table 4.**

The initial number of matches is gradually reduced over the years, until the CD4+ T-cell count drops below 200. At that time, the number of matches increases dramatically. The match number appears to closely correlate with the number of HIV RNA molecules in the patient's blood. For example, after treatment, the number of matches drops to zero, along with a reduction in the plasma HIV RNA number. Subsequently, after 10 yr of HIV infection, the number of matches begins to creep up again.

A possible explanation for this finding is that the presence of CDRH3 pentapeptides in the V3-loop reduces its antigenicity. Such mutant HIV would bind existing anti-HIV antibodies in the patient less effectively, becoming more pathogenic. Based on this observation, the use of amino acid or nucleotide sequences of V3-loop as a vaccine would not be very efficient.

Table 4
Longitudinal Study of HIV gp120 V3-Loop Sequence Variations

Sample	Months after Infection	Sequence of V3-loop determined	Matches in human CDRH3	CDR4+ T-cells	HIV RNA per mL of plasma
A1	0	10	6		230
A2	12	10	3		230
A2b	27	7	0	427	2,300
A3	42	5	0	277	230
A4	70	3	0	186	230
A5	94	12	21	156	23,000
treatment	97				
A6	110	12	0	248	2,300
A7	118	12	1	212	2,300

An effective vaccine would most likely be made from an area of the exposed surface that does not contain high variability, as indicated in **Fig. 3.** There are several segments of seven or more nearly invariant amino acid residues in HIV gp120, in contrast to influenza virus hemagglutinin. Nearly invariant residues are defined as those that occur more than about 95% of the time at a particular position *(1)*. They are located at the following positions (numbering including the precursor region) in the C1, C2, or C5 region of gp120:

Segment #	Position #	Sequence
I	4 to 14	WVTVYYGVPVW
II	23 to 30	LFCASDA
III	44 to 50	ACVPTDP
IV	225 to 231	PIPIHYC
V	261 to 267	VQCTHGL
VI	269 to 282	PVVSTQLLL-NGSL
VII	538 to 545	ELYKYKVV

Some of the adjacent residues occur more than 90% of the time. Furthermore, segments II and III and segments VI and V form disulfide bonds. Segment VI is only one residue away from segment V, and that residue is either K or R most of the time. Segment I is near the N-terminal and segment VII near the C-terminal, and they are physically located near each other in the folded structure of gp120 *(28)*. If these segments are indeed located on the surface of gp120, we may then suggest that segment I linked to segment VII—with linkers consisting of repeats of GGGGS, segment II disulfide bounded to segment

III, and segment IV S-S bounded to segment V joined to segment VI with an intervening residue of K or R—should be used as possible peptide vaccine candidates. Additional residues that occur more than 90% of the time may also be included in these segments, suggesting the following three possible peptides:

```
WVTVYYGVPVWGGGSGGGSDNWRSELYKYKVV,

              LFCASDAK
                 |
           WATHACVPTDP, and

            PIPIHYC
                  |
                 VQCTHGIKPVVSTQLLLNQSL.
```

In contrast, for influenza virus hemagglutinin amino acid sequences, no such segments of seven or more residues are found.

3. Future Directions

As previously discussed, during the past few years a substantial decline in the number of published sequences of proteins of immunological interest has occurred. With the shift in focus from brute-force data collection to in-depth analysis and "data mining" by various researchers, well-characterized data sets have become extremely important. Each entry in the database inherently contains a large amount of bioinformatic analysis such as alignment information, the relationship between gene sequence and protein sequence, and coding region designation. These relationships prove most valuable in allowing researchers to ask more intuitive, abstract questions than would be possible with most unaligned, raw sequence databases. We continue to locate, annotate, and align sequences found in the published literature. Periodically, the database and website are updated to reflect inclusion of the new data. Corrections of errors found in the sequence data by us and by database users are constantly made, ensuring the collection's accuracy. We continue to explore new ways of relating the database entries, such as incorporating links to journal abstracts, links to 3D structural information, and germline gene assignment.

We continue to create and develop software programs for performing various analyses of the data. We are in the process of converting many tools we have used into Java and adding graphical interfaces. Two major groupings of tools are currently being created: the first to update and extend the current entry retrieval tools (such as SeqhuntII), and the second to perform distribution analyses on entire groups of sequences (such as variability). Java tools for locating sequences based on pattern matching, length distribution of a specified region, positional examination of a codon or residue, and sequence length have been developed and are undergoing testing. Many of the studies we have performed on the database require tools for grouping and analyzing collections of sequences rather than each one individually. We are developing a Java interface for creating distributions based on position (used most frequently for calculating variability), region length (used in length distribution analyses), and sequence pattern (used in gene count estimations and various homology studies). Together, these powerful interfaces will allow researchers to quickly perform many complex bioinformatics studies on the aligned sequence data and combine their results.

4. Conclusion

The fundamental reason for creating and maintaining most sequence databases is to study and correlate a protein's primary sequence structure with its 3D structure. Although there are many proteins with known 3D structures, there are probably two orders of magnitude more proteins with known amino acid or nucleotide sequences. In the 1950s, Anfinsen proposed and summarized in his 1973 paper ***(29)*** that the primary sequence of a protein should determine its 3D folding. Unfortunately, we still do not know how to decipher this information.

In the long run, the Kabat Database must be self-sustained. However, the transition from a free NIH-supported database to a self-sustaining format will take time and continued investigator interest. For example, it is hoped that the rapid development of therapeutic antibody techniques, using chimeric or humanized approaches, will eventually lead to the *de novo* synthesis of designer antibodies. Thus, immunotherapy for cancers and viral infections may rely heavily on the *Kabat Database* collections.

We will also rely on users to suggest to us what basic immunological ideas, what computer programs, and which types kinds of structure and function information will be of importance for future studies in this central problem in biomedicine. This feedback from users is of primary importance to the existence of the *Kabat Database.*

References

1. Wu, T. T. and Kabat, E. A. (1970) An analysis of the sequences of the variable regions of Bence Jones proteins and myeloma light chains and their implications for antibody complementarity. *J. Exp. Med.* **132,** 211–250.
2. Kabat, E. A., Wu, T. T., and Bilofsky, H. (1976) *Variable Regions of Immunoglobulin Chains.* Bolt Beranek and Newman Inc., Cambridge, MA.
3. Kabat, E. A., Wu, T. T., and Bilofsky, H. (1979) *Sequences of Immunoglobulin Chains.* NIH Publication No. 80–2008, Bethesda, MD.
4. Kabat, E. A., Wu, T. T., Bilofsky, H., Reid-Miller, M., and Perry, H. (1983) *Sequences of Proteins of Immunological Interest.* NIH Publication No. 369–847, Bethesda, MD.

5. Kabat, E. A., Wu, T. T., Reid-Miller, M., Perry, H., and Gottesman, K. (1987) *Sequences of Proteins of Immunological Interest,* 4th ed., U. S. Govt. Printing Off. No. 165–492, Bethesda, MD.
6. Kabat, E. A., Wu, T. T., Perry, H., Gottesman, K., and Foeller, C. (1991) *Sequences of Proteins of Immunological Interest,* 5th ed., NIH Publication No. 91–3242, Bethesda, MD.
7. Hilschmann, N., and Craig, L. C. (1965) Amino acid sequence studies with Bence Jones proteins. *Proc. Natl. Acad. Sci. USA* **53,** 1403–1409.
8. Kabat, E. A. and Wu, T. T. (1971) Attempts to locate complementarity-determining residues in the variable portions of light and heavy chains. *Ann. NY Acad. Sci.* **190,** 382–393.
9. Kohler, G. and Milstein, C. (1975) Continuous cultures of fused cells secreting antibody of predefined specificity. *Nature* **256,** 495–497.
10. Amit, A. G., Mariussa, R. A., Phillips, S. E., and Poljak, R. J. (1986) Three-dimensional structure of antigen-antibody complex at 2.8 A resolution. *Science* **233,** 747–753.
11. Wu, T. T., Kabat, E. A., and Bilifsky, H. (1975) Similarities among hypervariable segments of immunoglobulin chains. *Proc. Natl. Acad. Sci. USA* **72,** 5107–5110.
12. Kabat, E. A., Wu, T. T., and Bilofsky, H. (1978) Variable region genes for immunoglobulin framework are assembled from small fragments of DNA—a hypothesis. *Proc. Natl. Acad. Sci. USA* **75,** 2429–2433.
13. Bernard, O., Hozumi, N., and Tonegawa, S. (1978) Sequences of mouse light chain genes before and after somatic changes. *Cell* **15,** 1133–1144.
14. Milstein, C. (1967) Linked groups of residues in immunoglobulin chains. *Nature* **216,** 330–332.
15. Early, P., Huang, H., Davis, M., Calame, K., and Hood, L. (1980) An Immunoglobulin heavy chain variable gene is generated from three segments of DNA: VH, DH, and JH. *Cell* **19,** 981–992.
16. Sakano, H., Maki, R., Kurosawa, Y., Roeder, W., and Tonegawa, S. (1980) Two types of somatic recombinations are necessary for the generation of complete heavy chain genes. *Nature* **286,** 676–683.
17. Baltimore, D. (1981) Gene conversion: some implications for immunoglobulin genes. *Cell* **24,** 592–594.
18. Reynaud, C., Anquez, V., Dahan, A., and Weill, J. (1985) A single rearrange event generates most of the chicken immunoglobulin light chain diversity. *Cell* **40,** 283–291.
19. Desiderio, S. V., Yancopoulos, G. D., Paskind, M., Thomas, E., Boss, M. A., Landau, N., et al. (1984) Insertion of N regions into heavy-chain genes is correlated with expression of terminal deoxytransferase in B cells. *Nature* **311,** 752–755.
20. Sleckman, B. P., Gorman, J. R., and Alt, F. W. (1996) Accessibility control of antigen-receptor variable-region gene assembly: role of cis-acting elements. *Annu. Rev. Immunol.* **14,** 459–481.
21. Kabat, E. A. and Wu, T. T. (1991) Indentical V-region amino acid sequences and segments of sequences in antibodies of different specificities: relative contributions

of VH and VL genes, minigenes and CDRs to binding of antibody combining sites. *J. Immunol.* **147,** 1709–1819.
22. Wu, T. T. (1994) From esoteric theory to therapeutic antibodies. *Appl. Biochem. Biotechnol.* **47,** 107–118.
23. Wu, T. T., Johnson, G., and Kabat, E. A. (1993) Length distribution of CDRH3 in antibodies. *Proteins* **16,** 1–7.
24. Wu, T. T. (2001) *Analytical Molecular Biology.* Kluwer Academic Publishers, Norwell, MA.
25. Wilson, M. R., Middleton, D., and Warr, G. W. (1988) Immunoglobulin heavy chain variable region gene evolution: structure and family relations of two genes and a pseudogene in a teleost fish. *Proc. Natl. Acad. Sci. USA* **85,** 1566–1570; and (1989) Erratum. *Proc. Natl. Acad. Sci. USA* **86,** 3276.
26. Johnson, G., Wu, T. T., and Kabat, E. A. (1995) SEQHUNT, a program to search aligned nucleotide and amino acid sequences, in *Antibody Engineering Protocols* (Paul, S., ed.), Humana Press, Totowa, NJ, pp. 1–15.
27. Janssens, W., Nkengasong, J., Heyndricks, L. van der Auwera, G., Vereecken, K., Coppens, S., et al. (1999) Intrapatient variability of HIV type I group O ANT70 during a 10-year follow-up. *AIDS Res. Hum. Retrovir.* **15,** 1325–1332.
28. Wyatt, R., Kwong, P. D., Desjardins, E., Sweet, R. W., Robinson, J., Hendrickson, W. A., et al. (1998) The antigen structure of HIV gp120 envelope glycoprotein. *Nature* **393,** 705–711.
29. Anfinsen, C. B. (1973) Principles that govern the folding of protein chains. *Science* **181,** 223–230.
30. Wu, T. T. and Kabat, E. A. (1992) Possible use of similar framework region amino acid sequences between human and mouse immunoglobulins for humanizing mouse antibodies. *Mol. Immunol.* **29,** 1141–1146.
31. Johnson, G. and Wu, T. T. (1997a) A method of estimating the numbers of human and mouse immunoglobulin V-genes. *Genetics* **145,** 777–786.
32. Johnson, G. and Wu, T. T. (1997b) A method of estimating the numbers of human and mouse T cell receptor for antigen alpha and beta chain V-genes. *Immunol. Cell Biol.* **75,** 580–583.
33. Johnson, G. and Wu, T. T. (1998a) Possible assortment of a1 and a2 regiuon gene segments in human MHC class I molecules. *Genetics* **149,** 1063–1967.
34. Johnson, G. and Wu, T. T. (1998b) Preferred CDRH3 lengths for antibodies with defined specificities. *Int. Immunol.* **10,** 1801–1805.
35. Johnson, G. and Wu, T. T. (2000a) Kabat database and its applications: 30 years after the first variability plot. *Nucleic Acids Res.* **28,** 214–218.
36. Johnson, G. and Wu, T. T. (2000b) Matching amino acid and nucleotide sequences of mouse rheumatoid factor CDRH3-FRH4 segments to other mouse antibodies with known specificities. *Bioinformatics* **16,** 941–943.
37. Johnson, G. and Wu, T. T. (2001) Kabat database and its applications: future directions. *Nucleic Acids Res.* **29,** 205–206.

3

IMGT, The International ImMunoGeneTics Information System®, http://imgt.cines.fr

Marie-Paule Lefranc

1. Introduction

The molecular synthesis and genetics of the immunoglobulin (IG) and T-cell-receptor (TR) chains are particularly complex and unique, as they include biological mechanisms such as DNA molecular rearrangements in multiple loci (three for IG and four for TR in human) located on different chromosomes (four in human), nucleotide deletions and insertions at the rearrangement junctions (or N-diversity), and somatic hypermutations in the IG loci (for review, *see* **refs.** ***1*****,*****2***). The number of potential protein forms of IG and TR is almost unlimited. Because of the complexity and large number of published sequences, data control and classification and detailed annotations are a very difficult task for the general databanks such as *EMBL, GenBank,* and *DDBJ* ***(3–5)***. These observations were the starting point of *IMGT, the International ImMunoGeneTics Information System*® (http://imgt.cines.fr) ***(6)***, created in 1989 by the Laboratoire d'ImmunoGénétique Moléculaire (LIGM), at the Université Montpellier II, CNRS, Montpellier, France.

IMGT is a high-quality knowledge resource and integrated information system that specializes in IG, TR, major histocompatibility complex (MHC), and related proteins of the immune system (RPI) of humans and other vertebrates. *IMGT* provides a common access to standardized data that include nucleotide and protein sequences, oligonucleotide primers, gene maps, genetic polymorphisms, specificities, and two-dimensional (2D) and three-dimensional (3D) structures. *IMGT* includes three sequence databases *(IMGT/LIGM-DB, IMGT/MHC-DB, IMGT/PRIMER-DB),* one genome database *(IMGT/GENE-DB),* one 3D structure database *(IMGT/3Dstructure-DB),* Web resources ("*IMGT* Marie-Paule page"), and interactive tools *(IMGT/V-QUEST, IMGT/JunctionAnalysis, IMGT/Allele-Align, IMGT/PhyloGene, IMGT/Gene-*

From: *Methods in Molecular Biology, Vol. 248: Antibody Engineering: Methods and Protocols*
Edited by: B. K. C. Lo © Humana Press Inc., Totowa, NJ

Search, IMGT/GeneView, IMGT/LocusView, IMGT/Structural Query). IMGT expertly annotated data and *IMGT* tools are particularly useful in medical research (repertoire in leukemias, lymphomas, myelomas, translocations, autoimmune diseases, and acquired immunodeficiency syndrome [AIDS]), therapeutic approaches, and biotechnology related to antibody engineering. *IMGT* is freely available at http://imgt.cines.fr.

2. *IMGT* Databases

The *IMGT* databases comprise:

1. Three sequence databases: i) *IMGT/LIGM-DB* is a comprehensive database of IG and TR nucleotide sequences from human and other vertebrate species, with translation for fully annotated sequences, created in 1989 by LIGM, Montpellier, France, on the Web since July 1995 ***(6–10)***. In July 2003, *IMGT/LIGM-DB* contained 74,387 nucleotide sequences of IG and TR from 105 species. ii) *IMGT/MHC-DB* is hosted at the European Bioinformatics Institute (EBI) and comprises a database of the human MHC allele sequences (*IMGT/MHC-HLA,* developed by Cancer Research, UK and Anthony Nolan Research Institute, London, UK), on the Web since December 1998 ***(11)***, databases of MHC class II sequences from nonhuman primates (*IMGT/MHC-NHP,* curated by BPRC, the Netherlands), and from felines and canines (*IMGT/MHC-FLA* and *IMGT/MHC-DLA,* on the Web since April 2002. iii) *IMGT/PRIMER-DB* is an oligonucleotide primer database for IG and TR, developed by LIGM, Montpellier in collaboration with EUROGENTEC, Belgium.
2. One genome database: *IMGT-GENE-DB* allows a query per gene name.
3. One three-dimensional (3D) structure database: *IMGT/3Dstructure-DB* provides the *IMGT* gene and allele identification and Colliers de Perles of IG and TR with known 3D structures, created by LIGM, on the Web since November 2001 ***(12)***. In July 2003, *IMGT/3Dstructure-DB* contained 623 atomic coordinate files.

In the following sections, we describe in more detail *IMGT/LIGM-DB,* which is the first and largest IMGT database.

2.1. IMGT/LIGM-DB *Data*

IMGT/LIGM-DB sequence data are identified by the *EMBL/GenBank/DDBJ* accession number. The unique source of data for *IMGT/LIGM-DB* is *EMBL,* which shares data with the other two generalized databases *GenBank* and *DDBJ.* Once the sequences are allowed by the authors to be made public, LIGM automatically receives IG and TR sequences by e-mail from EBI. After control by LIGM curators, data are scanned to store sequences, bibliographical references, and taxonomic data, and standardized *IMGT/LIGM-DB* keywords are assigned to all entries. Based on expert analysis, specific detailed annotations are added to *IMGT* flat files in a second step *(7)*.

Since August 1996, the *IMGT/LIGM-DB* content closely follows the *EMBL* one for the IG and TR, with the following advantages: *IMGT/LIGM-DB* does not contain sequences that have previously been wrongly assigned to IG and TR; conversely, *IMGT/LIGM-DB* contains IG and TR entries that have disappeared from the generalized databases [as examples: the L36092 accession number that encompasses the complete human TRB locus is still present in *IMGT/LIGM-DB*, whereas it has been deleted from *EMBL/GenBank/DDBJ* because of its too large size (684,973 bp); in 1999, *IMGT/LIGM-DB* detected the disappearance of 20 IG and TR sequences that had inadvertently been lost by *GenBank*, and allowed the recuperation of these sequences in the generalist databases].

2.2. IMGT/LIGM-DB *Interface and Data Distribution*

The *IMGT/LIGM-DB* Web interface allows searches according to immunogenetic-specific criteria, and is easy to use without any knowledge of a computing language. The interface allows the users to easily connect from any type of platform (PC, Macintosh, workstation) using freeware such as Netscape. All *IMGT/LIGM-DB* information is available through five search modules **(Fig. 1)**:

1. **Catalogue** (accession number, mnemonic, *EMBL* first reception date, sequence length, definition, *IMGT/LIGM-DB* annotation level;
2. **Taxonomy and characteristics** (species and classification level, nucleic acid type, "loci, genes or chains," functionality, structure, specificity, group, and subgroup);
3. **Keywords** (standardized keywords, selection of *IMGT* reference sequences—for human and mouse IG and TR);
4. **Annotation labels;**
5. **References** (authors, publication type, journal, year, title, MEDLINE reference number).

Selection is displayed at the top of the resulting sequences pages, so you can check your own queries *(9)* **(Fig. 2)**. You have the possibility to modify your request or to consult the results: you can decrease or increase the number of resulting sequences by adding new conditions, view details concerning the selected sequences, or search for sequence fragments (subsequences) that correspond to a particular label *(9)* **(Fig. 2)**. When selecting the "View" options, you can choose among nine possibilities **(Fig. 3)**: annotations, *IMGT* flat file, coding regions with protein translation, catalogue and external references, sequence in dump format, sequence in *FASTA* format, sequence with three reading frames, *EMBL* flat file, and *IMGT/V- QUEST.*

IMGT/LIGM-DB data are also distributed by anonymous FTP servers at CINES (ftp://ftp.cines.fr/IMGT/) and at EBI (ftp://ftp.ebi.ac.uk/pub/databases/imgt/), and from many SRS (Sequence Retrieval System) sites. *IMGT/LIGM-DB* can be searched by *BLAST* or *FASTA* on different servers (e.g., EBI, IGH, INFOBIOGEN, or Institut Pasteur).

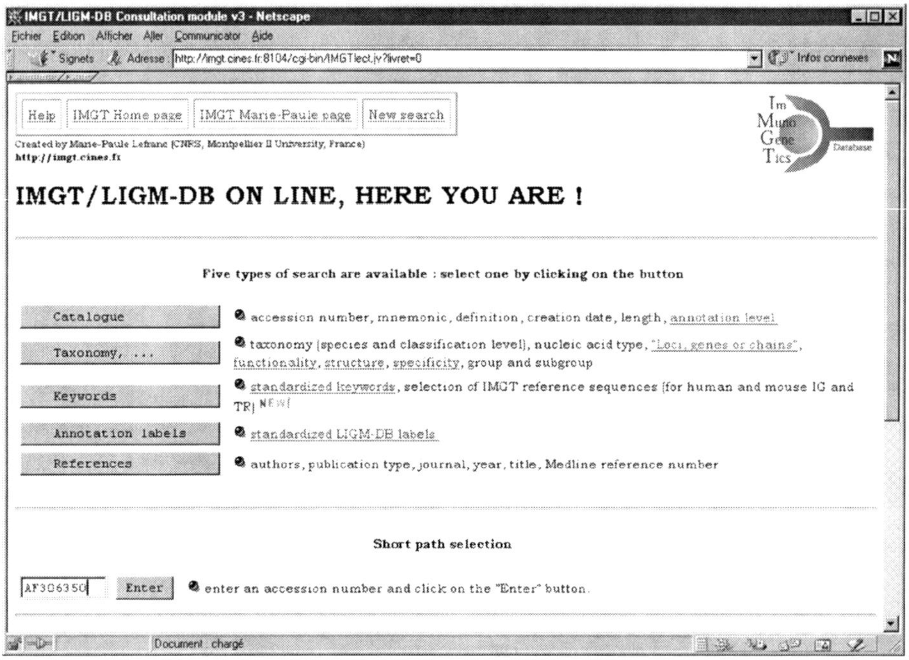

Fig. 1. *IMGT/LIGM-DB* search page (http://imgt.cines.fr). Five modules of search are available. Catalogue, Taxonomy and Characteristics, Keywords, Annotation labels, and References. These modules allow extensive and complex queries on immunoglobulin and T-cell-receptor sequences from human and other vertebrates. In July 2003, *IMGT/LIGM-DB* contained 74,837 sequences of IG and TR from 105 species. A short path selection allows a direct query with an accession number or a part of it. For example, "AF306350" will retrieve that sequence, whereas "AF306" will retrieve all sequences beginning with AF306.

3. *IMGT* Web Resources

IMGT Web resources ("*IMGT* Marie-Paule page") *(6)* comprise the following sections: "*IMGT* Scientific chart," "*IMGT* Repertoire," "*IMGT* Bloc-notes," "*IMGT* Education," "*IMGT* Aide-mémoire," and "*IMGT* Index."

3.1. IMGT *Scientific Chart*

The *IMGT* Scientific chart provides the controlled vocabulary and the annotation rules and concepts defined by *IMGT* ***(13)*** for the identification, the description, the classification, and the numeration of the IG and TR data of human and other vertebrates.

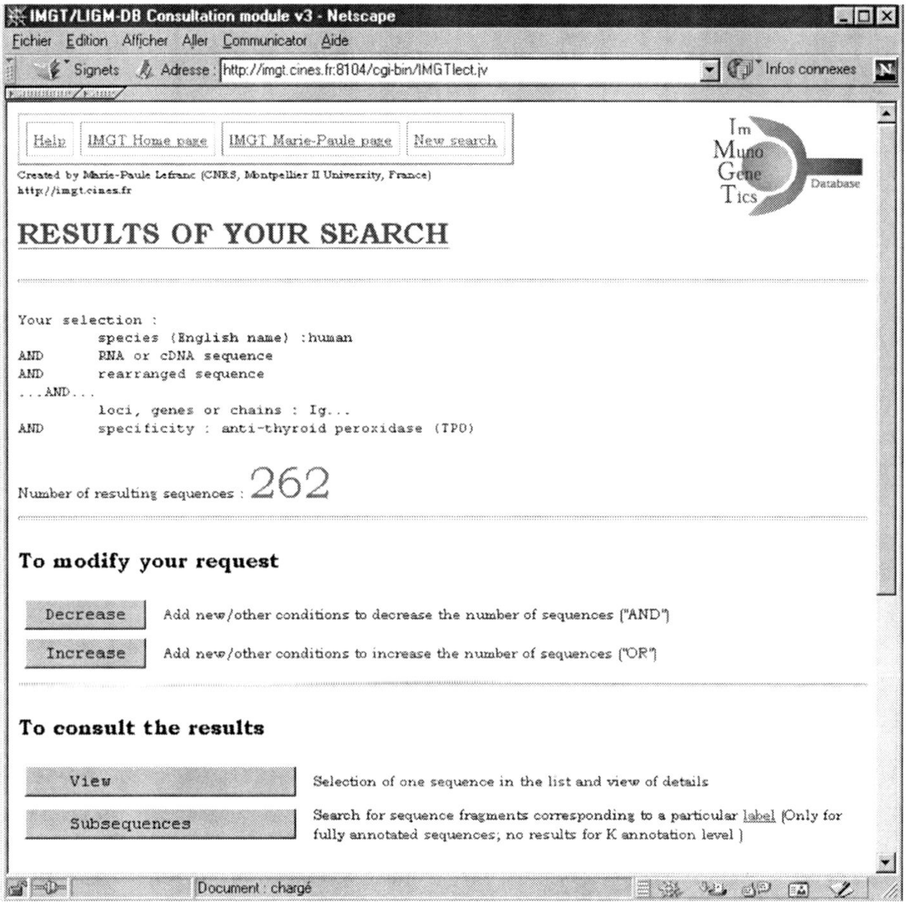

Fig. 2. Example of *IMGT/LIGM-DB* results of search (http://imgt.cines.fr). There are 262 resulting sequences for the query "human," "RNA or cDNA sequence," "rearranged sequence," "IG," and "anti-thyroid peroxidase (TPO)" specificity. The user can modify the request ("Decrease," "Increase") or consult the results ("View," "Subsequences").

3.1.1. Concept of Identification: Standardized Keywords

IMGT standardized keywords for IG and TR include the following: i) **General keywords:** indispensable for the sequence assignments, they are described in an exhaustive and nonredundant list, and are organized in a tree structure; ii) **Specific keywords:** they are more specifically associated to particularities of the sequences (e.g., orphon or transgene) or to diseases (e.g., leukemia, lymphoma, or myeloma) *(7)*. The list is not definitive, and new specific keywords can easily be added if needed. *IMGT/LIGM-DB* standardized keywords have been assigned to all entries.

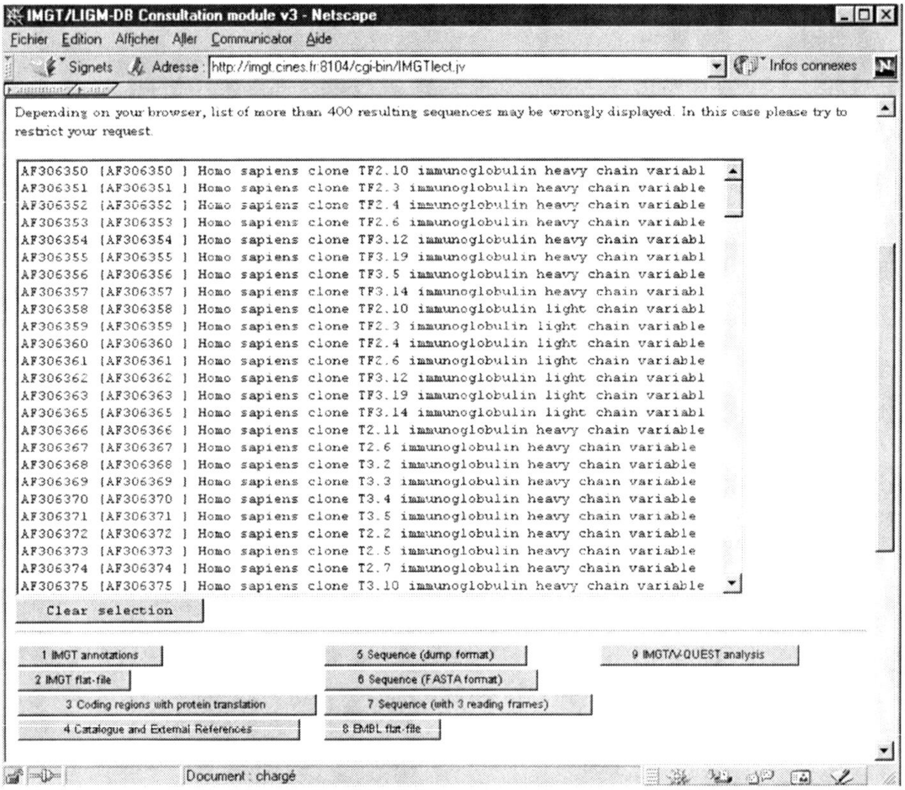

Fig. 3. Example of *IMGT/LIGM-DB* resulting screen for the "View" choice. The user clicks on a line corresponding to an accession number and can choose among nine possibilities (e.g., *IMGT* annotations, *IMGT* flat file, or coding regions with protein translation.)

3.1.2. Concept of Description: Standardized Sequence Annotation

One hundred and seventy-seven feature labels are needed to describe all structural and functional subregions that compose IG and TR sequences *(7)*, whereas only seven of them are available in *EMBL, GenBank,* or *DDBJ.* Annotation of sequences with these labels constitutes the main part of the expertise. Levels of annotation have been defined that allow the users to query sequences in *IMGT/LIGM-DB,* although they are not fully annotated *(7)*.

Prototypes represent the organizational relationship between labels and provide information on the order and expected length (in number of nucleotides) of the labels *(7,9)*.

3.1.3. Concept of Classification: Standardized IG and TR Gene Nomenclature

The objective is to provide immunologists and geneticists with a standardized nomenclature per locus and per species that will allow extraction and comparison of data for the complex B- and T-cell antigen-receptor molecules. The concepts of classification have been used to set up a unique nomenclature of human IG and TR genes, which was approved by *HGNC,* the HUGO (Human Genome Organization) Nomenclature Committee in 1999 ***(6)***. The complete list of the human IG and TR gene names ***(1,2,14–20)*** has been entered by the *IMGT* Nomenclature Committee in *GDB,* Toronto, and *LocusLink,* NCBI, United States, and is available from the *IMGT* site ***(6)***. *IMGT* reference sequences have been defined for each allele of each gene based on one or, whenever possible, several of the following criteria: germline sequence, first sequence published, longest sequence, mapped sequence ***(9,21)***. They are listed in the germline gene tables of the *IMGT* repertoire ***(22–29)***. The protein displays show translated sequences of the alleles (*01) of the functional or ORF genes ***(1,2,30,31)***.

3.1.4. Concept of Numerotation: the IMGT Unique Numbering

A uniform numbering system for IG and TR sequences of all species has been established to facilitate sequence comparison and cross-referencing between experiments from different laboratories whatever the antigen receptor (IG or TR), the chain type, or the species ***(32,33,41)***. This numbering results from the analysis of more than 5,000 IG and TR variable region sequences of vertebrate species from fish to humans. It takes into consideration and combines the definition of the framework (FR) and complementarity-determining region (CDR) ***(34)***, structural data from X-ray diffraction studies ***(35)***, and the characterization of the hypervariable loops ***(36)***. In the *IMGT* numbering, conserved amino acids from frameworks always have the same number, regardless of the IG or TR variable sequence, and whatever the species they come from. As examples: Cysteine 23 (in FR1-IMGT), Tryptophan 41 (in FR2- IMGT), Leucine 89 and Cysteine 104 (in FR3-IMGT). Tables and graphs are available on the *IMGT* web site at http://imgt.cines.fr and in **refs.** ***1,2***.

This *IMGT* unique numbering has several advantages:

1. It has allowed the redefinition of the limits of the FR and CDR of the IG and TR variable domains. The FR-IMGT and CDR-IMGT lengths themselves become crucial information that characterizes variable regions that belong to a group, a subgroup, and/or a gene.
2. Framework amino acids (and codons) located at the same position in different sequences can be compared without requiring sequence alignments. This is also true for amino acids that belong to CDR-IMGT of the same length.

3. The unique numbering is used as the output of the *IMGT/V-QUEST* alignment tool. The aligned sequences are displayed according to the *IMGT* numbering and with the FR-IMGT and CDR-IMGT delimitations.
4. The unique numbering has allowed a standardization of the description of mutations and the description of IG and TR allele polymorphisms ***(1,2)***. These mutations and allelic polymorphisms are described by comparison to the *IMGT* reference sequences of the alleles (*01) ***(8,9)***.
5. The unique numbering allows the description and comparison of somatic hypermutations of the IG *IMGT* variable domains.

By facilitating the comparison between sequences and allowing the description of alleles and mutations, the *IMGT* unique numbering represents a major step forward in the analysis of the IG and TR sequences of all vertebrate species ***(41)***. Moreover, it provides insight into the structural configuration of the variable domain and opens interesting views on the evolution of these sequences, since this numbering has been successfully applied to all the sequences belonging to the V-set of the immunoglobulin superfamily, including non-rearranging sequences in vertebrates (e.g., human CD4 and *Xenopus* CTXg1) and in invertebrates (e.g., *Drosophila* amalgam and *Drosophila* fasciclin II) ***(8,9,32,33,41)***.

3.2. IMGT *Repertoire*

IMGT Repertoire is the global Web Resource in immunogenetics for the IG, TR, MHC, and RPI of human and other vertebrates, based on the "*IMGT* Scientific chart." *IMGT* Repertoire provides an easy-to-use interface for carefully and expertly annotated data on the genome, proteome, polymorphism, and structural data of the IG, TR, MHC, and RPI ***(6)***. Only titles of this large section are quoted here. *Genome* data include chromosomal localizations, locus representations, locus description, germline gene tables, potential germline repertoires, lists of IG and TR genes, and links between *IMGT, HUGO, GDB, LocusLink,* and *OMIM,* correspondence between nomenclatures ***(1,2)***. *Proteome* and *polymorphism* data are represented by protein displays, alignments of alleles, tables of alleles, allotypes, particularities in protein designations, *IMGT* reference directory in *FASTA* format, correspondence between IG and TR chain and receptor *IMGT* designations ***(1,2)***. *Structural data* comprise 2D graphical representations designated as *IMGT* Colliers de Perles ***(1,2,6,8,9)***, FR-IMGT and CDR-IMGT lengths, and 3D representations of IG and TR variable domains ***(10,12)***. This visualization permits rapid correlation between protein sequences and 3D data retrieved from the *Protein Data Bank (PDB)*. Other data comprise: i) phages; ii) probes used for the analysis of IG and TR gene rearrangements and expression, and restriction fragment-length polymorphism (RFLP) studies; iii) data related to gene regulation and expression: promoters, primers, cDNAs, and reagent monoclonal antibodies (MAbs); iv) genes and clinical entities: translocations and inversions, humanized antibodies, MAbs with clinical indications; v) taxonomy of vertebrate species present in *IMGT/LIGM-DB;* vi) immunoglobulin superfamily: gene exon–intron organization, protein displays, Colliers de Perles, and 3D representations of V-LIKE and C-LIKE domains.

3.3. IMGT *Bloc-Notes*

The *IMGT* Bloc-notes provide numerous hyperlinks for the Web servers that specialize in immunology, genetics, molecular biology, and bioinformatics (e.g., associations, collections, companies, databases, immunology themes, journals, molecular biology servers, resources, societies, and tools) *(37)*.

3.4. IMGT *Education*

IMGT Education is a section that provides useful biological resources for students. It includes figures and tutorials (in English and/or in French) on the IG and TR variable and constant domain 3D structures, the molecular genetics of immunoglobulins, the regulation of IG gene transcription, B-cell differentiation and activation, and translocations.

3.5. IMGT *Aide-mémoire and IMGT Index*

IMGT Aide-mémoire provides easy access to information such as genetic code, splicing sites, amino acid structures, and restriction enzyme sites.

IMGT Index is a fast way to access data when information must be retrieved from different parts of the *IMGT* site. For example, "allele" provides links to the *IMGT* Scientific chart rules for the allele description, and to the *IMGT* Repertoire Alignments of alleles and Tables of alleles (http://imgt.cines.fr).

4. *IMGT* Interactive Tools

4.1. IMGT/V-QUEST *Tool*

4.1.1. Overview

IMGT/V-QUEST (V-QUEry and Standardization) (http://imgt.cines.fr) is an integrated software for IG and TR *(6)*. This tool is easy to use and analyzes an input IG or TR germline or rearranged variable nucleotide sequence (**Fig. 4**). *IMGT/V-QUEST* results comprise the identification of the V, D, and J genes and alleles and the nucleotide alignment by comparison with sequences from the *IMGT* reference directory (**Fig. 5**), the delimitations of the FR-IMGT and CDR-IMGT based on the *IMGT* unique numbering, the protein translation of the input sequence, the identification of the JUNCTION and the 2D Collier de Perles representation of the V-REGION. Note that *IMGT/V-*

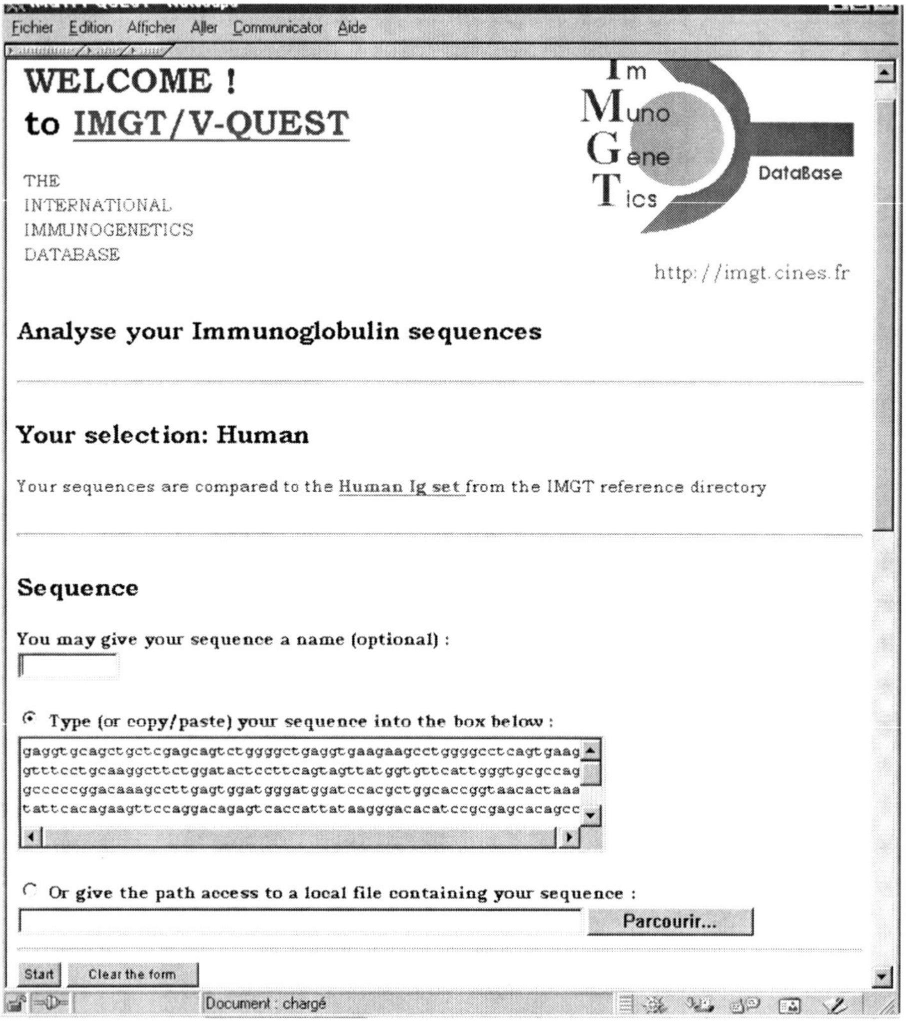

Fig. 4. *IMGT/V-QUEST* analysis for human immunoglobulin sequences (http://imgt.cines.fr). The user can type (or copy/paste) a sequence or give the path access to a local file.

QUEST does not work, or will give aberrant results, for pseudogenes with DNA insertions or deletions, partial sequences that are too short, sequences containing a cluster of V-GENEs, or sequences with 5′-untranslated regions (5′-UTR) or 3′-UTR that are too long. The set of sequences from the *IMGT* reference directory, used for *IMGT/V-QUEST,* can be downloaded in *FASTA* format from the *IMGT* site.

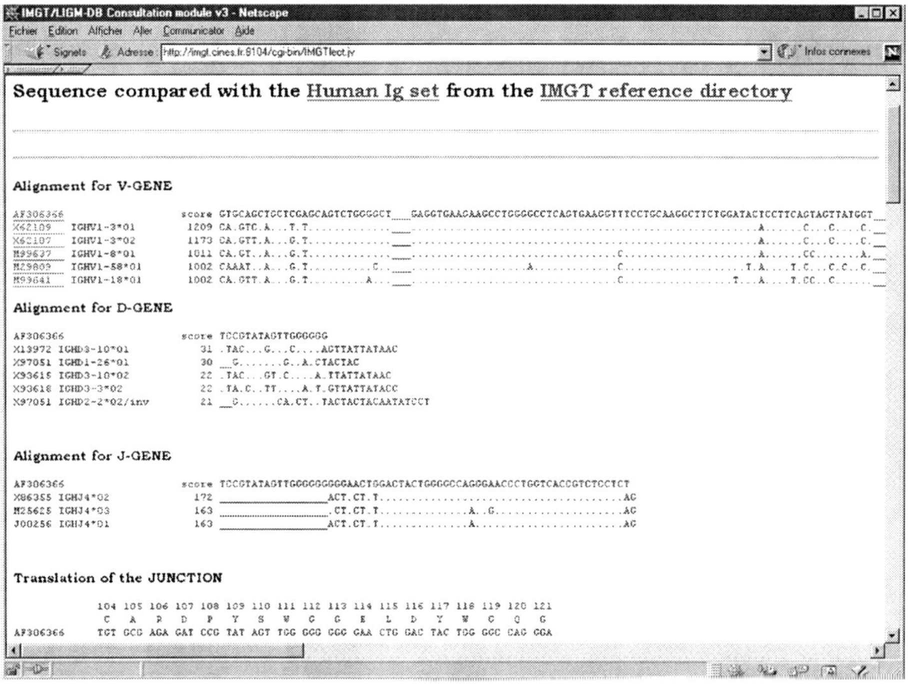

Fig. 5. *IMGT/V-QUEST* results for the gene and allele identification, and the translation of the JUNCTION. *IMGT/V-QUEST* compares the input germline or rearranged IG or TR variable sequences with the *IMGT/V-QUEST* reference directory sets. For example, the highest scores for the input AF306366 rearranged sequence allow the identification of IGHV1-3*01, IGHD3-10*01, IGHJ4*02 as the genes and alleles most likely to be involved in the V-D-J rearrangement. The CDR3-IMGT of that sequence is 13 amino acids (or codons) long (from position 105 to position 117). The JUNCTION extends from 2nd-CYS 104 to J-TRP 118 included. The translation is displayed from 2nd-CYS to the Phe/Trp-Gly-X-Gly (here, W-G-Q-G) motif included. The information provided by *IMGT/V-QUEST* [V and J gene and allele names, sequence of the JUNCTION (from 2nd-CYS 104 to J-PHE or J-TRP 118)] can be used in *IMGT/Junction-Analysis* for a confirmation of the D gene and allele identification and a more accurate analysis of the junction (*see* **Figs. 9** and **10**). Gene and allele names are according to the *IMGT* nomenclature ***(1,2,18–20)***.

4.1.2. IMGT/V-QUEST Reference Directory Sets

Depending on your selection in the *IMGT/V-QUEST* Search page (IG or TR, species), your sequence will be compared to a given *IMGT/V-QUEST* reference directory set. The *IMGT/V-QUEST* reference directory sets are constituted by sets of sequences that contain the V-REGION, D-REGION, and J-REGION alleles, isolated from the Functional and ORF allele *IMGT* reference sequences. By definition, these sets contain *one sequence for each allele.* Allele names of these sequences are shown in red in Alignments of alleles in the *IMGT* repertoire (http://imgt.cines.fr). Exceptionally, the *IMGT/V-QUEST* reference directory sets may include sequences isolated from pseudogene allele *IMGT* reference sequences [indicated with (P) following the allele name]. For sequence alignments, the *IMGT/V-QUEST* uses the *DNAPLOT* program, an alignment tool that is part of *IMGT,* developed by Hans-Helmar Althaus and Werner Müller (Institut für Genetik, Köln, Germany). Since 1997, the *IMGT/V-QUEST* developments have been implemented by Véronique Giudicelli (IMGT, LIGM, Université Montpellier II, Montpellier, France).

4.1.3. IMGT/V-QUEST Output

The *IMGT/V-QUEST* output comprises five different displays:

1. Alignment for the identification of the V-GENE, D-GENE, and J-GENE: the alignment in **Fig. 5** shows the input sequence aligned with the closest V-REGION, D-REGION, and J-REGION alleles, from the *IMGT/V-QUEST* reference directory sets. Dots represent identity, and dashes and lines are gaps. The alignments for the D-REGION and the J-REGION start from the end of the V-REGION. Note that the IGH D-REGIONs are not easily identified by *IMGT/V-QUEST,* and that it is recommended to use the *IMGT/JunctionAnalysis* tool (http://imgt.cines.fr) for the IGHD gene and allele identification.
2. Translation of the JUNCTION: the JUNCTION extends from 2nd-CYS 104 to J-PHE or J-TRP inclusive. J-PHE or J-TRP are easily identified for in-frame rearranged sequences and when the conserved Phe/Trp-Gly-X-Gly motif of the J-REGION is present. The translation of in-frame sequences is displayed starting from 2nd-CYS up to the Phe/Trp-Gly-X-Gly motif (inclusive), if the motif is present in your sequence **(Fig. 5)** or in the closest J-REGION, or up to the end of your sequence if the motif is not found. The length of the CDR3-IMGT of rearranged V-J-GENEs or V-D-J-GENEs is a crucial piece of information. It is the number of amino acids or codons from position 105 to J-PHE or J-TRP, noninclusive. Note that for an out-of-frame sequence, it is necessary to look at the nucleotide sequence to identify the codon that would have encoded J-PHE or J-TRP, in order to delimit, in 3′, the JUNCTION (codon included), or the CDR3-IMGT (codon not included). Some V-REGIONs have a Cysteine at position 103. In those cases, for a technical reason, the translation of the JUNCTION will start with "CC." Be aware that the first Cysteine corresponds to position 103, and is not part of the JUNCTION.
3. Alignment with FR-IMGT and CDR-IMGT delimitations: the sequences are shown with the *IMGT* unique numbering and with the *IMGT* framework region (FR-IMGT) and complementarity-determining region (CDR-IMGT) delimitations **(Fig. 6)**. Dashes indicate identical nucleotides. Dots indicate gaps according to the *IMGT* unique numbering. The resulting alignment shows the CDR3-IMGT of the germline V-REGION alleles of the *IMGT* reference directory. The CDR3-IMGT of

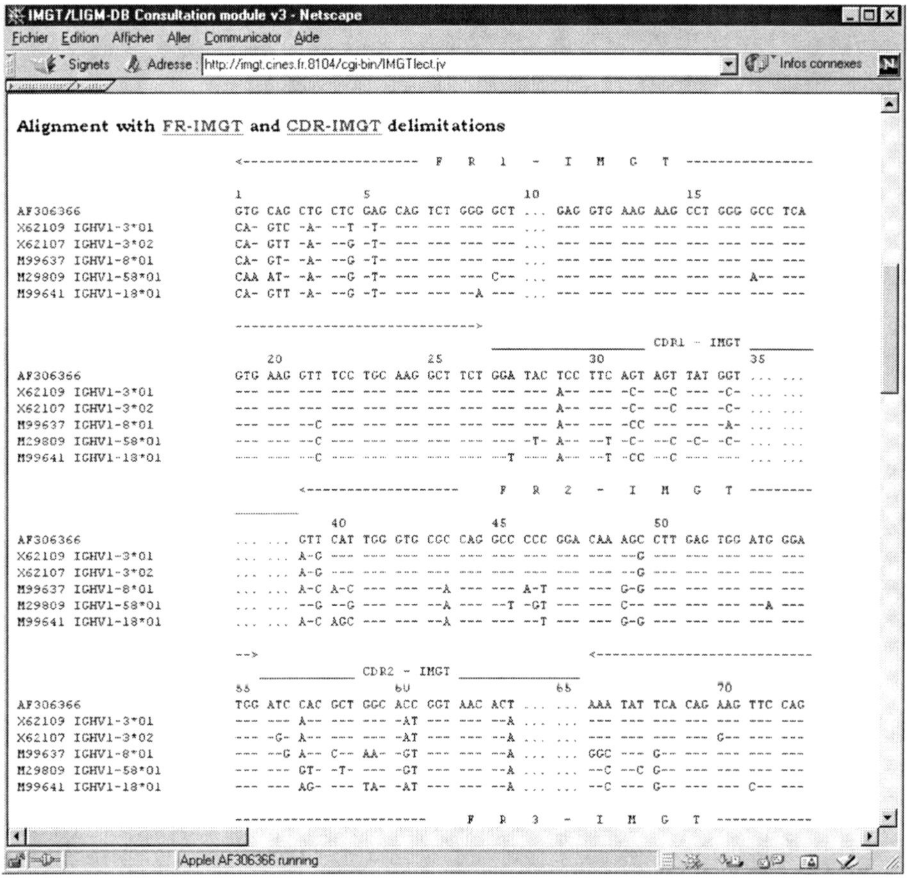

Fig. 6. *IMGT/V-QUEST* results for alignment with FR-IMGT and CDR-IMGT delimitations. The input nucleotide sequence (AF306366) is aligned with the five most similar sequences from the *IMGT* reference directory set (in that example, human IG). FR-IMGT and CDR-IMGT delimitations are according to the IMGT unique numbering *(32,33,41)*.

input rearranged sequences can be identified in the translation of the JUNCTION and in the translation of the input sequence.

4. Translation of the input sequence: the nucleotide sequence and deduced amino acid translation of the input sequence are shown with the FR-IMGT and CDR-IMGT delimitations (**Fig. 7**). The 3′ limit of the CDR3-IMGT of the input rearranged sequence is correctly identified if the conserved Phe/Trp-Gly-X-Gly motif of the J-REGION has been identified. If not, the 3′ limit of the CDR3-IMGT must be checked.
5. *IMGT/Collier de Perles* for the input sequence V-REGION: the *IMGT/Collier de Perles* 2D graphical representation (**Fig. 8**) is automatically generated by the *IMGT*

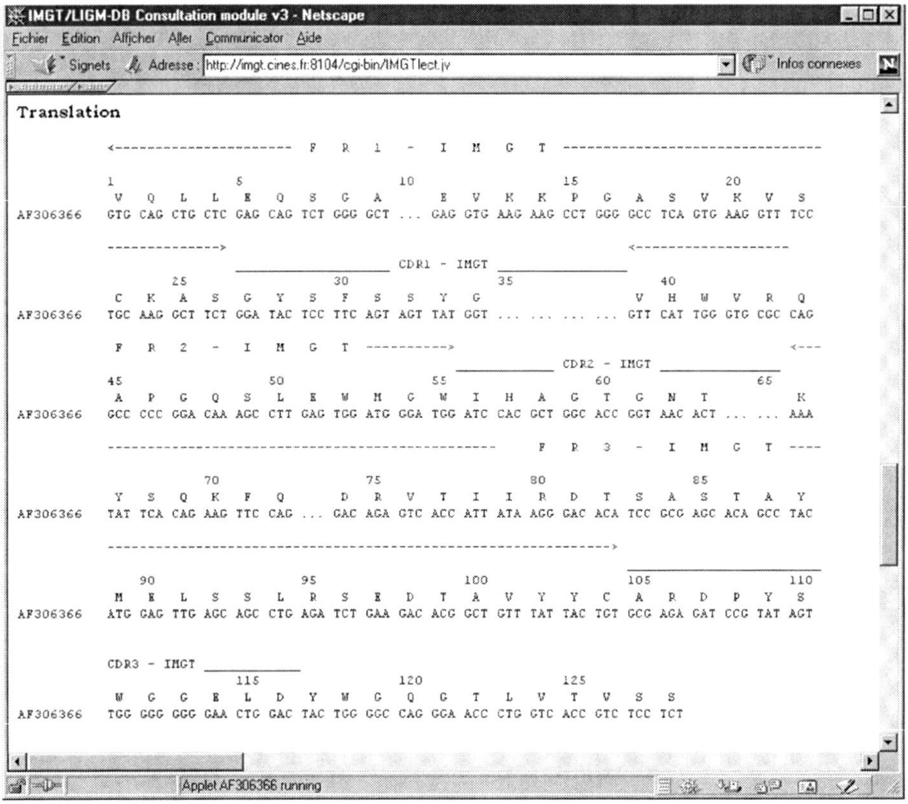

Fig. 7. *IMGT/V-QUEST* results for translation. The translation of the input nucleotide sequence is displayed with FR-IMGT and CDR-IMGT delimitations according to the *IMGT* unique numbering ***(32,33,41)***. CDR1-IMGT, CDR2-IMGT, and rearranged CDR3-IMGT lengths of the AF306366 V-DOMAIN sequence are 8, 8, and 13 amino acids (or codons) long, respectively, or [8.8.13] as described in the *IMGT* Scientific chart ***(41)***.

Collier de Perles program developed by Gérard Mennessier (Laboratoire de Physique Mathématique, Montpellier, France) and adapted for Java applet by Denys Chaume and Manuel Ruiz (IMGT, LIGM, Montpellier, France). Only a portion of the CDR3-IMGT of the rearranged sequence is shown. The length which is displayed corresponds to that of the longest germline CDR3-IMGT in the V-REGION set used. The representation of the CDR3-IMGT loop of rearranged sequences is in development.

4.2. IMGT/JunctionAnalysis *Tool*

4.2.1. Overview

IMGT/JunctionAnalysis (http://imgt.cines.fr) is a tool developed by Mehdi Yousfi (IMGT, LIGM, Montpellier, France), complementary to *IMGT/V-*

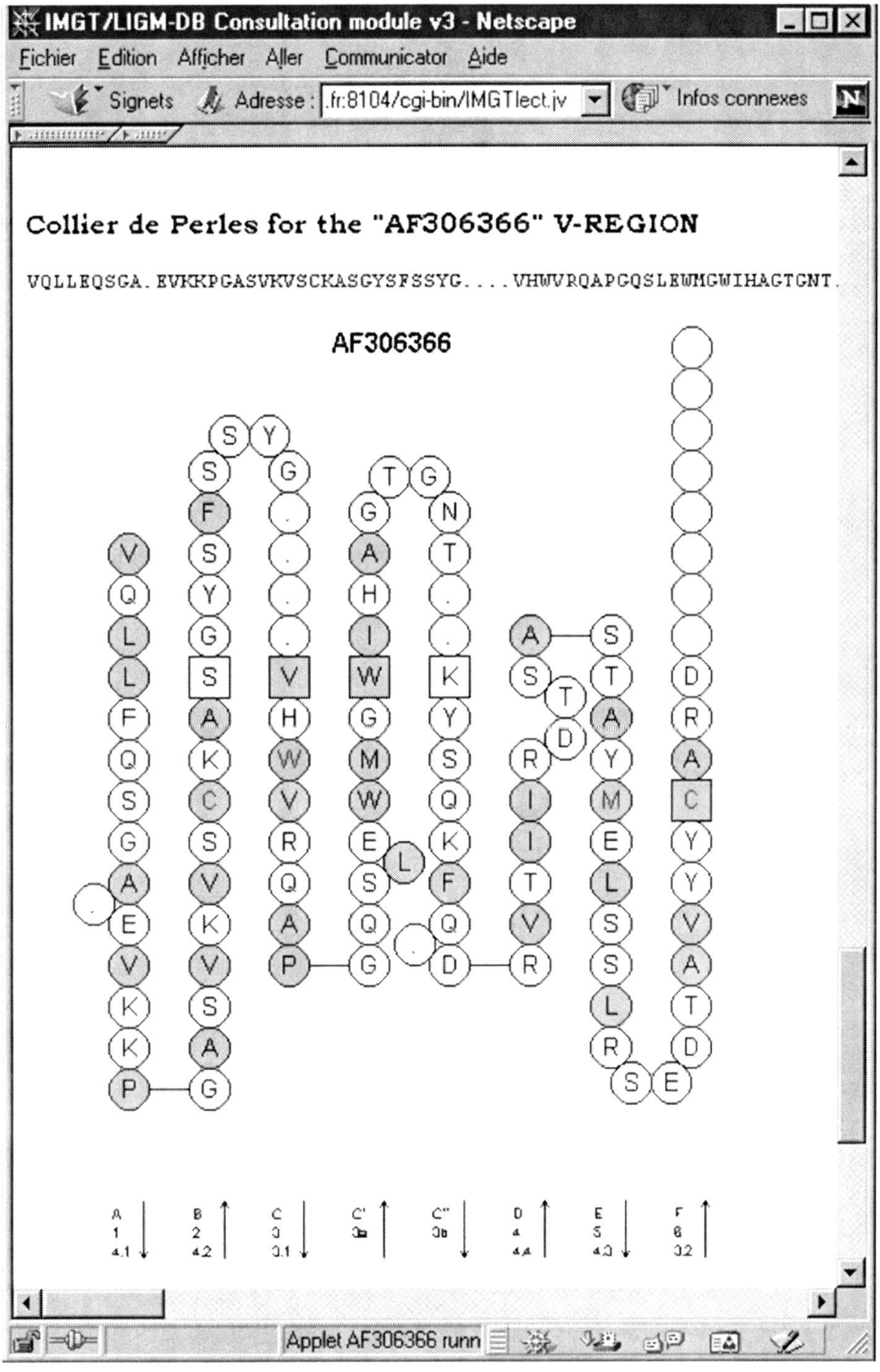

Fig. 8. *IMGT/V-QUEST* results for Collier de Perles. Positions with a dot correspond to missing positions according to the *IMGT* unique numbering ***(32,33,41)***. The CDR-IMGT are limited by amino acids shown in squares, which belong to the neighboring FR-IMGT.

QUEST, which provides a thorough analysis of the V-J and V-D-J junction of IG and TR rearranged genes **(Fig. 9)**. *IMGT/JunctionAnalysis* identifies the D-GENE and allele involved in the IGH, TRB, and TRD V-D-J rearrangements by comparison with the *IMGT* reference directory, and delimits precisely the P, N, and D regions *(1,2)*. Results from *IMGT/JunctionAnalysis* are more accurate than those given by *IMGT/V-QUEST* regarding the D-GENE identification. Indeed, *IMGT/JunctionAnalysis* works on shorter sequences (JUNCTION), and with a higher constraint because the identification of the V-GENE and J-GENE and alleles is a prerequisite to perform the analysis. Several hundreds of junction sequences can be analyzed simultaneously.

4.2.2. IMGT/JunctionAnalysis Input Information

Select Species: At present, only human sequences can be analyzed, but mouse sequences will follow **(Fig. 9)**.

Select Locus (IGH, IGK or IGL for the immunoglobulins): By default, the IGH search will yield results by comparing input data to the IGHD genes and alleles in direct and inverted orientations. A faster (but simplified!) search does not include "inversed" D genes and gives results by comparing input data to only the IGHD (*01) alleles **(Fig. 9)**.

Enter the junction sequences: The junction nucleotide sequences can be entered either directly in the reserved box by typing or by "copy/paste," or by a local file (click on "Browse" or type its full path in the reserved box). The required format is the *FASTA* format. Each junction nucleotide sequence must be preceded by the following information **(Fig. 10)**:

1. An identifier ("input"), with a ten-character maximum length. This identifier can be a sequence name, an accession number, or a clone name.
2. The name of the V-GENE and ALLELE according to the *IMGT* gene name nomenclature.
3. The name of the J-GENE and ALLELE according to the *IMGT* gene name nomenclature.

There is no limitation in the number of sequences to be analyzed in a single search. Sequences only need to be entered in the same format, starting a new line for each sequence:

>"input1," "V-GENE and ALLELE name," "J-GENE and ALLELE"
"nucleotide sequence (in uppercase or lowercase)"
>"input2," "V-GENE and ALLELE name," "J-GENE and ALLELE"
"nucleotide sequence (in uppercase or lowercase)"

Example:

>M62724, IGHV7-4-1*02, IGHJ4*02
tgtgcgagagaagatagcaatggctacaaaatatttgactactgg

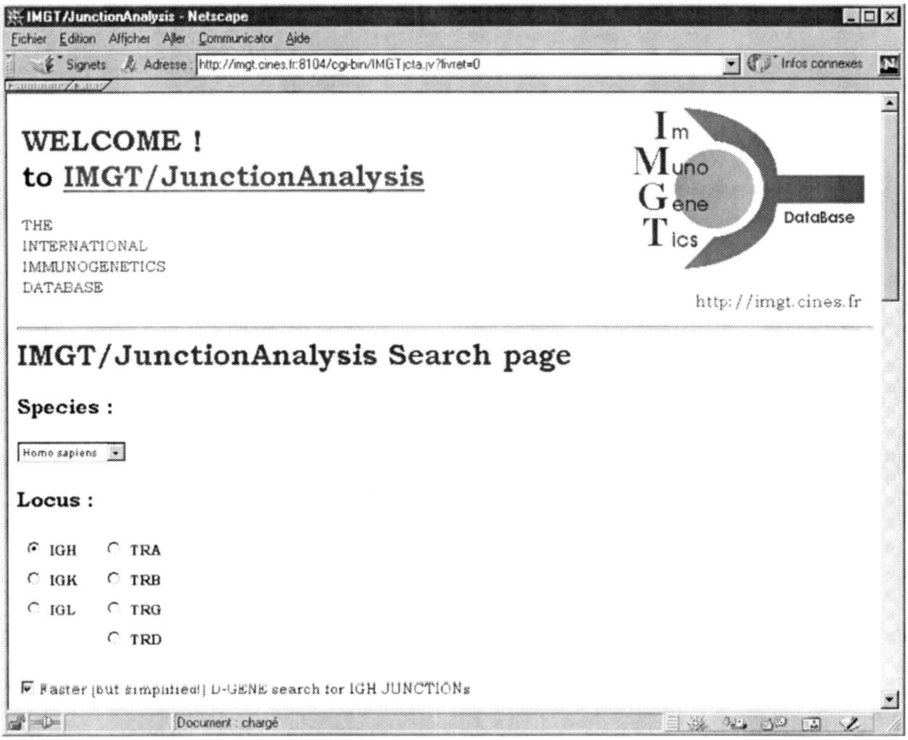

Fig. 9. *IMGT/JunctionAnalysis* search page (http://imgt.cines.fr).

>Z47269, IGHV1-69*06, IGHJ5*02
tgtgcgagagggggggctaaggtcgaatttttggagtggtttcatgggtactggttcgacccctgg

The junction nucleotide sequences must start with V-REGION 2nd-CYS codon and end with the J-REGION J-PHE or J-TRP codon. "V-GENE and ALLELE" and "J-GENE and ALLELE" are those obtained by querying *IMGT/V-QUEST.* If several alleles give the same score, select the most probable one.

If the V-GENE ALLELE or J-GENE ALLELE is unknown, the *IMGT/JunctionAnalysis* tool accepts a "?" character instead of the allele number (example: IGHV1-2*?) and will run the search against the allele (*01) by default. If there are several proposed V-GENEs and/or J-GENEs, the different V-GENE and ALLELE names and/or J-GENE and ALLELE names must be separated by the "/" character (example: IGHV1-2*01/IGHV1-3*?/IGHV1-18*02, IGHJ1*01/IGHJ2*01). The *IMGT/JunctionAnalysis* tool will run the search against the first V-GENE and ALLELE and J-GENE and ALLELE listed.

Select the output options: The analysis results can be either displayed or downloaded into a local file. The maximum number of characters per line allows the user to select the most appropriate format for viewing or printing.

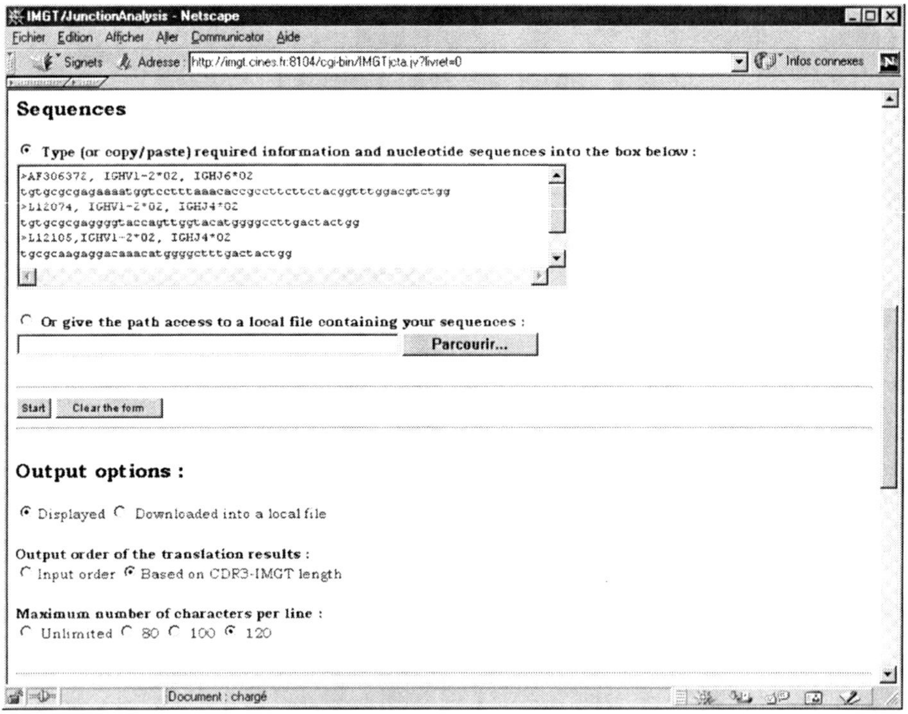

Fig. 10. *IMGT/JunctionAnalysis* input information page. Each junction sequence (from 2nd-CYS 104 to J-PHE or J-TRP 118) must be preceded by an identifier, a V-GENE gene and allele, and a J-GENE and allele. There is no limitation in the number of sequences to be analyzed in a single search.

The order of the translation results can be either according to the input or based on the CDR3-IMGT lengths **(Fig. 10)**.

4.2.3. IMGT/JunctionAnalysis Output

The *IMGT/JunctionAnalysis* output comprises:

1. Analysis of the JUNCTIONs: Results are displayed according to the selected output option (in the order of the sequence entries or based on the CDR3-IMGT lengths). Nucleotides of each region identified in a JUNCTION are displayed. Dots in V-REGION, D-REGION, and J-REGION indicate nucleotides deleted in the rearranged sequence, by comparison to the corresponding germline V-GENE, D-GENE, and J-GENE. N, N1, N2, N3, N4 indicate N-REGIONs. If there are several N-REGIONs, they are numbered from left to right. P indicates P-REGIONs (there is no numbering for the P-REGIONs). The information provided by the user in the *IMGT/Junction-Analysis* Search page is reported in three columns: **Input, V name** (*IMGT* V-GENE and ALLELE name), and **J name** (*IMGT* J-GENE and ALLELE name). The information which results from *IMGT/JunctionAnalysis* is displayed in the other columns:

D name: *IMGT* D-GENE and ALLELE name (for IGH junctions).
Vmut: Number of mutations in the "input" sequence from 2nd-CYS to the V-REGION 3′ end, by comparison to the corresponding germline allele sequence.
Dmut: Number of mutations in the D-REGION sequence identified by the *IMGT/JunctionAnalysis* tool, by comparison to the corresponding germline allele sequence.
Jmut: Number of mutations in the "input" sequence from the J-REGION 5′ end to J-PHE or J-TRP, compared to the corresponding germline allele sequence.
Ngc: Ratio of the number of g+c nucleotides to the total number of N region nucleotides.

2. Translation of the JUNCTIONs: Each junction nucleotide sequence is translated into an amino acid sequence **(Fig. 11)**. In the case of frameshifts, gaps indicated by one or two dots are inserted to maintain the J-REGION reading frame and to facilitate sequence comparison. Codons and amino acids are numbered according to the *IMGT* unique numbering for the V-DOMAINs. The numbering is made according to the longest JUNCTION obtained in the results. Note that gaps inserted in the other JUNCTIONs may split a D-REGION or a J-REGION because the gaps are localized at the top of the CDR3-IMGT loops and depend on the CDR3-IMGT length and not on the sequence alignment. "*" indicates a STOP-CODON. "#" indicates a frameshift. "+" and "–" at the end of the line indicates "in-frame" and "out-of-frame" junction, respectively.

4.3. *Other* IMGT *Tools*

IMGT/Allele-Align makes it possible to compare two alleles, highlighting the nucleotide and amino acid differences. *IMGT/PhyloGene* is an easy to use tool for phylogenetic analysis of IMGT standardized reference sequences. *IMGT/GeneSearch*, *IMGT/GeneView*, and *IMGT/LocusView* are tools for genome analysis of the IG, TR, and MHC genes and foci. *IMGT/StructuralQuery* is a tool for 3D structure analysis of the IG, TR, MHC, and RPI.

5. IMGT-ONTOLOGY and IMGT Interoperability

5.1. IMGT-ONTOLOGY

IMGT distributes high-quality data with an important incremental value added by the *IMGT* expert annotations, according to the rules described in the *IMGT* Scientific chart. *IMGT* has developed a formal specification of the terms to be used in the domain of immunogenetics and bioinformatics to ensure accuracy, consistency, and coherence in *IMGT.* This has been the basis of *IMGT-ONTOLOGY (13)*, the first ontology in the domain, which allows the immunogenetics knowledge to be maintained for all vertebrate species. Control of coherence in *IMGT* combines data integrity control and biological data evaluation *(38,39)*.

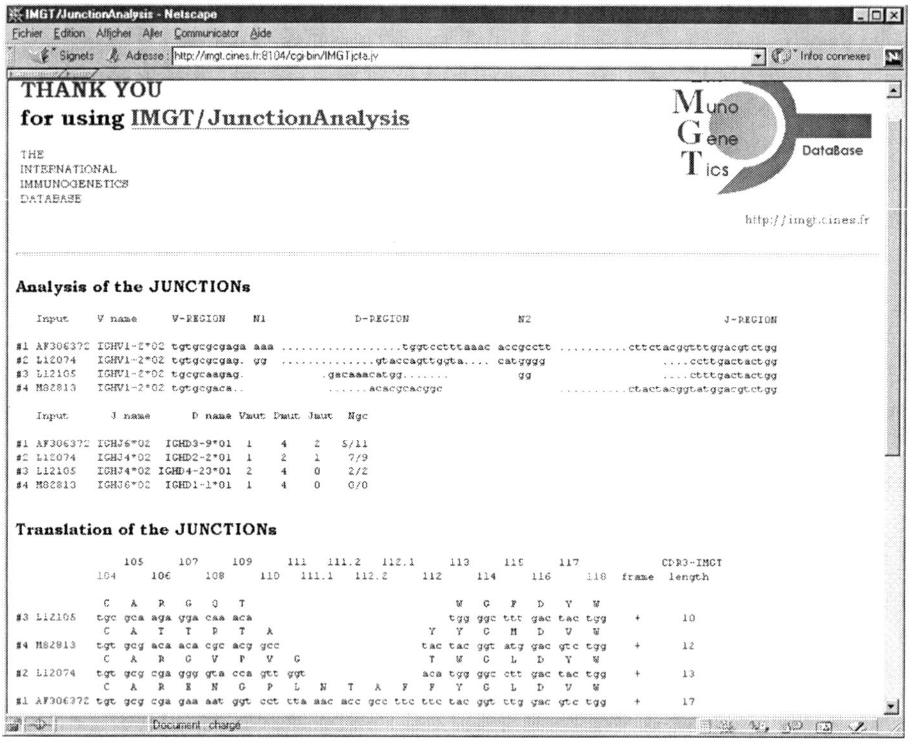

Fig. 11. *IMGT/JunctionAnalysis* results. The *IMGT/JunctionAnalysis* results comprise, for each junction, the identification of the D-GENE and allele, the identification of the P and N regions (e.g., N1 and N2) and their precise delimitations, and the junction translation. **Vmut, Dmut,** and **Jmut** correspond to the number of mutations in the input junction sequence by comparison to the germline allele sequences. **Ngc** is the ratio of the number of g + c nucleotides to the total number of nucleotides in the N regions. The CDR3-IMGT numbering is done according to the *IMGT* unique numbering for V-DOMAIN *(41)*. *IMGT/JunctionAnalysis* analyzes, in a single search, an unlimited number of junctions, provided that the *IMGT* V-GENE and ALLELE and the J-GENE and ALLELE names are identified in the input information page.

5.2. IMGT *Interoperability*

Since July 1995, IMGT has been available on the web at http://imgt.cines.fr. *IMGT* provides biologists with an easy-to-use and friendly interface. From January 2000 to July 2003, the *IMGT* WWW server at Montpellier was accessed by more than 210,000 sites. *IMGT* has an exceptional response, with more than 120,000 requests a month. Two-thirds of the visitors are equally distributed between the European Union and the United States. To facilitate the integration of *IMGT* data into applications developed by other laboratories, we have built an

Application Programming Interface to access the database and its software tools (*see* "*IMGT* Informatics page [API…]") ***(38)***. This API includes: a set of URL links to access biological knowledge data (keywords, labels, functionalities, and lists of gene names), a set of URL links to access all data related to one given sequence, a set of *JAVA*™ class packages to select and retrieve data from an appropriate *IMGT* server using an Object-oriented approach.

6. Conclusion

The information provided by *IMGT* is valuable to clinicians and biological scientists in general ***(40)***. *IMGT* is designed to allow a common access to all immunogenetics data, and a particular attention is given to the establishment of cross-referencing links to other databases that are relevant to the users of *IMGT.*

7. Citing *IMGT*

Authors who make use of the information provided by *IMGT* should cite **ref.** ***6*** as a general reference for the access to and content of *IMGT,* and quote the *IMGT* home page URL, http://imgt.cines.fr.

Acknowledgments

I thank Véronique Giudicelli, Denys Chaume, and Gérard Lefranc for their helpful discussions, and Nora Bonnet for typing the manuscript. I am deeply grateful to the IMGT team for its expertise and constant motivation. *IMGT* is funded by the European Union's 5th PCRDT programme (QLG2-2000-01287), the CNRS (Centre National de la Recherche Scientifique), the Ministère de l'Education Nationale, and the Ministère de la Recherche. Subventions have been received from ARC (Association pour la Recherche sur le Cancer), and the Région Languedoc-Roussillon.

References

1. Lefranc, M.-P. and Lefranc, G. (2001) The Immunoglobulin FactsBook. Academic Press, London, UK, pp. 1–458.
2. Lefranc, M.-P. and Lefranc, G. (2001) The T cell receptor FactsBook. Academic Press, London, UK, pp. 1–398.
3. Stoesser, G., Baker, W., Van den Broek, A., Camon, E., Garcia-Pastor, M., Kanz, C., et al. (2001) The EMBL nucleotide sequence database. *Nucleic Acids Res.* **29,** 17–21.
4. Benson, D. A., Karsch-Mizrachi, I., Lipman, D., Ostell, J., Rapp, B. A., and Wheeler, D. L. (2000) GenBank. *Nucleic Acids Res.* **28,** 15–18.
5. Tateno, Y., Miyazaki, S., Ota, M., Sugawara, H., and Gojobori, T. (2000) DNA Data Bank of Japan (DDBJ) in collaboration with mass sequencing teams. *Nucleic Acids Res.* **28,** 24–26.
6. Lefranc, M.-P. (2001) IMGT, the international ImMunoGeneTics database. *Nucleic Acids Res.* **29,** 207–209.

7. Giudicelli, V., Chaume, D., Bodmer, J., Müller, W., Busin, C., Marsh, S., et al. (1997) IMGT, the international ImMunoGeneTics database. *Nucleic Acids Res.* **25,** 206–211.
8. Lefranc, M.-P., Giudicelli, V., Busin, C., Bodmer, J., Müller, W., Bontrop, R., et al. (1998) IMGT, the international ImMunoGeneTics database. *Nucleic Acids Res.* **26,** 297–303.
9. Lefranc, M.-P., Giudicelli, V., Ginestoux, C., Bodmer, J., Müller, W., Bontrop, R., et al. (1999) IMGT, the international ImMunoGeneTics database. *Nucleic Acids Res.* **27,** 209–212.
10. Ruiz, M., Giudicelli, V., Ginestoux, C., Stoehr, P., Robinson, J., Bodmer, J., et al. (2000) IMGT, the international ImMunoGeneTics database. *Nucleic Acids Res.* **28,** 219–221.
11. Robinson, J., Malik, A., Parham, P., Bodmer, J. G., and Marsh, S. G. E. (2000) IMGT/HLA Database – a sequence database for the human major histocompatibility complex. *Tissue Antigens* **55,** 280–287.
12. Ruiz, M. and Lefranc, M.-P. (2002) IMGT gene identification and Colliers de Perles of human immunoglobulin with known 3D structures. Immunogenetics DOI 10.1007/s00251-001-0408-6. *Immunogenetics* **53,** 857–883.
13. Giudicelli, V. and Lefranc, M.-P. (1999) Ontology for Immunogenetics: IMGT-ONTOLOGY. *Bioinformatics* **12,** 1047–1054.
14. Lefranc, M.-P. (2000) Nomenclature of the human immunoglobulin genes. Current Protocols in Immunology. J. Wiley and Sons, New York, NY, **40** (Suppl.) A.1P.1–A.1P.37.
15. Lefranc, M.-P. (2000) Nomenclature of the human T cell Receptor genes. Current Protocols in mmunology. J. Wiley and Sons, New York, NY, **40,** (Suppl.) A.1O.1–A.1O.23.
16. Lefranc, M.-P. (2000) Locus maps and genomic repertoire of the human Ig genes. *The Immunologist* **8,** 80–87.
17. Lefranc, M.-P. (2000) Locus maps and genomic repertoire of the human T-cell receptor genes. *The Immunologist* **8,** 72–79.
18. Lefranc, M.-P. (2001) Nomenclature of the human immunoglobulin heavy (IGH) genes. *Exp. Clin. Immunogenet.* **18,** 100–116.
19. Lefranc, M.-P. (2001) Nomenclature of the human immunoglobulin kappa (IGK) genes. *Exp. Clin. Immunogenet.* **18,** 161–174.
20. Lefranc, M.-P. (2001) Nomenclature of the human immunoglobulin lambda (IGL) genes. *Exp. Clin. Immunogenet.* **18,** 242–254.
21. Lefranc, M.-P. (1998) IMGT (ImMunoGeneTics) Locus on Focus. A new section of Experimental and Clinical Immunogenetics. *Exp. Clin. Immunogenet.* **15,** 1–7.
22. Pallarès, N., Frippiat, J. P., Giudicelli, V., and Lefranc, M.-P. (1998) The human immunoglobulin lambda variable (IGLV) genes and joining (IGLJ) segments. *Exp. Clin. Immunogenet.* **15,** 8–18.
23. Barbié, V. and Lefranc, M.-P. (1998) The human immunoglobulin kappa variable (IGKV) genes and joining (IGKJ) segments. *Exp. Clin. Immunogenet.* **15,** 171–183.
24. Pallarès, N., Lefebvre, S., Contet, V., Matsuda, F., and Lefranc, M.-P. (1999) The human immunoglobulin heavy variable (IGHV) genes. *Exp. Clin. Immunogenet.* **16,** 36–60.

25. Ruiz, M., Pallarès, N., Contet, V., Barbié, V., and Lefranc, M.-P. (1999) The human immunoglobulin heavy diversity (IGHD) and joining (IGHJ) segments. *Exp. Clin. Immunogenet.* **16,** 173–184.
26. Folch, G. and Lefranc, M.-P. (2000) The human T cell receptor beta variable (TRBV) genes. *Exp. Clin. Immunogenet.* **17,** 42–54.
27. Scaviner, D. and Lefranc, M.-P. (2000) The human T cell receptor alpha variable (TRAV) genes. *Exp. Clin. Immunogenet.* **17,** 83–96.
28. Scaviner, D. and Lefranc, M.-P (2000) The human T cell receptor alpha joining (TRAJ) genes. *Exp. Clin. Immunogenet.* **17,** 97–106.
29. Folch, G. and Lefranc, M.-P. (2000) The human T cell receptor beta diversity (TRBD) and beta joining (TRBJ) genes. *Exp. Clin. Immunogenet.* **17,** 107–114.
30. Scaviner, D., Barbié, V., Ruiz, M., and Lefranc, M.-P. (1999) Protein displays of the human immunoglobulin heavy, kappa and lambda variable and joining regions. *Exp. Clin. Immunogenet.* **16,** 234–240.
31. Folch, G., Scaviner, D., Contet, V., and Lefranc, M.-P. (2000) Protein displays of the human T cell receptor alpha, beta, gamma and delta variable and joining regions. *Exp. Clin. Immunogenet.* **17,** 205–215.
32. Lefranc, M.-P. (1997) Unique database numbering system for immunogenetic analysis. *Immuno. Today* **18,** 509.
33. Lefranc, M.-P. (1999) The IMGT unique numbering for Immunoglobulins, T cell receptors and Ig-like domains. *The Immunologist* **7,** 132–136.
34. Kabat, E. A., Wu, T. T., Perry, H. M., Gottesman, K. S., and Foeller, C. (1991) Sequences of proteins of immunological interest. National Institute of Health Publications, Washington, D.C., pp. 91–3242.
35. Satow, Y., Cohen, G. H., Padlan, E. A., and Davies, D. R. (1986) Phosphocholine binding immunoglobulin Fab McPC603. *J. Mol. Biol.* **190:** 593–604.
36. Chothia, C. and Lesk, A. M. (1987) Canonical structures for the hypervariable regions of immunoglobulins. *J. Mol. Biol.* **196,** 901–917.
37. Lefranc, M.-P. (2000) Web sites of interest to immunologists, in *Current Protocols in Immunology,* J. Wiley and Sons, New York, NY, pp. A.1J.1–A.1J.33.
38. Giudicelli, V., Chaume, D., and Lefranc, M.-P. (1998) IMGT/LIGM-DB: a systematized approach for ImMunoGeneTics database coherence and data distribution improvement. Proceedings of the Sixth International Conference on Intelligent Systems for Molecular Biology, ISBM-98, 59–68.
39. Giudicelli, V., Chaume, D., Mennessier, G., Althaus, H. H., Müller, W., Bodmer, J., et al. (1998) IMGT, the international ImMunoGeneTics database: a new design for immunogenetics data access. Proceedings of the Ninth World Congress on Medical Informatics, MEDINFO' 98 (Cesnik, B. et al., eds.), IOS Press, Amsterdam, The Netherlands, pp. 351–355.
40. Lefranc, M.-P. (2002) IMGT, the international ImMunoGeneTics database: a high-quality information system for comparative immmunogenetics and immunology. *Dev. Comp. Immunol.* **26,** 697–705.
41. Lefranc, M.-P., Pommié, C., Ruiz, M., Giudicelli, V., Foulquier, E., Truong, L., et al. (2003) IMGT unique numbering for immunoglobulin and T cell receptor variable domains and Ig superfamily V-like domains. *Dev. Comp. Immunol.* 27, 55–77.

4

Antibody Variable Regions

Toward a Unified Modeling Method

Nicholas Whitelegg and Anthony R. Rees

1. Introduction

Predicting the structure of the antibody variable region from sequence has been the focus of considerable research since the work of Kabat and Wu *(1)*, Padlan et al. *(2)*, Stanford and Wu *(3)*, and Feldmann et al. *(4)*. Following the essentially "homology"-based predictions of these early approaches, methods were developed that introduced more rule-based procedures exemplified by our own work *(5–11)* and that of Chothia, Lesk, and colleagues *(12–17)* who, building on the observations of Kabat et al. *(18,19)* and Padlan and Davies *(20)*, developed the concept of canonical classes for certain of the variable region complementarity-determining region (CDRs). Since then, attention has focused on the more difficult problem of non-canonical CDRs, of which the antibody heavy-chain CDR3 (hereafter referred to as H3) is the most unique.

Antibody modeling offers an advantage over protein modeling in general because only the Fv has to be modeled, and the constant region is conserved. In addition, the majority of the Fv itself, the framework, is highly conserved in structure between different antibodies, and can be modeled using the most sequence-homologous known framework *(11)*. For CDR modeling, five of the six CDRs (all except H3) frequently fall into one of between one and ten *canonical* classes, a set for each CDR. Members of a canonical class all have approximately the same backbone conformation. This is determined by the loop length and the presence of a number of key residues, both in the CDR and the framework, which hold the CDR in a particular conformation by hydrogen bonding, electrostatic, and/or hydrophobic interactions. In order to model an unknown CDR, the sequence is examined, the appropriate canonical class assigned, and the most sequence-homologous known CDR is used. For each loop except L2, a few examples fall outside existing canonical classes, and,

From: *Methods in Molecular Biology, Vol. 248: Antibody Engineering: Methods and Protocols*
Edited by: B. K. C. Lo © Humana Press Inc., Totowa, NJ

along with the H3 loop, must be modeled in other ways. One question that must be asked is whether these examples are real "outliers" or whether they are members of additional canonical classes for which only one member has been observed thus far. This is one of the questions that is investigated here.

To date, no strictly "canonical" classification has been possible for CDR H3. The difficulty of modeling the H3 loop arises from the extensive variability in both sequence and structure among different antibodies. There are essentially three approaches to the prediction of H3 conformations: *knowledge-based* methods, such as database searching, in which the closest matching database loop [either from antibodies, or from the entire Protein Data Bank (PDB) ***(21)***] in sequence and length is used as the model—e.g., as in ABGEN ***(22)***—or *ab initio* methods, such as the CONGEN conformational search ***(23)***, or a combination of both ***(8)***. More recent analysis suggests that eventually, knowledge-based methods will provide the degree of accuracy required for most applications of modeling, such as antibody humanization ***(24,25)***. However, at present, a combination of empirical and *ab initio* methods is likely to be necessary when accuracies approaching those of medium resolution X-ray structures are required.

The use of such methods inevitably requires a *screening* step, since choices between a number of different possible structures must be made. This screening may be carried out using some sort of energy function ***(26)***, or by knowledge-based methods in which, for example, the suitability of a sequence for each constructed conformation is evaluated (e.g., as in threading algorithms such as THREADER) ***(27)***. The relative success of such screening methods in antibody modeling is described later in this chapter.

1.1. Do Sequence-Structure Rules Exist for CDR H3?

Recent studies have shown that there are indeed sequence-structure rules, largely for the C-terminal region of the H3 loop ***(11)***. The first extensive analysis was carried out by Shirai et al. ***(28)***, who observed that the C-terminal conformation fell into two conformational groups:

1. "kinked," in which residue H3:C-2 (*see* **Subheading 6.** for specific H3 nomenclature rules) points inward and H3:C-1 outward, and
2. "extended," in which a standard beta-strand extended conformation is assumed, with residue H3:C-2 pointing outward and H3:C-1 inward, in contrast to the "kinked" group.

A number of rules ***(28)*** were formulated for deciding which conformation would be adopted, and the majority of structures were classed as "kinked." These can be summarized as follows: if Asp is present at H3:C-1, either the kinked or extended conformation is formed depending on the presence or absence respectively of a positively charged group at H3:N-1. If Asp is not present at H3:C-1, the kinked conformation is formed.

Morea et al. ***(29)*** have also proposed these two subtypes for H3 loops, using the same set of rules; however, they subdivided the "extended" class into two subtypes, determined by interactions between H3 and conserved aromatic residues of H1. They also suggest that the "sharpness" or "shallowness" of the kink is determined by whether or not the H3:L3 interface is sterically hindered. If it is, the residue at H3:C-3 points toward L2, forming a "sharp" kink, whereas if it is not, it points toward L3, forming a "shallow" kink. However, our studies of an updated set of structures reveal no such correlation.

The majority of existing structures conform to the rules of Shirai et al. and Morea et al. There were three exceptions, which were explained by prolines distorting the H3 loop [Protein Databank (PDB) access code: 1mam], positively charged residues in the N-terminal half of the H3 loop, allowing formation of the kink where an extended structure would be expected (1igi), and differing structures in the free and complexed antibodies (1hil). Neither group offered a clear classification of the center of the loop, except that if a hairpin structure was formed, the subclass of beta-turn formed was determined by the rules of Sibanda et al. ***(30)***.

Shirai et al. have recently extended their investigations ***(31)*** because more structures have become available in the PDB. Their findings have reinforced the kinked/extended rules, and one or two additional rules have emerged. First, more structures are known containing charged residues in (either) the N-terminal half of the loop (or) the residue immediately before the L2 loop. These new structures would be expected to be extended on the basis of sequence, but are actually kinked (5/6), leading to the proposition that if charged residues are in such positions, a kinked rather than an extended structure will be formed.

Secondly, the class of kink has been subdivided into three types: T(rans), C(is), and G(auche), based on the value of the pseudo-dihedral angle formed by the C-alpha atoms of the residues H3:C-4, H3:C-3, H3:C-2 and H3:C-1. Shirai et al. ***(31)*** have proposed that, for steric reasons, if there is a tryptophan in position H3:C-3, the C-type kink is always formed, whereas if there is a glycine in this position, the G-type kink is always formed. This holds true based on the evidence thus far, with all eight structures with a Gly at this position forming a G kink and all four structures with a Trp forming the C kink. However, the converse—that there *must* be a Gly or Trp at these positions to form the G or C kinks respectively—is *not* true. Thus, there are no sequence rules to determine whether a structure will form the most common (T) kink type, and several G or C kinks are sequentially unexplained. Shirai et al. ***(31)*** have further attempted to classify H3 loops into structures that form beta-ladders in the center of the loop, and those that are more disordered. They propose that prolines in certain positions of the loop cause the central section to be completely disordered, whereas aromatic residues at certain positions result in the formation of beta-ladders. The former is true in five of six cases, and the latter in 12 out of 13 (the exception is a very long loop of 17 residues). However, once again, the converse is *not* true: most beta-ladder structures do not have aromatics in the given positions, and six disordered structures do not have prolines in the correct positions. Furthermore, this classification into beta-ladders and disordered structures is questionable: few structures show well-defined beta ladders. Our own detailed structural examination suggests that the structures classified as beta-ladder (for example, the 10-residue loop of 1dba) do not appear to have any more beta-ladder character than others classified as disordered (for example, the 10-residue loops of 1for and 1igi). It appears that for the moment, the beta-ladder rules are too poorly defined to be of any real use. Finally, Shirai et al. proposed that the presence and type of a beta-turn is determined by the rules of Sibanda et al. ***(30)***. Yet again, although when the residues required are present that type of turn will form (but only for H3 loops of 10 residues or less), there are many structures with beta-turns but without the required residues.

In a separate investigation, Oliva et al. ***(32)*** have proposed a further H3 classification. Cluster analysis was performed on the observed H3 structures, from which a number of structural groups emerged. Attempts were made to identify common sequence motifs in each group that would determine the structure. However, few clear, concise motifs emerged, and those that did required broadening as new structures in the database arrived. An additional problem with this classification was that each group contained only a small number of structures (typically two or three), in some of which kinked, extended, and differing length structures were clustered together because the center of the loops occupied similar conformational spaces.

In summary, the investigations reviewed here have provided a defined set of rules for the conformation of the C-terminal of H3 loops in terms of kinked or extended conformations, and some useful information as to whether the G or C subtype of kink will form. This information is clearly useful for antibody modeling, and indeed is exploited by us in our updated, web-based implementation of the AbM algorithm ***(33)***. However, there are too many exceptions to the other proposed rules to be able to construct anything resembling a clear, canonical-style set of rules for H3 conformation. Furthermore, published studies thus far have consistently missed, or only covered to a limited extent, the role of intra-H3 or H3-other CDR side chain-side chain interactions in CDR conformation. For the non-H3 CDRs, this has been shown to be very important in canonical class definitions ***(12)***. These interactions are also very important for H3.

In this chapter we describe the progress in determining a more detailed classification of CDRs, we analyze and present results on side chain conformational modeling and its effect on the accuracy of CDR prediction, particularly CDR H3, and, finally, we propose effective methods for screening candidate H3 conformations. The details of the algorithms and the protocols for modeling an antibody using the web-based algorithm, WAM, are detailed at http://antibody.bath.ac.uk. These new developments result in a modeling protocol that exhibits considerable improvements over other antibody modeling methods.

2. Modeling Approach

2.1. Reclassification of Canonical Loops

For the most recent non-redundant set of antibody structures available from the PDB at the time of writing, L1, L3, H1, and H2 CDRs were grouped into length classes. The conformations were examined (Insight-II, MSI Inc.), any clear outliers were noted, and possible reasons for unusual conformations were determined. In addition, new clusters of conformations were catalogued, and their sequences and interactions were examined for possible new canonical classes. For CDRs that belonged to existing classes on conformational grounds, but did not conform to the existing rules, the exceptional residues were noted, and a decision was made regarding whether the definitions should be widened to include them. The criteria used were conservation of residue type and frequency of occurrence of the atypical residue(s).

2.2. Canonical and CDR-H3 Conserved Side Chains

For the same structure set used in the canonical analysis and each existing canonical class of CDRs L1, L2, L3, H1, and H2, the chi-1 values at each position were analyzed. The data readily separated out into positions in which the chi-1 was entirely random, or where the vast majority of examples exhibited conserved chi-1 angles, for either all or specific residue types. A residue position was conserved if 85% of observed chi-1s were within 20 degrees of the mean, with at least ten observed examples, or all the examples were within 20 degrees of the mean with at least five observed examples. A small number of positions that narrowly failed to meet these criteria (nine observed examples, of which one was an outlier) were also included, particularly if they were in canonical key residue positions, or were conserved in other classes of that loop that differed only in length.

A similar approach was used for the conserved side chains at the C-terminus of H3. Since the "kink" was a conserved feature, the entire set of "kinked" conformations were taken, the chi-1s of the four C-terminal residues were analyzed, and the same criteria were used to determine whether each position was conserved. For the H3-conserved side chains, the chi-2 angles were also analyzed.

2.3. H3 Backbone Analysis

A set of 70 non-redundant crystal structures of antibodies was used for the analysis of H3 loops **(Table 1)**. Selected structures were required to have a resolution below 3.0 Å with the absence of any structural uncertainties. The H3 loops were grouped into structural classes by fitting members of each length group onto each other, followed by examination in Insight-II. Any apparent similarities were subsequently verified by calculating the global RMSD between each H3 loop. When similarities were observed, a search was made for possible canonical-type features, such as hydrogen bond, salt bridge, and hydrophobic interactions, using both visual examination for identification and the *CONTACT* program (N. Whitelegg, unpublished results) to analyze and list probable interactions of those types. For exceptions in a length class that exhibited a similar conformation to the remaining members, the same methods were used to search for interactions that might be responsible for the unusual conformation. Figures were generated with MOLSCRIPT *(36)* and Raster3D *(37)*.

2.4. Modeling Protocol

The methods for modeling an antibody variable region from sequence are now associated with a web-based service, WAM, which is free to academic users. It can be found at http://antibody.bath.ac.uk. A typical procedure is presented in **Subheading 5.** Given this existing service, this chapter does not follow the style of others in this volume by describing detailed procedures within the text. However, to understand the theoretical basis for some of these procedures, a detailed analysis of CDR canonical and non-canonical conformations follows. The results of this analysis are used within the WAM algorithm, but are not detailed here. Of course, readers may use the detailed descriptions of canonical structures and the H3 "rules" in a purely manual fashion.

3. CDR Conformations

3.1. Reclassification of Canonical Loops

Our new findings on canonical structures, together with a summary of the rules for each loop, are presented here. For more details of the existing classes, refer to the studies of Chothia and colleagues *(12–17)*. Note that the numbering scheme for some of the classes differs from that used by Chothia and colleagues, but where differences exist they are clearly indicated. Note also that CDR residues are indicated using a different notation here: for example, L1:N + 1 indicates the residue immediately following the N-terminal residue of L1, and L3:C-2 is the residue two before the L3 C-terminal residue. Residue H2:C + 13 describes the framework residue 13 residues after the H2 C-terminus, and L:N + 1 describes the framework residue following the N-terminal residue of the entire light chain.

Table 1
Sequences and Lengths of the CDR-H3 Loops Used in the Analysis

PDB Code	H3 Length	H3 Sequence	PDB Code	H3 Length	H3 Sequence
1psk	5	KSFDY	1ay1	10	YYYGYWYFDV
1hkl	5	YYGIY	1cfv	10	LNYAVYGMDY
1ind	5	HRFVH	1clz	10	GLDDGAWFAY
1ggb	5	EGYIY	1nbv	10	DQTGTAWFAY
1gpo	5	WHGDY	1kb5	10	SRTDLYYFDY
1cr9	5	DLMDY	1clo	10	DRGLRFYFDY
1nld	5	RGGDF	1igf	10	YSSDPFYFDY
1sm3	6	VGQFAY	1rmf	10	GGWLLLSFDY
1cgs	7	GYSSMDY	1bln	10	YYRYEAWFAS
1mlb	7	GDGNYGY	1eap	10	SYYGSSYVDY
1tet	7	RSWYFDV	1aif	10	RPLFYYAVDY
1ghf	7	VEAGFDY	1a6v	11	YDYYGSSYFDY
4fab	7	SYYGMDY	1fvc	11	WGGDGFYAMDY
1mrc	7	LRGYFDY	2h1p	11	RDSSASLYFDY
3hfl	7	GNYDFDG	1frg	11	RERYDEKGFAY
1vfa	8	ERDYRLDY	1igc	11	WGNYPYYAMDV
1fgn	8	DNSYYFDY	1mcp	11	NYYGSTWYFDV
1a6t	8	RDDYYFDF	1fpt	11	DFYDYDVGFDY
1mim	8	DYGYYFDF	1ncb	11	GEDNFGSLSDY
1kem	8	WGSYAMDY	1axs	11	GHSYYFYPGDY
1mam	8	DPYGPAAY	12e8	11	GHDYDRGRFPY
1plg	8	GGKFAMDY	1ad9	11	EKTTYYYAMDY
1cic	8	GLAFYFDH	1ejo	11	RAFDSDVGFAS
1b2w	8	GFLPWFAD	1nqb	11	YDYYGSSYFDY
2fbj	9	LHYYGYNAY	1hil	11	RERYDENGFAY
1igt	9	HGGYYAMDY	1igm	12	HRVSYVLTGFDS
1ar1	9	HEYYYAMDY	1vge	12	DPYGGGKSEFDY
1jhl	9	DDNYGAMDY	6fab	12	SEYYGGSYKFDY
2jel	9	VMGEQYFDV	1iai	12	DGYYFNYYAMDY
7fab	9	NLIAGGIDV	1dee	12	VKFYDPTAPNDY
1nfd	9	AGRFDHFDY	1osp	12	SRDYYGSSGFAF
1mfb	9	GGHGYYGDY	1ap2	12	REVYSYYSPLDV
1for	10	SGNYPYAMDY	1jrh	12	RAPFYGNHAMDY
1dba	10	GDYVNWYFDV	1mpa	12	IYFDYADFIMDY
1igi	10	SSGNKWAMDY	1dsf	12	SPIYYDYAPFTY
1a4j	10	AERLRRTFDY			

Table 2
Ten-Residue L1 Loops that Structurally Belong to Canonical Class 1 But Are Not Classed as Such by the Existing Rules, Showing the Differences in Residue Type From Those Required

Residues	L:N + 1	L1:N + 1	L1:N + 5	L1:C-1	L2:C + 15
Required for class	I	A	V	L,M	Y
1a6t	S	P	V	M	F
1clo	T	A	V	I	Y
1mim	I	A	R	M	Y
1psk	I	A	V	I	Y

3.2. CDR-L1

The existing classification *(17)* defines five definitive canonical classes for kappa chains, plus five indefinite classes: four lambda, and one kappa. New structures solved since that classification have clarified some of the definitions. The five existing classes—class 1 (10 residues), 2 (17 residues), 3 (11 residues), 4 (16 residues), and 8 (15 residues) (referred to by Chothia and colleagues as K1 to K5, respectively)—are all held in place by a hydrophobic interaction made by buried residue L1:N + 5, which contacts other hydrophobics at L1:C-1, L:N + 1 and L2:C + 15, as well as a Ser or Ala at L1:N + 1.

3.2.1. Class 1 (10 Residues)

An examination of the entire set of 10-residue L1s has expanded the allowed range of residues that make the hydrophobic interaction. Of the ten structures, four—1a6t, 1clo, 1mim, and 1psk—do not fit into the existing definition, yet on structural grounds, they are members of the class. **Table 2** shows that much of the sequence variation involves homologous residues: Phe rather than Tyr for 1a6t L2:C + 15, Ile rather than Met for 1clo and 1psk L1:C-1, and Pro rather than Ala for 1a6t L1:N + 1. However, there are three non-conserved substitutions: Ser and Thr rather than Ile for 1a6t and 1clo L:N + 1, respectively, and Arg rather than Val for 1mim L1:N + 5. On this basis, it was decided to widen the definition of this class to include loops containing Pro at L1:N + 1, Ile at L1:C-1, and Phe at L2:C + 15, but to exclude those loops containing the non-conserved substitutions.

3.2.2. Class 2 and New Class 6 (11 Residues)

Of the 59 occurrences of 11-residue L1s, only 40 fit the canonical sequence rules for class 2, whereas all but two are members of the class on conformational grounds. The two exceptions—8fab and 1nfd—both contain a Pro at residue

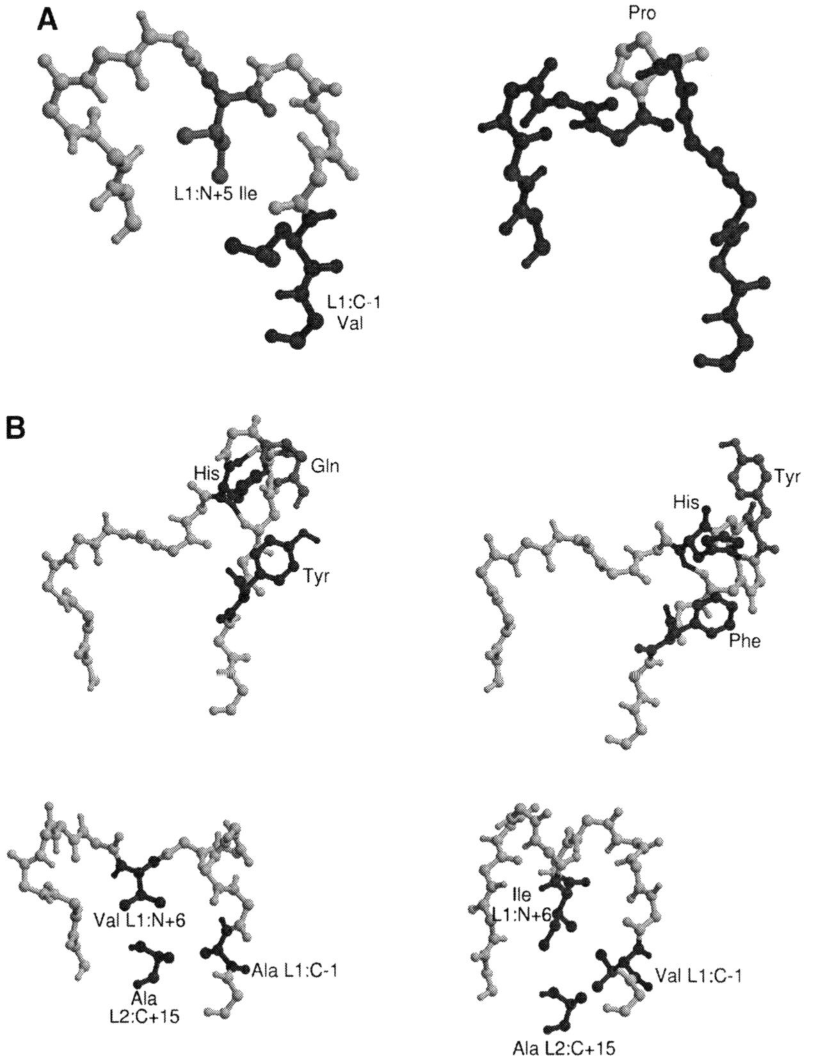

Fig. 1A. The contrasting conformations of 10-residue L1 loops of class 1 (1vfa, *left*) and class 6 (8fab, *right*).

Fig. 1B. **Top:** 16-residue L1 loops. Left: 1dba, an example of a 16-residue L1 loop with the defining class 4 interaction; right, 1axt, an example of a 16-residue L1 loop that cannot form the class 4 interaction as a result of steric effects. **Bottom:** the contrasting conformations of 14-residue L1 loops of class 9 (1gig, *left*) and class 10 (7fab, *right*).

L1:N + 5 rather than the normal large hydrophobic, thus distorting the usual structure (Al-Lazikani et al. ***[17]*** also place 8fab in a separate class). To explain this unusual structure, we define here a new class, class 6 (*see* **Fig. 1A**).

Of the remaining 17 structures that are members of the class on conformational but not sequence grounds, widening of the range of allowed residues at L:N + 1 and L1:C-1 from Ile and Leu respectively to either Ile, Val, or Leu in either position, returns 16 of the 17 structures to class 2 membership. The remaining member—1ghf—shows an unusual substitution from Ala to Glu at residue L1:N + 1, and is excluded.

3.2.3. Class 3 (17 Residues)

Of the seven 17-residue L1s, only one fails to meet the canonical sequence criteria, but passes the conformational test. This exception, 43c9, results from a substitution of Ile by Val at residue L:N + 1. It was decided to add this conservative substitution to the range of allowed residues for this class.

3.2.4. Class 4 (16 Residues)

Of the 32 instances of 16-residue L1s, all but four are conformational members of this class, but only 16 fit the canonical rules. As with class 2, all but one L1 could be returned to this class by expanding the definitions to include conserved substitutions, with one exception. 1kem has Phe rather than Ser at L1:N + 1, a substitution that slightly perturbs the structure.

The four conformational outliers—1bln, 4fab, 1axt and 1cfv—are all different, and therefore are not classifiable. By sequence, three of the four show differences (Thr, Gln, or Tyr) from the normally conserved Asn (occasionally Asp or Ser) at residue L1:C-6, whereas in the remaining L1– 1cfv—the otherwise absolutely conserved Gly at L1:C-5 is substituted by Arg, preventing formation of the canonical conformation on steric grounds.

Structural examination of the canonical members reveals that Asn at L1:C-6 forms hydrogen bonds with one or both of L1:C-2 (normally Tyr) and L1:C-8 (normally His) **(Fig. 1B)**. In those instances when one of these two is substituted by a residue unable to hydrogen bond, L1:C-6 is able to form the other hydrogen bond and maintain the structure; furthermore, Asp and Ser at L1:C-6 could also make the interaction. However, when L1:C-6 is one of Thr, Gln, or Tyr (in the outliers) it is impossible to form this interaction **(Fig. 1B)**. Therefore, in addition to widening the range of hydrophobic residues allowed for class 4, we have proposed a restriction of residue L1:C-6 to Asn, Asp or Ser, and L1:C-5 to Gly.

3.2.5. Class 5 (13 Residues)

The only one member of this class—2fb4—has a lambda light chain. The conformation is unlike other lambda light chain L1s of 14 residues, but, as with other class 5 members, contains a buried residue at L1:N + 5. Until other 13-residue L1s are observed, we propose to retain this class, first suggested by Al-Lazikani et al. *(17)*.

3.2.6. Class 7 (12 Residues)

Three 12-residue L1 structures (1aif, plus the new structures 15c8 and 1cf8) exhibit very similar structures, and also feature the typical hydrophobic interaction of L1s (*see* **Subheading 3.2.**; Andrew Henry, personal communication). Therefore, on this basis, it was decided to define a new class for 12-residue L1s, determined by hydrophobics at L1:N + 5, L1:C-1, L:N + 1 and L2:C + 15, as well as a Ser or Ala at L1:N + 1.

3.2.7. Class 8 (15 Residues)

There are currently six 15-residue L1s. All conform to the "K5" class ***(17)***, both on conformational and sequence grounds.

3.2.8. Class 9 and Class 10 (14 Residues)

A number of new 14-residue L1 structures have recently been defined, and all of these are from lambda light chains, making a total of nine structures. Again, they exhibit the hydrophobic interaction typical of L1s, but involve L1:N + 6 rather than L1:N + 5. It can be seen from **Table 3A** that on sequence grounds they fall into two clear categories (1aqk and 7fab, and the remainder), based on the residue type at L1:N + 1, L1:N + 6 and L1:C-1. On conformational grounds, they also separate into two groups corresponding to the sequence groups. The reason for the differences in conformation appears to be sterically driven, arising from residue size differences at L1:C-1, and L1:N + 6 (**Fig. 1B,** bottom). The conservation of sequence and structure in the main group of 7, together with the fact that two 14-residue classes corresponding to the classes outlined here were also proposed by Al-Lazikani et al. ***(17)***, leads us to define a new class 9 and a tentative new class 10 **(Table 3B)**.

3.3. CDR L2

The overwhelming majority of CDR-L2 loops follow the one existing canonical structure described by Chothia and colleagues ***(12–14)***. Currently, only one or two L2 loops fail to fit this classification, but the low incidence of nonconformist structures does not allow additional classes to be defined with any degree of confidence.

3.4. CDR-L3

Existing canonical classes for CDR-L3 ***(12–17)*** are present for eight- (class 3) and nine- (classes 1 and 2) residue loops. They are conformationally similar to hairpin loops, but with perturbations that depend on the positions of proline residues. A Gln residue at L3:N + 1 also plays a role in holding the conformation through hydrogen bonds to backbone atoms. The L3 classes have been substantially revised, as described here.

Table 3A
Residues Involved in the Hydrophobic Interaction in 14-Residue L1s, Showing the Differing Residue Types at L1:N+6 and L1:C-1 for 1aqk and 7fab and the Remainder

	L1:N+1	L1:N+6	L1:C-1	L2:C+15
1a6v	S	V	A	A
1aqk	**G**	**I**	**V**	**A**
1gig	S	V	A	A
1ind	S	V	A	A
1mfb	S	V	A	A
1ngq	S	V	A	A
1sm3	S	V	A	A
1yuh	S	V	A	A
7fab	**G**	**I**	**V**	**A**

Table 3B
Proposed Definition of New 14-Residue L1 Canonical Classes, Based on the Differing Residue Patterns and Conformation in 1aqk and 7fab Compared to Examples in Table 3A

	Class 9 (all except 1aqk,7fab)	Class 10 (tentative) (1aqk,7fab)
L1:N+1	S	G
L1:N+6	V	I
L1:C-1	A	V
L2:C+15	A	A

3.4.1. Class 1 (9 Residues)

This is the most populated class, and is determined by a Gln at L3:N + 1 and Pro at L3:C-2. All 65 structures that fit the canonical rules have the required conformation (broadly), and similarly, all structures with the required conformation fit the rules.

3.4.2. Nine-Residue Loops Outside Class 1

The existing Chothia classes include a class 2 for nine-residue L1s, which has an altered conformation as the Pro is at L3:C-3 rather than L3:C-2. However, an examination of the ten nine-residue L3s not in class 1, listed here, reveals that only one structure thus far—2fbj—fits the rules and the remainder—all lambda light chains—have different sequence patterns (**Table 4A**). A new group of six structures with virtually identical sequence exists for nine-residue L3s. Examination of the conformations reveals that these six had almost identical conformation, and therefore should be considered to be a new

Table 4A
Nine-Residue L3s Outside Canonical Class 1, Showing Six Examples (in Bold) That Exhibit a New Hydrophobic Interaction Not Seen in Previous Examples

2fbj	QQWTYPLIT
1a6v	**ALWYSNHWV**
1ind	**ALWYSNLWV**
1gig	**ALWYSNHWV**
1ngq	**ALWYSNHWV**
1sm3	**ALWYSNHWV**
1yuh	**ALWYSNHWV**
7fab	QSYDRSLRV
8fab	QAWDNSASI
1mfb	ALWCNNHWI

Table 4B
Definition of New Hydrophobic L3 Canonical Class 6

L3:N+1	L
L3:N+2	W
L3:N+3	Y
L3:C-1	W
L:N+3	V
L1:N+5	V
L1:C-2	Y
H2:N-3	W

canonical class. Examination of the interactions reveals that unlike other L3 canonical classes, but like L1 and H1 classes, they are held in place by hydrophobic interactions, of which three exist. First, the Leu at L3:N + 1 is surrounded by the fourth residue in the light-chain framework (Val), L1:N + 6 (Val), and L3:N + 3 (Tyr). Second, the Trp at L3:N + 2 is surrounded by Tyr L1:C-2 and the other L3 Trp (L3:C-1). Finally, Trp L3:C-1 is surrounded by the first L3 Trp as well as the framework Trp immediately preceding H2 (H2:N-3), and in five of six cases, the hydrophobic H3 residue at H3:C-2 that contributes to its characteristic "kink" *(28)*. In the other case—1yuh—the H3 has an unusual kink conformation, because of a Pro at H3:C-3, and the Trp packs against the Pro instead of the H3:C-2 residue. Therefore, a new L3 canonical class can be designated (**Table 4B; Fig. 2,** top).

It is notable that all six structures exhibit the new L1 14-residue canonical class 9, which contains the key Val and Tyr residues described here. The other three nine-residue L3 structures—7fab, 8fab, and 1mfb—contain only some of

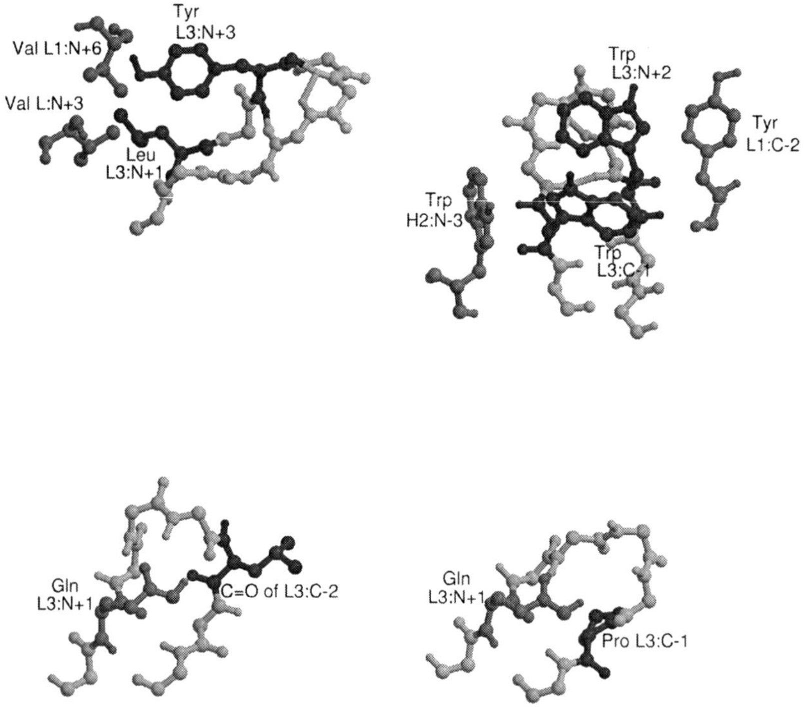

Fig. 2. L3 loop interactions. **Top:** two views of the hydrophobic interactions of new class 6 as seen in an example structure (1gig); **bottom:** the differing conformations of eight-residue L3s of new canonical class 4 (1a0q, *left*) and existing class 3 (3hfl, *right*).

the key residues for the new class, and therefore are perturbed from the canonical conformation.

3.4.3. Eight-Residue Loops

The existing Chothia classes ***(12,17)*** define only one canonical class for eight-residue L3s, determined by the Gln at L3:N + 1 and a Pro at L3:C. However, of the five eight-residue L3s with known structure, only one—3hfl—exhibits this sequence pattern, and the remainder show a different sequence and conformation (**Table 5; Fig. 2,** bottom).

The remaining loops are virtually identical in conformation. They are simple hairpin loops, with a conformation determined by hydrogen bonds from the Gln at L3:N + 1, as in class 1. Thus, it can be said that 3hfl is in fact the outlier with the Pro perturbing the structure from the norm for eight-residue L3s, and we can divide eight-residue loops into two canonical classes: the existing class 3 (3hfl) and new class 4, determined by Gln L3:N + 1 and the absence of Pro at L3:C-1.

Table 5
Sequences of Eight-Residue L3 Loops

3hfl	QQWGRNPT
1a0q	LQYYNLRT
1c5d	LQYGNLYT
1eap	LQYYNLRT
1jrh	QQYWSTWT

3.4.4. Seven-Residue Loops

Al-Lazikani et al. ***(17)*** suggested a seven-residue class. The new eight-residue class 5, with the Gln as the only important determining residue (no Pro is needed to determine the conformation) lends support to the definition of a new L3 canonical class 5 for the two seven-residue L3s determined thus far, which also lack a Pro yet contain the Gln and are very similar to each other in conformation (1dfb: QQYNSYS, and 1mim: HQRSSYT).

3.4.5. Ten- and 11-Residue Loops

Apart from the six canonical classes for L3, four longer loops are also seen. These are the 10-residue loops of 1aqk and 1ar1, and the 11-residue loops of 1nfd and 2fb4. Unfortunately, both the 10-residue and 11-residue examples are dissimilar, so classes cannot be defined.

3.5. CDR-H1

Existing Chothia classes for CDR-H1 cover a 10-residue class 1—a slightly perturbed version of class 1 designated class 2—and an 11-residue class 3. Both class 1 and class 3 are determined by hydrophobic contacts made by H1:N + 1 and H1:N + 3, whereas class 2 contains polar residues at H1:N + 1, which perturbs the conformation slightly, although it is still within 2 Å of the average class 1 conformation.

3.5.1. Classes 1 and 2 (10 Residues)

3.5.1.1. Conformation

Of 116 10-residue H1s, in general, 113 structurally fit into class 1 or 2. As Chothia and colleagues have suggested previously, there are minor differences in individual members of the main structural group, caused by minor differences in sequence and in the other CDRs. The three members of class 2—3hfm, 1gpo, and 1osp (included by widening the dependence on Asp or Thr at L1:N + 1 to Asp, Thr or Glu)—do not noticeably stand out in conformation from the rest, and all members of class 1 and 2 are within 1.5–2.0 Å of each

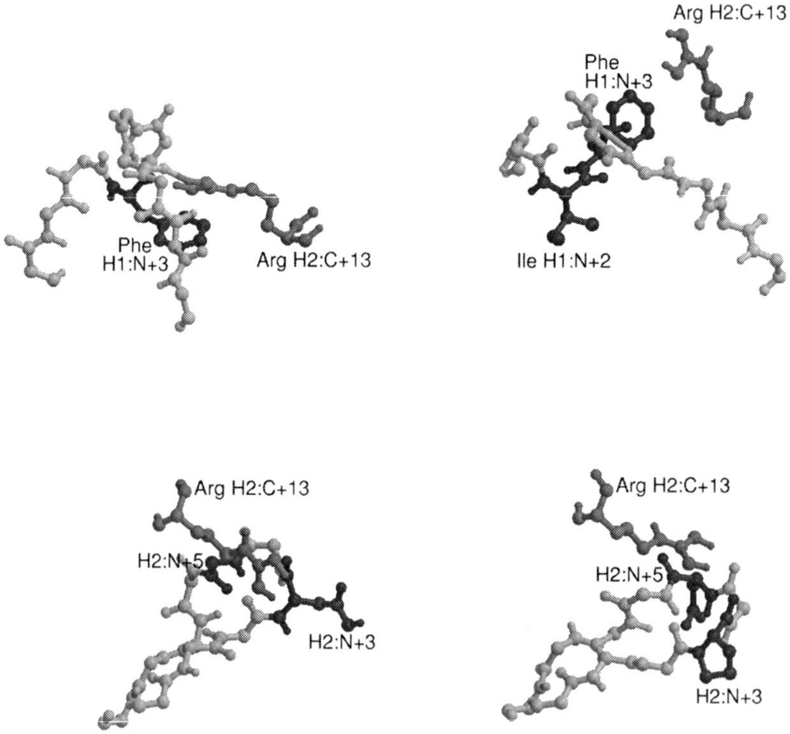

Fig. 3. The unusual H1 and H2 conformations of 1b2w, compared to a normal structure (1igt). The larger residue at H2:N + 5 in 1b2w *(bottom right)* compared to 1igt *(bottom left)* would cause a steric clash if the local backbone conformation and the conformation of Arg H2:C + 13 adopted the typical class 3 interaction; therefore H2:C + 13 adopts an alternative placing, which in turn causes the H2 to adopt a class 2 rather than the expected class 3 conformation. In addition, the differing H2:C + 13 in 1b2w also influences the H1 conformation *(top right)* compared to the normal *(top left)*.

other by locally fitted root-mean-square deviation (RMSD), although they are more similar to each other than to the majority of class 1 members. Three structures (1nfd, 1b2w, and 1igi) are structurally distinct from the main group. 1nfd has a very different sequence from the norm, with an Asp at H1:N + 3 compared to a hydrophobic in all but one other examples, whereas the difference in 1b2w is restricted to the N-terminal and results from a combination of two effects. First, Ile H1:N + 2 (relatively unusual; normally a hydrophilic residue is found here) is buried in the normal position of Phe H1:N + 3, which has a different conformation in order to retain packing with Arg H2:C + 13, which in turn shows an unusual conformation because of features in CDR-H2 **(Fig. 3)**. Both these unusual features (the Ile and the unusual Arg conformation) appear to be essential, as structures with just one form the normal H1 10-residue N-terminal conformation. The conformation of 1igi is very different from the remainder—a slack, open loop—but no explanation for the difference has been determined. It could be, as with the CDR-H3 of 1hil, that the normal canonical conformation is formed on complexation.

3.5.1.2. SEQUENCE

A total of 17 structures, which on conformational grounds belong to the main 10-residue group, are not classified as canonical by the existing rules. 1osp has been mentioned here, and of the others, 12 could be included by removing the dependency on a positively charged residue at H3:C-3. A wide range of residues are observed at this position over all the class 1 members, including Gly, Ser, Ile, Ala, and Gln, so it would appear that this is not a determining residue. Three more are included by widening the residues at position H:N + 1 to include Leu and Ile and H1:C-1 to include Phe, leaving only one structure, 1eap, still sequentially lying outside the class. This is because of the presence of Ser rather than a hydrophobic at H1:N + 3, and for reasons already mentioned in the L1 discussion, it was decided not to include Ser in the range of allowed residues at this position.

3.5.2. Class 3 (11 Residues)

A total of five 11 residue H1 loops have been observed (1baf, 1c12, 1cf8, 1f58, and 1ay1). All of these have the canonical conformation and fit the canonical rules.

3.5.3. Class 4 (12 Residues)

This is another class proposed by Al-Lazikani et al. ***(17)***, formed by inserting two residues into the standard CDR-H1 conformation (the 10-residue class 1) rather than one as in class 3. Examination of the four examples (1acy, 1ggi, 1jrh, and 2hmi) reveals that all have an almost identical conformation, and like the other classes, they are held in place by the two hydrophobic interactions at H1:N + 1 and N + 3. Therefore, a new class 4 has been defined for 12-residue H1s, with the same residue dependencies as for classes 1 and 3.

3.6. CDR-H2

There are four canonical CDR-H2 conformations—one of 9 residues, one of 12 residues, and two of 10 residues. The conformations are determined by residue types required for hairpin loops, and by the residue at framework position H2:C + 13, which packs against the H2 loop.

3.6.1. Class 1 (9 Residues)

All of the 23 nine-residue H2s show the canonical conformation. However, two fail to match the sequence requirements. 7fab shows Val at position H2:C +

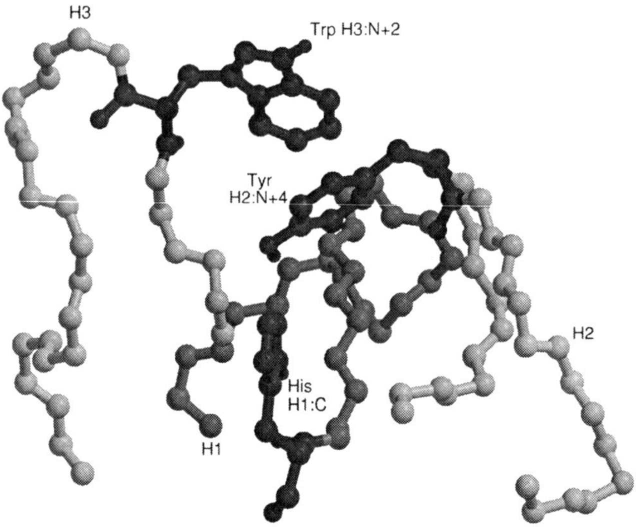

Fig. 4. The unusual conformation of the H2 loop of 1rmf showing the underlying interaction.

13 rather than Arg, Lys, or Ile. Since Val is similar in type to Ile, it was decided to widen the allowed types here to include Val. 1f58 shows Ala at H2:N + 5 rather than Gly or Asp, residues favored in turns. Since this is the first observation of Ala at this position, it was decided not to include it as an allowed residue type for the class, although if it is observed in subsequent structures, it may be included.

3.6.2. Classes 2 and 3 (10 Residues)

As discussed by Chothia et al. ***(13)***, the main determinant for whether a 10-residue loop belongs to class 2 or 3 is the residue type at position H2:C + 13. In class 3, this residue is Arg, which forces the loop residue H2:N + 3 onto the surface. In class 2, this residue is smaller (Ala, Leu, or Thr), leaving a cavity in the space the Arg would normally occupy, which is instead filled by residue H2:N + 3.

3.6.2.1. Conformation

A single structure—1rmf—belongs neither to class 2 or class 3 conformationally, with the N-terminal end of the H2 pointing outward toward H3. This was found to be stabilized by a "sandwich" aromatic interaction, between the Tyr at residue H2:N + 4, the His at the C-terminal of H1, and a Trp at H3:N + 2 (**Fig. 4**). The Tyr and His are frequently observed in these positions, whereas the H3 Trp is the only incidence in 10-residue H3s at this position, so it can be concluded that the Trp is the residue forcing the conformation.

Additionally, two structures that fit the class 3 rules conformationally belong to class 2; these are discussed here.

3.6.2.2. SEQUENCE

All 22 structures in class 3 on conformational grounds fit the sequence requirement of an Arg at framework residue H2:C + 13, yet of 66 structures in class 2, 33 do not fit the sequence-structure rules. A total of 23 of these show Val at H2:C + 13 rather than Ala, Leu, or Thr, a clear indication that Val is among the range of smaller (than Arg) residues allowed at this position. Two structures each show Asp, Glu, or Val at residue H2:N + 6, compared to the stipulated Gly or Ser. The requirement for Gly and Ser is related to requirements for turns ***(12,13)***; however, these are preferred rather than required residues. Because there was more than one observation of each, the range of allowed residues was widened to include Asp, Glu, and Val. One structure (1dba) shows Ile rather than Pro, Thr, or Ala at H2:N + 3, and one more (1mpa) Arg at H2:N + 6; however, the canonical rules were not extended to cover these because there has only been a single observation of each. Nonetheless, the variation at H2:N + 6 reinforces the idea that the most important determining residue is the framework residue H2:C + 13, rather than any of the loop residues themselves. Finally, however, two structures (1b2w and 1vge)—which sequentially belong to class 3, with an Arg at this position—conformationally belong to class 2, which shows smaller residues at this position. This is because the Arg points outward rather than into the cavity, resulting in less steric hindrance than there would have been otherwise, and in turn allowing H2:N + 3—the smaller residues Ala and Pro—to be buried, leading to the class 2 conformation (**Fig. 3**). However, the reason for the differing Arg conformation in these two cases could not be determined, although it appears to be a result of steric reasons as a result of a larger than normal (Asp, Thr vs Gly, Ser) residue at H2:N + 5. (The Asp in 1b2w appears to be unable to form a salt-bridge interaction with the Arg on geometric grounds). In addition, in 1b2w, this influences the CDR-H1 conformation (**Fig. 3**). Therefore, it was decided to modify the rules for class 3 so that if Arg is present at H2:C + 13, but a residue other than Gly or Ser (or, by inference, Ala) is found at H2:N + 5, the class 2 rather than class 3 conformation is formed.

3.6.3. Class 4 (12 Residues)

Of a total of 14 twelve-residue H2s, 12 conformationally belong to this class. Of the two outliers, one—1nfd—has a completely different sequence, whereas the other—1aif—features a conserved substitution of the key residue Tyr by Phe at H2:C-3. Chothia and colleagues ***(15)*** indicate that the role of the Tyr involves a nonpolar packing interaction against the Arg at H2:N + 13, something that Phe should also be able to do. One possible explanation for the difference is that, although no actual hydrogen bonds are formed, polar interactions between the Tyr-OH and the backbone CO and NH groups in the preceding (apex) residues may play a role in holding the apex conformation, something suggested by examination in Insight-II. Phe cannot do this, and the energetics could therefore favor the observed conformation, away from the Arg and the apex of the loop, packing with Ile H2:N + 1 instead. This in turn would influence the loop apex conformation, the region in which 1aif differs.

Of the 12 with the required conformation, two do not meet the sequence requirements. 1ce1 features Val rather than Arg at the key framework residue H2:C + 13, whereas 1dlf features His rather than Tyr at H2:C-3. Since Val has been shown to be a replacement for Arg in class 1, and His can perform the function of Tyr in terms of the polar interactions suggested here, both these were deemed to be acceptable substitutions to incorporate into the canonical rules.

3.7. Conserved Side Chains

3.7.1. Canonical CDR-Conserved Side Chains

Examination of the chi-1 angles of side chains of canonical loops of known structure has revealed that side chains at certain positions in each class are conserved in conformation, with respect to chi-1 (**Table 6**). Although certain canonical key residues are included in this list, others vary more between structures in terms of chi-1 (although their general position is obviously similar). By contrast, certain other residues, which may be involved in other roles such as solubilizing the antibody, do have conserved chi-1s.

3.7.2. H3-Conserved Side Chains

The investigation of the "kinked" H3 structures conducted in this work has led to the discovery of the fact that the chi-1s of the residues forming the "kink" are conserved (for definition, *see* **Subheading 2.2.**). The conservation is independent of the length of the loop; almost all examples of "kinked" loops conserved the chi-1s of the four C-terminal residues (**Table 7**).

The small number of outliers are not dependent on the length of the loop, or whether or not the H3 falls into one of the well-defined structural classes presented in the preceding section, so it was deemed safe to use the entire set of "kinked" loops for determination of conservation.

There is also some degree of conservation in chi-2 for specific residue types only. **Table 8** illustrates residues for which at least 64% are within 30 degrees of the mean.

3.8. H3 Backbone Classification: An Improved Description of H3 Loops

The analysis of CDR-H3 loops has yielded some interesting results, not covered in the work of others, and this is presented below for each length H3.

Table 6
Residues of Canonical Loops with Conserved Chi-1 Values

Loop and class	Residue	Residue type	Loop and class	Residue	Residue type	Loop and class	Residue	Residue type
L1,1	3	S	L1,10	7	I		8	W
L1,2	6	I		13	V		9	I
	7	Y	L2,1	4	T		9	V
	9	Y		6	D		10	[EHNY]
	10	L		6	I	H1,2	7	[FY]
	10	V		6	[FHY]		9	W
L1,3	6	L	L3,1	1	F	H1,3	2	Y
	7	[FL]		4	[HWYNF]		4	I
L1,4	3	S		5	H		5	T
	6	L		6	Y		11	[NH]
	7	V		7	[FWY]	H2,1	1	Y
	8	N	L3,4	1	[LQ]		2	I
	10	H		3	Y		3	S
	13	T		4	[YW]		3	W
	14	[YFH]		5	[NS]		5	D
	15	[LMF]		8	T	H2,2	2	[IFLV]
	16	E	L3,5	1	[QH]		3	[DNY]
	16	H		2	Q		5	Y
L1,7	11	*	L3,6	2	L		6	T
L1,8	5	S		3	W	H2,3	1	T
	6	*		7	[HL]		2	I
	13	F		8	W	H2,4	1	F
	15	*	H1,1	2	[LIYF]		2	I
L1,9	2	S		4	F		3	R
	7	*		7	[FY]		4	N
	11	N		7	N		9	Y
	12	[YH]		8	N			

The residue numbering is a local numbering for each loop, where 1 indicates the N-terminal residue. Note that when more than one type of conserved residue is shown for a position on separate lines, the chi-1s are different for each residue type, whereas where residues are grouped together, they all have chi-1s within the same range

3.8.1. Five-Residue Loops

Because of their small size, five-residue H3s do not show well-defined "kinked" or "extended" C-terminals. Nevertheless, they do show either the "kinked" H3:C-2 carbonyl/Trp H3:C + 1 side chain interaction, or, in one case

Table 7
Conserved chi-1 Angles at the C-Terminus of H3 Loops

Residue	Number conserved	% conserved	Chi-1 (approx)	Residue types not conserved[i]
H3:C-3	21/24	87.5	60	Trp (3 examples) have chi-1 of –60, rather than 60
H3:C-2	All	100	–60	
H3:C-1	30/33	90.9	60	
H3:C	33/35	94.3	–60	Val not conserved

[i] These residue types are not included in the calculation of percentage conserved, or in the number of conserved totals.

Table 8
Conserved chi-2 Angles at the C-Terminus of H3 Loops

Residue and type	Number within 30 degrees of the mean	Percentage within 30 degrees of the mean	Mean chi-2 (of this group only)
H3:C-3 (Y)	9/14	64	85
H3:C-2 (YFN)	22/27	81	–59
H3:C-2 (MLI)	9/12	75	176
H3:C-1 (D)	22/33	67	–11
H3:C (YF)	29/33	88	–82

(1nld), the "extended" Asp H3:C-1 side chain/Trp H3:C + 1 side chain interaction. Furthermore, unlike the longer loops, they separate into reasonably well-defined structural subgroups that can be determined by sequence. For this reason, they are treated separately from H3 loops of longer lengths. The global RMS deviations between the structures are shown in **Table 9.** Three well-defined structural groups, determined by sequence, are revealed **(Fig. 5)** and described here.

3.8.1.1. No Asp at H3:C-1 or Asp at H3:C-1 and Arg at H3:N-1

This is the rule which, for longer loops, defines a "kinked" structure. Five structures—1ggi, 1ind, 3fct, 1aj7, and 1nld—show this sequence pattern, and as **Table 9** shows, all except 1nld are close in conformation. 1ind is slightly different, as it is perturbed slightly by an interaction between His H3:C and Ser H3:N-1. A further structure, 1psk, shows this conformation although it has an Asp and no Arg; the Lys at H3:N takes the place of the Arg in forming the salt-bridge with the Asp.

Table 9
Backbone, Global RMSD Between Five-Residue H3 Loops

	1gpo	1dqq	1nld	1aj7	3fct	1ggi	1psk	1ind
1dqq	0.4	–	–	–	–	–	–	
1nld	3.2	3.0	–	–	–	–	–	
1aj7	2.2	2.1	2.5	–	–	–	–	
3fct	1.5	1.4	2.1	1.3	–	–	–	
1ggi	2.4	2.4	2.4	1.2	1.3	–	–	
1psk	2.6	2.5	2.8	1.0	1.7	1.3	–	
1ind	2.4	2.3	3.3	1.3	1.9	2.2	1.9	
1cr9	5.5	5.4	3.2	4.4	4.4	3.8	4.4	5.3

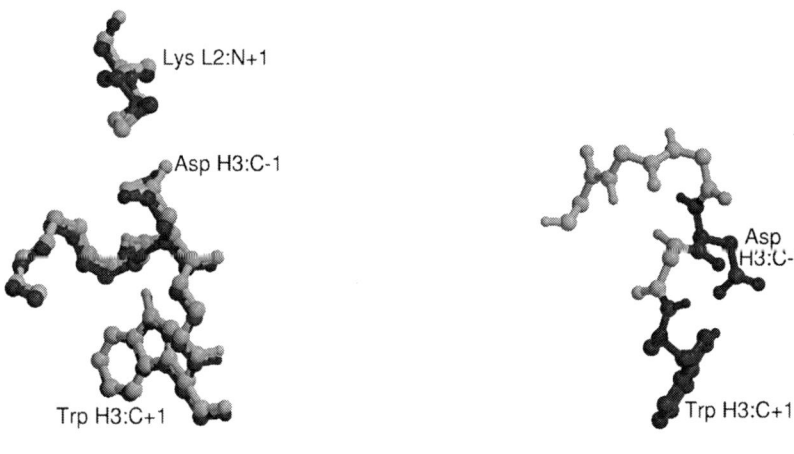

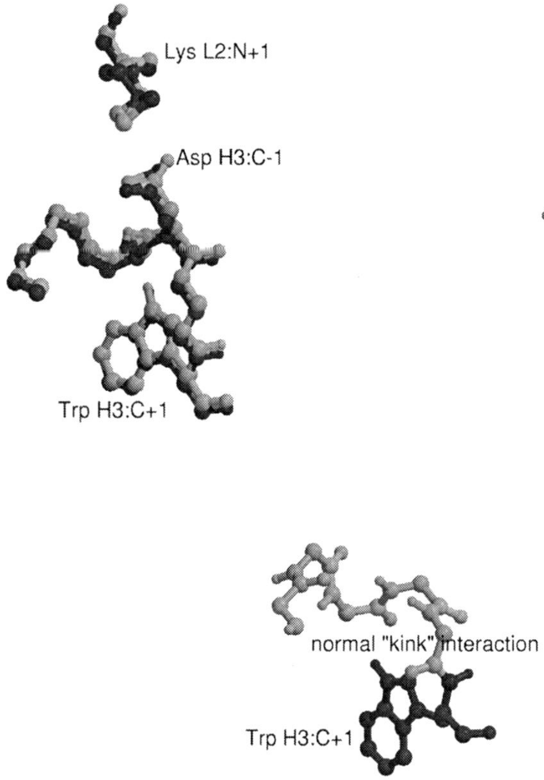

Fig. 5. Five-residue H3 conformations. **Top left;** 1gpo and 1dqq (kinked because of interaction with L2); **top right;** 1cr9 (extended); **bottom;** 1ggi (typical kinked).

The reason that 1nld does not show the typical conformation (instead, the "extended" Asp H3:C-1/Trp H3:C + 1 interaction is formed) is unclear. One possibility is that because of the small size of the loop, it is difficult for the Arg H3:N-

1 to form the salt-bridge with the Asp. 3fct also shows Arg H3:N-1 and Asp H3:C-1, but has a second interaction between Arg H3:N and Asp H3:N + 1, which may "tip the balance" in favor of the "kinked" interaction pattern, unlike 1nld.

3.8.1.2. Asp at H3:C-1 and no Arg at H3:N-1 plus Lys at L2:N-1

The Asp and no Arg sequence pattern would normally result in the "extended" interaction pattern. However, in the two five-residue structures (1gpo and 1dqq) with this pattern, the "kinked" interaction of the H3:C-2 carbonyl with the Trp H3:C + 1 is shown, as the Asp salt-bridges with Lys L2:N-1. Lys is not always seen at this position, and evidence for its role in H3 conformation is provided by the five-residue H3 of 1cr9, which exhibits an Asp at H3:C-1, but no Lys here and no Arg at H3:N-1. This forms an "extended"-type conformation. More support for this rule comes from structure 3hfm, which at 3.0 Å, only just fails the resolution criterion used in this analysis. This structure has an Asp at H3:C-1 and a Lys at L2:N-1, and shows the same conformational class as 1gpo and 1dqq.

3.8.1.3. Asp at H3:C-1 and no Lys at L2:N-1

One structure (1cr9) shows an Asp at H3:C-1 but, unlike the group described here (**Subheading 3.8.1.2.**), has no Lys at L2:N-1. Thus, the Asp interacts with Trp H3:C + 1 in the normal "extended" manner, and the structure shows an extended C-terminal.

3.8.1.4. Summary

Based on the evidence thus far, it can be said with reasonable confidence that the structural class described in **Subheading 3.7.1.2.** will lead to the conformation shown in 1gpo, 1dqq, and 3hfm. Because of the problem of 1nld, the likelihood of the structural class described in **Subheading 3.7.1.1.** forming the conformation shown in 1ggi, 1aj7, 3fct, and 1psk is less, but we can say with reasonable confidence that if there is no Arg at H3:N-1 and no Asp at H3:C-1, the conformation will form. If, as in 1ind, there is a His at H3:C and a Ser or similar-size hydrogen-bonding residue (e.g., Thr or Asn) at H3:N-1, there will be a slight perturbation of the type seen in 1ind.

3.8.2. Six- to Twelve-Residue Loops

3.8.2.1. Six-Residue Loops

There is just one kinked six-residue H3: 1sm3. This shows the typical U-shaped conformation found in the main group of seven- to nine-residue kinked H3s.

3.8.2.2. Seven-Residue Loops

These loops fall into three classes, a kinked group (1cgs, 1tet, 1mlb, 1ghf), an extended group (4fab, 1mrc), and an irregular structure, 3hfl. The kinked/

Table 10
Backbone, Global RMSD Between Seven-Residue Kinked H3 Loops

	1cgs	1tet	1mlb
1cgs	–	–	–
1tet	1.6	–	–
1mlb	1.6	1.8	–
1ghf	1.0	1.4	1.7

Table 11
Backbone, Global RMSD Between Eight-Residue Kinked H3 Loops

	1vfa	1fgn	1mim	1a6t	1kem	1b2w	1plg
1vfa	–	–	–	–	–	–	–
1fgn	1.4	–	–	–	–	–	–
1mim	1.2	0.8	–	–	–	–	–
1a6t	0.9	1.5	1.3	–	–	–	–
1kem	2.3	2.1	2.5	2.7	–	–	–
1b2w	2.1	2.7	2.3	1.9	4.0		
1plg	3.7	4.2	3.6	3.4	5.5	2.0	–
1cic	2.9	3.2	2.7	2.6	4.4	1.4	1.2

extended rules apply to the first two groups. In 3hfl, the irregular structure (in which a kink would be expected) is caused by the Asp H3:C-1 residue forming a saltbridge with residue L2:N-4, unusually an Arg. This forces the carbonyl of H3:C-2 away from Trp H3:C + 1, preventing the kink from being formed. The kinked group is fairly conserved, as the global RMSD **(Table 10)** show.

3.8.2.3. Eight-Residue Loops

Again, there are two broad classes for eight-residue loops: a kinked class, covering most structures (1vfa, 1fgn, 1mim, 1a6t, 1kem, 1b2w), a single non-kinked, irregular example (1mam), and two structures with a "weak" kink (1plg, 1cic). All should form the kink by the Shirai rules, however, the two prolines in 1mam distort the structure *(28)*.

a. Kinked **Table 11** shows that all examples exhibiting a normal kink, except 1kem, are highly conserved in conformation (**Fig. 6,** left). 1kem (**Fig. 6,** center) has its loop pointing more "upright" than the others. Examination on Insight-II has revealed that this is because of the Trp residue at H3:N that

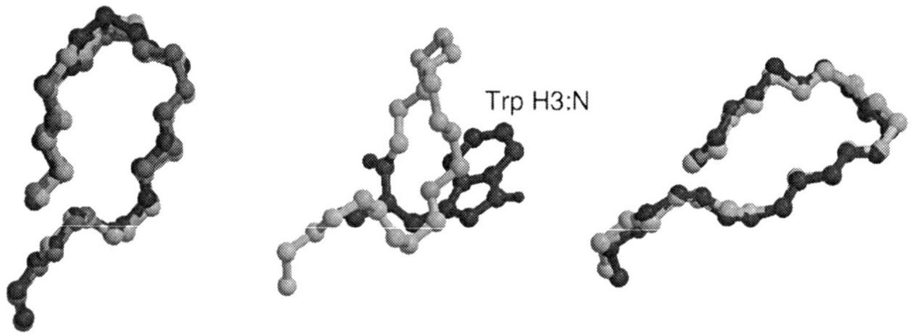

Fig. 6. Eight-residue H3 conformations. **Left:** the superimposed "typical" conformations of 1vfa, 1fgn and 1a6t; **center:** the unusual conformation of 1kem; **right:** the "weak kink" conformations of 1plg and 1cic.

blocks the normal conformation. Kim et al. ***(34)*** have also made this observation.1b2w also has a slightly differing conformation in the center of the loop (residues Pro H3:N + 3 and Trp H3:N + 4).

b. Weakly Kinked Two structures, 1plg and 1cic (**Fig. 6,** right), are very similar in conformation (global backbone RMSD 1.2 Å), and show a "weak" kink. Kim et al. ***(34)***, through the results of dynamics simulations, have speculated that Gly at the N-terminal (common to both) could be important. Our own investigations suggest that this could indeed be true for eight- and nine-residue loops, as the nine-residue structure 1mfb also has a "weak" kink and is the only example with Gly at the N-terminal. Furthermore, the eight-residue H3 loop 1b2w is similar in conformation to 1plg and 1cic, if not quite as similar as they are to each other (2.0-Å RMSD from 1plg; 1.5 Å RMSD from 1cic), and it too has Gly at H3:N and a "weak" kink, although its apex differs somewhat, because of the large Trp H3:N + 4 side chain and the Pro H3:N + 3. However, glycines at the N-terminal for other lengths do not exhibit "weak" kinks; it thus appears to be a steric effect (because Gly allows more freedom in the backbone than other residue types), depending on loop length.

3.8.2.4. Nine-Residue Loops

With nine-residue loops, the majority are kinked. 1bbd forms an extended structure by the rules of Shirai et al., yet 2fbj, 1igt, 1jhl, 2jel, 7fab, and 1nfd all form true kinks. As already seen, one structure, 1mfb, forms a "weak" kink, probably because of the presence of Gly at H3:N.

a. Main Kinked Group **Table 12** shows that three of the structures, 2fbj, 1igt, and 1jhl, are all conserved in structure. The H3 forms a standard "U-shaped loop" shape (*see* **Fig. 7**), more open than hairpin loops.

Table 12
Backbone, Global RMSD Between Nine-Residue Kinked H3 Loops

	2fbj	1igt	1jhl	1ar1	2jel	7fab	1nfd
2fbj	–	–	–	–	–	–	–
1igt	1.4	–	–	–	–	–	–
1jhl	1.3	1.3	–	–	–	–	–
1ar1	1.6	1.5	1.6	–	–	–	–
2jel	2.4	2.6	2.4	3.0	–	–	–
7fab	2.3	2.6	2.4	2.7	2.4	–	–
1nfd	4.9	5.3	4.6	4.8	3.2	4.5	–
1mfb	1.8	2.1	2.2	2.0	2.4	2.6	4.5

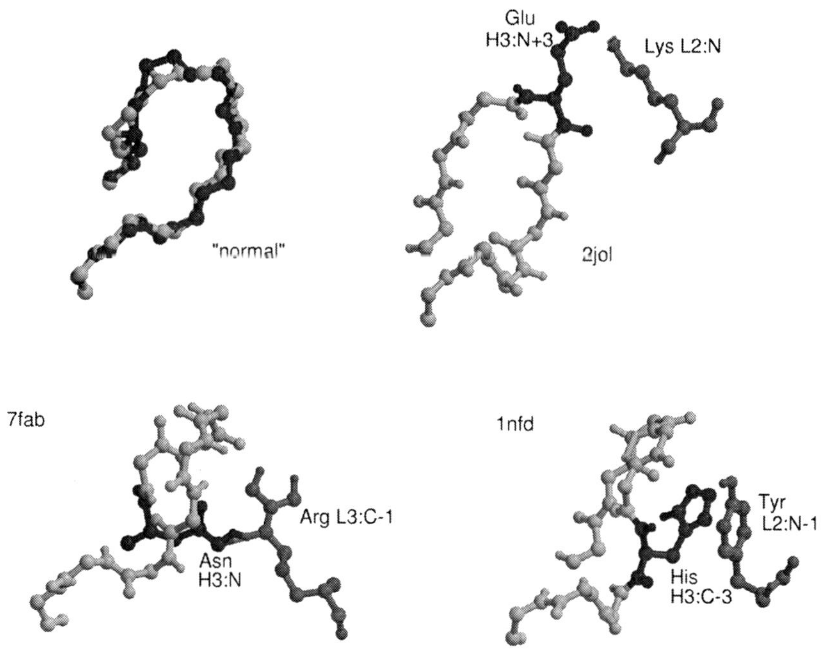

Fig. 7. Nine-residue H3 conformations. **Top Left:** the superimposed "typical" conformations of 1igt and 1jhl; **top right:** 2jel; **bottom left:** 7fab; **bottom right:** 1nfd.

b. Kinked Structures with Differences in the Center 2jel and 7fab both form kinks, but show differences in the center of the loop. 7fab, uniquely, has an Arg in the L3 loop at position L3:C-1, which forms a hydrogen bonding interaction with the side chain of Asn H3:N and with the carbonyl of residue H3:N + 2, forming an unusual "bulge" in the N-terminus that subsequently affects the center. In 2jel, the difference is not so great, but a salt-bridge between the Lys at residue L2:N and the Glu at residue H3:N + 3 perturbs the center of the loop. This is the only H3 for which this is possible in terms of sequence. 2jel and 7fab are also rather "tighter" loops than the main group **(Fig. 7)**.

Of all the true kinked structures, 1nfd **(Fig. 7)** is the most unusual. This has some similarities with 1kem in that the histidine at position H3:C-3 occupies the position that the apex would normally occupy, forcing a more "upright" conformation. This is further stabilized by hydrogen bonds between the histidine ring and the backbone, and an aromatic interaction between the histidine and the Tyr immediately before L2.

3.8.2.5. Ten-Residue Loops

There are a large number of ten-residue H3 loops, and the majority are kinked. Two (1eap and 1aif) form extended C-terminals, in accordance with the rules of Shirai et al. The kinked structures break up into different groups, depending on the conformation of the center of the loop, as shown in **Table 13.**

a. "U-shaped" Structures This is the most frequent group **(Fig. 8)** with four definite members (1for, 1igi, 1dba, and 1a4j) and four additional members close in structure to those five, though slightly different (1ay1, 1bln, 1clz, and 1cfv). The conformation is a U-shaped loop, rather like hairpin loops but lacking the even beta-structure (*see* **Fig. 8**), although 1bln, 1clz, and 1cfv are closer to a hairpin structure than the remainder. All satisfy the Shirai "kink" rules except for 1igi and 1a4j, which have a Asp at H3:C-1 but no Arg or Lys at H3:N-1. In both cases, the Arg or Lys at H3:N + 4 takes the place of the N-terminal residue in forming the salt-bridge.

Within the five close structures, the RMSD values suggest a further separation into three groups: 1for and 1igi, 1dba and 1a4j, and 1ay1. In the latter, the bulky Trp at H3:C-4, and the large number of aromatics in general, indicate that there is more of a steric hindrance problem, restricting the space available to the conformation. For the other two groups, sequence distinctions are unclear except that at position H3:N + 1, 1for and 1igi have Gly and Ser, whereas 1dba and 1a4j have Asp and Glu. However, no distinguishing interaction could be determined.

b. "Lid and finger" Structures The three structures of 1igf, 1kb5 and 1clo are all fairly—although not very—similar (RMSD from each other 2–2.5 Å). They have a more perturbed structure, with a "lid"-like centre with a "finger" projecting upward **(Fig. 8)**. However, examination of the interactions does not reveal a single defining interaction: in 1igf, the "finger" appears to be held in place by residue Tyr H3:C-3 hydrogen bonding with Ser H3:N + 1 and the carbonyl of Asp H3:N + 3; in 1kb5 it appears to be the side chains of Asp H3:C-1 with Arg H3:N + 2 and Tyr H3:C-3; and in 1clo it appears to be the side chain of Asp H3:N with the N-H groups of H3:N + 2 and H3:C-3.

Table 13
Backbone, Global RMSD Between Kinked 10-Residue H3 Loops

	1for	1dba	1igi	1a4j	1ay1	1cfv	1clz	1nbv	1kb5	1clo	1igf	1rmf	1bln
1for	–	–	–	–	–	–	–	–	–	–	–	–	–
1dba	1.4	–	–	–	–	–	–	–	–	–	–	–	–
1igi	1.1	1.4	–	–	–	–	–	–	–	–	–	–	–
1a4j	1.3	0.9	1.3	–	–	–	–	–	–		–	–	–
1ay1	1.7	2.0	1.9	1.3	–	–	–	–	–	–	–	–	–
1cfv	1.8	1.8	1.7	1.4	1.8	–	–	–	–	–	–	–	–
1clz	2.0	2.3	1.8	1.9	2.4	1.6	–	–	–	–	–	–	–
1nbv	2.5	2.3	2.2	2.1	2.4	2.0	2.0	–	–	–	–	–	–
1kb5	3.0	2.3	2.9	2.8	3.4	3.4	3.9	4.1	–	–	–	–	–
1clo	2.7	2.7	2.6	2.8	3.2	3.2	3.5	4.1	2.1	–	–	–	–
1igf	3.0	2.7	2.7	3.1	3.5	2.9	3.4	3.8	2.3	2.1	–	–	–
1rmf	2.9	3.1	3.2	2.6	2.2	2.8	2.7	2.3	4.9	4.8	5.0	–	–
1bln	1.7	1.8	1.5	1.5	1.9	1.7	1.2	1.7	3.4	3.2	3.2	2.6	–

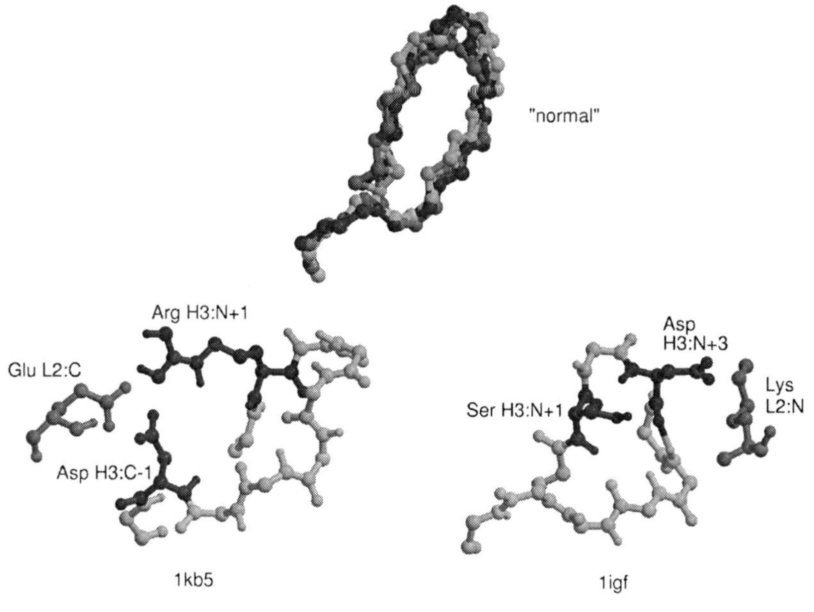

Fig. 8. Ten-residue conformations. **Top:** the superimposed "typical" conformations of 1dba, 1a4j and 1for; **bottom left:** 1kb5; **bottom right:** 1igf.

Each of these three structures also form unusual salt-bridges with other CDRs, which could stabilize the structures. In 1kb5, Arg H3:N + 1 interacts with Glu L2:C; a Glu at this position is unusual. In 1igf, Asp H3:N + 3 interacts

Table 14
Backbone, Global RMSD Between Kinked 11-Residue H3 Loops

	1a6v	1fvc	2h1p	1frg	1igc	1mcp	1fpt	1ncb	1axs	12e8	1ad9	1nqb
1a6v	–	–	–	–	–	–	–	–	–	–	–	
1fvc	2.3	–	–	–	–	–	–	–	–	–	–	
2h1p	2.6	2.1	–	–	–	–	–	–	–	–	–	
1frg	2.3	2.1	2.1	–	–	–	–	–	–	–	–	
1igc	2.6	1.8	2.6	2.4	–	–	–	–	–	–	–	
1mcp	2.5	2.1	2.3	2.6	1.8	–	–	–	–	–	–	
1fpt	2.2	3.0	3.0	1.8	3.7	3.7	–	–	–	–	–	
1ncb	5.3	5.2	4.2	3.2	5.9	6.0	4.5	–	–	–	–	
1axs	4.2	5.0	5.9	5.2	5.4	5.7	4.2	7.4	–	–	–	
12e8	3.0	3.2	2.3	4.5	3.8	3.6	2.4	3.3	5.9	–	–	
1ad9	2.0	2.1	3.3	2.5	2.5	3.0	2.4	6.4	3.8	4.1	–	
1nqb	2.5	2.3	1.5	2.0	2.9	2.3	3.0	4.1	5.5	2.4	3.2	
1ejo	3.0	2.3	2.3	1.6	3.1	3.1	1.3	4.6	4.6	2.6	3.1	2.0

with Lys L2:N; this is the only instance in the 10-residue H3 loops where this combination is found. In 1clo, Arg H3:N + 4 (unusual) interacts the Glu H2:C.

c. Irregular Structures 1nbv shares a broad similarity to the U-shaped class, but differs in conformation around Thr H3:C-5. This appears to be the result of Lys L2:N occupying a region of space slightly closer to the H3 loop than other examples with a Lys here, which was in turn a result of the apex of the L1 loop (around residues L1:N + 8 to L1:N + 11) occupying a slightly different position than most members of L1 canonical class 4. One further kinked structure (1rmf) does not fit well into either class, which is a result of the aromatic interaction formed by the Trp with the Tyr in H2 (*see* **Subheading 3.6.**).

3.8.2.6. 11-Residue Loops

All known structures have a kinked C-terminal. As the RMSD values (**Table 14**) show, the 11-residue loops do not usually show the same degree of conservation found in smaller loops, although one group of three with a well-defined sequence/structure relationship has been determined. This is described first.

d. A Potential Canonical Class for Eleven-Residue H3s

With the discovery of a new 11-residue H3 structure (1ejo), a possible canonical-type structure can be defined. This is very similar (1.3-Å RMSD) to another 11-residue H3 (1fpt), and both are moderately similar (1.8-Å RMSD) to a third, 1frg. This compares to a typical RMSD of 2–2.5Å among the more diffuse complete set of eleven-residue kinked H3 structures.

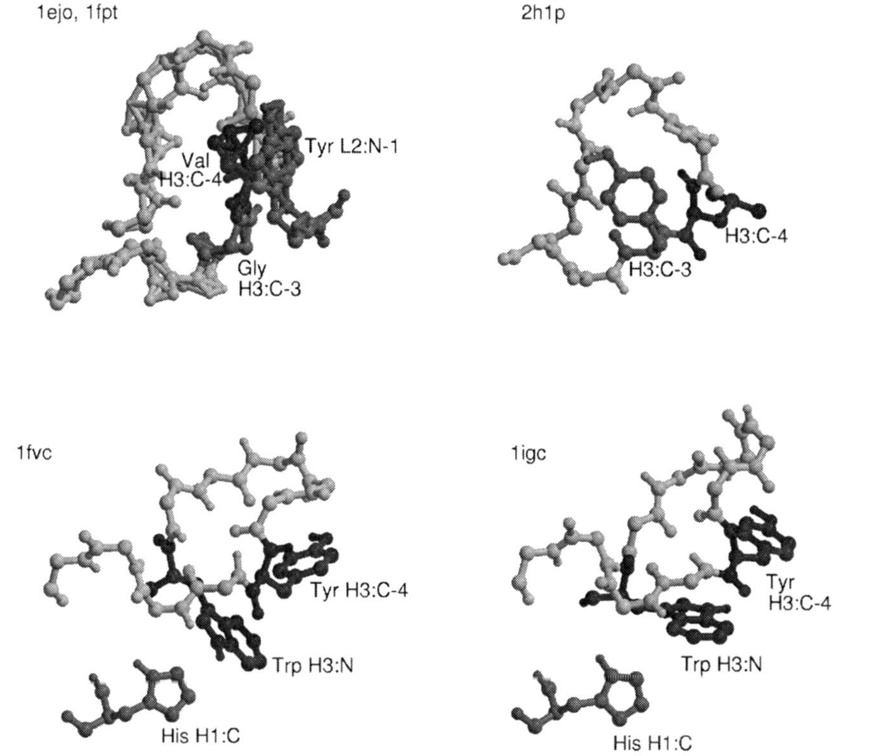

Fig. 9. **Left:** the defining interaction in the 11-residue H3 loops of 1ejo and 1fpt compared to a structure which does not exhibit the interaction or conformation (2h1p); **right:** the defining interaction of the 11-residue H3 loops of 1fvc and 1igc.

1ejo RAFDSD**VGF**AS
1fpt DFYDYD**VGF**DY
1frg RERYDE**KGF**AY

Considering the sequences, both 1ejo and 1fpt contain the motif VGF from residue H3:C-4 to residue H3:C-2. Examination of different 11-residue H3 structures on Insight-II reveals that the residue H3:C-4 normally points into the interior, and the residue H3:C-3 (normally non-Gly, and often large) sticks upward (although it is still buried because of the loop apex). In 1ejo and 1fpt, Val H3:C-4 instead sticks upward, packing against the Tyr immediately before the L2, and occupying the space left by the presence of Gly, rather than a larger residue, at H3:C-3. This influences the loop conformation, particularly in this region (**Fig. 9**). In 1frg, there is a Lys rather than Val, but similar behavior is observed (if not exactly the same conformation because of the differing properties of Lys compared to Val).

Table 15
Backbone, Global RMSD Between Kinked 12-Residue H3 Loops

	1igm	1vge	6fab	1iai	1jrh	1osp	1dee
1igm	–	–	–	–	–	–	
1vge	1.7	–	–	–	–	–	
6fab	4.9	5.7	–	–	–	–	
1iai	5.8	6.7	1.7	–	–	–	
1jrh	4.6	5.3	2.1	2.2	–	–	
1osp	2.5	3.6	3.5	4.1	3.0	–	
1dee	2.1	2.9	3.7	4.4	3.0	1.6	
1dsf	5.7	6.1	4.7	4.3	3.2	6.5	4.4

3.8.2.7. Other 11-Residue Kinked H3s

Aside from the above, only three pairs (1fvc/1igc, 1igc/1mcp, and 2h1p/1nqb) have an RMSD from each other of less than 2 Å, and then only just (except for the 1.5 Å RMSD between 2h1p and 1nqb). Most structures generally occupy the same conformational space (2.6 Å or less), although there are also some markedly irregular structures; 1axs, 1ncb, and 12e8.

One common interaction reveals itself. The low RMSD pair 1fvc/1igc share similar sequences (WGGDGFYAMDY, WGNYPYAMDY) and form some common hydrophobic interactions not observed in other 11-residue H3 loops (**Fig. 9**). Although the orientation of the ring differs somewhat, the unusual Trp at H3:N packs against a His at H1:C in both cases (His has moderate frequency of occurrence here), and also against a Tyr at H3:C-4. A further possible stabilizing interaction is between the Tyr at H3:N + 5 and the Tyr at L2:C-1. Tyr is fairly unusual at this L2 position.

One structure, 1hil, has a poorly defined kink. However, upon binding antigen, the true kink is formed, so the rules of Shirai et al. are not violated *(28)*.

3.8.2.8. Twelve-Residue Loops

Like the 11-residue H3s, the 12-residue kinked loops show more variation than those of shorter lengths (**Table 15**), although some common interactions and conformations in regions of the loop other than the kink have been observed.

a. Common Conformations Examination on Insight-II has revealed that 1dee, 1osp, and 1igm all share a particular conformation of the N-terminal four residues (1osp is shown in **Fig. 10**). 1dee and 1osp are similar at the apex, whereas 1igm differs. Although 1igm and 1vge exhibit differing N-terminal conformations, they share a similar conformation in the remainder of the loop from the apex to the C-terminal (1igm is shown in **Fig. 10**).

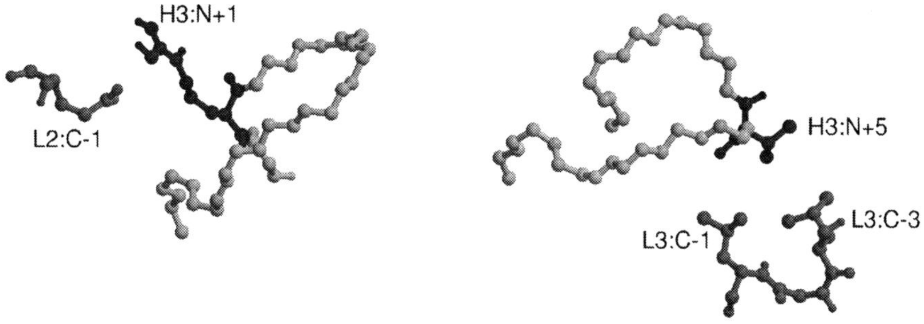

Fig. 10. **Left:** an example (1osp) of the common conformation and interaction of the N-terminal of the 12-residue H3s of 1osp, 1dee, and 1igm; **right:** an example (1igm) of the common conformation and interaction of the apex-to-C-terminal region of the 12-residue H3s of 1igm and 1vge.

1dee	VKFYDPTAPNDY
1osp	SRDYYGSSGFAF
1igm	HRVSYVLTGFDS
1vge	DPYGGGKSEFDY

Considering the sequences, 1dee, 1osp, and 1igm all contain an Arg or Lys at H3:N + 1. In all three cases, this interacts with a Glu or Gln at L2:C-1, and it is this interaction that defines the conformation of the N-terminal four residues. The interaction defining the common apex-to-kink conformation in 1igm and 1vge is more poorly defined. It is best seen in 1igm, where the Val at H3:N + 5 packs with hydrophobics at L3:C-3 and L3:C-1. The only other 12-residue H3 structure with hydrophobics at both of these L3 positions is 1vge, which has a Gly at H3:N + 5 rather than Val. Although the L3 residues appear to influence the backbone, this is rather speculative, and is limited to 12-residue H3s, as no other H3 length exhibits a conformation that is dependent on the L3:C-3 and C-1 residues.

b. Other Structures Two more structures—6fab and 1iai—form a "hairpin-like" U-shaped loop shape, although the similarity is not as strong as in the loop subregions of 1osp/1dee/1igm, and 1igm/1vge, described here. Both of these have a Tyr at H3:C-4, which influences the conformation on steric grounds. At present, no other 12-residue kinked H3 structure has a Tyr or Phe here, so whether this is a canonical feature is uncertain. Another point of interest is that there are two structures that do not form the C-terminal expected by the kinked/extended rules. 1jrh is predicted not to be kinked because the Arg at H3:N rather than H3:N-1 is predicted to form a salt-bridge with Asp H3:C-128. However, the Arg at H3:N instead hydrogen bonds with His H3:C-4, leaving Asp

H3:C-1 to bond with Arg H3:N-1 leading to a kinked conformation. This Arg/His interaction also occupies the space that a 6fab/1iai-like conformation would occupy, forcing the bulge at H3:N + 2. By contrast, 1ap2 is predicted to be kinked by the Shirai rules, but forms an irregular C-terminus. This appears to be a result of the Arg H3:N-1 residue that forms a salt-bridge with Asp H1:C-3 instead of Asp H3:C-1—the H1 residue is, unusually, charged. Additionally, the H3 loop is in close contact with the H3 loop of the second molecule in the unit cell. Finally, 1mpa is—as expected—extended, and 1dsf is kinked but with a unique apex conformation arising from a salt-bridge between Asp H3:N + 5 and two Lys residues at L1:N-4 and L2:N; the only instance of a 12-residue H3 with this combination.

3.8.3. Longer Loops

There is more chance of flexibility in solution in the center of the H3 for loops of lengths greater than 12, and none of the examples showed any similarity other than the kink, so they were not examined for contacts.

4. Conclusions

4.1. Canonical Reclassification

Observations have shown that, by making conserved expansions of the range of allowed residues for canonical conformations, the vast majority of structures can be brought into the canonical classes, leading to accurate models for CDRs L1 to H2 in the majority of structures. With an increased number of structures in the PDB, some new canonical classes were also defined. Explanations for the few exceptions—usually the result of differing sequence or unusual residues in the other CDRs—have been postulated. A small number of structures show the canonical conformation, yet exhibit non-conserved residues in the key positions. Therefore, it may be that only one of the range of interactions which hold a conformation together is required. However, because of the dangers of allowing non-canonical structures to be classified as canonical, these non-conserved mutations were not brought into the range of allowed residues, unless two or more observations of them have been documented.

4.2. Conserved Side Chain Conformations

The results indicated here have shown that for certain residue positions and residue types, there is conservation in chi-1 angles for both canonical loops and the kink region of CDR-H3. This information could be used to improve antibody modeling—the accuracy of the side chains at those positions will be improved, and, as a result, any backbones subsequently built will be less prone to error because of the elimination of clashes with incorrect side chains.

4.3. H3 Structural Investigations

A set of rules for the conformation of five-residue H3s, which works in all but one of nine structures examined, has been drawn up. These loops are too small to form well-defined "kinked" or "extended" structural features, yet all but one show the interactions that define the "kinked" structure *(28)*. The structures are subdivided by further electrostatic or hydrogen-bonding interactions.

For H3 lengths 7–10, the "kinked" H3 loops by default appear to have a preference for a U-shaped hairpin-like loop, with exceptions being caused by specific interactions or steric effects caused by unusual residue types. In particular, there is support for the effect of Gly at the N-terminal *(34)* in producing a "weak" kink, but only for eight- and nine-residue H3s. A preference for hairpin-like loops was described by Shirai et al.; they suggested that aromatics at certain loop positions (H3:N + 2 and H3:C-3) sharpened the hairpin definition. However, no correlation with aromatics at those positions could be found in the current investigation, and, indeed, in contrast to the results of Chothia and Lesk *(12)*, no sequence determinants could be found for these semi-conserved (RMSD between examples typically between 0.5 and 2.0 Å) structures. Thus, the rule appears to be "default U-shaped loop, unless there are unusual interactions or steric effects." Once more structures become available, other examples of the unusual structures for each length may be found that will define canonical classes, but as yet, there is not enough information here to be used in an improved modeling algorithm.

The 11-residue loops generally vary in conformation so that only broad classifications can be made, although one very distinct group and one less distinct pair of common structures do emerge. The 12-residue loops are also variable, although two possible rules for sections of the N-terminal and apex have been determined.

4.4. Postscript: Recent H3 Loops

Since the above analysis was carried out, the structures of a number of kinked CDR-H3 loops of lengths 8–12 (2pcp[8], 1c5d[9], 1ct8[9], 1sbs[12], 1ce1[10], 1dlf[10], and 1a3l[10]) have been described. These are all outlying structures from the main group for the corresponding length, although for most of the 8–10-residue examples, any large differences are restricted to the apex of the CDR. As with the outliers discussed here, most of the differences result from interactions caused by unusual residues, either within the H3 loop itself or within an adjacent CDR. For example, in 1dlf, the presence of Pro residues may be responsible for the perturbations observed. Since the conformations and the interactions are not the same as the other outliers previously described for this length class, they are not discussed in detail here; however, they reinforce the conclusions reached is this section.

5. A Practical Case Study Using WAM

A worked example of a typical WAM modeling run is presented in the following. The following sequence (CDRs in bold) is aligned (see http://antibody.bath.ac.uk/) and entered into the interactive online interface.

V_L DVVMTQIPLSLPVNLGDQASISC**RASQSISNN**
LNWYLQKPGQSPKLLMY**YATSLAD**GVPDRF
SGSGTDFTLKISRVEAEDLGIYFC**QHFWSTPK**
TFGGGTKLEIKRA

V_H RVQLLESGAELMKPGASVQISCKAT**FEFAKY**
WMSWVKERPGHGLEWIG**YIANAGGSTY**YR
EKFKGKATFTRDTSSNTAYMQLSSLTSEDSA
VYYCTR**SGNYAMDY**WGQGTSVTVS

5.1. Framework and Canonicals

The first stage is to determine:

1. the most sequence-homologous framework for each of the light and heavy chains;
2. the canonical class of each loop;
3. the most sequence-homologous database loop of that class.

For the example here, the most homologous light-chain framework is *1dba,* and for the heavy chain it is *1cgs.*

The canonical structures are as follows:

CDR	*Sequence*[a]	Structure used[b]	Class[c] (or H3 type)
L1	RASQSISN.... .NLN	1mlb (RASQSISN....NLH)	2
L2	YATSLAD	1fgn (YATSLAD)	1
L3	QHFWS. .TPKT	1vfa (QHFWS. .TPRT)	1
H1	GFEFAK. .YWMS	2fbj (GFDFSK. .YWMS)	1
H2	YIANAGG. .STY	1igt (YISNGGG. .STY)	3
H3	(R)SGNYA.........MDY	–	Kinked

[a] For H3, the residue immediately preceding the CDR (R) is also shown, since it determines the presence of the kink.

[b] Most sequence-homologous member of the appropriate class. Side chain positions with conserved chi-1 values between individual members of the class are highlighted.

[c] According to Chothia and Lesk *(12)*, and subsequent publications by Chothia and colleagues.

Thus, the light-chain framework is built using that of 1dba, re-sequencing residues using maximum-overlap if necessary; and the heavy-chain framework is built using that of 1cgs. The two chains are combined into a single model by fitting on an average beta-barrel generated from the known Fv structures. The frameworks of loops L1 to H2 are deleted and rebuilt using the canonicals from

1mlb, 1fgn, 2fbj, 1igm, and 1igt, respectively. Finally, the conformationally conserved canonical side chains, described elsewhere is this chapter, are built using maximum-overlap re-sequencing.

5.2. The H3 Loop: Construction

The eight-residue H3 loop in this instance is preceded by Arg, and features Asp at the residue preceding the C-terminal, specifying a kinked conformation *(31)*. Therefore, CAMAL (the combined database/conformational search method, described in **ref.** ***33***) is used—in this instance with C-alpha to C-alpha distance constraints derived from kinked H3 structures only. For this example, the database search generated 40 hits, and after rebuilding the apex using CONGEN, a total of 8,400 conformations were produced. On clustering, this was reduced to 69.

5.3. The H3 Loop: Final Screen

This is an eight-residue kinked CDR, and therefore, the RMSD screen is the method of choice because it is a relatively short loop and the known structures have a well-defined profile. Since side chains are required to be in place for this screen, side chains must be built on *all* the clustered conformations. As CONGEN is a quicker method than dead-end elimination, this is the default algorithm here, although the user may alter this if it is decided that a slight time penalty can be taken in order to achieve a somewhat more theoretically sound model. (Note that if the RMSD screen was used, the screening requires only the backbone conformation, and therefore only the final backbone, selected by the screen, would require side chain construction. Therefore dead-end elimination would be the default method). All non-conserved side chains (H3 and canonical CDRs) are built simultaneously using CONGEN, which iteratively places side chains at each residue until an acceptable global minimum is reached according to energy convergence. After side chain construction, each conformation is energy-minimized [by steepest descent, using the Valence Force Field (VFF)] to relieve steric clashes, and the accessibility screen performed on each model to select the final conformation, which is that most closely matching the typical accessibility profile of eight-residue kinked H3s.

The final model is then made available as a PDB file for visualization by any standard graphics software package.

6. Appendix: CDR and Residue Naming Conventions

6.1. CDR Definitions

The definitions of the CDRs used here are those formerly used by this group *(7,8,11,33,35)*.

6.1.1. L1

The L1 loop is defined to start at the residue immediately after the first conserved cysteine of the light chain, and finish at the residue immediately before the conserved tryptophan beyond.

6.1.2. L2

The L2 loop is defined to start 15 residues beyond the conserved tryptophan mentioned here, and finish 32 residues before the conserved cysteine mentioned here.

6.1.3. L3

The L3 loop is defined to start at the residue immediately after the second conserved cysteine of the light chain, and finish at the residue immediately before the conserved FG pair beyond.

6.1.4. H1

The H1 loop is defined to start four residues after the first conserved cysteine of the heavy chain, and finish at the residue immediately before the conserved tryptophan residue beyond.

6.1.5. H2

The H2 loop is defined to start 15 residues beyond the conserved tryptophan mentioned here, and finish 37 residues before the conserved cysteine mentioned here.

6.1.6. H3

The H3 loop is defined to start three residues beyond the second conserved cysteine of the heavy chain, and finish at the residue immediately before the conserved WG pair beyond.

6.2. Residue Naming

In order to provide a unified, unambiguous nomenclature for loop residues, a special convention has been used here. The general form is:

LL:T+x

where LL represents the loop name (e.g., L1 or H3) or the chain identity (L or H); T represents the terminal (N or C); and x the displacement from the given terminal. Thus, L1:N + 1 represents the residue following the L1 N-terminal residue; L3:C-1 represents the residue preceding the L3 C-terminal residue, H3:N-1 represents the framework residue preceding the H3 C-terminal and L:N + 1 represents the residue following the light-chain N-terminal.

References

1. Kabat, E. A. and Wu, T. T. (1972) Construction of a three-dimensional model of the polypeptide backbone of the variable region of kappa immunoglobulin light chains. *Proc. Natl. Acad. Sci. USA* **69,** 960–964.
2. Padlan, E. A., Davies, D. R., Pecht, I., Givol, D., and Wright, C. (1976) Model-building studies of antigen binding sites: the hapten-binding site of MOPC-315. *Cold Spring Harbor Symp. Quant. Biol.* **41,** 627–637.
3. Stanford, J. M. and Wu, T. T. (1981) A predictive method for determining possible three-dimensional foldings of immunoglobulin backbones around antibody combining sites. *J. Theor. Biol.* **88,** 421–439.
4. Feldmann, R. J., Potter, M., and Glaudemans, C.P.J. (1981) A hypothetical space-filling model of the variable regions of the galactan binding myeloma immunoglobulin J539. *Mol. Immunol.* **18,** 683–698.
5. Darsley, M. J. and Rees, A. R. (1985) Three distinct epitopes within the loop region of hen egg lysozyme defined with monoclonal antibodies. *EMBO J.* **4,** 383–392.
6. de la Paz, P., Sutton, B. J., Darsley, M. J., and Rees, A. R. (1986) Modelling of the combining sites of three anti-lysozyme monoclonal antibodies and of the complex between one of the antibodies and its epitope. *EMBO J.* **5,** 415–425.
7. Martin, A.C.R., Cheetham, J. C., and Rees, A. R. (1989) Modelling antibody hypervariable loops: a combined algorithm. *Proc. Natl. Acad. Sci. USA* **86,** 9268–9272.
8. Martin, A.C.R., Cheetham, J. C., and Rees, A. R. (1991) Molecular modelling of antibody combining sites. *Methods Enzymol.* **203,** 121–153.
9. Pedersen, J. T., Searle, S.M.J., Henry, A. H., and Rees, A. R. (1992) Antibody modelling: Beyond homology. *Immunomethods* **1,** 126–136.
10. Rees, A. R., Martin, A.C.R., Webster, D., Cheetham, J. C., and Roberts, S. (1990) Antibody combining sites: prediction and design. *Biophys. J.* **57,** A384.
11. Rees, A. R., Searle, S.M.J., Henry, A. H., Pedersen, J. T., and Whitelegg, N.R.J. (1996) Antibody combining sites: structure and prediction. In Sternberg M.J.E. (ed.), *Protein Structure Prediction* (1st ed.), Oxford University Press, 141–172.
12. Chothia, C. and Lesk, A. M. (1987) Canonical structures for the hypervariable loops of immunoglobulins. *J. Mol. Biol.* **196,** 901–917.
13. Chothia, C., Lesk, A. M., Tramontano, A., Levitt, M., Smith-Gill, S. J., Air, G., et al. (1989) Conformations of immunoglobulin hypervariable regions. *Nature* **342,** 877–883.
14. Chothia, C., Lesk, A. M., Gherardi, E., Tomlinson, I. M., Walter, G., Marks, J. D., et al. (1992) Structural repertoire of the human Vh segments. *J. Mol. Biol.* **227,** 799–817.
15. Tramontano, A., Chothia, C., and Lesk, A. M. (1990) Framework residue 71 is a major determinant of the position and conformation of the second hypervariable region in the Vh domains of immunoglobulins. *J. Mol. Biol.* **215,** 175–182.
16. Tomlinson, I. M., Cox, J.P.L., Gherardi, E., Lesk, A. M., and Chothia, C. (1995) The structural repertoire of the human V-kappa domain. *EMBO J.* **14,** 4628–4638.
17. Al-Lazikani, B., Lesk, A. M., and Chothia, C. (1997) Standard conformations for the canonical structures of immunoglobulins. *J. Mol. Biol.* **273,** 927–948.

18. Wu, T. T. and Kabat, E. A. (1970) An analysis of the sequences of the variable regions of Bence-Jones proteins and myeloma light chains and their implications for antibody complementarity. *J. Exp. Med.* **132,** 211–250.
19. Kabat, E. A., Wu, T. T., and Bilofsky, H. (1977) Unusual distributions of amino acids in complementarity determining (hypervariable) segments of heavy and light chains of immunoglobulins and their possible roles in specifity of antibody combining sites. *J. Biol. Chem.* **252,** 6609–6616.
20. Padlan, E. A. and Davies, D. R. (1975) Variability of three-dimensional structure in immunoglobulins. *Proc. Natl. Acad. Sci. USA* **72,** 819–823.
21. Berman, H. M., Westbrook, J., Feng, Z., Gilliland, G., Bhat, T. N., Weissig, H., et al. (2000) The Protein Data Bank. *Nucleic Acids Res.* **28,** 235–242.
22. Mandal, C., Kingery, B. D., Anchin, J. M., Subramaniam, S., and Linthicum, D. S. (1996) ABGEN: a knowledge based automated approach for antibody structure modelling. *Nat. Biotechnol.* **14,** 323–328.
23. Bruccoleri, R. E. and Karplus, M. (1987) Prediction of the folding of short polypeptide segments by uniform conformational sampling. *Biopolymers* **26,** 137–168.
24. Reichmann, L., Clark, M., Waldmann, H., and Winter, G. (1988) Reshaping human antibodies for therapy. *Nature* **332,** 323–327.
25. Roguska, M. A., Pedersen, J. T., Henry, A. H., Searle, S.M.J., Roja, C. M., Avery, B., et al. (1996) A comparison of two murine monoclonal antibodies humanised by CDR-grafting and variable domain resurfacing. *Protein Eng.* **9,** 895–904.
26. Dauber-Osguthorpe, P., Roberts, V. A., Osguthorpe, D. J., Wolff, J., Genest, M., and Hagler, A. T. (1988) Structure and energetics of ligand-binding to proteins: Eschericia coli dihydrofolate reductase/trimethoprim, a drug-receptor system. *Proteins* **4,** 31–47.
27. Miller, R. J., Jones, D. J., and Thornton, J. M. (1996) Protein fold recognition by sequence threading: tools and assessment techniques. *FASEB J.* **10,** 171–178.
28. Shirai, H., Kidera, A., and Nakamura, H. (1996) Structural classification of CDR-H3 in antibodies. *FEBS Lett.* **399,** 1–8.
29. Morea, V., Tramontano, A., Rustici, M., Chothia, C., and Lesk, A. M. (1998) Conformation of the third hypervariable region in the Vh domain of antibodies. *J. Mol. Biol.* **275,** 269–294.
30. Sibanda, B. L., Blundell, T. L., and Thornton, J. M. (1988) Conformation of beta-hairpins in protein structures: a systematic classification with applications to modelling by homology, electron-density fitting and protein engineering. *J. Mol. Biol.* **206,** 759–777.
31. Shirai, H., Kidera, A., and Nakamura, H. (1999) H3-rules: identification of CDR-H3 structures in antibodies. *FEBS Lett.* **455,** 188–197.
32. Oliva, B., Bates, P. A., Querlo, E., Aviles, F.X., and Sternberg, M.J.E. (1998) Automated classification of antibody complementarity determining region 3 of the heavy chain (H3) loops into canonical forms and its application to protein structure prediction. *J. Mol. Biol.* **279,** 1193–1210.
33. Whitelegg, N. R. J. and Rees, A. R. (2000) WAM—an improved algorithm for modelling antibodies on the Web. *Protein Eng.* **13,** 819–824.

34. Kim, S. T., Shirai, H., Nakajima, N., Higo, J., and Nakamura, H. (1999) Enhanced conformational diversity search of CDR-H3 in antibodies: role of the first CDR-H3 residue. *Proteins* **37,** 683–696.
35. Whitelegg, N.R.J. (1998) PhD Thesis, "Molecular Modelling of Antibody Combining Sites", University of Bath, UK.
36. Kraulis, P. J. (1991) MOLSCRIPT: a program to produce both detailed and schematic plots of protein structures. *J. Appl. Cryst.* **24,** 946–950.
37. Merritt, E. A. and Bacon, D. J. (1997) Raster3D—Photorealistic molecular graphics. *Methods Enzymol.* **277,** 505–524.

5

Studying Antibody Conformations by Ultracentrifugation and Hydrodynamic Modeling

Stephen E. Harding, Emma Longman, Beatriz Carrasco, Alvaro Ortega, and Jose Garcia de la Torre

1. Introduction

Intact immunologically active antibodies are too large for high-resolution structural analysis by nuclear magnetic resonance (NMR) spectroscopy. The flexibility between the Fab and Fc domains at the hinge region also makes structure determination by X-ray crystallography generally difficult, except for a limited number of cases (antibodies that possess short hinge regions, hinge-deleted mutants, or those stabilized by complexation with antigen). Hydrodynamic methods—such as measurement of the translational frictional ratio by ultracentrifugation or dynamic light scattering—have no such limitations. However, two caveats exist. First, the structures obtained are only low-resolution models. Second, hydrodynamic parameters, such as the frictional ratio, are often at least as sensitive to the volume of the particle as to conformation. This is tricky, since this volume includes a significant contribution from any solvent associated with the antibody; and this "solvation" or "hydration" is a complex dynamic process. Nonetheless, if the hydration problem can be adequately dealt with, useful information about domain orientation can be obtained, and in particular how orientations differ between different antibody classes and subclasses, and between wild-type and mutant or engineered structures. This information may ultimately assist with strategies for antibody engineering, particularly if we can understand which sequences are responsible for "compact" as opposed to open structures. Domain orientation and flexibility are intimately linked with an antibody's ability to crosslink antigen and bind to the effector system, and are also related to its stability *(1)*.

By combining the measured frictional ratio data (from measurements of sedimentation or diffusion behavior) of the individual domains and of the

From: *Methods in Molecular Biology, Vol. 248: Antibody Engineering: Methods and Protocols*
Edited by: B. K. C. Lo © Humana Press Inc., Totowa, NJ

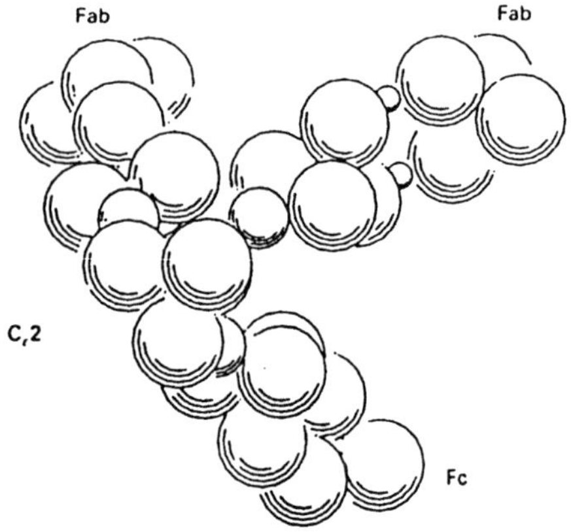

Fig. 1. The cusp-shape conformation of IgE was first discovered using hydrodynamic methods in 1990 (from **ref. 3**).

intact antibody, with information calculated from the known crystallographic structures of the domains, the possible conformations—in terms of the domain orientation—of intact antibody molecules can be given, *without ambiguities through hydration.* This combination of hydrodynamics and crystallography was introduced recently as "crystallohydrodynamics" *(2)*. An earlier form of this approach was reported in 1990 with the first demonstration that the antibody IgE was cusp-shaped in solution *(3)* (**Fig. 1**).

Essentially, the time-averaged apparent hydration (expressed as the amount of water associated with the protein on a mass/mass basis) is dealt with by comparing hydrodynamic properties with the crystal structures of those antibody forms with crystal structures that are known. This can then be used to represent the hydration of other structures with conformations that are being sought. In the case of **ref.** ***2*** "those antibody forms" were the known structures of the Fab′ and Fc domains of IgG. In the case of the earlier **ref.** ***3***, "those antibody forms" were the hingeless mutant antibody IgGMcg, one of the few intact (although not immunologically active) antibodies with structures that have been solved.

This chapter focuses on the measurement and use of the translational frictional ratio for representing the conformation of antibodies and other multidomain structures in solution, in terms of the low-resolution orientation of the domains.

1.1. The Sedimentation Coefficient, Frictional Ratio, and the Perrin "Frictional Ratio due to Shape"

1. In simple terms of boundary sedimentation, the sedimentation coefficient, *s*, (second, or Svedberg units, S, in which 1 S = 10^{-13}s) of a macromolecule is defined as

the velocity of sedimentation, v (cm/s), of a macromolecule per unit centrifugal field (cm/s^2):

$$s = v/\omega^2 r \tag{1}$$

where ω is the angular velocity (=2π.rpm/60) in rads/s and r (cm) is the radial position of the sedimenting *boundary.* NB: The full description of the change in concentration with time and radius across the centrifugal cell is given by a much more complicated differential relation known as the Lamm equation, which also takes into account boundary spreading caused by diffusion (*see* **ref. 4**). The Lamm equation forms the basis of modern software considered here.

2. The Svedberg unit, S, is often used instead of seconds: $1S = 10^{-13}$ s. Increase in mol wt tends to increase *s*. Increase in asymmetry or hydration tends to decrease *s*.
3. The sedimentation coefficient of a macromolecule measured in an arbitrary buffer solution or solvent will depend on the intrinsic properties of the macromolecule as well as on the viscosity and density of the solution it is sedimenting through. We therefore normalize to standard solvent conditions—namely, the density and viscosity of water at 20.0°C (*see* **ref. 4**):

$$s_{20,w} = \{(1 - \bar{v} \cdot \rho_{20,w})/(1 - \bar{v} \cdot \rho_{T,b})\} \cdot \{\eta_{T,b}/\eta_{20,w}\} \cdot s_{T,b} \tag{2}$$

where $s_{T,b}$ is the measured sedimentation coefficient in the buffer and at the temperature used; $s_{20,w}$ is the corresponding corrected value at 20.0°C in water; $\rho_{T,b}$ and $\eta_{T,b}$ are the densities and viscosities of the buffer; $\rho_{20,w}$ and $\eta_{20,w}$ are the corresponding values at 20.0°C in water; $\bar{v}$ is the partial specific volume, which can be reasonably estimated from the amino acid and carbohydrate content: the program *SEDNTERP* will do this (*see* **Subheading 1.2.3.**).
4. The sedimentation coefficient is corrected for non-ideality effects by extrapolation to zero concentration to give $s^o_{20,w}$ according to the relation:

$$s_{20,w} = s^o_{20,w}\,(1 - k_s c) \tag{3}$$

where k_s is known as the concentration dependence parameter or the Gralén coefficient (mL/g).
5. $s_{20,w}$ depends not only on the conformation but the mol wt and (hydrated) volume of the macromolecule. We need a parameter that can be explicitly given in terms of conformation alone. The first step is to combine sedimentation coefficient with the mol wt to give the *translational frictional ratio* (f/f_o) (the ratio of the frictional coefficient of the macromolecule to a spherical particle of the same mass and hydrated volume):

$$(f/f_o) = M(1 - \bar{v} \cdot \rho_{20,w})/[N_A 6\pi\eta_{20,w} s^o_{20,w}\,\{(3\bar{v} \cdot M)/(4\pi N_A)\}^{1/3}] \tag{4}$$

where M is the mol wt (g/mol) and N_A is Avogadro's number (6.02×10^{23} mol^{-1}).
6. The final step is to define a "frictional ratio due to shape" or "Perrin function" *P,* by combining the frictional ratio with the hydrated volume of the macromolecule

$$P = (f/f_o) \cdot \{1 + (\delta_{app}/\bar{v} \cdot \rho_{20,w})\}^{-1/3} \quad (5)$$

where δ_{app} is the apparent hydration. As stated in the previous section, hydration is a complicated, dynamic process. For the purpose of hydrodynamic modeling represents a time-average, and δ_{app} represents an "apparent" hydration because it considers other non-shape contributions toward the frictional ratio.

7. The P value obtained from experimental data in this way can then be compared with the P values computed from models of the antibody shape—using, for example, the computer program SOLPRO as described under **Subheading 1.2.3.**

1.2. Choice of Modeling Methods: Ellipsoids, Beads, and Shells

1.2.1. Ellipsoids

The simplest representation of the shapes of proteins in solution is as an ellipsoidal shell (**Fig. 2**). An ellipsoid is a symmetric structure, and there are three types. There is the general "tri-axial" ellipsoid with three unequal axes, a>b>c, with a,b,c the semi-axial dimensions. Its shape is specified by two axial ratios (a/b, b/c). The general triaxial ellipsoid is symmetric about the *center.*

There are two simpler ellipsoids that are symmetric about an *axis:* as in the situation when two of the three axes are forced to be equal. The prolate ellipsoid has a > b = c—e.g., it has two equal shorter axes and one longer axis, and its shape is specified by the axial ratio a/b. The oblate ellipsoid has a = c > b, and its shape is also specified by a single axial ratio a/b. The prolate ellipsoid resembles a rugby ball, whereas the oblate ellipsoid resembles a "smartie" or an "M&M" shape. In the limit a = b = c, the shape is a sphere. The advantage of ellipsoids is that their hydrodynamic properties can be calculated exactly *(5)*; the disadvantage is that they can only be applied directly to reasonably symmetrical structures. In fact, antibody Fab domains approximate a prolate ellipsoid and Fc domains approximate an oblate ellipsoid. However, intact immunologically active antibodies are not symmetric, and cannot be represented by a single ellipsoid structure. Unfortunately, the hydrodynamic properties of three ellipsoids joined together cannot be calculated. Fortunately, however, the hydrodynamic properties of arrays of spheres can be calculated to a high degree of accuracy.

1.2.2. Beads and Shells

The hydrodynamic and other solution properties of macromolecules or particles arbitrary shapes can be calculated using a methodology that is known as bead modeling (*see,* for example, **ref.** *6*). This approach has two variants *(2)*. One of them is bead modeling in a strict sense, in which the particle is represented as an assembly of spherical elements of arbitrary shape, with the only

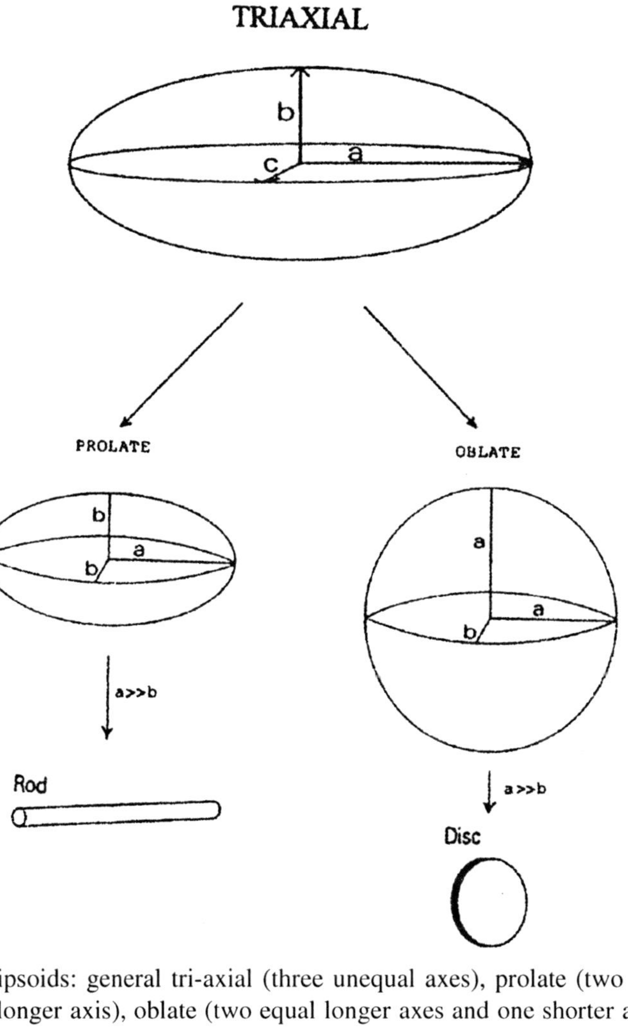

Fig. 2. Ellipsoids: general tri-axial (three unequal axes), prolate (two equal short axes and one longer axis), oblate (two equal longer axes and one shorter axis). A Fab domain can be approximated by a prolate ellipsoid, an Fc domain by an oblate. In the extremes a>>b (not antibodies), a prolate approaches a rod, an oblate approaches a disk.

condition that the overall size and shape of the particle must be as close as possible to that of the particle. The other variant is shell modeling, in which the *surface* of the particle is represented by a shell of small elements ("minibeads"); the results are extrapolated to a zero minibead radius. This procedure is slightly more computing-intensive than bead modeling, but it is hydrodynamically more rigorous and avoids problems (such as bead overlapping) that had usually plagued the use of the first method. **Figs. 3–6** have been

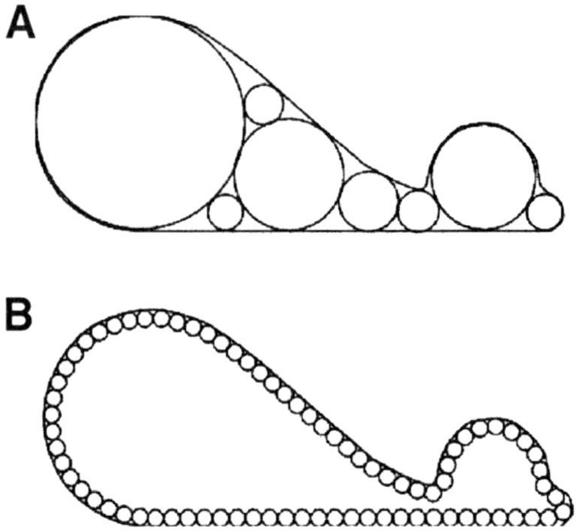

Fig. 3. Schematic representation of **(A)** a bead model in a strict sense and **(B)** a shell model.

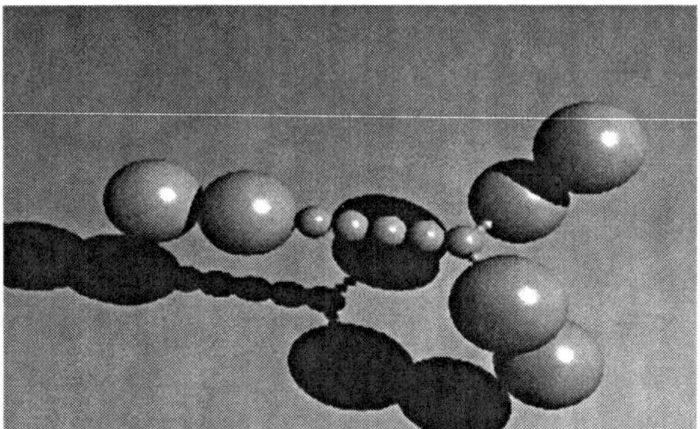

Fig. 4. A bead model for the human antibody molecule IgG3.

included to illustrate both conventional bead and bead-shell models for various proteins (and a glycoprotein in the case of IgG3).

1.2.3. Bead and Shell Modeling Software

We now describe the basic aspects of those programs that can be most useful for hydrodynamic modeling of antibodies and related molecules ***(7–13)***. *See* **Note 1** for details of how to access the software described.

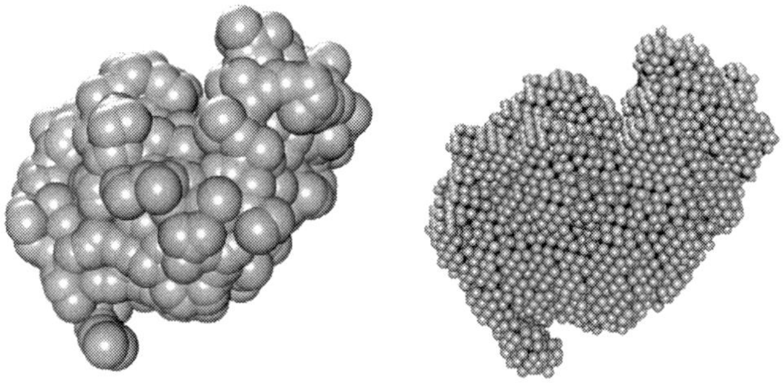

Fig. 5. **Left:** A primary bead model of lysozyme, showing bead overlap. **Right:** A shell model, produced by HYDROPRO.

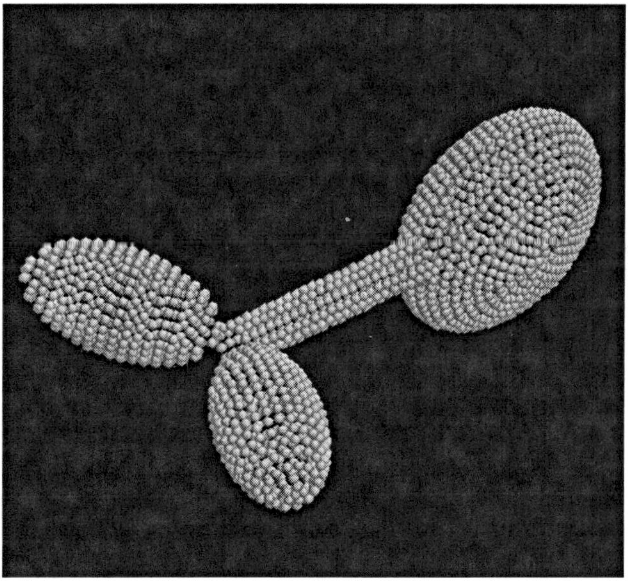

Fig. 6. A shell model for an IgG3 molecule, generated with HYDROSUB showing the two prolate Fab domains, the oblate Fc subunit and the long (nearly rodlike) hinge. The molecule is the same as that in **Fig. 4:** note the difference between a bead model and a shell model.

1. HYDRO is a program for the calculation of hydrodynamic coefficients and other solution properties of rigid macromolecules, colloidal particles, etc., that employs bead models in the strict sense, as mentioned previously. The user of HYDRO has to first build the bead model, which will be specified as a list of Cartesian coordinates and radii of the beads. An example of a bead model (for rat IgE) has already been given, in **Fig. 1.** Another (for human IgG3) is given in **Fig. 4.**

2. HYDROPRO computes the hydrodynamic properties of rigid macromolecules (e.g., globular proteins and small nucleic acids) from their atomic-level structure, as specified by the atomic coordinates taken from a Protein Data Bank (PDB) file supplied by the user, from which the proper hydrodynamic model is built by the program itself. This program employs the shell-modeling method. In the primary hydrodynamic model, each non-hydrogen atom is represented by a sphere of radius approx 3 Å. In order to avoid the inconveniences associated with bead overlapping, this model is replaced by a shell model, in which small "minibeads" are placed at the surface of the primary hydrodynamic model. HYDROPRO calculates the basic hydrodynamic properties: translational diffusion coefficient, sedimentation coefficient, intrinsic viscosity, and relaxation times, along with the radius of gyration. Optionally, HYDROPRO computes also other solution properties such as the co-volume (related to the second virial coefficient) and scattering related properties such as the angular dependence of scattering intensities and the distribution of intramolecular distances.
3. HYDROSUB is intended for the calculation of solution properties of a macromolecular or supramolecular structure modeled as an assembly of subunits that have ellipsoidal (prolate, spherical, or oblate) and/or cylindrical shapes. The subunits are represented using a variation of the bead-shell method, which is particularly suitable for axially symmetric structures. The program builds the shell model, stacking rings of tangential minibeads. The radius of the rings is fixed for cylinders and decreases from the equator to the poles in the case of ellipsoids. This is done internally by the program; the user just has to specify, for each subunit, the coordinates of their centers and the polar angles of their symmetry axis.
4. SOLPRO. The results from HYDRO, HYDROPRO, and HYDROSUB are the primary solution properties: e.g., hydrodynamic coefficients, radius of gyration, and co-volume. SOLPRO *(11,12)* is an ancillary program, whose input is one of the output files produced by HYDRO, HYDROPRO, or HYDROSUB. This program takes the values of the solution properties and combines them to obtain a number of size-independent shape quantities such as the P function defined in **Equation 5.** The user should use this program according to the supplied instructions.
5. SEDNTERP. This is a general support program for researchers using analytical ultracentrifugation, which provides a very convenient interface for calculating buffer densities, viscosities from buffer/colvent composition data. It will also estimate macromolecular partial specific volumes—and in the case of proteins, extinction coefficients, the isoelectric pH and the valency on a protein for a given pH, all from chemical composition data.

1.3. Workflow Summary of Hydrodynamic Modeling

1. Measure the sedimentation coefficient, and check to determine whether it has been affected by aggregation (boundary shape, is the mol wt consistent with the value from the sequence?).
2. Calculate the experimental frictional ratio from the sedimentation coefficient and (sequence) mol wt.

3. Calculate the experimental Perrin function *P* from the frictional ratio and the appropriate estimate for the (time-averaged) hydration.
4. Construct candidate models using HYDROSUB or HYDRORO and calculate their *P* values using SOLPRO.

2. Materials

2.1. Analytical Ultracentrifugation

1. Beckman XL-I or XL-A ultracentrifuge.
2. Optical system: Ultraviolet (UV) absorption optics at 280 nm.
3. Rotor (conventionally a four-hole or eight-hole) for Beckman XL-I or XL-A analytical ultracentrifuge.
4. Ultracentrifuge cells: double sector. Multichannel may be used for sedimentation equilibrium, but not velocity.
5. Solvent: choose as appropriate. To keep problems of hydrodynamic or thermodynamic non-ideality to a minimum (by minimizing polyelectrolyte or charge effects), we suggest a minimum ionic strength of 0.1 *M*.
6. Sample requirements:
   a. Sedimentation velocity: at least four concentrations in the range 0.2–1.0 mg/mL (0.3–0.4 mL each), unless reversible self-association phenomena (*see* **Note 2**) are present and need to be probed, in which case a wider concentration range may be appropriate.
   b. Sedimentation equilibrium: one concentration should suffice, at ~0.4 mg/mL (0.2 mL), unless, there is a complication through self-association phenomena (*see* **Note 2**).

2.2. Computer

The hydrodynamic modeling programs run on a variety of platforms, including Microsoft DOS/Windows and Linux for PCs. Thus, the computational work can be carried out on inexpensive personal computers. With fast Pentium or AMD PC processors, the calculations run with a speed similar to that of more expensive machines. We recommend the use of a PC with the second- or third- fast Pentium processor (which optimize the performance/cost ratio), and a moderate amount of fast memory. At the time of writing this chapter, we are using a DELL Dimension 8200, but any computer with the following requirements should be adequate:

1. Pentium IV 2 GHz (is not the fastest Pentium, but the performance/cost ratio is rather good).
2. Memory: RDRAM is recommended. This is not common, but is very good for the matrix calculations that are within the HYDRO programs. An alternative is "DDR" memory. Although 256 Mb is usually sufficient, we advise 512 Mb.
3. A medium-range graphics card.

3. Methods

3.1. Experimental Methods

3.1.1. Sedimentation Velocity in the Analytical Ultracentrifuge

This is used to provide a check on the homogeneity and monomeric state of the antibody. It involves placing the solution and a reference solvent in a specialized ultracentrifuge cell that has end windows transparent to visible and UV light. An optical system (either UV-absorption- or refratometric-based) can be used to record the distribution of solute concentration as a function of radial position and time. The balanced cell is placed in a special rotor. Sedimentation velocity speeds typically range from 30,000–50,000 rpm, depending on the size and shape of the macromolecular system being analyzed. The primary parameter that results from a sedimentation velocity experiment is the sedimentation coefficient. After consulting the manufacturer's operation instructions and guidelines, for measurements on antibodies we suggest the following protocol:

1. Select a rotor speed of ~50,000 rev/min (**Note:** With analytical ultracentrifugation, researchers almost always refer to rotor settings in terms of revolutions per min rather than the equivalent *g*-force. This does not create any ambiguities, because at the time of writing there is only one commercial manufacturer, and all rotors are of standard size (just differing in the numbers of holes for cells).
2. Choose a temperature: 20.0°C is normally used, unless reproducibility of physiological conditions is necessary or there are particular temperature effects on conformation or stability you wish to probe. If the antibody is relatively fragile, choose 4°C.
3. An experimental run lasts for approx 3–4 h. This should allow the protein to sediment through most of the cell.
4. Choose the appropriate software for analysis. We recommend one of the three packages: DCDT+ ***(14)***, SVEDBERG (15), or SEDFIT ***(4)***. All three are downloadable from the internet (*see* **Note 1** and **ref. *16***), SEDFIT is free, whereas DCDT+ and SVEDBERG require a nominal charge. All three come with comprehensive instructions and help files. They all follow the change in the whole concentration vs radial position with time (old methods used to follow just the sedimenting boundary), which facilitates extraction of the sedimentation coefficient to a high degree of precision.
5. Check the homogeneity (**Fig. 7A**): there should only be one (macromolecular) component. If there are multiple components present, select the component which corresponds to the monomeric antibody: you may wish to consult a hydrodynamics expert (*see* **Note 3**).
6. Extract the sedimentation coefficient $s_{20,w}$. The subscripts 20,w, refer to corrected values in "standard solvent conditions," namely the viscosity and density of water at 20°C. The most recent versions of the three packages DCDT+, SVEDBERG, and SEDFIT provide that correction for you: the packages ask you for the buffer/solvent details you used in your experiment, and the partial specific volume.

If you don't know the latter, then take the uncorrected sedimentation coefficient from these programs and then use the program SEDNTERP (**ref.** ***16,17***; *see* **Note 1**) to correct it—you need to enter the PDB data file for the protein and it will evaluate the partial specific volume from the amino acid (and carbohydrate) content. You can then enter the buffer details and $s_{20,w}$ will then be calculated.

7. Extrapolate $s_{20,w}$ to zero concentration, (**Fig. 7B**), to give $s^o_{20,w}$.
8. Also extract the translational diffusion coefficient $D_{20,w}$ (cm^2/s): the software packages referred to here also estimate this for you, from the rate of boundary spreading, and extrapolate this to zero concentration. Extrapolate $D_{20,w}$ to zero concentration to give $D^o_{20,w}$ (**Fig. 7C**).
9. From $s^o_{20,w}$ and $D^o_{20,w}$ calculate the mol wt from the Svedberg equation.

$$M_w = \{s^o_{20,w}/D_{20,w}\} \cdot \{RT/(1-\bar{v} \cdot \rho_{20,w})\} \quad (6)$$

where R is the gas constant. Check that this is within experimental error (+/– 5%) of the mol wt for the monomeric antibody from its sequence. If this is confirmed by further checks from sedimentation equilibrium studies (**Subheading 3.1.2.**), you can then proceed with some confidence using the $s^o_{20,w}$ and the sequence mol wt for modeling.

3.1.2. Sedimentation Equilibrium in the Analytical Ultracentrifuge

This is used to provide a further check on the mol wt of the antibody. Lower speeds are chosen compared with sedimentation velocity. At these lower speeds, the effects of sedimentation are matched by the effects of diffusion, and instead of a boundary forming, an equilibrium distribution of solute is obtained in the ultracentrifuge cell, with a lower concentration at the air/solution interface and a higher concentration at the base of the cell. Because the system is at equilibrium, shape or frictional effects do not affect the distribution, which is entirely dependent on molecular mass (provided that non-ideality effects are allowed for). After consulting the manufacturer's operation instructions and guidelines for measurements on antibodies, we suggest the following protocol:

1. Select a rotor speed of ~10,000 rev/min and the same solvent and temperature conditions as described in **Subheading 3.1.1.**
2. Choose a temperature: 20.0°C is normally used, unless reproducibility of physiological conditions is necessary or there are particular temperature effects on conformation or stability you wish to probe. If the antibody is relatively fragile, choose 4°C.
3. An experimental run lasts at least overnight. This is because of the long period required for sedimentation and diffusive processes to reach equilibrium.
4. Choose the appropriate software for analysis. We recommend the package MSTARA *(18)*, which gives the mol-wt average for all the macromolecular components across the entire radial range in the cell, not just a selected region (**Fig. 7D**). This is also downloadable from the web *(16)* (*see* **Note 1**) and is free. If the value

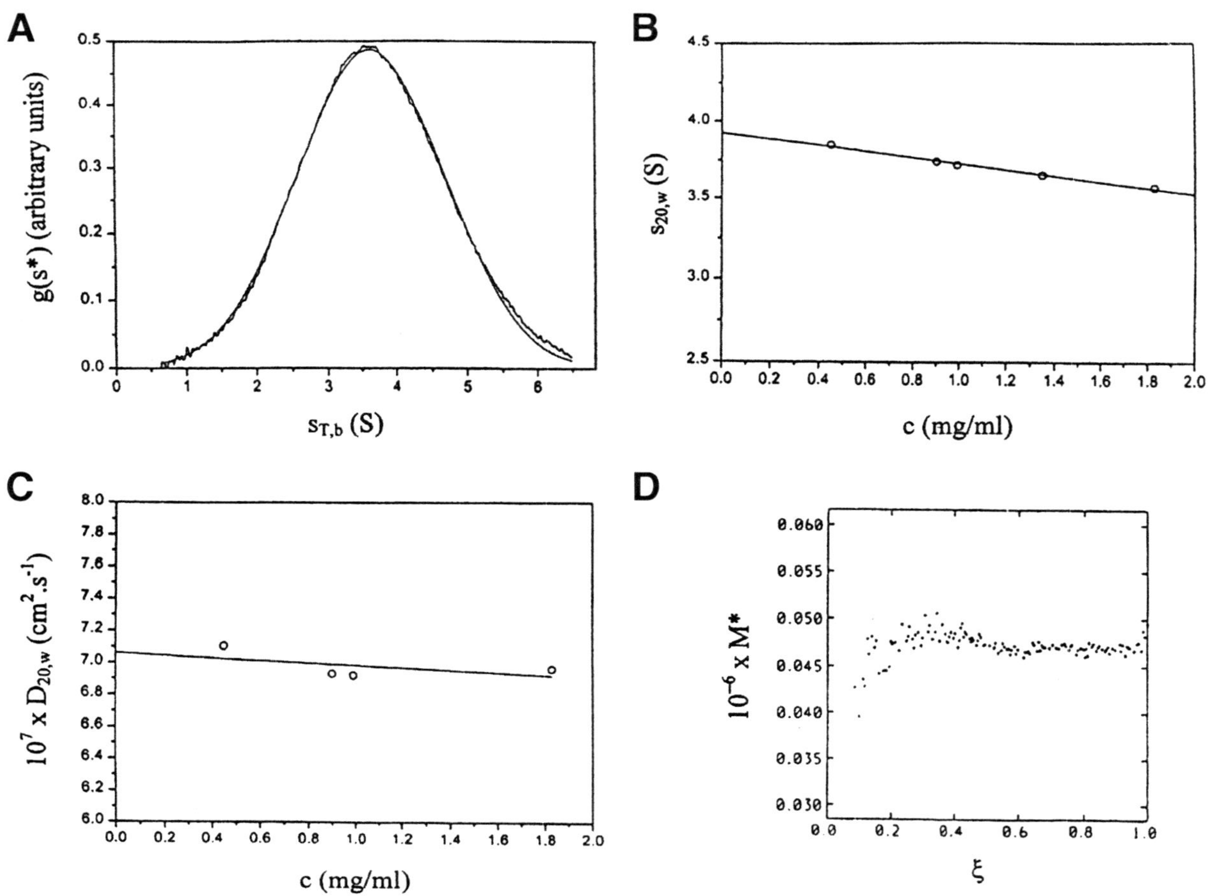

A
g(s*) (arbitrary units)
s_T,b (S)
B
s_20,w (S)
c (mg/ml)
C
10^7 x D_20,w (cm^2.s^-1)
c (mg/ml)
D
10^-6 x M*
ξ

Fig. 7. Sedimentation analysis of IgGFab′. **(A)** Sedimentation velocity DCDT+ plot for IgGFab' at a loading concentration of 1.10 mg/mL. Rotor speed = 49,000 rpm, temperature = 20.0°C. *g(s*)* is the apparent (e.g., not corrected for diffusion or non-ideality) distribution of sedimentation coefficients. The shape of the peak gives information of homogeneity (this sample is monodisperse). The center of the peak gives $s_{T,b}$; the width, for a monodisperse peak, gives $D_{T,b}$. **(B)** Gralén plot to obtain $s^o_{20,w}$. The line fitted is to **Equation 3.** **(C)** Plot of $D_{20,w}$ vs c to obtain $D^o_{20,w}$. **(D)** Extraction of M from sedimentation equilibrium using MSTAR analysis. M* is an operational point average mol wt whose value extrapolated to the bottom (ξ=1) of the ultracentrifuge = the weight average mol wt of all the macromolecular components in the centrifuge cell M. ξ is simply a normalized radial displacement squared parameter = $(r^2 - a^2)/(b^2 - a^2)$, with r=radial distance from the rotor center, and a,b are the corresponding distances of the air/solution meniscus and the cell bottom, respectively.

for M (a weight average) is also within experimental error of the mol wt of the monomeric antibody from its sequence, you can use with complete confidence the $s^o_{20,w}$ and the sequence mol wt for modeling.

3.1.3. Calculation of the Frictional Ratio and the "Frictional Ratio due to Shape"

1. Obtain the sedimentation coefficient $s^o_{20,w}$ for your antibody from the analytical ultracentrifuge as described in **Subheading 3.1.1.**
2. Obtain the sequence mol wt, M.
3. Use *s*, M, and the partial specific volume ($\bar{v}$) to calculate the frictional ratio f/f_o of the antibody according to **Equation 4.** In that formula, the value you use for the density and viscosity of solvent should be that of water at 20°C, since you have corrected your sedimentation coefficient to $s^o_{20,w}$. The program ELLIPS_PRIME *(20)* (*see* **Note 1**) can perform this calculation for you.
4. f/f_o cannot be used directly to model the shape, since it also depends also on the volume of the macromolecule: this volume has to take into account "swelling" through association of water/solvent with the protein. Although this association is a complicated dynamic process, for the purpose of overall shape modeling, a time-averaged "hydration" value, δ will suffice **(Subheading 3.2.1.)**. Then, the "frictional ratio due to shape," *P* can be estimated **(Equation 5)**, and compared with values computed from the shape models using HYDROSUB and SOLPRO as described in the following section.

3.2. Modeling Methods

3.2.1. δ-Value

1. Taking a so-called average protein value of 0.3–0.4 is inadequate for antibodies, primarily because they are glycoproteins and sugar residues show a higher affinity for water association than amino acid residues do.

2. δ can be estimated from hydrodynamic data, providing that the crystal structures of the Fab (or Fab′) and Fc domains are also known (this is the case for human IgG and IgE). This procedure involves taking the crystal structure for the domains and obtaining the corresponding value of *P* using bead modeling. Then, this is compared with the experimental value for f/f_o for the domains (from the sedimentation coefficient and mol wt), and δ can be obtained. An average δ for the intact antibody can then be estimated from a simple average:

$$\delta_{app} = \{2 \times \delta_{app}(\text{Fab}') + 1 \times \delta_{app}(\text{Fc})\}/3 \tag{7}$$

where δ_{app} means "apparent" hydration in the sense it takes into account other non-shape contributions towards the frictional ratio. For human IgG a value of ~0.59 has been obtained ***(12)***.

3.2.2. Surface Shell Representation of the Fab and Fc Domains and Hinge: HYDROSUB and SOLPRO

P values for domain orientations of the Fabs and Fc in an antibody (taking into account the hinge) can be calculated using the program HYDROSUB with SOLPRO. However, before this can be done, a way of specifying i) the shapes of the domains; ii) relative length of any hinge; and iii) the positions of the domains relative to each other needs to be achieved. This may be the most laborious part of the modeling procedure.

1. The Protein Data Bank (PDB) file is taken for the domains and a surface ellipsoid is fitted using the program ELLIPSE used by X-ray crystallographers ***(21)***. This provides the shape of the ellipsoid in terms of the two axial ratios a/b, b/c (**Fig. 8**).
2. Conveniently, the crystal structure of Fab approximates well to a "prolate" ellipsoid (two equal shorter axes, and one longer axis), and the Fc domain approximates an oblate ellipsoid (two equal long axes, and one shorter axis) as described in **Subheading 1.1.** and illustrated in **Fig. 2.**
3. The equivalent axial ratio (a/b) is then given for the Fab and Fc domains (in terms of a prolate and oblate ellipsoid, respectively) (**Fig. 9**).
4. Once the axial ratios are known, in preparation for piecing together the domains, we need to know the individual values of a and b for each domain: this can be done by expanding the ellipsoid fit to the surface of the crystal structures during the application of the ELLIPSE routine. A surface bead model to each ellipsoid domain is then generated using HYDROSUB. We are now ready to piece the molecule together.

*3.2.3. Piecing the Molecule Together to Calculate P for the Intact Antibody (see **Notes 3–6**)*

1. To generate the shell model for the pieced together antibody the user just has to specify, for the two Fabs and the Fc, the relative coordinates of their centers and the angles that define the orientation of the symmetry axis (**Fig. 10**), together, when appropriate, with the relative length of the hinge.

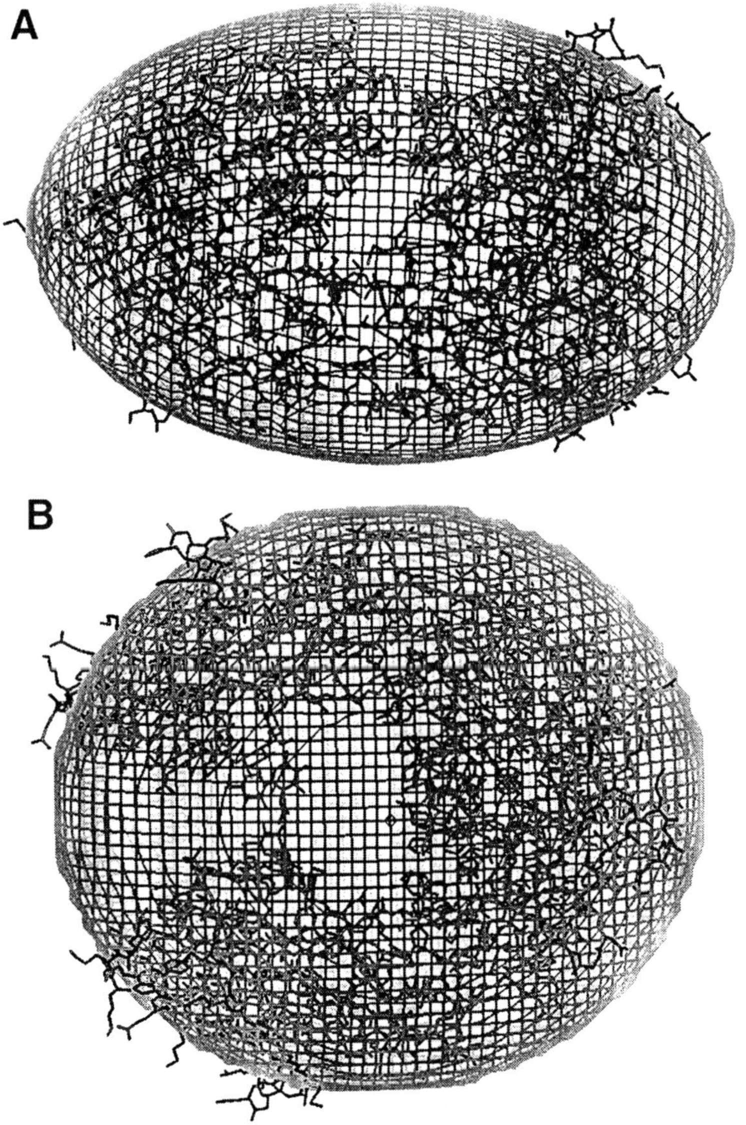

Fig. 8. General triaxial shape fits to the surface of the crystal structures of (**A**) IgGFab′ and (**B**) IgGFc using the ELLIPSE algorithm *(21)*. Instructions for fitting an ellipsoidal surface to the crystal structure of an antibody domain can be found in **ref.** ***2***.

2. The coordinates and angles of each subunit are given in the main input data file for HYDROSUB (for more details on the convention for angles, file format, etc, see the HYDROSUB user guide). SOLPRO is subsequently run to calculate *P* and all

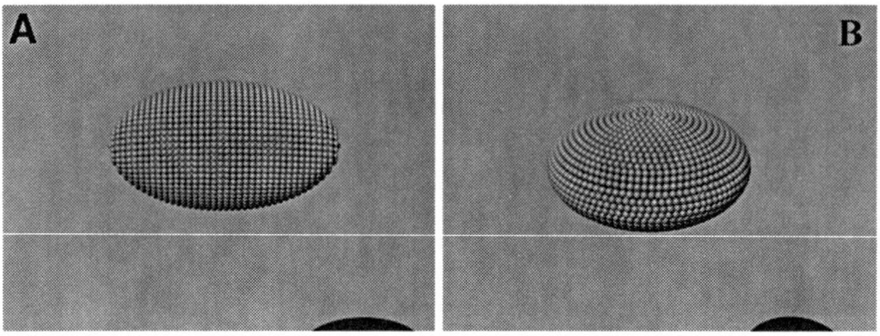

Fig. 9. Bead shell models for **(A)** the equivalent prolate ellipsoid fit to **Fig. 3A** for IgGFab and **(B)** the equivalent oblate ellipsoid fit to **Fig. 3B** for IgGFc. The procedure for generating the bead coordinates is described in the documentation for HYDROSUB.

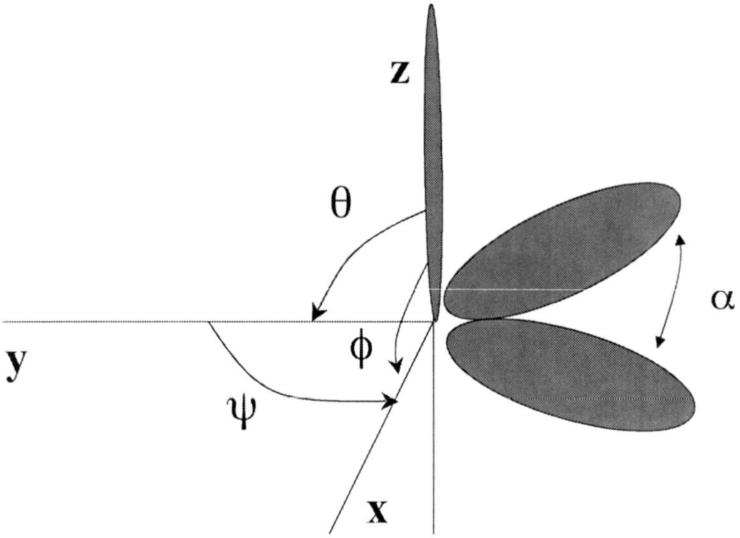

Fig. 10. Orientation angles between Fab′ and Fc domains relative to an xyz Cartesian frame. α: angle between Fab domains; θ: non-coplanarity angle of the Fc domain to the Fab's. $\theta = 90°$ corresponds to a planar structure; φ: twist angle (swivel of the Fc face); ψ: bend angle (bending up of Fc toward one of the Fab arms).

the other shape functions. A value of P is returned for a given arrangement of the domains and hinge length. The value of P does not depend on the absolute dimensions of the domains or hinge, merely on their dimensions relative to each other—e.g., the overall shape.

3. The values of P for various domain orientations are compared with the experimental value obtained from **Subheading 3.1.** The most appropriate models are selected by matching the P values for the models with the experimental P value after taking into account experimental error. **Fig. 11** gives some examples.

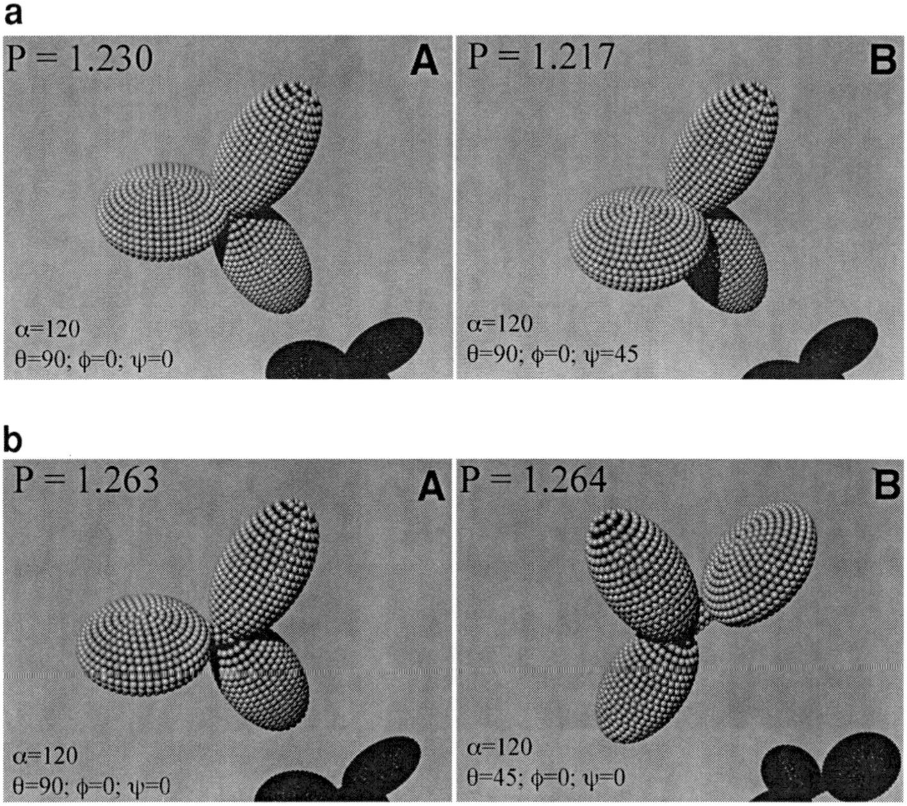

Fig. 11. Examples of domain representations for IgG subclasses of human antibody **(a)** Open Y-models for human IgG2 and IgG4. Experimental $P = 1.22 \pm 0.03$ for IgG2 and 1.23 ± 0.02 for IgG4. Experimentally, it is impossible to distinguish these two subclasses on the basis of sedimentation properties alone. Model A ($P = 1.230$): coplanar Y shape; Model B ($P = 1.217$) distorted (non-coplanar) Y. Compact models with the Fab arms folded down yield P values that are too small compared with the experimental values, and are ruled out. **(b)** Open Y-models for human IgG1. Small cylindrical hinge is necessary to model the experimental value of $P = 1.26 \pm 0.03$. Model A, Coplanar hinged Y ($P = 1.263$); Model B, Non-coplanar hinged Y ($P = 1.264$). **(c)** Coplanar T and Y-models for the hingeless mutant IgGMcg. The experimental value of $P = (1.23 \pm 0.03)$ is best satisfied by either Model A (coplanar Y shape, with twist angle, $\varphi = 90$, $P = 1.215$), or with Model B, a coplanar T-shape, with twist $\varphi = 0$ and $P = 1.194$. Although immunologically inactive, this hingeless mutant is one of the few intact antibodies whose crystal structures are known: that structure is consistent with the planar hydrodynamic models shown here. **(d)** Bead shell model for human $(Fab)_2$. Experimental Perrin function $P = (1.23 \pm 0.02)$. Modeled $P = 1.208$ (linear arrangement of domains: other arrangements yield lower values). *(Figure continues.)*

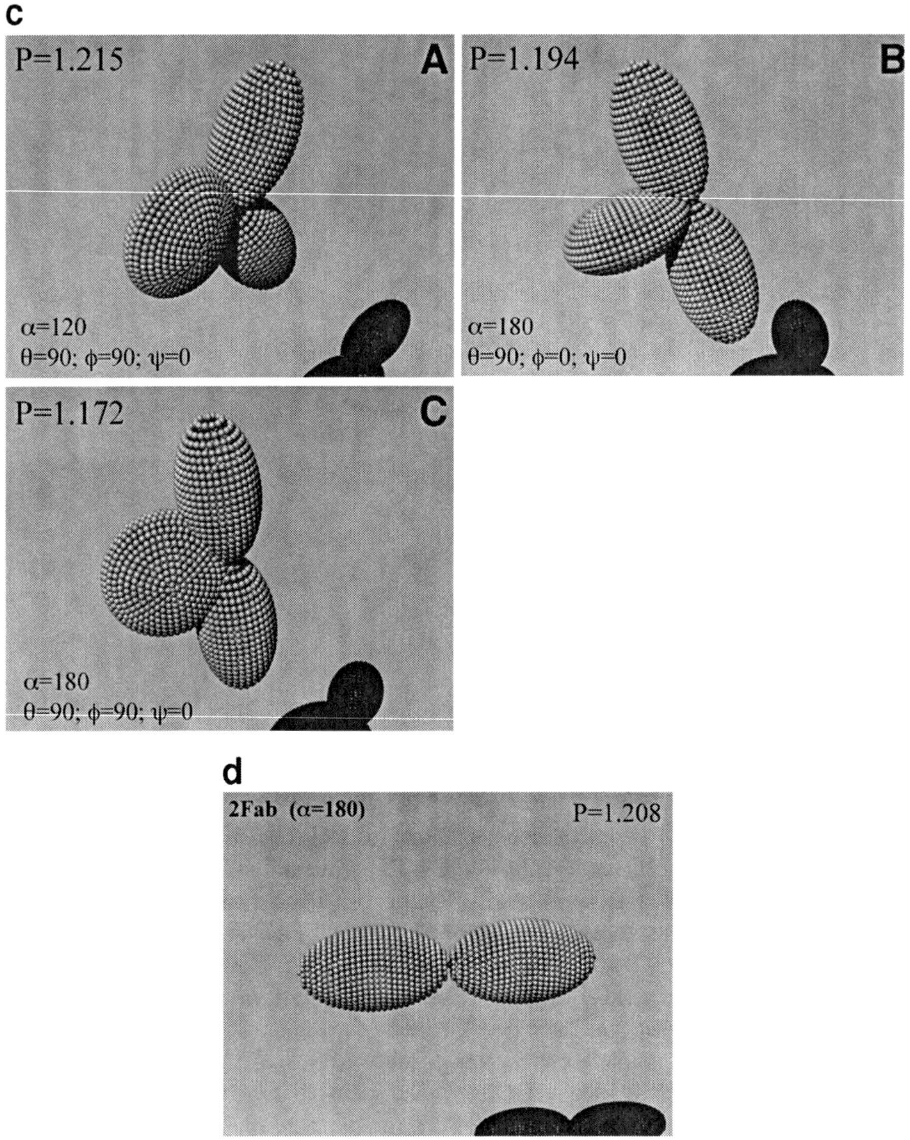

Fig. 11. *(Continued.)*

4. Notes

1. HYDRO, HYDROPRO, HYDRONMR, and SOLPRO (executables for several platforms, sample files and user guides) are available from the Internet web site http://leonardo.fcu.um.es/macromol. ELLIPS_PRIME is available from http://www.nottingham.ac.uk/ncmh/unit/method.html#Software; MSTAR and SEDNTERP are available from http://www.bbri.org/RASMB/rasmb.html.

2. The symptoms of this are that the measured sedimentation coefficient increases with increase in concentration; and that the measured mol wt > sequence molecular weight, and which also increases with concentration.
3. Experimental advice can be obtained from SEH (steve.harding@nottingham.ac.uk). For specific inquiries with regards the running of HYDROSUB, HYDROPAR, and SOLPRO, please contact JGT (jgt@um.es). General problems over the ultracentrifuge you can post your enquiry to rasmb@server1.bbri.org, where a host of experts will be willing to help.
4. It will be clear from **Fig. 11** that several models can give equally good fits to experimental values for *P*, the frictional ratio resulting from shape. This is known as the uniqueness problem: using just *P* as a hydrodynamic modeling parameter limits the detail that can be specified. For example, in the case of IgG1 we can distinguish open from compact conformations, but to be more specific as to which type of open conformation in terms of specific domain angles more information is required. This is the subject of ongoing research: researchers are currently investigating the virtue of additional information from viscosity, fluorescence anisotropy decay methods, and solution X-ray scattering.
5. Other sources of hydrodynamic information are intrinsic viscosity and rotational hydrodynamic parameters. The equivalent shape parameter to *P* from the intrinsic viscosity is the Einstein-Simha viscosity increment *ν*, which has a value of 2.5 for spheres and greater for other shapes *(22)*. It is a much more sensitive function of shape than *P*, but until recently, much higher concentrations have been required for its measurement (>5 mg/mL). A new generation of viscometer based on a measurement of the pressure difference between flow of a solution and flow of the corresponding pure solvent offers the possibility of measurements at much lower concentrations (for example, *see* **ref.** *22*). Fluorescence anisotropy relaxation times (including the harmonic mean) are also much more sensitive to shape than *P*, but have the disadvantage of requiring the presence of a suitable chromophore which does not rotate with respect to the rest of the molecule. The range of hydrodynamic methods that could be used is considered in **ref.** *6*.
6. Bead models can also be calculated from solution X-ray scattering data. X-ray scattering is more sensitive to shape than the translational frictional ratio, but again there are problems, namely in the form of irradiation damage to the protein: this is particularly serious for antibodies because of the extra susceptibility of the hinge region. Neutron scattering is a further alternative, but sources are not readily accessible, and the hydration corresponding to an antibody in aqueous solution may be quite different in a solution of deuterium oxide.

Acknowledgment

Studies at the University of Murcia are funded by Ministerio de Ciencia y Tecnologia, Spain (Proyect BQU2000-0229 and predoctoral fellowship awarded to A. Ortega). The work of the NCMH is supported by the Biotechnology and Biological Sciences Research Council and the Engineering and Physical Sciences Research Council, UK.

References

1. Burton, D. R. (1990) The conformation of antibodies, in *Fc Receptors and the Action of Antibodies* (Metzger, H. ed.), American Society for Microbiology, Washington, DC, pp. 31–54.
2. Carrasco, B., Garcia de la Torre, J., Davis, K. G., Jones, S., Athwal, D., Walters, C., et al. (2001) Crystallohydrodynamics for solving the hydration problem for multi-domain proteins: open physiological conformations for IgG. *Biophys. Chem.* **93,** 181–196.
3. Davis, K. G., Glennie, M., Harding, S. E., and Burton, D. R. (1990) A model for the solution conformation of rat IgE. *Biochem. Soc. Trans.* **18,** 935–936.
4. Schuck, P., Perugini, M. S., Gonzales, N. R., Howlett, G. J., and Schubert, D. (2002) Size-distribution analysis of proteins by analytical ultracentrifugation: strategies and application to model systems. *Biophys. J.* **82,** 1096–1111.
5. Harding, S. E. (1994) Determination of macromolecular homogeneity, shape and interactions using sedimentation velocity analytical ultracentrifugation. *Methods in Molecular Biology* (Jones, C., Mulloy, B. and Thomas, A. H. eds.), Humana Press, Totowa, NJ, pp. 61–73.
6. Harding, S. E. (1995) On the hydrodynamic analysis of macromolecular conformation. *Biophys. Chem.* **55,** 69–93. *See also* Harding, S. E., Horton, H., and Cölfen, H, (1997) The ELLIPS suite of macromolecular conformation algorithms. *Eur. Biophys. J.* **25,** 347–359.
7. Carrasco, B. and Garcia de la Torre, J. (1999) Hydrodynamic properties of rigid particles. Comparison of different modelling and computational strategies. *Biophys. J.* **76,** 3044–3057.
8. Garcia de la Torre, J., Navarro, S., Lopez Martinez, M. C., Diaz, F. G., and Lopez Cascales, J. (1994) HYDRO. A computer software for the prediction of hydrodynamic properties of macromolecules. *Biophys. J.* **67,** 530–531.
9. Garcia de la Torre, J., Huertas, M. L., and Carrasco, B. (2000) Calculation of hydrodynamic properties of globular proteins from their atomic-level structure. *Biophys. J.* **78,** 719–730.
10. Garcia de la Torre, J. (2001) Building hydrodynamic bead-shell models for rigid bioparticles of arbitrary shape. *Biophys. Chem.* **94,** 265–274.
11. Garcia de la Torre, J. and Carrasco, B. (2002) Hydrodynamic properties of rigid macromolecules composed of ellipsoidal and cylindrical subunits" *Biopolymers* **63,** 163–167.
12. Garcia de la Torre, J., Carrasco, B., and Harding, S. E. (1997) SOLPRO: theory and computer program for the prediction of SOLution PROperties of rigid macromolecules and bioparticles. *Eur. Biophys. J.* **25,** 361–372.
13. Garcia de la Torre, J., Harding, S. E., and Carrasco, B. (1999) Calculation of NMR relaxation, covolume and scattering properties of bead models using the SOLPRO computer program. *Eur. Biophys. J.* **28,** 119–132.
14. Philo, J. S. (2000) A method for directly fitting the time derivative of sedimentation velocity data and an alternative algorithm for calculating sedimentation coefficient distribution functions. *Analytical Biochem.* **279**, 151–163.

15. Philo, J. S. (1997) An improved function for fitting sedimentation velocity data for low-molecular-weight solutes. *Biophys. J.* **72,** 435–444.
16. http://www.bbri.org/RASMB/rasmb.html
17. Laue, T. M., Shah, B. D., Ridgeway, T. M., and Pelletier, S. L. (1992) Computer-aided interpretation of analytical sedimentation data for proteins, in *Analytical Ultracentrifugation in Biochemistry and Polymer Science* (Harding S. E., Rowe A. J. and Horton J. C. eds.), Royal Society of Chemistry, Cambridge, UK, pp. 90–125.
18. Harding, S. E. (1994) Determination of absolute molecular weights using sedimentation equilibrium analytical ultracentrifugation. *Methods in Molecular Biology* (Jones, C., Mulloy, B., and Thomas, A. H. eds.), Humana Press, Totowa, NJ, pp. 75–84.
19. Cölfen, H. and Harding, S. E. (1997) MSTARA and MSTARI: interactive PC algorithms for simple, model independent evaluation of sedimentation equilibrium data. *Eur. Biophys. J.* **25,** 333–346.
20. http://www.nottingham.ac.uk/ncmh/unit/method.html#Software
21. Taylor, W. R., Thornton, J. M., and Turnell, W. G. (1983) An ellipsoidal approximation of protein shape. *J. Mol. Graphics* **1,** 30–38.
22. Harding, S. E. (1997) The intrinsic viscosity of biological macromolecules. Progress in measurement, interpretation and application to structure in dilute solution. *Prog. Biophys. Mol. Biol.* **68,** 207–262.

II

Antibody-Lead Generation

6

PCR Cloning of Human Immunoglobulin Genes

James D. Marks and Andrew Bradbury

1. Introduction

One of the first steps in an antibody-engineering project is the isolation of the immunoglobulin heavy (V_H)- and light (V_L)-chain variable-region genes that encode the binding domains of an antibody. This is best accomplished using the polymerase chain reaction (PCR). Before PCR, cloning antibody genes was a laborious process, requiring the creation and screening of genomic or cDNA libraries. PCR has streamlined the process considerably, simplifying tasks such as the isolation of V_H and V_L genes from hybridomas, and allowing entirely new approaches such as generation of large antibody fragment gene repertoires from immunized or naïve hosts that can be displayed on filamentous phage (antibody phage display; *see* Chapter 8) *(1)* or in other display technologies (e.g., ribosome display; *see* Chapter 9).

For PCR amplification, first-strand cDNA is obtained by reverse transcription of mRNA with constant region primers [primers that anneal in the C_H gene ($C_\gamma 1$, $C_\gamma 2$, $C_\gamma 3$, $C_\gamma 4$, C_μ, C_ε, C_α, or C_δ) or the C_L gene (C_κ or C_λ)]. Since the sequences of the constant domain exons are known *(2)*, primer design is straightforward. For PCR amplification, primers are usually designed so that either Fab, Fv, or single-chain Fv (scFv) antibody fragments can be created. For the creation of Fabs, the V_H-C_H1 (fd) and light chain (V_L-C_L) are amplified. Design of primers that anneal to the 3′ end of these genes is straightforward, since the constant regions have been sequenced *(2)*. For Fv or scFv, only the rearranged V_H and V_L are amplified. Design of PCR primers for the 3′ end of rearranged murine *(3)* or human *(1)* V_H and V_L gene is also straightforward, since primers can be based on the J gene segments, which have been sequenced. These primers can also be used for first-strand cDNA synthesis instead of constant region primers. A number of groups have also designed sets of "universal" V-gene primers containing internal or appended restriction

From: *Methods in Molecular Biology, Vol. 248: Antibody Engineering: Methods and Protocols*
Edited by: B. K. C. Lo © Humana Press Inc., Totowa, NJ

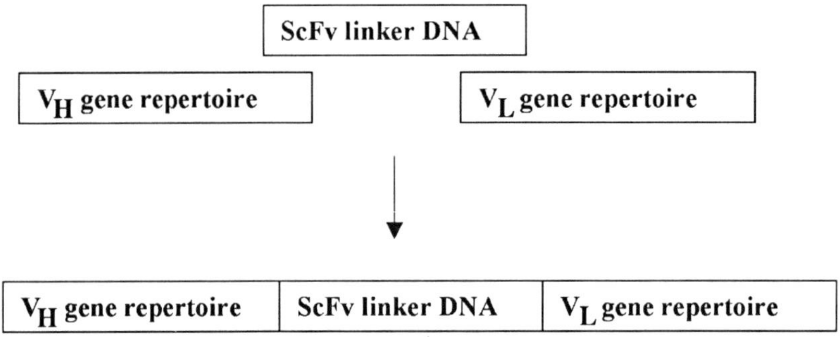

Fig. 1. Construction of scFv gene repertoires using splicing by overlap extension (PCR assembly). For three-fragment assembly, the linker is created in such a way that it contains a region of homology to all 3′ J_H and 5′ V_L genes. This anchors the assembly and ensures that the linker remains at the correct length and is important when the linker is repetitive, such as with the $(Gly_4Ser)_3$ scFv linker. If a non-repetitive linker is used, a two-fragment assembly can be carried out with the region of overlap between the two V-genes containing an overlapping region of the linker.

sites that are suitable for amplification of murine *(4–8)*, human *(1,9–11)*, chicken *(12)*, and rabbit *(13)* V-genes. These "universal" primers are based on the extensive published V-region sequences, and permit PCR amplification of most V-genes.

Fv, Fab, and scFv genes are typically constructed by the sequential cloning of V_H (V_H-C_H1) and V_L (V_L-C_L) genes *(14,15)*. Alternatively, PCR splicing by overlap extension (PCR assembly) of V_H and V_L genes has been used to construct scFv genes *(1,3)* (**Fig. 1**). With PCR assembly, only two restriction sites are required to clone the scFv. These can be appended to the 5′ and 3′ end of the scFv gene cassette, and expression systems have been described in which octanucleotide cutters can be utilized *(1)*. This markedly decreases the likelihood that a restriction enzyme will cut internally in the V-gene. Sequential cloning requires four restriction sites and increases the chances of restriction enzymes cutting internally *(16)*.

In this chapter, protocols are described for PCR amplification of human V_H and V_L gene repertoires from human peripheral-blood lymphocytes (PBLs). These protocols can be adapted to PCR amplification of V_H and V_L genes from hybridomas or immunized mice by substituting murine primers *(4–8,17)*. We also describe how the human repertoires can be used to create naïve scFv phage antibody libraries using splicing by overlap extension and cloning into the vector pHEN1 *(18)*. The same protocols can also be used to construct libraries from immunized humans or patients with diseases for which an immune response is mounted. Alternatively, different primer sets can be used to generate libraries from immunized rodents or hybridomas *(4–8,17)*.

The steps required to construct scFv phage antibody libraries are:

1. Isolation of RNA from a V-gene source (**Subheading 3.1.**)
2. First-strand cDNA synthesis (**Subheading 3.2.**)
3. V-region PCR and construction of a scFv gene repertoire (**Subheadings 3.3.–3.6.**)
4. Restriction digestion and ligation of the scFv gene repertoire and transformation of *E. coli* phage display vector and scFv gene repertoire (**Subheadings 3.7–3.10.**)

2. Materials

2.1. RNA Isolation from PBLs (see Note 1)

1. 30-mL Corex tubes, acid-washed and silated.
2. Ficoll (Amersham Biosciences).
3. Ice-cold phosphate-buffered saline (PBS): 154 m*M* NaCl, 8.1 m*M* Na_2HPO_4, 1.9 m*M* NaH_2PO_4, pH 7.4.
4. Lysis buffer: 5 *M* guanidine monothiocyanate, 10 m*M* ethylenediaminetetraacetic acid (EDTA), 50 m*M* Tris-HCl, pH 7.5, 1 m*M* dithiothreitol (DTT), filtered through a Millipore 0.45-μm filter.
5. 4 *M* and 3 *M* lithium chloride, autoclaved.
6. RNA solubilization buffer: 0.1% sodium dodecyl sulfate (SDS), 1 m*M* EDTA, 10 m*M* Tris-HCl, pH 7.5, autoclaved.
7. 3 *M* sodium acetate, pH 4.8, diethylpyrocarbonate (DEPC)-treated (add 0.2 mL DEPC/100 mL solution) and autoclaved.
8. DEPC-treated water (0.2 mL DEPC/100 mL water), autoclaved.
9. Phenol equilibrated in 10 m*M* Tris-HCl, pH 8.0 (Sigma; Cat. #P4557).
10. Chloroform: Isoamyl alcohol (24:1) (Sigma; Cat. #C0549).
11. 100% ethanol at –20°C.

2.2. First-Strand cDNA Synthesis of Human Ig Heavy- and Light-Chain Genes

1. 10X RT buffer: 1.4 *M* KCl, 500 m*M* Tris-HCl, pH 8.1 at 42°C, 80 m*M* $MgCl_2$.
2. RNAsin ribonuclease inhibitor (Promega).
3. AMV reverse transcriptase (Promega).
4. 70% ethanol.
5. 100 m*M* DTT.
6. 20X dNTPs: 5 m*M* each of deoxyadenosine triphosphate (dATP), deoxyguanosine triphosphate (dGTP), deoxycytidine triphosphate (dCTP), and deoxythymidine triphosphate (dTTP).
7. IgG, IgM, κ and λ chain constant region primers, or random hexamer primers (10 pmol/μL) (*see* **Table 1,** item 1).
8. Silated 1.5-mL centrifuge tubes.
9. RNA (from **Subheading 3.1.**).

Table 1
Primers for the Creation of Human scFv Libraries

1. Primers for first-strand cDNA synthesis (Subheading 3.2.)	
Human Heavy-Chain Constant Region Primers	
HuIgG1–4CH1FOR	5′-GTC CAC CTT GGT GTT GCT GGG CTT-3′
HuIgMFOR	5′-TGG AAG AGG CAC GTT CTT TTC TTT-3′
Human κ-Chain Constant Region Primer	
$HuC_{\kappa}FOR$	5′-AGA CTC TCC CCT GTT GAA GCT CTT-3′
Human λ-Chain Constant Region Primer	
HuCλFOR	5′-TGA AGA TTC TGT AGG GGC CAC TGT CTT-3′
2. Primers for primary amplifications of V_H, V_{κ}, and V_{λ} genes (Subheading 3.3.)	
Human V_H Back Primers	
HuVH1aBACK	5′-CAG GTG CAG CTG GTG CAG TCT GG-3′
HuVH2aBACK	5′-CAG GTC AAC TTA AGG GAG TCT GG-3′
HuVH3aBACK	5′-GAG GTG CAG CTG GTG GAG TCT GG-3′
HuVH4aBACK	5′-CAG GTG CAG CTG CAG GAG TCG GG-3′
HuVH5aBACK	5′-GAG GTG CAG CTG TTG CAG TCT GC-3′
HuVH6aBACK	5′-CAG GTA CAG CTG CAG CAG TCA GG-3′
Human V_{κ} Back Primers	
HuVκ1aBACK	5′-GAC ATC CAG ATG ACC CAG TCT CC-3′
HuVκ2aBACK	5′-GAT GTT GTG ATG ACT CAG TCT CC-3′
HuVκ3aBACK	5′-GAA ATT GTG TTG ACG CAG TCT CC-3′
HuVκ4aBACK	5′-GAC ATC GTG ATG ACC CAG TCT CC-3′
HuVκ5aBACK	5′-GAA ACG ACA CTC ACG CAG TCT CC-3′
HuVκ6aBACK	5′-GAA ATT GTG CTG ACT CAG TCT CC-3′
Human V_{λ} Back primers	
HuVλ1BACK	5′-CAG TCT GTG TTG ACG CAG CCG CC-3′
HuVλ2BACK	5′-CAG TCT GCC CTG ACT CAG CCT GC-3′
HuVλ3aBACK	5′-TCC TAT GTG CTG ACT CAG CCA CC-3′
HuVλ3bBACK	5′-TCT TCT GAG CTG ACT CAG GAC CC-3′
HuVλ4BACK	5′-CAC GTT ATA CTG ACT CAA CCG CC-3′
HuVλ5BACK	5′-CAG GCT GTG CTC ACT CAG CCG TC-3′
HuVλ6BACK	5′-AAT TTT ATG CTG ACT CAG CCC CA-3′
Human J_H Forward Primers	
HuJH1–2FOR	5′-TGA GGA GAC GGT GAC CAG GGT GCC-3′
HuJH3FOR	5′-TGA AGA GAC GGT GAC CAT TGT CCC-3′
HuJH4–5FOR	5′-TGA GGA GAC GGT GAC CAG GGT TCC-3′
HuJH6FOR	5′-TGA GGA GAC GGT GAC CGT GGT CCC-3′
Human J_{κ} Forward Primers	
HuJκ1FOR	5′-ACG TTT GAT TTC CAC CTT GGT CCC-3′
HuJκ2FOR	5′-ACG TTT GAT CTC CAG CTT GGT CCC-3′
HuJκ3FOR	5′-ACG TTT GAT ATC CAC TTT GGT CCC-3′
HuJκ4FOR	5′-ACG TTT GAT CTC CAC CTT GGT CCC-3′
HuJκ5FOR	5′-ACG TTT AAT CTC CAG TCG TGT CCC-3′

Table 1 *(Continued)*

Human J_λ Forward Primers	
HuJλ1FOR	5′-ACC TAG GAC GGT GAC CTT GGT CCC-3′
HuJλ2–3FOR	5′-ACC TAG GAC GGT CAG CTT GGT CCC-3′
HuJλ4–5FOR	5′-ACC TAA AAC GGT GAG CTG GGT CCC-3′
3. Primers to create scFv linker DNA (Subheading 3.4.)	
Reverse J_H primers	
RHuJH1–2	5′-GCA CCC TGG TCA CCG TCT CCT CAG GTG G-3′
RHuJH3	5′-GGA CAA TGG TCA CCG TCT CTT CAG GTG G-3′
RHuJH4–5	5′-GAA CCC TGG TCA CCG TCT CCT CAG GTG G-3′
RHuJH6	5′-GGA CCA CGG TCA CCG TCT CCT CAG GTG C-3′
Reverse V_κ for scFv linker	
RHuVκ1aBACKFv	5′-GGA GAC TGG GTC ATC TGG ATG TCC GAT CCG CC-3′
RHuVκ2aBACKFv	5′-GGA GAC TGA GTC ATC ACA ACA TCC GAT CCG CC-3′
RHuVκ3aBACKFv	5′-GGA GAC TGC GTC AAC ACA ATT TCC GAT CCG CC-3′
RHuVκ4aBACKFv	5′-GGA GAC TGG GTC ATC ACG ATG TCC GAT CCG CC-3′
RHuVκ5aBACKFv	5′-GGA GAC TGC GTG AGT GTC GTT TCC GAT CCG CC-3′
RHuVκ6aBACKFv	5′-GGA GAC TGA GTC AGC ACA ATT TCC GAT CCG CC-3′
Reverse V_λ for scFv linker	
RHuVλBACK1Fv	5′-GGC GGC TGC GTC AAC ACA GAC TGC GAT CCG CCA CCG CCA GAG-3′
RHuVλBACK2Fv	5′-GCA GGC TGA GTC AGA GCA GAC TGC GAT CCG CCA CCG CCA GAG-3′
RHuVλBACK3aFv	5′-GGT GGC TGA GTC AGC ACA TAG GAC GAT CCG CCA CCG CCA GAG-3′
RHuVλBACK3bFv	5′-GGG TCC TGA GTC AGC TCA GAA GAC GAT CCG CCA CCG CCA GAG-3′
RHuVλBACK4Fv	5′-GGC GGT TGA GTC AGT ATA ACG TGC GAT CCG CCA CCG CCA GAG-3′
RHuVλBACK5Fv	5′-GAC GGC TGA GTC AGC ACA GAC TGC GAT CCG CCA CCG CCA GAG-3′
RHuVλBACK6Fv	5′-TGG GGC TGA GTC AGC ATA AAA TTC GAT CCG CCA CCG CCA GAG-3′
4. Primers with appended restriction sites for reamplification of scFv gene repertoires (Subheadings 3.5. and 3.6.)	
HuVH1aBACKSfi	5′-GTC CTC GCA ACT GCG GCC CAG CCG GCC ATG GCC CAG GTG CAG CTG GTG CAG TCT GG-3′

(continues)

Table 1 *(Continued)*

HuVH2aBACKSfi	5′-GTC CTC GCA ACT GCG GCC CAG CCG GCC ATG GCC CAG GTC AAC TTA AGG GAG TCT GG-3′
HuVH3aBACKSfi	5′-GTC CTC GCA ACT GCG GCC CAG CCG GCC ATG GCC GAG GTG CAG CTG GTG GAG TCT GG-3′
HuVH4aBACKSfi	5′-GTC CTC GCA ACT GCG GCC CAG CCG GCC ATG GCC CAG GTG CAG CTG CAG GAG TCG GG-3′
HuVH5aBACKSfi	5′-GTC CTC GCA ACT GCG GCC CAG CCG GCC ATG GCC CAG GTG CAG CTG TTG CAG TCT GC-3′
HuVH6aBACKSfi	5′-GTC CTC GCA ACT GCG GCC CAG CCG GCC ATG GCC CAG GTA CAG CTG CAG CAG TCA GG-3′
HuJκ1FORNot	5′-GAG TCA TTC TCG ACT TGC GGC CGC ACG TTT GAT TTC CAC CTT GGT CCC-3′
HuJκ2FORNot	5′-GAG TCA TTC TCG ACT TGC GGC CGC ACG TTT GAT CTC CAG CTT GGT CCC-3′
HuJκ3FORNot	5′-GAG TCA TTC TCG ACT TGC GGC CGC ACG TTT GAT ATC CAC TTT GGT CCC-3′
HuJκ4FORNot	5′-GAG TCA TTC TCG ACT TGC GGC CGC ACG TTT GAT CTC CAC CTT GGT CCC-3′
HuJκ5FORNot	5′-GAG TCA TTC TCG ACT TGC GGC CGC ACG TTT AAT CTC CAG TCG TGT CCC-3′
Hu Jλ1FORNot	5′-GAG TCA TTC TCG ACT TGC GGC CGC ACC TAG GAC GGT GAC CTT GGT CCC-3′
Hu Jλ2–3FORNot	5′-GAG TCA TTC TCG ACT TGC GGC CGC ACC TAG GAC GGT CAG CTT GGT CCC-3′
Hu Jλ4–5FORNot	5′-GAG TCA TTC TCG ACT TGC GGC CGC ACY TAA AAC GGT GAG CTG GGT CCC-3′

The primers above correspond to those used to create libraries described in **refs. *1,19*.** Although they may not amplify all V genes, they have produced excellent libraries. They are adapted for cloning into pHEN1.

2.3. Amplification of Human V_H and V_L Genes from First-Strand cDNA

1. Vent DNA polymerase and 10X buffer (New England Biolabs).
2. 20X dNTPs: 5 m*M* each of dATP, dGTP, dCTP, and dTTP.
3. V-region specific forward (3′) (J_HFOR, J_κ FOR, and J_λ FOR) and back (5′) (V_H BACK, V_κ BACK, and V_λ BACK) PCR primers (**Table 1,** item 2).
4. PCR thermocycler.
5. Agarose, DNA-grade and high melting (Fisher Biotech; Cat. #BP164-500).
6. 10X TBE buffer: dissolve 108 g of Tris base and 55 g of boric acid in 900 mL of water, add 40 mL of 0.5 *M* EDTA, pH 8.0, and bring vol to 1 L.
7. 100-bp ladder DNA mass markers (New England Biolabs; Cat. #N3231S).

8. Geneclean kit (Bio 101, Inc.).
9. First-strand cDNA (from **Subheading 3.2.**).

2.4. Preparation of ScFv Linker DNA

1. PCR reagents and equipment as described in **Subheading 2.3.**
2. Vector containing an scFv with the $(Gly_4Ser)_3$ linker (1 µg)—available from the corresponding author (JDM).
3. $RHuJ_H$, $RHuV_\kappa BACK$ and $RHuV_\lambda BACK$ primers at 10 pmol/µL (**Table 1,** item 3).
4. Agarose, DNA-grade and high melting.
5. 10X TBE buffer (*see* **Subheading 2.3.**).
6. 100-bp ladder DNA mass markers.

2.5. PCR Assembly of V_H and V_L Genes with ScFv Linker to Create ScFv Gene Repertoires

1. PCR reagents and equipment as described in **Subheading 2.3.**
2. Geneclean kit.
3. V-region specific forward ($HuJ_\kappa FORNot$ and $HuJ_\lambda FORNot$) and back ($HuV_H BACKSfi$) PCR primers (**Table 1,** item 4).
4. ScFv linker DNA (from **Subheading 3.4.**).
5. Agarose, DNA-grade and high melting.
6. 10X TBE buffer (see **Subheading 2.3.**).
7. 100-bp ladder DNA mass markers.

2.6. Re-Amplification of ScFv Gene Repertoires to Append Restriction Sites for Cloning

1. PCR reagents and equipment as described in **Subheading 2.3.,** except substitute Taq polymerase and buffer for Vent polymerase and buffer (*see* **Note 2**).
2. V-region specific forward (3′) ($HuJ_\kappa FORNot$ and $HuJ_\lambda FORNot$) and back (5′) ($HuV_H BACKSfi$) PCR primers **(Table 1, item 4.)**.
3. Assembled scFv repertoires (from **Subheading 3.5.**).
4. Wizard PCR purification kit (Promega).

2.7. Restriction Digestion of ScFv Gene Repertoires and pHEN1 Phage Display Vector

1. *Nco*I and *Not*I restriction enzymes and NEB3 buffer (New England Biolabs).
2. 100X acetylated BSA solution (New England Biolabs, supplied with NotI).
3. 37°C incubator or heat block with heated lid.
4. Geneclean kit.
5. pHEN1 DNA ***(18)***, prepared by cesium chloride purification.
6. Reamplified scFv repertoires (from **Subheading 3.6.**).

2.8. Ligation of ScFv and Vector DNA

1. T4 DNA ligase and 10X ligation buffer (New England Biolabs).
2. Phenol/chloroform (Sigma).

3. 100% and 70% ethanol at –20°C.
4. Digested vector and scFv repertoires from **Subheading 3.7.**
5. TE buffer: 10 m*M* Tris-HCl, pH 8.0, 1 m*M* EDTA.

2.9. Generation of Phage Antibody Libraries

1. Electrocompetent *E. coli* strain TG1 (*see* **Subheading 3.10.** for preparation).
2. 37°C incubator.
3. Electroporator and 0.2-cm path cuvets (BioRad Gene Pulser™; or equivalent).
4. 16°C water bath.
5. SOC media: dissolve 20 g of bacto-tryptone, 5 g of yeast extract, and 0.5 g of NaCl in 950 mL of distilled water; add 10 mL of 250 m*M* KCl, adjust to pH 7.0 with 5 *N* NaOH and make up to 1 L with distilled water; autoclave the solution; when it is cool to the touch, add 5 mL of 2 *M* $MgCl_2$ and 20 mL of 1 *M* glucose.
6. 100 and 150 mm TYE/amp/glu agar plates (100 µg/mL ampicillin, 2% glucose): dissolve 16 g of bacto-tryptone, 10 g of yeast extract, 5 g of NaCl, and 15 g of bacto agar in distilled water and adjust the volume to 1 L. Autoclave the solution, and when it is cool to the touch, add 1 mL of 100 mg/mL ampicillin and 100 mL of 20% glucose. Mix and pour into plates.
7. Bio-assay dishes (243 × 243 × 18 mm, 530 cm^3 area) (Nunc).
8. 2X TY media: dissolve 16 g of bacto-tryptone, 10 g of yeast extract, and 5 g of NaCl in 1 L of distilled water; autoclave the solution.
9. 2X TY/amp/glu media: 2X TY media containing 100 µg/mL ampicillin and 2% glucose.
10. 50% glycerol.

2.10. Preparation of Electrocompetent E. coli TG1

1. *E. coli* TG1 strain (Stratagene).
2. Sorvall RC5 centrifuge (or similar), GS3 rotor (both prechilled to 4°C).
3. Prechilled 500-mL polypropylene centrifuge tubes.
4. 2-L baffled flasks.
5. Minimal media agar plate: 10.5 g of K_2HPO_4, 4.5 g of KH_2PO_4, 1 g of $(NH_4)_2SO_4$, 0.5 g of sodium citrate· $2H_2O$, 15 g of agar with water to 1 L; autoclave to sterilize and allow to cool until the flask is hand-hot, and then add 1 mL of 1 *M* $MgSO_4$, 0.5 mL of 1% vitamin B1 (thiamine), and 10 mL of 20% dextrose.
6. 2X TY media (*see* **Subheading 2.9.**).
7. 1 *M* HEPES, pH 7.4.
8. 1 M HEPES, pH 7.4 in 10% glycerol.
9. 10% glycerol.

3. Methods

3.1. RNA Isolation from PBLs (see Note 3)

1. Collect fresh human blood and separate white blood cells over Ficoll immediately.
2. Wash the PBLs 3× with ice-cold PBS.

3. Collect the PBLs in a 50-mL plastic centrifuge tube and add 7 mL of lysis buffer (will lyse up to 5×10^8 PBLs), and vortex vigorously to lyse the cells (*see* **Note 4**).
4. Add 7 vol (49 mL) of 4 *M* lithium chloride and incubate at 4°C for 15–20 h (overnight).
5. Transfer the suspension to 30-mL Corex tubes and centrifuge at 6,800*g* in a swinging bucket rotor at 4°C for 2 h.
6. Pour off the supernatant and wipe the lips of the tubes with Kimwipes. Pool the pellets by resuspending them in 3 *M* lithium chloride (approx 15 mL). Centrifuge the resuspended pellets at 6,800*g* for 1 h.
7. Pour off the supernatant and dissolve the pellets in 2 mL of RNA solubilization buffer. Freeze the suspension thoroughly at –20°C.
8. Thaw the suspension by vortexing for 20 s every 10 min for 45 min.
9. Extract once with an equal volume of phenol and once with an equal volume of chloroform.
10. Precipitate the RNA by adding 1/10 vol of 3 *M* sodium acetate, pH 4.8 and 2 vol of –20°C ethanol. Mix the solution thoroughly and incubate at –20°C overnight.
11. Centrifuge the RNA at 23,500*g* in a swinging bucket rotor for 30 min. Suspend the pellet in 0.2 mL of DEPC-treated water. Transfer the dissolved DNA to a 1.5-mL microcentrifuge tube and reprecipitate it by adding 1/10 vol of 3 *M* sodium acetate, pH 4.8, and 2 vol of –20°C ethanol. Store the RNA as an ethanol-precipitate until ready to use.

3.2. First-Strand cDNA Synthesis of Human Ig Heavy- and Light-Chain Genes (see Notes 5 and 6)

1. For first-strand cDNA synthesis, prepare the following reaction mixture in a silated 1.5-mL microcentrifuge tube (*see* **Note 7**):

10X RT buffer	5 µL
20X dNTPs	2.5 µL
100 m*M* DTT	5 µL
HuIgMFOR primer	2 µL
HuC_κ FOR primer	2 µL
HuC_λ FOR primer	2 µL
RNAsin	80 U (2 µL)

All primers are 10 pmol/µL

2. Take an aliquot (1–4 µg) of RNA in ethanol, place in a sterile 1.5-mL microcentrifuge tube, and centrifuge it at maximum speed for 5 min in a microcentrifuge. Wash the pellet once with 70% ethanol, dry it, and resuspend in 27 µL of DEPC-treated water.
3. Heat the solution to 65°C for 3 min to denature the RNA, quench it on ice for 2 min, and add it to the first-strand reaction mixture. Add 12.5 U (2.5 µL) of AMV reverse transcriptase and incubate the solution at 42°C for 1 h.
4. Boil the cDNA reaction mixture for 3 min, centrifuge it for 5 min in a microcentrifuge, transfer the cDNA-containing supernatant to a new silated tube, and use it immediately for PCR amplification of V-genes (*see* **Subheading 3.3.**).

3.3. Amplification of Human V_H and V_L Genes from First-Strand cDNA

1. Make up 50-µL PCR reaction mixes in 0.5-mL microcentrifuge tubes containing:

Millipore water	31.5 µL
10X Vent polymerase buffer	5.0 µL
20X dNTPs	2.5 µL
Forward primer(s) (*see* **Notes 8–10**)	2.0 µL
Back primer(s) (*see* **Notes 8–10**)	2.0 µL
cDNA reaction mix (from **Subheading 3.2.**)	5.0 µL

2. Heat the reactions to 94°C for 5 min in a thermal cycling block. If this does not have a heated lid, add one drop of light mineral oil to cover the reaction.
3. Add 2 U (2.0 µL) of Vent DNA polymerase.
4. Cycle 30× to amplify the V-genes at 94°C for 1 min, 60°C for 1 min, and 72°C for 1 min.
5. Purify the PCR fragments (V_H: ~350 bp; V_L: ~330 bp) by electrophoresis on a 1.5% agarose gel, extract from the gel using the Geneclean kit according to the manufacturer's instructions. Resuspend each product in 20 µL of water. Determine the concentration of the DNA by analysis on a 1% agarose gel compared to markers of known size and concentration.

3.4. Preparation of ScFv Linker DNA

1. Make up a master mix for amplification in 0.5-mL microcentrifuge tubes. The master mix is shown for the V_H-V_κ linker. For the V_H-V_λ linker master mix, $n = 29$ (*see* **Note 11**).

	Individual reaction	Master mix ($n = 25$)
Water	37 µL	925 µL
10X Vent buffer	5.0 µL	125 µL
20X dNTPs	2.5 µL	62.5 µL
Vent DNA polymerase	0.5 µL	12.5 µL

2. Place into 24 × 0.5-mL tubes, 2 µL each, of RHuJ$_H$ and RHuV$_\kappa$ primers for each of the 24 possible combinations of RHuJ$_H$ and RHuV$_\kappa$ primers. For the V_H-V_λ linker, use the RHuV$_\lambda$ primers instead of RHuV$_\kappa$ primers, and make up 28 tubes. If this does not have a heated lid, add one drop of light mineral oil to cover the reaction.
3. Put 50 µL of the master mix into a negative control tube.
4. Add 1.0 µL of the scFv gene template (~1 ng) per sample to the master mix.
5. Add 45 µL of the master mix to each of the 24 tubes generated in **step 2.**
6. Preheat the PCR block to 94°C.
7. Cycle 25× to amplify the linker sequence at 94°C for 1 min, 55°C for 1 min, and 72°C for 1 min.
8. Purify the PCR fragments by electrophoresis on a 2.0% agarose gel in TBE, excise the PCR product (~750 bp), extract it using the Geneclean kit, and resuspend the DNA in 50 µL of water.

3.5 PCR Assembly of V_H and V_L Genes with ScFv Linker to Create ScFv Gene Repertoires (see Notes 12 and 13)

1. Set up the following PCR mixtures in 0.5-mL microcentrifuge tubes. Make up two "reaction tubes," one containing the V_κ repertoire DNA and one containing the V_λ repertoire DNA.

	Reaction Tube	Control 1	Control 2
scFv linker DNA (100 ng/µL)	1.0 µL	1.0 µL	
V_H repertoire DNA (100 ng/µL)	3.5 µL		1.0 µL
V_L repertoire DNA (V_κ or V_λ; 100 ng/µL)	3.0 µL		1.0 µL
Water		6.5 µL	5.5 µL

2. To each tube add:

Water	33 µL
10X Vent buffer	5.0 µL
20X dNTPs	2.5 µL

3. Heat the reaction tubes to 94°C for 5 min in a thermal cycling block. If this does not have a heated lid, add one drop of light mineral oil to cover the reaction.
4. Add 2 U (2 µL) of Vent DNA polymerase.
5. Cycle 7× at 94°C for 1 min, 65°C for 1 min, and 72° C for 2 min to randomly join the fragments.
6. After 7 cycles, hold the temperature at 94°C while adding 1 µL of each set of flanking primers (an equimolar mixture of the six V_HBACKSfi primers and an equimolar mixture of the five J_κFORNot or three J_λFORNot primers, yielding a final total concentration of each primer mix of 10 pmol/µL).
7. Cycle 25× to amplify the fragments at 94°C for 1 min, 60°C for 1 min, and 72°C for 1 min.
8. Analyze 3 µL of the PCR products on a 1.5% agarose gel in TBE to determine success of the splicing. The assembled scFv should be 0.8–0.9 kbp, and there should be no product of this size in the negative control reactions.
9. Purify the assembled gene repertoires by electrophoresis on a 1.0% agarose gel in TBE, followed by extraction from the gel using the Geneclean kit. Resuspend each product in 20 µL of water. Determine the DNA concentration by analysis on a 1% agarose gel compared to markers of known size and concentration.

3.6. Re-Amplification of ScFv Gene Repertoires to Append Restriction Sites for Cloning (see Note 14)

1. Make up two 50-µL PCR reaction mixes (one for the V_H-V_κ scFv repertoire and one for the V_H-V_λ scFv repertoire) in 0.5-mL microcentrifuge tubes containing:

Water	37.5 µL
10X Taq buffer	5.0 µL
20X dNTPs	2.5 µL
Forward primer (*see* **Note 15**)	2.0 µL
Back primer (*see* **Note 15**)	2.0 µL
scFv gene repertoire (~10ng)	1.0 µL

2. Heat the reaction mixes to 94°C for 5 min in a thermal cycling block. If this does not have a heated lid, add one drop of light mineral oil to cover the reaction.
3. Add 5 U (1.0 µL) of Taq DNA polymerase.
4. Cycle 25× to amplify the scFv genes at 94°C for 1 min, 55°C for 1 min, and 72°C for 1 min.
5. Purify the PCR product using the Wizard PCR purification kit (*see* **Note 16**).

3.7. Restriction Digestion of the ScFv Gene Repertoires and pHEN1 Phage Display Vector

1. Make up two 100-µL reaction mixes to digest the scFv repertoires, one for the V_H-V_κ scFv repertoire and one for the V_H-V_λ scFv repertoire:

scFv DNA (1 to 4 µg)	50 µL
Water	33 µL
10X NEB3 buffer	10 µL
100X acetylated BSA	1.0 µL
*Nco*I (10 U/µL)	3.0 µL
NotI (10 U/µL)	3.0 µL

2. Incubate the reactions at 37°C overnight. The use of a 37°C incubator or heat block with heated lid prevents evaporation. Otherwise, add a layer of mineral oil (Sigma) as used for PCRs.
3. Gel-purify the gene repertoire on a 1% agarose gel and extract the DNA using the Geneclean kit. Resuspend the product in 30 µL of water. Determine the DNA concentration by analysis on a 1% agarose gel with markers of known size and concentration.
4. Prepare the vector by digesting 4 µg of cesium chloride-purified pHEN 1 in a reaction exactly as described previously for the scFv gene repertoire, but substituting plasmid DNA for scFv DNA (*see* **Note 17**).
5. Purify the digested vector DNA on a 0.8% agarose gel. Extract the cut vector from the gel using the Geneclean kit the same way as for the scFv insert. Resuspend the product in 30 µL of water. Determine the DNA concentration by analysis on a 1% agarose gel compared with markers of known size and concentration.

3.8. Ligation of ScFv and Vector DNA

1. Make up two 100-µL ligation mixtures as described here, one for the V_H-V_κ scFv repertoire and one for the V_H-V_λ scFv repertoire, as well as the two control reactions (*see* **Note 18**). The controls determine how much of the vector is uncut (Control 2) and how much is cut once and can be religated in the absence of insert (Control 1). The proportions are maintained, and the number of colonies obtained (for the same ligation volume) should be less than 10% of the library:

	Libraries	Control 1	Control 2
10X ligation buffer	10 µL	0.5	0.5
Millipore water	32 µL	2.3	2.5
Digested pHEN 1 (100 ng/µL)	40 µL	2	2
ScFv gene repertoire (100 ng/µL)	14 µL		
T4 DNA ligase (400 U/µL)	4 µL	0.2	

2. Incubate at 16°C overnight.
3. Increase the volume of the library reaction mix to 200 µL with TE buffer. Extract the DNA with an equal volume of phenol/chloroform. Ethanol-precipitate the DNA and wash the pellet twice in 70% ethanol.

3.9. Generation of Phage Antibody Libraries

1. Set the electroporator at 200 ohms (resistance), 25 µF (capacitance), and 2.5 kilovolts.
2. Thaw the electrocompetent bacteria on ice or use freshly prepared electrocompetent cells. Place the electroporation cuvet, cuvet holder, cells, and DNA on ice.
3. Mix 80 ng of purified, salt-free DNA with 50 µL of bacteria. Incubate the mixture on ice for 1 min.
4. Place the mixture in a 0.2-cm path electroporation cuvet, taking care to avoid leaving air bubbles between the electrodes. Tap the cuvet gently to transfer all liquid to the bottom of the cuvet. Move quickly to keep the cuvet cold.
5. Place the cuvet into the electroporator, making sure the sides of the cuvet are dry (to avoid arcing). Pulse and immediately add 1.0 mL of SOC media and mix. The time constant should be in the 4.5- to 5.5-ms range. If the time constant is less than 4.3 ms, repeat the DNA precipitation (**Subheading 3.8**, **step 3**).
6. Grow the bacteria from each electroporation in 1 mL of SOC media at 37°C for 1 h, shaking at 250 rpm.
7. Combine all the electroporation cultures and plate serial dilutions onto 100 mm TYE/amp/glu plates to determine the size of the library (e.g., the total number of independent clones).
8. Centrifuge the remaining bacterial culture at 2,000*g* at 4°C for 10 min. Resuspend the pellet in 500 µL for every four electroporations and plate 125-µL aliquots onto 150 mmm TYE/amp/glu plates or plate 500 µL on 530 cm^2 Bio-assay dishes containing TYE/amp/glu. Incubate the plates overnight at 30°C.
9. Scrape the bacteria from the plates by adding 10 mL of 2X TY/amp/glu media to the 150-mm dishes containing 15% glycerol (sterilized by filtration through 0.2-µm filter). Store the library stock at –70°C (*see* **Notes 19** and **20**).

3.10. Preparation of Electrocompetent E. coli TG1

1. Inoculate a minimal media plate with *E. coli* TG1, and incubate overnight at 37°C.
2. Inoculate 15 mL of 2X TY in a 100-mL flask with *E. coli* TG1 from the minimal plate. Shake the culture overnight at 37°C (*see* **Note 21**).
3. Inoculate two 2-L baffled flasks containing 500 mL of 2X TY with 5 mL of the overnight TG1 culture. Grow in a 37°C shaker for ~1.5 h, until the A_{600} is approx 0.5.
4. Transfer the bacteria to four 500-mL prechilled centrifuge bottles (~250-mL/bottle). Balance the bottles and keep them on ice for 20 min.
5. Centrifuge the bottles at 3,000*g*, 4°C for 15 min. Prechill all rotors before use.
6. Discard the supernatant and resuspend the cells in each bottle in the original volume (~250 mL) of ice-cold 1 m*M* HEPES solution, pH 7.0. All solutions should be made fresh.

7. Centrifuge the bottles again at 3,000*g*, 4°C for 15 min.
8. Resuspend the cells in one-half the original volume (~125 mL) of ice-cold 1 m*M* HEPES, pH 7.0, now using two centrifuge bottles.
9. Centrifuge the bottles again at 3,000*g*, 4°C for 15 min.
10. Resuspend all the cells in total volume of 20 mL of 1 m*M* HEPES, pH 7.0, in 10% glycerol.
11. Centrifuge the bottles again at 3,000*g*, 4°C for 15 min.
12. Prepare an ethanol/dry ice bath, if the cells are to be stored rather than used fresh.
13. Resuspend the cells in a total volume of 2 mL 10% glycerol.
14. Use the cells fresh, or freeze aliquots of cells using the dry ice bath (*see* **Note 22**). Use the cells promptly after testing efficiency (within 1 wk) (*see* **Note 23**).

4. Notes

1. Use disposable plasticware when possible. All glassware, including Corex tubes, should be baked overnight at 180°C. Use separate reagents (phenol, chloroform, or ethanol) for RNA work.
2. Any brand of Taq polymerase in its appropriate buffer can be used.
3. A number of easy-to-use kits are available for RNA generation from vendors such as Promega. We have also found that these kits make high-quality RNA for repertoire cloning.
4. The same protocol can be used to prepare RNA from hybridoma cells or from mouse spleens by using these cells as starting material. For spleens, we pass the dissected spleen through a fine wire mesh to separate the cells from matrix.
5. The following protocols are for first-strand synthesis of human V_H and V_L repertoires using constant region primers. The sequences of the primers are given in **Table 1.** It is not necessary to prepare polyadenylated RNA to prepare cDNA: total RNA works very well. Immune libraries should be created using IgG-based primers, and naïve libraries are best prepared using IgM primers to ensure maximal diversity. This protocol can be adapted for first-strand synthesis from hybridomas or immunized mice by substituting murine-specific constant region primers *(4–8)*.
6. This protocol has been successfully used in our laboratories. There are also many kits available that are suitable for cDNA synthesis.
7. The heavy-chain primer used in this protocol is for naïve library generation. For immune libraries, use an IgG-specific primer.
8. V_H, V_κ, and V_λ genes are amplified in separate PCRs using the appropriate back and forward primers. Back primers are an equimolar mixture of the six V_HBACK, six V_κBACK, or seven V_λBACK primers (final total concentration of primer mixture equals 10 pmol/µL), corresponding to the different Kabat V-gene families *(2)*. Similarly, forward primers are an equimolar mixture of the four J_HFOR, five J_κFOR, or three J_λFOR primers (final total concentration of primer mixture equals 10 pmol/µL). Alternatively, amplifications can be done with each back primer individually with the mixture of forward primers. This usually biases the library slightly toward the V-genes that are less highly expressed in the PBL population.

```
  G  G  G  G  S  G  G  G  G  S  G  G  G  G  S
GGTGGAGGCGGTTCAGGCGGAGGTGGTTCTGGCGGTGGCGGATCG
```

Fig. 2. The $(Gly_4Ser)_3$ scFv linker. Many phage antibody libraries use the 15-amino acid Gly-Ser[$Gly_4Ser)_3$] linker to connect V_H and V_L in the order V_H–V_L. It has been shown that linkers of this length can have a tendency to form dimeric scFv, in which the V_H of one scFv pairs with the V_L of another scFv. By increasing the length of the linker to 20 amino acids, this tendency can be reduced, although not eliminated.

9. These primers have been validated as generating diverse repertoires and high-quality, diverse libraries *(1,9,19)*. Alternative sets of primers exist for generating human repertoires *(10,11)*.
10. To amplify V-genes from hybridomas or from RNA prepared from immunized mice, substitute murine V-region primers.
11. The linker is generated with primers that anneal to the 5′ and 3′ ends of the cloned linker and have overhangs that encode 24 nucleotides of perfect complementarity with the various J_H (5′ end) or V_L (3′ end) genes (**Fig. 1** and **2**). Since there are four different J_H primers and six different V_κ primers, a total of 24 separate PCRs are required to generate the V_H-V_κ linker. Similarly, since there are four different J_H primers and seven different V_λ primers, a total of 28 PCR separate PCRs is required to generate the V_H-V_λ linker.
12. A number of ways exist to combine V_H and V_L gene repertoires into either scFv or Fab repertoires that can be displayed on phage or in other display systems. One of the simplest (and first described) methods is to use splicing by overlap extension (PCR assembly) (**Fig. 1**). In the protocol provided here, V_H and V_L gene repertoires are combined with linker DNA that encodes the $(Gly_4Ser)_3$ peptide used to connect V_H and V_L in one embodiment of the scFv (**Fig. 2**). The linker DNA overlaps the 3′ end of the V_H gene and the 5′ end of the V_L gene, and is synthesized from a template containing the scFv gene using complementary primers of the J_H, V_κ, and V_λ primers (*see* **Subheading 3.4.** and **Table 1**).
13. As an alternative to splcing by overlap extension, V_H and V_L genes can be TA-cloned (useful after amplifying from monoclonal hybridomas) or sequentially cloned into a number of described vector systems using appropriate restriction sites *(20,21)*. Obviously, different primers would be required for PCR that contained these restriction sites.
14. For library creation, the spliced scFv gene repertoires must be reamplified to append restriction sites (NcoI and NotI) to permit cloning into the phagemid vector pHEN1 *(18)*.
15. Forward primer is an equimolar mix of the five J_κFORNot or three J_λFORNot, at a final concentration of 10 pmol/µL. Back primer is an equimolar mix of the six V_HBACKSfi primers, at a final concentration of 10 pmol/µL. The sequences of these primers are provided in **Table 1, item 4.**
16. In order to clone the scFv gene repertoires into pHEN1, they must be digested with the restriction enzymes NcoI and NotI. We have found that the repertoires can be

digested by these enzymes simultaneously with good efficiency in NEB3 buffer. Typically, we digest 1–4 μg of repertoire in 100 μL reaction volumes. We ligate approx 1.5 μg of repertoire DNA into 4 μg of vector DNA in a vol of 100 μL. Ligation mixtures are "cleaned up" by phenol/chloroform extraction followed by ethanol-precipitation. After resuspension in water, the ligation mixture is used to transform electrocompetent *E. coli* TG1 (*see* **Subheading 3.10.** for electrocompetent TG1 preparation).

17. Efficient digestion is important because a small amount of undigested vector leads to a very large background of non-recombinant clones. Use of vector DNA prepared by techniques other than cesium chloride will yield lower transformation efficiencies and higher background.
18. The amounts of DNA indicated are chosen to give the theoretical optimal molar ratio of insert to vector of 2:1. Given that the ratio of sizes of scFv (800 bp) to vector (4,500 bp) is approx 1:6, this translates into a ratio of insert to vector of 1:3 in weight terms.
19. Typical transformation efficiencies are $1–10 \times 10^9$/μg of pUC19, and $1–100 \times 10^7$/μg of ligated vector with insert. The background of ligated vector without insert should be less than 10% of the ligated vector with insert. This will depend upon the efficiency of the ligation reaction and the quality of the vector used—see controls 1 and 2 in **Subheading 3.8.**
20. Ligations and electroporations should be continued until there are at least 10^9 separate colonies for naïve libraries, and 10^7 separate colonies for immune libraries.
21. It is critical that all glass and plastic equipment, medium, and buffer are absolutely soap- and detergent-free (it is best to use a set of dedicated equipment that is washed only with water and autoclaved).
22. Cells can be quick-frozen in 25–200 μL aliquots and stored at –80°C, with an approx 10-fold decrease in transformation efficiency.
23. Test that cells are free of contaminant plasmids, phagemids or phages, by growing on appropriate antibiotic plates and in top agar with lawn bacteria.

References

1. Marks, J. D., Hoogenboom, H. R., Bonnert, T. P., McCafferty, J., Griffiths, A. D., and Winter, G. (1991) By-passing immunization: Human antibodies from V-gene libraries displayed on phage. *J. Mol. Biol.* **222,** 581–597.
2. Kabat, E. A., Wu, T. T., Perry, H. M., Gottesman, K. S., and Foeller, C. (1991) *Sequences of Proteins of Immunological Interest.* US Department of Health and Human Services, US Government Printing Office.
3. Clackson, T., Hoogenboom, H. R., Griffiths, A. D., and Winter, G. (1991) Making antibody fragments using phage display libraries. *Nature* **352,** 624–628.
4. LeBoeuf, R. D., Galin, F. S., Hollinger, S. K., Peiper, S. C., and Blalock, J. E. (1989) Cloning and sequencing of immunoglobulin variable-region genes using degenerate oligodeoxyribonucleotides and polymerase chain reaction. *Gene* **82,** 371–377.

5. Iverson, S. A., Sastry, L., Huse, W. D., Sorge, J. A., Benkovic, S. J., and Lerner, R. A. (1989) A combinatorial system for cloning and expressing the catalytic antibody repertoire in Escherichia coli. *Cold Spring Harbor Symp. Quant. Biol.* **1,** 273–281.
6. Sastry, L., Alting, M. M., Huse, W. D., Short, J. M., Sorge, J. A., Hay, B. N., et al. (1989) Cloning of the immunological repertoire in Escherichia coli for generation of monoclonal catalytic antibodies: construction of a heavy chain variable region-specific cDNA library. *Proc. Natl. Acad. Sci. USA* **86,** 5728–5732.
7. Kettleborough, C. A., Saldanha, J., Ansell, K. H., and Bendig, M. M. (1993) Optimzation of primers for cloning libraries of mouse immunoglobulin genes using the polymerase chain reaction. *Eur. J. Immunol.* **23,** 206–211.
8. Orum, H., Andersen, P. S., Oster, A., Johansen, L. K., Riise, E., Bjornvad, M., et al. (1993) Efficient method for constructing comprhensive murine Fab antibody libraries displayed on phage. *Nucleic Acids Res.* **21,** 4491–4498.
9. Marks, J. D., Tristrem, M. Karpas, A., and Winter, G. (1991) Oligonucleotide primers for polymerase chain reaction amplification of human immunoglobulin variable genes and design of family-specific oligonucleotide probes. *Eur. J. Immunol.* **21,** 985–991.
10. Persson, M. A., Caothien, R. H., and Burton, D. R. (1991) Generation of diverse high-affinity human monoclonal antibodies by repertoire cloning. *Proc. Natl. Acad. Sci. USA* **88,** 2432–2436.
11. Burton, D. R., Barbas, C. F., Persson, M.A.A., Koenig, S., Chanock, R. M., and Lerner, R. A. (1991) A large array of human monoclonal antibodies to type 1 human immunodefficiency virus from combinatorial libraries of asymptomatic seropositive individuals. *Proc. Natl. Acad. Sci. USA* **88,** 10,134–10,137.
12. Davies, E., Smith, J., Birkett, C., Manser, J., Anderson-Dear, D., and Young, J. (1995) Selection of specific phage-display antibodies using libraries derived from chicken immunoglobulin genes. *J. Immunol. Methods* **186,** 125–135.
13. Lang, I., Barbas, C. R., and Schleef, R. (1996) Recombinant rabbit Fab with binding activity to type-1 plasminogen activator inhibitor derived from a phage-display library against human alpha-granules. *Gene* **172,** 295–298.
14. Bird, R. E., Hardman, K. D., Jacobson, J. W., Johnson, S., Kaufman, B. M., Lee, S. M., et al. (1988) Single-chain antigen-binding proteins. *Science* **242,** 423–426.
15. Huston, J. S., Levinson, D., Mudgett, H. M., Tai, M. S., Novotny, J., Margolies, M. N., et al. (1988) Protein engineering of antibody binding sites: recovery of specific activity in an anti-digoxin single-chain Fv analogue produced in Escherichia coli. *Proc. Natl. Acad. Sci. USA* **85,** 5879–5883.
16. Chaudhary, V. K., Batra, J. K., Gallo, M. G., Willingham, M. C., FitzGerald, D. J., and Pastan, I. (1990) A rapid method of cloning functional variable-region antibody genes in Escherichia coli as single-chain immunotoxins. *Proc. Natl. Acad. Sci. USA* **87,** 1066–1070.
17. Orlandi, R., Gussow, D. H., Jones, P. T., and Winter, G. (1989) Cloning immunoglobulin variable domains for expression by the polymerase chain reaction. *Proc. Natl. Acad. Sci. USA* **86,** 3833–3837.

18. Hoogenboom, H. R., Griffiths, A. D., Johnson, K. S., Chiswell, D. J., Hudson, P., and Winter, G. (1991) Multi-subunit proteins on the surface of filamentous phage: methodologies for displaying antibody (Fab) heavy and light chains. *Nucleic Acids Res.* **19,** 4133–4137.
19. Sheets, M. D., Amersdorfer, P., Finnern, R., Sargent, P., Lindqvist, E., Schier, R., et al. (1998) Efficient construction of a large nonimmune phage antibody library: the production of high-affinity human single-chain antibodies to protein antigens. *Proc. Natl. Acad. Sci. USA* **95,** 6157–6162.
20. Barbas, C. F., Kang, A. S., Lerner, R. A., and Benkovic, S. J. (1991) Assembly of combinatorial antibody libraries on phage surfaces: The gene III site. *Proc. Natl. Acad. Sci. USA* **88,** 7978–7982.
21. de Haard, H. J., van Neer, N., Reurs, A., Hufton, S. E., Roovers, R. C., Henderikx, P., et al. (1999) A large non-immunized human Fab fragment phage library that permits rapid isolation and kinetic analysis of high affinity antibodies. *J. Biol. Chem.* **274,** 18,218–18,230.

7

Antibody Humanization by CDR Grafting

Benny K. C. Lo

1. Introduction

The goal of antibody humanization is to engineer a monoclonal antibody (MAb) raised in a nonhuman species into one that is less immunogenic when administered to humans. The development of this technology had transformed the stagnant state of antibody therapeutics in the 1980s, when the major obstacle was the human anti-murine antibody (HAMA) response *(1, 2)*. The HAMA response occurred in up to 50% of patients upon the administration of murine hybridoma-derived antibodies *(3)* and severely compromised the safety, efficacy, and biological half-life of these reagents. In addition, murine antibody constant regions are inefficient in directing suitable human immune effector functions for therapeutic effects. Efforts to produce human antibodies by hybridoma technology *(4)* and Epstein-Barr virus (EBV)-mediated B-lymphocyte transformation *(5)* have met with limited success, but their widespread application is hampered by the lack of robust human hybridoma fusion partners and the instability of EBV-transformed clones, respectively (*see* **ref.** ***6–8*** for review). Consequently, humanization technology was well-placed to exploit the plethora of murine MAbs against a variety of disease targets and turn them into effective clinical reagents.

Two methods were developed in the mid-1980s as attempts to reduce the immunogenicity of murine antibodies: chimerization *(9–11)* and humanization by CDR grafting *(12–14)* (*see* **Fig. 1A**). In mouse/human chimeric antibodies, the immunogenic murine constant domains (66% of total sequence) are replaced by the human counterpart to reduce the likelihood of eliciting the HAMA response. At the same time, intact murine variable domains are preserved to maintain the intrinsic antigen-binding affinity. Effector functions of chimeric antibodies may also be programmed by choosing constant domains of different human immunoglobulin isotypes *(15, 16)*. In the clinic, chimeric anti-

From: *Methods in Molecular Biology, Vol. 248: Antibody Engineering: Methods and Protocols*
Edited by: B. K. C. Lo © Humana Press Inc., Totowa, NJ

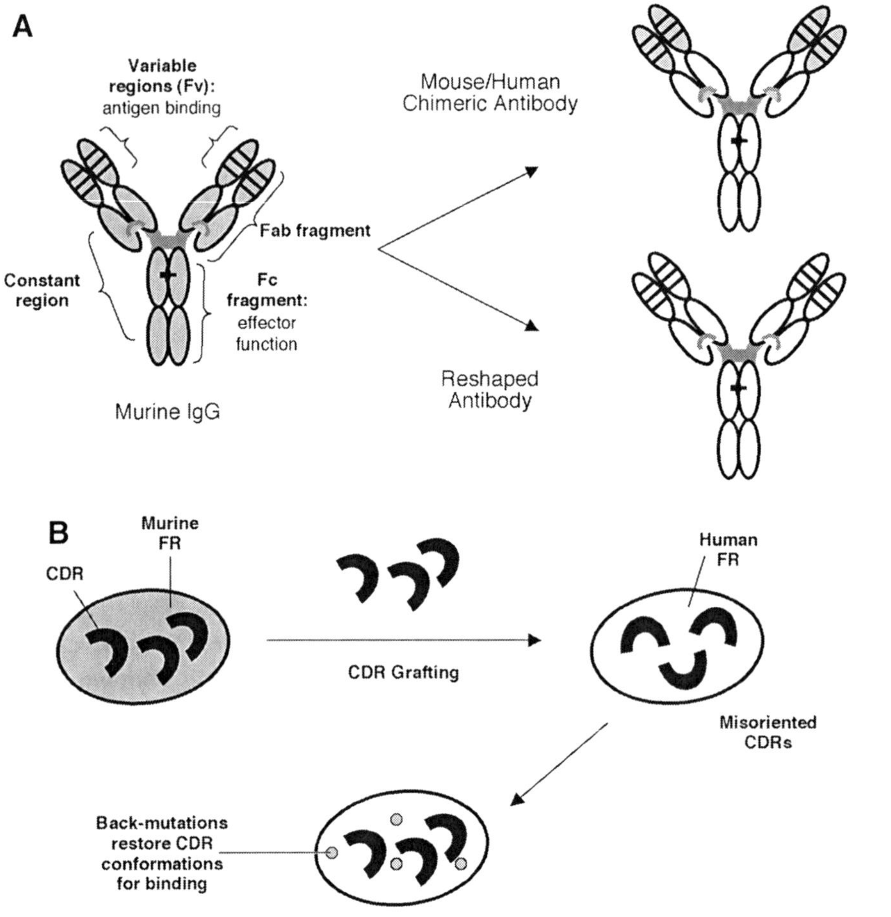

Fig. 1. Antibody humanization and CDR Grafting. (**A**) Mouse/human chimeric antibodies retain the entire Fv regions from the murine antibody (66% human). Reshaped antibodies, on the other hand, only retain the CDRs and a few framework residues from the murine antibody (90–95% human), making them minimally immunogenic. (Shaded ovals: murine sequences; blank ovals: human sequences; black lines across ovals: murine CDRs). (**B**) The transplantation of murine CDRs onto a human framework (FR) often lead to suboptimal orientations of these binding loops. Consequently, critical murine framework residues need to be re-introduced as back-mutations to restore the optimum CDR conformations for antigen binding.

bodies have been successfully used in the treatment of Crohns disease (Remicade®), non-Hodgkin's lymphoma (Rituxan®), and renal transplant rejection (Simulect®). Although the chimerization of some murine antibodies led to the disappearance of the HAMA response, others remained immunogenic ***(3)***, partly because of the presence of murine epitopes in the Fv. As an alternative approach, Dr. Greg Winter and colleagues at the Medical Research Council, Cambridge pioneered the complementarity-determining region (CDR) grafting technology ***(12–14)*** in which the antigen-binding specificity of a murine antibody is transferred to a human antibody by replacing the CDR loops and selected variable-region framework residues. The reshaped, humanized antibody only retained the essential binding elements from the murine antibody (5–10% of total sequence) and was predicted to be minimally immunogenic. Indeed, the in vivo tolerance and efficacy of many humanized antibodies have been shown to be more favorable than murine antibodies. Clinical trials of four such antibodies—Hu2PLAP ***(17)***, hMN-14 ***(18)***, CAMPATH-1H ***(19)***, and HuOKT3 ***(20)***—demonstrated minimal or no detectable HAMA response after intravenous administration, thus establishing the potential for therapeutic use. A number of humanized antibodies have recently reached the market for diseases such as cancer (Herceptin®, Myelotarg®, Campath®), allograft rejection (Zenapax®) and viral infection (Synagis®) (*see* **ref. *21*** for review). These successes provide further evidence of the power of CDR grafting and antibody-based reagents in complementing traditional chemotherapeutic drugs.

Despite the development of molecular display technologies (Chapters 8 and 9) and transgenic animals (Chapter 10) for the generation of fully human antibodies, CDR grafting remains an attractive and proven strategy for turning well-characterized, highly specific murine antibodies into clinical reagents. The vast amounts of data often available for the parent murine antibody should greatly facilitate the movement of the humanized antibody from the laboratory to the clinic. Furthermore, the grafting of naturally evolved CDRs onto human antibody frameworks eliminates the need for the generation of a completely new human antibody with methods such as in vitro library selection, affinity maturation, or further immunization of transgenic animals.

1.1. Structural Analysis of the Murine Fv Region

The key to a successful CDR grafting experiment lies in the preservation of the murine CDR conformations in the reshaped antibody for antigen binding. This can benefit from a detailed analysis of the structure and sequence of the murine Fv. The antibody Fv region comprises variable domains from the light chain (V_L) and heavy chain (V_H) and confers antibodies with antigen-binding specificity and affinity. The variable domains adopt the immunoglobulin fold in which two antiparallel β-sheet framework scaffolds support three hypervariable CDRs. Structural variation between antibodies is dependent on the length, sequence, conformation, and relative disposition of the CDRs, and the pairing of CDRs from repertoires of V_L and V_H sequences. These parameters create the unique topography of each combining site and form the basis for specific antigen recognition. Although antigen binding is predominantly regulated by CDR residues, framework residues may also contribute, either through direct antigen contact *(22)* or indirectly through packing with CDR residues. The implication for CDR grafting is that it may be necessary to revert one or more human framework residues to the murine equivalents in the reshaped antibody (so-called framework back-mutations; *see* **Fig. 1B**) to restore the native biomolecular environment for antigen binding *(13)*.

The fact that grafted murine CDRs are able to function on human antibody frameworks is primarily made possible by the highly conserved structure of antibody framework regions and the limited repertoire of backbone conformations in five of the six hypervariable regions. These canonical structures, described by Dr. Cyrus Chothia and colleagues *(23–26)*, were identified for CDRs 1, 2, and 3 of V_L and CDRs 1 and 2 of V_H. Each conformation is dependent on the interactions of a few conserved residues at specific sites, both within the CDRs and the framework regions. Murine CDRs are thus expected to adopt the same backbone conformation when the basic sequence requirements for the corresponding canonical structures are met by the human frameworks. Moreover, the inherent flexibility of the antigen-binding site may accommodate small imperfections from loop grafting in an induced-fit fashion upon antigen binding *(14,* and references herein). Such conformational self-correction mechanisms have been proposed as the basis for binding, even when considerable differences exist between the antigen-free structures of the murine and reshaped binding sites *(27–29)*. On the whole, these properties of the antibody Fv make CDR grafting a versatile process, providing that the key framework requirements are satisfied.

Framework residues that participate in antigen binding can often be identified by inspecting a three-dimensional (3D) model of the murine Fv for potential framework–CDR and framework–antigen interactions. Ideally, an X-ray crystallographic structure of the antigen–Fv complex is used in such an analysis. If this is unavailable, a predicted structure from molecular modeling is usually sufficient. Molecular models can be generated for both the murine and reshaped forms of the Fv to locate critical framework residues and evaluate the suitability of the human acceptor frameworks. Additional analyses are performed by sequence alignment with other antibodies to reveal unique sequence features. These insights into the murine Fv structure are invaluable for the selection of human acceptor frameworks and, if necessary, the introduction of framework back-mutations to optimize antigen binding.

1.2. Choice of Human Acceptor Frameworks

With the blueprint of critical murine framework residues from structural analysis, database searching is next performed to short-list suitable human antibody frameworks to support the murine CDRs. The inclusion of these critical framework residues increases the chance of success in producing a functional reshaped antibody. Depending on the framework selection strategy, one or more such residues may be incorporated into the native acceptor sequence or introduced subsequently by framework mutagenesis. The five major framework selection strategies are summarized here:

1. Fixed frameworks: the same set of human acceptor frameworks is used regardless of sequence homology to the murine antibody. This is particularly advantageous when the chosen frameworks are well-characterized in terms of expression levels and availability of solved crystal structures. Early humanization experiments were performed with frameworks from the human myeloma proteins REI *(30)* and NEWM *(31)* for V_L and V_H, respectively *(12–14,32)*.
2. Homology: in this approach, human antibodies sharing maximum sequence homology to the murine antibody are selected to provide the acceptor frameworks *(33–36)*. Generally speaking, the use of homologous human frameworks is least likely to distort CDR conformations in the reshaped antibody. This strategy also minimizes the need for framework mutagenesis. However, one possible drawback is that the use of individual antibody sequences may include somatic hypermutations, thereby creating potentially immunogenic epitopes.
3. Subgroup consensus: the Kabat database *(37)* catalogs antibody sequences into subgroups of different variable-region families. The consensus sequence of each Kabat subgroup is compiled from the most common residue at all the framework positions. In this strategy, consensus from the human subgroups most homologous to the murine antibody is selected for CDR grafting *(38–42)*. The use of consensus frameworks may reduce the likelihood of immunogenicity as somatic hypermutations are minimized except to the extent that consensus sequences are artificial and may themselves contain unnatural elements that are immunogenic.
4. Germline: the most homologous human immunoglobulin germline sequences *(43–45)* are selected as the frameworks. Since germline sequences originate from un-rearranged immunoglobulin genes, such frameworks should therefore be free from idiosyncratic mutations and are minimally immunogenic. With the availability of the complete human immunoglobulin germline gene locus, this strategy is gaining popularity *(46,47)*.
5. Veneering: also known as "variable region resurfacing," veneering has developed from the view that protein epitopes recognized by antibodies mainly reside on the solvent-exposed surface; murine residues buried inside the protein fold are therefore predicted to be minimally antigenic. Subsequently, all buried antibody framework positions are kept murine while solvent-exposed positions are humanized *(48,49)*. This preserves all buried packing residues in the reshaped antibody that are potentially important for CDR conformations and minimizes the likelihood of antigenic response developed against the solvent-exposed regions. It remains to be seen whether the buried murine residues are immunogenic through the display of degraded antibody fragments as major histocompatibility complex (MHC) class II complexes on antigen-presenting cells, thus inducing a T-cell response.

Human frameworks for V_L and V_H are often selected from two different antibodies to achieve higher homologies for each chain. This is enabled by the conserved V_L/V_H interface that permits chain pairing and the formation of the antigen-binding site between variable domains from different antibodies. It is even possible to maximize homology by the use of hybrid frameworks for each variable domain *(50)*. Furthermore, the exact location of the framework regions can also vary depending on whether "Kabat CDRs" (**ref.** *37*; original Kabat definition of antigen-binding loops based on sequence variability) or "Chothia loops" (**ref.** *23*; Chothia definition based on structural loops) are used for grafting. Although these strategies have all resulted in functional reshaped antibodies, the nature of the human framework regions is likely to be structurally and immunologically distinct. It should therefore be emphasized that reshaped antibodies represent a *broad* class of CDR-grafted molecules, and the exact strategy chosen may carry significantly different consequences in the immunogenicity of the engineered molecule.

1.3. Introduction and Analysis of Framework Back-Mutations

The use of carefully selected human frameworks in CDR grafting maximizes the probability of retaining antigen-binding affinity in the reshaped antibody. This was well-exemplified by the anti-mucin antibody HMFG1 *(51)* and the anti-CD56 antibody N901 *(52)*, in which the reshaped antibody retains almost full binding affinity to the target antigen immediately after CDR grafting. However, these examples represent the ideal scenarios and are a minority among the many successfully reshaped antibodies. Often, antigen-binding affinity is greatly reduced or abolished after loop grafting *(38,39,53)*, and is restored only after the re-introduction of murine framework residues as back-mutations. The likelihood of producing a functional antibody usually increases with the number of back-mutations made on the CDR-grafted structure. However, one should aim to limit the number of back-mutations to a minimum, thereby keeping the reshaped antibody as "human" as possible and reducing the possibility of immunogenicity.

The identification of critical framework back-mutations truly represents the art of CDR grafting, as it is frequently the most difficult and unpredictable step in the process. Although the many reshaping experiments available in the literature provide a "rough guide" (see the website "Humanization by Design," http://people.cryst.bbk.ac.uk/~ubcg07s for an excellent database of published humanized antibodies), it is not possible to generalize back-mutations. A number of framework positions are known to be critical and buried residues in close proximity to the CDRs and the antigen-binding surface are generally more important than those that are solvent-exposed and distant from the binding site. Different strategies have been developed to derive the crucial combination of back-mutations to restore antigen binding in a timely manner. They mainly involve the stepwise or simultaneous incorporation of one or more back-mutations in the reshaped variable domains followed by their evaluation on antigen binding. The conventional approach relies on the use of a structural model to guide the introduction of strategic back-mutations. Alternatively, candidates for back-mutations are included in a combinatorial library of variable domains displayed on filamentous bacteriophage ***(54,55)***. Variable domain sequences in the library are synthesized so that both the murine and human residues are represented at selected positions for back-mutation. This allows the simultaneous analysis of a much larger pool of mutants. The best combinations of back-mutations are swiftly identified by panning the phage library for antigen binding, followed by sequencing of the clones of highest affinity.

1.4. Summary

The humanization of murine antibodies by CDR grafting is a major protein-engineering task. The transfer of antigen-binding loops from a naturally evolved murine framework to a different, artificially chosen human framework must be performed so that the native loop conformations are retained for antigen binding. Meanwhile, careful consideration should be given during the design process to minimize the immunogenic potential of the humanized antibody by limiting residual murine elements. Additional measures such as specificity determining residues grafting (**ref.** ***56***; *see* also Chapter 21), removal of immunogenic epitopes ***(57)***, use of constant regions that do not activate effector functions ***(58)***, and the generation of immunological tolerance ***(59)*** may also be taken. It should be noted that immunogenicity is a complex issue; intrinsic factors such as protein sequence and structure are likely to be as important as extrinsic ones (e.g., dosage, route, and schedule of administration, and patient's immune status). Thus, it is very difficult to predict the immunogenicity of a therapeutic antibody accurately. Only favorable results from clinical studies can confirm the success of a humanization experiment.

In this chapter, protocols are described for the CDR grafting of murine antibodies using germline human antibody frameworks and Kabat CDRs (*see* **Fig. 2**). Since many of the procedures involve standard molecular biology manipulations, emphasis is placed on the design strategy; cross-references to other chapters in this volume are provided where appropriate. Finally, the protocols are written purely from an academic perspective; if the reader is interested in developing a humanized antibody for commercial applications, professional advice should be sought because of the numerous patents covering all aspects of this technology.

2. Materials

2.1. Homology Modeling and Framework Selection

1. Silicon Graphics workstation (SGI, CA).
2. Molecular graphics software (e.g., Insight II; Accelrys, CA).

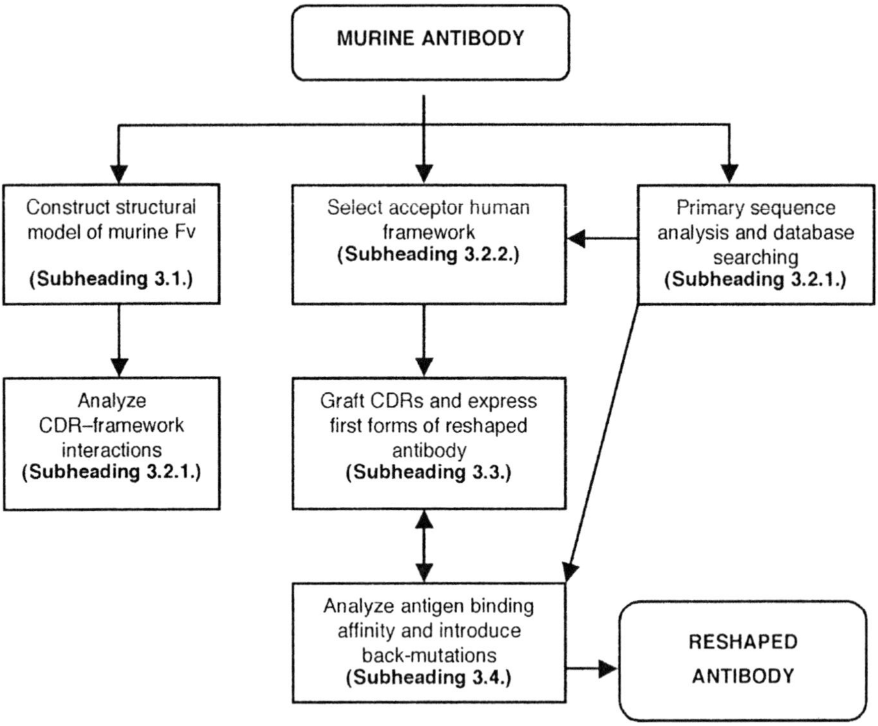

Fig. 2. CDR grafting strategy

3. Homology modeling software:
   a. WAM (**ref. *60***; accessible at http://antibody.bath.ac.uk);
   b. SWISS-MODEL (**ref. *61***; accessible at http://www.expasy.org);
   c. INSIGHT-HOMOLOGY (Accelrys); or
   d. COMPOSER (Tripos, MO).
4. SeqTest (accessible at http://www.bioinf.org.uk/abs/ – under "Sequences").
5. Kabat Database online (**ref. *62***; accessible at http://www.kabatdatabase.com) and printed ***(37)*** editions (*see* also Chapter 2).
6. Genedoc (**ref. *63***; downloadable from http://www.psc.edu/biomed/genedoc).
7. V-BASE (accessible at http://www.mrc-cpe.cam.ac.uk/vbase).
8. IMGT, the International ImMunoGeneTics Information System (**ref. *64***; accessible at http://imgt.cines.fr; *see* also Chapter 3).

2.2. CDR Grafting and Antibody Expression

1. GCG Wisconsin Package (Accelrys).
2. Thermocycling block for polymerase chain reaction (PCR).
3. PCR reagents: Geneamp 10X PCR buffer with 15 m*M* $MgCl_2$ (Applied Biosystems); AmpliTaq DNA polymerase (Applied Biosystems); 10 m*M* deoxynucleotide

triphosphate (dNTP) mix (Sigma); high-performance liquid chromatography (HPLC)-purified oligonucleotide primers; sterile, deionized water (H_2O).
4. Low melting-point agarose.
5. Ethidium bromide stock solution 10 mg/mL, store at 4°C.
6. Tris-borate EDTA (TBE) buffer: 90 m*M* Tris-borate, 2 m*M* EDTA in H_2O, autoclave at 121°C for 15 min and store at room temperature.
7. Agarose gel-running apparatus and ultraviolet (UV) transilluminator.
8. Qiaquick gel extraction kit, PCR purification kit, and Qiaprep miniprep kit (Qiagen).
9. TOPO TA cloning kit and One Shot TOP10 chemically competent *E. coli* (Invitrogen).
10. LB-ampicillin agar plates: suspend 40 g of Luria agar (Sigma) in 1 L of H_2O, boil to dissolve, autoclave at 121°C for 15 min, cool to 50°C, add ampicillin to 100 µg/mL, and dispense into Petri dishes. Store at 4°C for up to 3 wk.
11. LB medium: dissolve 25 g of Miller's LB broth (Sigma) in 1 L of H_2O, autoclave at 121°C for 15 min.
12. Restriction endonucleases, T4 DNA ligase, and associated reaction buffers (New England Biolabs).
13. Antibody expression vectors (*see* **Note 1**).
14. Sterile 0.45-µm filters.
15. Sodium azide stock solution 20% (w/v), store at 4°C.

2.3. Binding Analysis and Framework Back-Mutations

1. Nunc Maxisorp 96-well plates (Nalgen Nunc) and adhesive plate sealers.
2. Enzyme-linked immunosorbent assay (ELISA) plate shaker and reader.
3. Phosphate-buffered saline (PBS): 8 m*M* Na_2HPO_4, 1 m*M* KH_2PO_4, 160 m*M* NaCl, 3 m*M* KCl, pH 7.3, autoclaved.
4. PBST: PBS containing 0.05% (v/v) polyoxyethylenesorbitan monolaurate (Tween 20).
5. Blocking buffer: PBS containing 1% (w/v) bovine serum albumin (BSA).
6. ELISA buffer: PBS containing 0.2% (w/v) BSA and 0.02% (v/v) Tween 20.
7. Human serum IgG1/κ immunoglobulins (Sigma).
8. Goat anti-human IgG (Fc-specific), affinity-purified (Jackson Immunoresearch).
9. Goat anti-human κ-chain horseradish peroxidase (HRP) conjugate, affinity-purified (Sigma).
10. Goat anti-human IgG γ-chain HRP conjugate (Sigma).
11. TMB substrate solution (Pierce).
12. 1 *M* H_2SO_4.

3. Methods

3.1. Homology Modeling of the Murine Variable Region

The highly homologous nature of antibody Fv makes it convenient to build molecular models using homology modeling methods ***(65,66)***. A number of well-established antibody modeling approaches based on homology ***(67–69)*** and conformational searching ***(70–72)*** algorithms have been described (*see* **Note 2**).

Computer software incorporating these methods is also available (e.g., WAM, SWISS-MODEL, INSIGHT-HOMOLOGY, and COMPOSER; see **Subheading 2.1.**). As a first step in CDR grafting, construct a structural model of the murine variable region using one of these programs (if a crystal structure is not available). The reader is encouraged to consult Chapter 4 of this volume for an in-depth discussion on antibody modeling, and Chapter 1 for a list of general molecular graphics and modeling software accessible on the worldwide web.

3.2. Human Acceptor Framework Selection

3.2.1. Structural and Sequence Analysis of Murine Variable Region

1. Assign Kabat numbering *(37)* to the murine Fv using the program SeqTest and determine the CDR canonical structures for L1, L2, L3, H1, and H2 using **Table 1.**
2. Inspect a 3D model of the murine Fv (crystal structure or molecular model from **Subheading 3.1.**) for potential framework-CDR and framework-antigen interactions. Focus initially on five groups of critical framework residues with special significance (**Table 2**). Take note of framework residues that are in the immediate vicinity of the six CDRs and those on the antigen-binding surface.
3. Determine the Kabat subgroup of the V_L and V_H sequences using the Subgroup tool from Kabat Database online. Proceed to the printed edition of the Kabat Database and manually align the Fv sequence with the subgroup. For each framework residue that differs from the consensus sequence, evaluate its degree of uniqueness within the subgroup and take note of the rare framework residues. These could have arisen from somatic hypermutation, and may contribute to antigen-binding affinity.
4. Load the Fv sequences into sequence alignment programs such as Genedoc, highlight the critical framework residues from **steps 2** and **3** on the Fv sequence and use as the blueprint for short-listing suitable human frameworks for CDR grafting.

3.2.2. Framework Selection

1. Search V-BASE or IMGT for the closest human immunoglobulin germline sequences to the murine V_L and V_H (*see* **Note 3**).
2. Align the translated germline sequences to the sequence blueprint compiled under **Subheading 3.2.1.**
3. Compare the short-listed sequences to the murine Fv in terms of framework sequence identity and the presence of critical framework residues highlighted in the sequence blueprint.
4. Choose, as the human acceptor frameworks, the V_L and V_H germline sequences that share the highest sequence identity with the murine frameworks, especially at the critical framework positions.
5. **Optional:** Construct a molecular model of the reshaped Fv using the murine CDRs and the chosen germline framework sequence (*see* **Subheading 3.1.**). Access the quality of the chosen frameworks by comparing the CDR conformations on the murine and reshaped Fv (*see* **Note 4**). Select alternative acceptor frameworks if necessary.

Table 1
CDR Canonical Structures and Sequence Requirements

A number of CDR conformations have been observed with five of the six antibody hypervariable regions. Residue positions (Kabat numbering; **ref.** *37*) required to maintain the different canonical structures are shown together with the most commonly observed amino acid(s). Residues from the framework regions are shown in bold (reproduced with permission from **ref.** *73*).

CDR	Canonical class (loop length)	Residues important for canonical conformation: Kabat position [most common amino acid(s)]
L1	1 (10)[a]	**2**(I), 25(A), 30(V), 33(M, L), **71**(Y)
	2 (11)	**2**(I), 25(A), 29(V, I), 33(L), **71**(F,Y)
	2 (12)[b]	**2**(I, N), 25(A), 28(V, I), 33(L), **71**(F,Y)
	3 (17)	**2**(I), 25(S), 27B(V, L), 33(L), **71**(F)
	4 (15)[b]	**2**(I), 25(A), 27B(V), 33(M), **71**(F)
	4 (16)	**2**(V, I), 25(S), 27B(I, L), 33(L), **71**(F)
L2	1 (7)	**48**(I), 51(A, T), 52(S, T), **64**(G)
L3	1 (9)	90(Q, N, H), 95(P)
	2 (9)[a]	90(Q), 94(P)
	3 (8)	90(Q), 95(P)
H1	1 (5)	**24**(A, V, G), **26**[c](G), **27**[c](F, Y), **29**[c](F), 34(M, W, I), **94**(R, K)
	2 (6)	**24**(V, F), **26**[c](G), **27**[c](F, Y, G), **29**[c](I, L), 35(W, C), **94**(R, H)
	3 (7)	**24**(G, F), **26**[c](G), **27**[c](G, F, D), **29**[c](L, I, V), 35A(W, V), **94**(R, H)
H2	1 (16)	55 (G, D), **71**(V, K, R)
	2 (17)	52 (P, T, A), 55(G, S), **71**(A, T, L)
	3 (17)	54(G, S, N), **71**(R)
	4 (19)	54(S), 55(Y), **71**(R)
	5 (18)	52(Y), 54(K), 55(W), **71**(P)

[a] These canonical structures have only been observed in murine antibodies, but not in human antibodies.

[b] Approximately 20% of murine and 25% of human antibodies have 13 residues in canonical structure 2 or 14 residues in canonical structure 4. These minor variations in loop size result in changes at the tip of the L1 loop, but do not significantly alter loop conformation ***(24)***.

[c] Residues at H26–H30 are defined as framework residues in the Kabat system, but structurally they are part of the H1 loop as defined by Chothia et al. ***(24)***.

3.3. CDR Grafting and Expression of Reshaped Antibodies

Having finalized the acceptor framework sequences for the murine CDRs, the encoding DNA for the reshaped Fv is assembled by PCR. V_L and V_H sequences can be constructed in the form of "expression cassettes," which include additional upstream and downstream sequences to facilitate antibody

Table 2
Important Framework Positions for Antigen Binding

In addition to CDR residues, five classes of framework residues frequently influence antigen binding directly or indirectly. The actual framework positions (Kabat numbering; **ref.** *37*) or sequence motifs for these residues are shown.

Description	Significance	Framework positions	Reference
Vernier zone	A platform of residues directly underneath the CDRs	L2, L4, L35, L36, L46–L49, L64, L66, L68, L69, L71, L98; H2, H27–H30, H47–H49, H67, H69, H71, H73, H78, H93, H94, H103	*(74)*
Canonical residues	Maintenance of CDR canonical structures	See **Table 1** (bold residues)	See **Table 1** (bold residues)
V_L N-terminal residues	Form a contiguous surface with the antigen-binding pocket	L1–L6	*(75)*
Chain-packing residues	Residues buried at the V_L/V_H packing interface; may affect the relative disposition of CDRs	L34, L36, L38, L44, L46, L87, L89, L91, L96, L98; H35, H37, H39, H45, H47, H91, H93, H95, H100[a], H103	*(76)*
Putative N-/O-glycosylation sites	Carbohydrates may affect positioning of antigen in the binding pocket	See footnote[b,c]	

[a] The residue immediately N-terminal to residue H101 is a packing residue. The actual numbering of this residue varies with the number of loop insertions in CDR-H3.

[b] N-glycosylation consensus sequence: N-X-(S/T), where X is any amino acid *(77)*.

[c] No consensus exists for O-glycosylation, but four sequence motifs were found to predict O-glysolation *(78,79)*:

- X-P-X-X: at least one X is T.
- T-X-X-X: at least one X is T.
- X-X-T-X: at least one X is R or K.
- S-X-X-X: at least one X is S.

expression. The exact content of the cassettes will depend on the chosen vector and expression host, and whether the Fv is expressed as a full IgG or a single-chain Fv (scFv) fragment (*see* **Note 1**). Here, a general method (modified from **ref.** *73*) is described for the PCR assembly of the reshaped Fv expression cassettes using overlapping oligonucleotides. Further details on antibody expression in *E. coli* and mammalian systems can be found in Chapters 14 and 15,

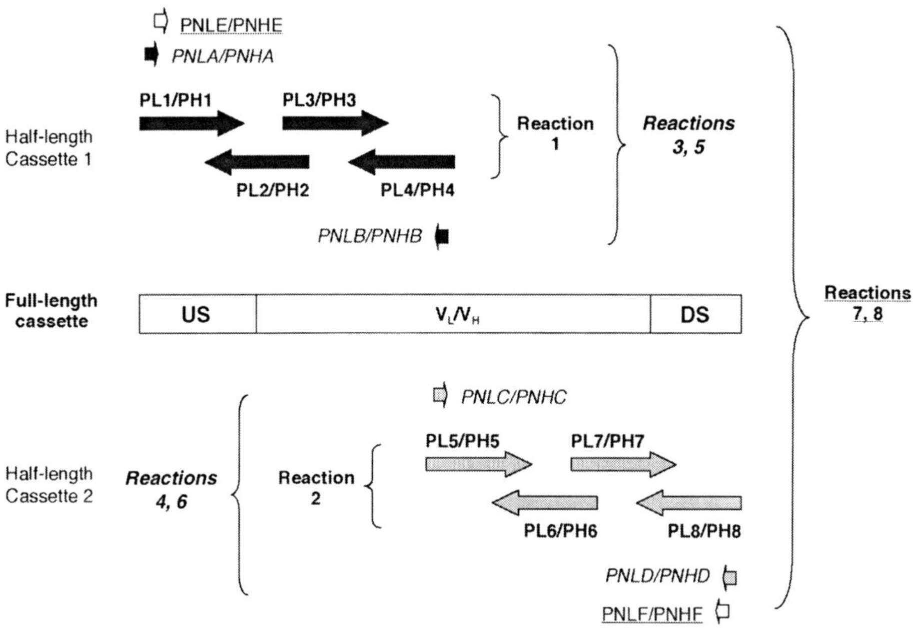

Fig. 3. Construction of reshaped Fv expression cassettes. The reshaped V_L and V_H expression cassettes are each produced by eight overlapping primers (block arrows) in eight PCRs. Overlaps between the megaprimers (PL1–8, PH1–8) are approx 24 bp. US = upstream accessory sequences, DS = downstream accessory sequences (tailored for individual expression vector systems).

respectively. A discussion on the design of monovalent and multivalent antibody fragments is also given in Chapter 12.

3.3.1. Design and Construction of the Reshaped Fv Expression Cassettes

1. Combine the nucleotide sequences of the chosen acceptor frameworks for V_L and V_H (from **Subheading 3.2.2.**) and the corresponding murine CDRs to form the reshaped Fv sequence. Append appropriate restriction sites, up- and downstream sequences as necessary for the chosen expression vector system.
2. Check the expression cassettes for the presence of internal restriction sites and stop codons. Remove these by alternative codon usage where necessary.
3. To construct the reshaped Fv cassettes, design eight partially complementary DNA megaprimers (approx 80 nucleotides each) and six nested primers (24 nucleotides each) covering the entire length of the V_L and V_H cassettes as illustrated in **Fig. 3.**
4. Using the Stemloop program in the GCG Wisconsin Package, check each megaprimer for secondary structure formation. Avoid stemloops with melting temperatures of above 40°C by alternative codon usage.

5. Set up two sequential sets of four PCRs for the assembly of half-length V_L and V_H expression cassettes (*see* **Note 5**):
   a. Assembly PCRs (V_L reactions 1, 2 and V_H reactions 1, 2):

5 μL	Geneamp 10X PCR buffer with 15 m*M* $MgCl_2$
1 μL	10 m*M* dNTP mix
2.5 pmol each	PL1, 2, 3, 4 (V_L reaction 1); PL5, 6, 7, 8 (V_L reaction 2); PH1, 2, 3, 4 (V_H reaction 1); PH5, 6, 7, 8 (V_H reaction 2) (*see* **Fig. 3**)
2.5 U	AmpliTaq DNA polymerase
+ H_2O to 50 μL	

   Thermocycle: 94°C for 5 min (hot start); 8 cycles of 94°C for 2 min and 72°C for 5 min; and final extension at 72°C for 10 min.

   b. Amplification PCRs (V_L reactions 3, 4, and V_H reactions 3, 4):

5 μL	Geneamp 10X PCR buffer with 15 m*M* $MgCl_2$
1 μL	10 m*M* dNTP mix
5 μL	PCR product from V_L reaction 1 (V_L reaction 3), V_L reaction 2 (V_L reaction 4), V_H reaction 1 (V_H reaction 3) or V_H reaction 2 (V_H reaction 4)
40 pmol each	PNLA and PNLB (V_L reaction 3)
	PNLC and PNLD (V_L reaction 4)
	PNHA and PNHB (V_H reaction 3)
	PNHC and PNHD (V_H reaction 4)
2.5 U	AmpliTaq DNA polymerase
+ H_2O to 50 μL	

   Thermocycle: 94°C for 5 min (hot start); 20 cycles of 94°C for 1.5 min, 64°C for 1.5 min, 72°C for 2.5 min; and final extension at 72°C for 10 min.

6. Visualize reaction products from amplification PCRs on a 1% (w/v) agarose gel containing 0.5 μg/mL ethidium bromide, in 1X TBE buffer.
7. Gel-purify bands of the correct size using the Qiaquick gel extraction kit (V_L: ~225 bp, V_H: ~240 bp; approximate guide only, the exact fragment lengths will depend on the individual antibody sequence and the positioning of primers).
8. Clone half-length expression cassettes into plasmid vector using the TOPO TA cloning kit. Check integrity of constructs by DNA sequencing.
9. Re-amplify the half-length expression cassettes from clones harboring the correct sequence with PCR (V_L reactions 5, 6, and V_H reactions 5, 6):

5 μL	Geneamp 10X PCR buffer with 15 m*M* $MgCl_2$
1 μL	10 mM dNTP mix
40 pmol each	PNLA and PNLB (V_L reaction 5)
	PNLC and PNLD (V_L reaction 6)
	PNHA and PNHB (V_H reaction 5)
	PNHC and PNHD (V_H reaction 6)
2.5 U	AmpliTaq DNA polymerase

   + *E. coli* colony stab of the corresponding cloned half-length expression cassettes
   + H_2O to 50 μL

   Thermocycle: as described in **step 5b.**

10. Visualize reaction products on an agarose gel and gel-purify bands of the correct size (V_L: ~225 bp, V_H: ~240 bp) using the Qiaquick gel extraction kit.
11. Set up two sequential sets of two PCRs to construct the final reshaped expression cassettes:
    a. Assembly PCRs (V_L reaction 7 and V_H reaction 7):

5 µL	Geneamp 10X PCR buffer with 15 m*M* $MgCl_2$
1 µL	10 m*M* dNTP mix
5 µL each	Gel-purified products from V_L reactions 5, 6 (V_L reaction 7) or, V_H reactions 7, 8 (V_H reaction 7)
2.5 U	AmpliTaq DNA polymerase
+ H_2O to 50 µL	

    Thermocycle: as described in **step 5a.**

    b. Amplification PCRs (V_L reaction 8 and V_H reaction 8):

5 µL	Geneamp 10x PCR buffer with 15 m*M* $MgCl_2$
1 µL	10 m*M* dNTP mix
5 µL	PCR product from V_L reaction 7 (V_L reaction 8) or V_H reaction 7 (V_H reaction 8)
40 pmol each	PNLE and PNLF (V_L reaction 8) PNHE and PNHF (V_H reaction 8)
2.5 U	AmpliTaq DNA polymerase
+ H_2O to 50 µL	

    Thermocycle: as described in **step 5b.**

12. Visualize reaction products on an agarose gel. Gel-purify appropriate-sized bands (V_L: ~450 bp, V_H: ~480 bp) and clone the expression cassettes into plasmid vector using the TOPO TA cloning kit. Check the integrity of the constructs by DNA sequencing.

3.3.2. Construction of the Murine Fv Expression Cassettes

The murine Fv should be expressed in parallel with the reshaped Fv as a control for binding analysis **(Subheading 3.4.)**. If the IgG format is chosen, it can be expressed with human antibody constant domains as a mouse/human chimeric antibody. For the scFv format, it can be simply expressed as a murine scFv fragment (*see* **Note 1**).

1. Append appropriate restriction sites, up- and downstream accessory sequences to the murine Fv sequence with PCR (*see* **Note 6**).
2. Visualize reaction products on an agarose gel, gel-purify bands of the correct size (V_L: ~450 bp, V_H: ~480 bp) and clone the expression cassettes into plasmid vector using the TOPO TA cloning kit. Check the integrity of constructs by DNA sequencing.

3.3.3. Antibody Expression

1. Digest the cloned reshaped and murine Fv expression cassettes with appropriate restriction endonucleases and ligate into the chosen expression vector(s). Arrange

the inserts in such a way so that four initial antibody constructs can be expressed (*see* **Note 7**):
a. Murine V_L + Murine V_H
b. Reshaped V_L + Reshaped V_H
c. Murine V_L + Reshaped V_H
d. Reshaped V_L + Murine V_H
2. Clone the ligated vectors into *E.coli* and confirm the ligated sequences by DNA sequencing.
3. Transform/transfect these vectors into the expression host of choice and express the four initial antibody constructs.
4. Pass the antibody-containing media through a 0.4-μm filter, add sodium azide to 0.02% (w/v), and store at 4°C.

3.4. Binding Analysis and Framework Back-mutations

The binding activity of the reshaped antibody is accessed using ELISA. A quantification ELISA is first performed to measure antibody concentration in the expression medium; this is followed by a binding assay for the cognate antigen. Sample protocols are described here for the analysis of IgG antibodies; appropriate reagents can be substituted for the analysis of scFv fragments (*see* **Note 8**).

3.4.1. Antibody Quantification ELISA

1. Coat a Nunc Maxisorp 96-well plate with 1 μg/mL affinity-purified goat anti-human IgG (Fc-specific) in PBS (100 μL/well). Cover with an adhesive plate sealer and incubate at 4°C overnight (*see* **Note 9**).
2. Wash wells 4× with 200 μL/well of PBST. Block nonspecific binding sites by adding 200 μL/well of blocking buffer and incubate at room temperature with shaking for 2 h.
3. Discard the blocking buffer. Add, in duplicates, 100 μL/well of standardized concentrations (1 mg/mL and nine additional 1:2 serial dilutions) of purified human serum IgG1/κ in ELISA buffer to the assay plate (Rows A–B). (*see* **Notes 10** and **11**)
4. Add 100 μL/well of antibody-containing medium and nine 1:2 serial dilutions of each in ELISA buffer to the assay plate (Rows C–H). (*see* **Note 11**).
5. Incubate plate at 37°C with shaking for 1 h.
6. Wash wells 4× with 200 μL/well of PBST.
7. Prepare a 1:5,000 dilution of affinity-purified goat anti-human κ-chain horseradish peroxidase (HRP) conjugate in ELISA buffer. Add to the assay plate at 100 μL/well and incubate at 37°C with shaking for 1 h.
8. Wash wells 4× with 200 μL/well of PBST.
9. Add 100 μL/well of TMB substrate solution to the assay plate and incubate at room temperature for 5–10 min, until sufficient color development is reached.
10. Stop color development by adding 50 μL/well of 1 *M* H_2SO_4 and read end-point absorbance values at 450 nm on an ELISA plate reader.

11. Construct a curve with mean absorbance values from the calibration standards (OD at 450 nm vs log [concentration]). Estimate sample IgG concentrations using absorbance values corresponding to the linear portion of the calibration curve.

3.4.2. Antibody Immunoreactivity ELISA

1. Coat a Nunc Maxisorp 96-well plate with 10 µg/mL antigen solutions in PBS (100 µL/well; *see* **Note 12**). Include the specific antigen(s) bearing the murine antibody epitope and several unrelated antigens to probe for nonspecific binding. Cover with an adhesive plate sealer and incubate at 4°C overnight.
2. Wash wells 4× with 200 µL/well of PBST. Block nonspecific binding sites by adding 200 µL/well of blocking buffer and incubate at room temperature with shaking for 2 h.
3. Discard the blocking buffer. Add antibody-containing medium to the assay plate at 100 µL/well (*see* **Note 13**).
4. Incubate plate at 37°C with shaking for 1 h.
5. Wash wells 4× with 200 µL/well of PBST.
6. Prepare a 1:4,000 dilution of affinity-purified goat anti-human IgG γ-chain HRP conjugate in ELISA buffer. Add to the assay plate at 100 µL/well and incubate at 37°C with shaking for 1 h.
7. Wash wells 4× with 200 µL/well of PBST.
8. Read plate as described under **Subheading 3.4.1., steps 9–10.**

3.4.3. Introduction and Analysis of Framework Back-Mutations

The expression and binding analysis of the four initial antibody constructs (**Subheading 3.3.3.**) should reveal whether back-mutations are required for V_L, V_H, or both variable domains of the reshaped antibody. Using these results as the basis, the following procedure describes the optimization of framework back-mutations by guided selection, a strategy that enables the individual analysis of each of the V_L and V_H mutants before the best ones are selected to form the final reshaped antibody (*see* **Fig. 4**).

1. Select the reshaped variable domain that resulted in a larger drop in antigen-binding affinity for initial mutagenesis. Align this with the sequence blueprint containing the critical murine framework residues (**Subheading 3.2.1.**) to identify which of these are absent from the human acceptor frameworks.
2. Construct hybrid Fvs in which the reshaped mutants are paired with the opposite murine domain. This can be conveniently achieved by mutating vectors constructed in **Subheading 3.3.3., steps 1c** and **1d,** using standard directed mutagenesis techniques (*see* Chapter 18). For example, if back-mutations are introduced to the reshaped V_L domain, prepare the following hybrid Fvs (*see* **Note 14**):
   a. Murine V_H-Reshaped V_L containing all Vernier zone back-mutations;
   b. Murine V_H-Reshaped V_L containing all canonical residue back-mutations;
   c. Murine V_H-Reshaped V_L containing all chain packing back-mutations;

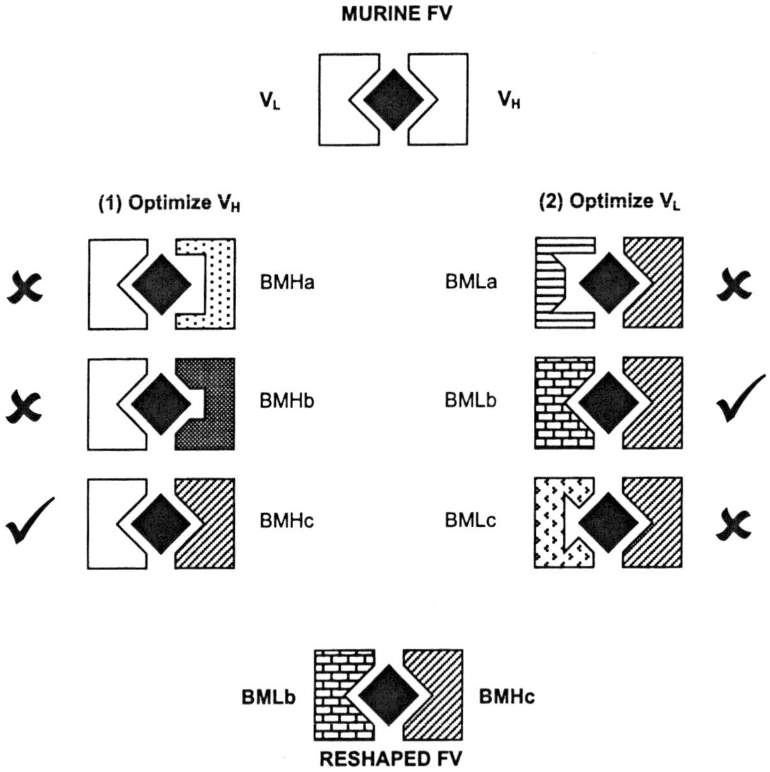

Fig. 4. Optimization of framework back-mutations by guided selection. The best combination of back-mutations can be identified by screening a series of hybrid antibodies for binding to antigen (◆). This figure illustrates the optimization of V_H back-mutations followed by V_L ones (the order can be reversed if appropriate, see under **Subheading 3.4.3.**). The murine V_L is initially expressed in combination with various reshaped V_H mutants (BMHx). The mutant V_L that gives the highest level of specific antigen binding is then used for the optimization of reshaped V_L mutants (BMLx), resulting in the final reshaped Fv.

d. Murine V_H-Reshaped V_L containing all V_L N-terminal back-mutations;
e. Murine V_H-Reshaped V_L containing all rare framework position back-mutations;
f. Murine V_H-Reshaped V_L containing all O- and N-glycosylation back-mutations;
g. Murine V_H-Reshaped V_L containing all of the above back-mutations.

In addition, prepare the following as controls:

a. Murine V_H Reshaped V_L containing no back-mutations; and
b. Murine V_H-Murine V_L.

3. Express these constructs in the expression system of choice (*see* **Note 1**). Perform quantification and immunoreactivity ELISAs as described under **Subheadings 3.4.1.** and **3.4.2.**

4. Compare specific antigen-binding signals of mutants 2a to 2g with those of mutants 2h and 2i. Evaluate immunoreactivity data to narrow down beneficial back-mutations that restore binding to a level comparable to that of the murine antibody (mutant 2i). Perform further mutagenesis as appropriate to derive the optimum set of back-mutations that fully reconstitutes specific antigen binding (*see* **Note 15**).
5. Repeat **steps 1–4** for the opposite reshaped domain, using the first optimized reshaped domain as a guide. The reshaped mutant that gives maximum antigen binding is chosen as the final reshaped antibody.
6. Characterize the reshaped antibody in terms of its epitope specificity (Chapter 24) and binding affinity (Chapters 25–27), as required. A continuous source of the reshaped antibody (as IgG or scFv) can be established by the stable transfection of a mammalian cell line (e.g., Chinese hamster ovary cells or mouse myeloma NS0 cells) with suitable expression vectors (*see* Chapter 15).

4. Notes

1. For the expression of antibodies as full IgGs, eukaryotic vectors that provide the light- and heavy-chain constant domains are available from AERES Biomedical, London, UK (SuperVector Expression System) and Lonza Biologics, NH, USA (pCON vectors). For expression as single-chain Fv fragments (scFv), common prokaryotic vectors that allow periplasmic protein expression can be used (*see* Chapters 12 and 14 for further details). The supplied literature for these expression systems should indicate the sequence requirements for the expression cassettes (**Subheadings 3.3.1.** and **3.3.2.**).
2. In the homology approach, a structural template(s) is chosen based on sequence homology and length. Conflicts between amino acid side-chains are then resolved by energy minimization. This approach is especially useful for the β-sheet framework and CDR loops with known canonical structures (*see* **Subheading 1.1.**). In contrast, the conformational search approach seeks to build CDR conformations *ab initio* by generating a large number of theoretically possible conformations by varying the dihedral angles of the peptide backbone. The predicted loop conformation is obtained after energetic evaluation of the conformers. This is useful for CDRs that do not belong to any of the canonical classes and those that are poorly modeled by homology alone.
3. V-BASE and IMGT are online databases featuring human immunoglobulin germline sequences. A searchable interface is provided by both databases (V-BASE: DNAPLOT; IMGT: IMGT-VQUEST) for nucleotide sequences. Translation of the germline genes into amino acid sequences is also possible. *See* Chapter 3 for further information on the IMGT database.
4. Although this model provides an early indication on the ability of the chosen germline frameworks to support the CDRs in similar conformations for binding, it should only be taken as an approximate guide, as conformational correction mechanisms have been shown to adjust suboptimal CDR conformations upon binding to antigen (*see* **Subheading 1.1.**).
5. The Fv expression cassettes are initially synthesized as half-length inserts and sequenced before the two halves are stitched together. This allows the early detec-

tion of PCR artifacts, which are likely at the numerous primer junctions, and avoids the subsequent need to sequence a large number of clones for the correct full-length cassette sequence.

6. Design overhanging primers that are partially complementary to the 5′- and 3′-ends of the murine V_L and V_H gene for the introduction of additional sequences.
7. The shuffling of the murine and reshaped V_L/V_H domains and subsequent analyses of the hybrid constructs will reveal whether the reshaped antibody retains the binding affinity and specificity of the murine antibody and, if not, the CDR grafting of which domain (V_L, V_H, or both) has resulted in the change in binding characteristics.
8. Possible capturing and detection agents include protein L (for kappa light chains) and anti-peptide tag antibodies (e.g., for hexahistidine, c-myc tags).
9. Coated plates can be sealed and stored at 4°C for up to 4 wk.
10. Purified immunoglobulins of a different isotype should be used if the reshaped/chimeric antibodies are not of the IgG1/κ isotype.
11. Serial dilutions can be made conveniently in the wells using a multichannel pipettor.
12. Antigen concentration (10–100 μg/mL) and coating buffer should be optimized if necessary.
13. Antibody supernatants should be diluted in ELISA buffer to equivalent concentrations for direct comparison.
14. Back-mutations of the same type are introduced initially as a group. The expression of these mutants with the opposite guide domain will provide an indication on the location of beneficial back-mutations. If a positive effect is seen within a particular group, further mutants containing different combinations of single back-mutations can be made to dissect out individual contributions. Priorities for individual mutations should be decided largely from results of structural analysis (**Subheading 3.2.1.**). After the analysis of the initial set of mutants, back-mutations of different types can then be combined to reveal any synergistic and/or antagonistic effects on antigen binding. If binding affinity cannot be restored to an acceptable level after back-mutating the critical framework positions, further back-mutations can be introduced at the remaining framework positions, again guided by the structural model of the murine Fv.
15. One should aim to achieve the maximum antigen-binding signal with the fewest number of back-mutations to keep the reshaped antibody as human as possible.

Acknowledgments

The author thanks Dr. Greg Winter for suggestions on this manuscript and the Croucher Foundation, Hong Kong for the award of a personal research fellowship.

References

1. Schroff, R. W., Foon, K. A., Beatty, S. M., Oldham, R. K., and Morgan, A. C. Jr, (1985) Human anti-murine immunoglobulin responses in patients receiving monoclonal antibody therapy. *Cancer Res.* **45,** 879–885.

2. Shawler, D. L., Bartholomew, R. M., Smith, L. M., and Dillman, R. O. (1985) Human immune response to multiple injections of murine monoclonal IgG. *J. Immunol.* **135,** 1530–1535.
3. Khazaeli, M. B., Conry, R. M., and LoBuglio, A. F. (1994) Human immune response to monoclonal antibodies. *J. Immunother.* **15,** 42–52.
4. Caroll, W. L., Thielemans, K., Dilley, J., and Levy, R. (1986) Mouse × human heterohybridomas as fusion partners with human B cell tumours. *J. Immunol. Methods* **89,** 61–72.
5. Borrebaeck, C.A.K. (1989) Strategy for the production of human, monoclonal antibodies using in vitro activated B cells. *J. Immunol. Methods* **123,** 157–165.
6. Carson, D. A. and Freimark, B. D. (1986) Human lymphocyte hybridomas and monoclonal antibodies. *Adv. Immunol.* **38,** 275–311.
7. Cole, S. P., Campling, B. G., Atlaw, T., Kozbor, D., Roder, J. C. (1984) Human monoclonal antibodies. *Mol. Cell. Biochem.* **62(2),** 109–120.
8. Niedbala, W. G. and Stott, D. I. (1998) A comparison of three methods for production of human hybridomas secreting autoantibodies. *Hybridoma* **17,** 299–304.
9. Boulianne, G. L., Hozumi, N., and Shulman, M. J. (1984) Production of functional chimeric mouse/human antibodies. *Nature* **312,** 643–646.
10. Morrison, S. L., Johnson, M. J., Herzenberg, L. A., and Oi, V. T. (1984) Chimaeric human antibody molecules: mouse antigen binding domains. *Proc. Natl. Acad. Sci. USA* **82,** 6851–6855.
11. Neuberger, M. S., Williams, G. T., and Fox, R. O. (1984) Recombinant antibodies possessing novel effector functions. *Nature* **312,** 604–608.
12. Jones, P. T., Dear, P. H., Foote, J., Neuberger, M. S., and Winter, G. (1986) Replacing the complementarity-determining regions in a human antibody with those from a mouse. *Nature* **321,** 522–525.
13. Riechmann, L., Clark, M., Waldmann, H., and Winter, G. (1988) Reshaping human antibodies for therapy. *Nature* **332,** 323–327.
14. Verhoeyen, M., Milstein, C., and Winter, G. (1988) Reshaping human antibodies: grafting an antilysozyme activity. *Science* **239,** 1534–1536.
15. Liu, A. Y., Robinson, R. R., Murray, E. D. Jr, Ledbetter, J. A., Hellstrom, I., and Hellstrom, K. E. (1987) Production of a mouse-human chimeric monoclonal antibody to CD20 with potent Fc-dependent biologic activity. *J. Immunol.* **139,** 3521–3526.
16. Dyer, M.J.S., Hale, G., Hayhoe, F.G.J., and Waldmann, H. (1989) Effects of CAMPATH-1 antibodies in vivo in patients with lymphoid malignancies: influence of antibody isotype. *Blood* **73,** 1431–1439.
17. Kalofonos, H. P., Kosmas, C., Hird, V., Snook, D. E., and Epenetos, A. A. (1994) Targeting of tumours with murine and reshaped human monoclonal antibodies against placental alkaline phosphatase: immunolocalisation, pharmacokinetics and immune response. *Eur. J. Cancer* **30A,** 1842–1850.
18. Sharkey, R. M., Juweid, M., Shevitz, J., Behr, T., Dunn, R., Swayne, L. C., et al. (1995) Evaluation of a complementarity-determining region-grafted (humanized) anti-carcinoembryonic antigen monoclonal antibody in preclinical and clinical studies. *Cancer Res.* **55,** 5935s–5945s.

19. Rebello, P. R., Hale, G., Friend, P. J., Cobbold, S. P., and Waldmann, H. (1999) Anti-globulin responses to rat and humanized CAMPATH-1 monoclonal antibody used to treat transplant rejection. *Transplantation* **68,** 1417–1420.
20. Woodle, E. S., Xu, D., Zivin, R. A., Auger, J., Charette, J., and O'Laughlin, R. (1999) Phase I trial of a humanized, Fc receptor nonbinding OKT3 antibody huOKT3gamma1(Ala-Ala) in the treatment of acute renal allograft rejection. *Transplantation* **68,** 608–616.
21. Vaswani, S. K. and Hamilton, R. G. (1998) Humanized antibodies as potential therapeutic drugs. *Ann. Allergy Asthma Immunol.* **81,** 105–115.
22. Mian, I. S., Bradwell, A. R., and Olson, A. J. (1991) Structure, function and properties of antibody binding sites. *J. Mol. Biol.* **217,** 133–151.
23. Chothia, C. and Lesk, A. M. (1987) Canonical structures for the hypervariable regions of immunoglobulins. *J. Mol. Biol.* **196,** 901–917.
24. Chothia, C., Lesk, A. M., Tramontano, A., Levitt, M., Smith-Gill, S. J., Air, G., et al. (1989) Conformations of immunoglobulin hypervariable regions. *Nature* **342,** 877–883.
25. Chothia, C., Lesk, A. M., Gherardi, E., Tomlinson, I. M., Walter, G., Marks, J. D., et al. (1992) Structural repertoire of the human V_H segments. *J. Mol. Biol.* **227,** 799–817.
26. Tramontano, A., Chothia, C., and Lesk, A. M. (1990) Framework residue 71 is a major determinant of the position and conformation of the second hypervariable region in the V_H domains of immunoglobulins. *J. Mol. Biol.* **215,** 175–182.
27. Holmes, M.A,. Buss, T. N., and Foote, J. (1998) Conformational correction mechanisms aiding antigen recognition by a humanized antibody. *J. Exp. Med.* **187,** 479–485.
28. Cheetham, G.M.T., Hale, G., Waldmann, H., and Bloomer, A. C. (1998) Crystal structures of a rat anti-CD52 (CAMPTAH-1) therapeutic antibody Fab fragment and its humanized counterpart. *J. Mol. Biol.* **284,** 85–99.
29. Fan, Z. C., Shan, L., Goldsteen, B. Z., Guddat, L. W., Thakur, A., Landolfi, N. F., et al. (1999) Comparison of the three-dimensional structures of a humanized and a chimeric Fab of an anti-gamma-interferon antibody. *J. Mol. Recognit.* **12,** 19–32.
30. Epp, O., Colman, P., Fehlhammer, H., Bode, W., Schiffer, M., Huber, R., et al. (1974) Crystal and molecular structure of a dimer composed of the variable portions of the Bence-Jones protein REI. *Eur. J. Biochem.* **45,** 513–524.
31. Saul, F. A., Amzel, L. M., and Poljak, R. J. (1978) Preliminary refinement and structural analysis of the Fab fragment from human immunoglobulin new at 2.0Å resolution. *J. Biol. Chem.* **253,** 585–597.
32. Tempest, P. R., Bremner, P., Lambert, M., Taylor, G., Furze, J. M., Carr, F. J., et al. (1991) Reshaping a human monoclonal antibody to inhibit human respiratory syncytial virus in vivo. *Biotechnology* **9,** 266–271.
33. Queen, C., Schneider, W. P., Selick, H. E., Payne, P. W., Landolfi, N. F., Duncan, J. F., et al. (1989) A humanized antibody that binds to the interleukin 2 receptor. *Proc. Natl. Acad. Sci. USA* **86,** 10,029–10,033.
34. Co, M. S., Deschamps, M., Whitley, R. J., and Queen, C. (1991) Humanized antibodies for antiviral therapy. *Proc. Natl. Acad. Sci. USA* **88,** 2869–2873.

35. Co, M. S., Avdalovic, N. M., Caron, P. C., Avdalovic, M. V., Scheinberg, D. A., and Queen, C. (1992) Chimeric and humanized antibodies with specificity for the CD33 antigen. *J. Immunol.* **148,** 1149–1154.
36. Co, M. S., Yano, S., Hsu, R. K., Landolfi, N. F., Vasquez, M., Cole, M., et al. (1994) A humanized antibody specific for the platelet integrin gpIIb/IIIa. *J. Immunol.* **152,** 2968–2976.
37. Kabat, E. A., Wu, T. T., Perry, H. M., Gottesman, K. S., and Foeller, C. (1991) Sequences of proteins of immunological interest, 5th ed. BETHESDA: US Department of Health and Human Services.
38. Carter, P., Presta, L., Gorman, C. M., Ridgway, J. B., Henner, D., Wong, W. L., et al. (1992) Humanization of an anti-p185HER2 antibody for human cancer therapy. *Proc. Natl. Acad. Sci. USA* **89,** 4285–4289.
39. Presta, L. G., Lahr, S. J., Shields, R. L., Porter, J. P., Gorman, C. M., Fendly, et al. (1993) Humanization of an antibody directed against IgE. *J. Immunol.* **151,** 2623–2632.
40. Couto, J. R., Christian, R. B., Peterson, J. A., and Ceriani, R. L. (1995) Designing human consensus antibodies with minimal positional templates. *Cancer Res.* **55,** 5973s–5977s.
41. Werther, W. A., Gonzalez, T. N., O'Connor, S. J., McCabe, S., Chan, B., Hotaling, T., et al. (1996) Humanization of an anti-lymphocyte function-associated antigen (LFA)-1 monoclonal antibody and reengineering of the humanized antibody for binding to rhesus LFA-1. *J. Immunol.* **157,** 4986–4995.
42. O'Connor, S. J., Meng, Y. G., Rezaie, A. R., and Presta, L. G. (1998) Humanization of an antibody against human protein C and calcium-dependence involving framework residues. *Protein Eng.* **11,** 321–328.
43. Tomlinson, I. M., Walter, G., Marks, J. D., Llewlyn, M. B., and Winter, G. (1992) The repertoire of human germline V(H) sequences reveals about 50 groups of V(H) segments with different hypervariable loops. *J. Mol. Biol.* **227,** 776–798.
44. Cox, J. P., Tomlinson, I. M., and Winter, G. (1994) A directory of human germ-line V kappa segments reveals a strong bias in their usage. *Eur. J. Immunol.* **24,** 827–836.
45. Williams, S. C., Frippiat, J. P., Tomlinson, I. M., Ignatovich, O., Lefranc, M. P., and Winter, G. (1996) Sequence and evolution of the human germline V lambda repertoire. *J. Mol. Biol.* **264,** 220–232.
46. Johnson, S., Oliver, C., Prince, G. A., Hemming, V. G., Pfarr, D. S., Wang, S. C., et al. (1997) Development of a humanized monoclonal antibody (MEDI-493) with potent in vitro and in vivo activity against respiratory syncytial virus. *J. Infect. Dis.* **176,** 1215–1224.
47. Caldas, C., Coelho, V. P., Rigden, D. J., Neschich, G., Moro, A. M., and Brigido, M. M. (2000) Design and synthesis of germline-based hemi-humanized single-chain Fv against the CD18 surface antigen. *Protein Eng.* **13,** 353–360.
48. Padlan, E. A. (1991) A possible procedure for reducing the immunogenicity of antibody variable domains while preserving their ligand-binding properties. *Mol. Immunol.* **28,** 489–498.

49. Roguska, M. A., Pedersen, J. T., Keddy, C. A., Henry, A. H., Searle, S. J., Lambert, J. M., et al. (1994) Humanization of murine monoclonal antibodies through variable domain resurfacing. *Proc. Natl. Acad. Sci. USA* **91,** 969–973.
50. Ohtomo, T., Tsuchiya, M., Sato, K., Shimizu, K., Moriuchi, S., Miyao, Y., et al. (1995) Humanization of mouse ONS-M21 antibody with the aid of hybrid variable regions. *Mol. Immunol.* **32,** 407–416.
51. Verhoeyen, M. E., Saunders, J. A., Price, M. R., Marugg, J. D., Briggs, S., Broderick, E. L., et al. (1993) Construction of a reshaped HMFG1 antibody and comparison of its fine specificity with that of the parent mouse antibody. *Immunology* **78,** 364–370.
52. Roguska, M. A., Pedersen, J. T., Henry, A. H., Searle, S. M., Roja, C. M., Avery, B., et al. (1996) A comparison of two murine monoclonal antibodies humanized by CDR-grafting and variable domain resurfacing. *Protein Eng.* **9,** 895–904.
53. Shearman, C. W., Pollock, D., White, G., Hehir, K., Moore, G. P., Kanzy, E. J., et al. (1991) Construction, expression and characterization of humanized antibodies directed against the human alpha/beta T cell receptor. *J. Immunol.* **147,** 4366–4373.
54. Rosok, M. J., Yelton, D. E., Harris, L. J., Bajorath, J., Hellstrom, K.-E., Hellstron, I., et al. (1996) A combinatorial library strategy for the rapid humanization of anticarcinoma BR96 Fab. *J. Biol. Chem.* **271,** 22,611–22,618.
55. Baca, M., Presta, L. G., O'Connor, S. J., and Wells, J. A. (1997) Antibody humanization using monovalent phage display. *J. Biol. Chem.* **272,** 10,678–10,684.
56. De Pascalis, R., Iwahashi, M., Tamura, M., Padlan, E. A., Gonzales, N. R., and Santos, A. D. (2002) Grafting of "abbreviated" complementarity-determining regions containing specificity-determining residues essential for ligand contact to engineer a less immunogenic humanized monoclonal antibody. *J. Immunol.* **169,** 3076–3084.
57. Mateo, C., Lombardero, J., Moreno, E., Morales, A., Bombino, G., Coloma, J., et al. (2000) Removal of amphipathic epitopes from genetically engineered antibodies: production of modified immunoglobulins with reduced immunogenicity. *Hybridoma* **19,** 463–471.
58. Gilliland, L. K., Walsh, L. A., Frewin, M. R., Wise, M. P., Tone, M., Hale, G., et al. (1999) Elimination of the immunogenicity of therapeutic antibodies. *J. Immunol.* **162,** 3663–3671.
59. Routledge, E. G., Falconer, M. E., Pope, H., Lloyd, I. S., and Waldmann, H. (1995) The effect of aglycosylation on the immunogenicity of a humanized therapeutic CD3 monoclonal antibody. *Transplantation* **60,** 847–853.
60. Whitelegg, N. R. and Rees, A. R. (2000) WAM: an improved algorithm for modeling antibodies on the WEB. *Protein Eng.* **13,** 819–824.
61. Guex, N., Diemand, A., and Peitsch, M. C. (1999) Protein modeling for all. *Trends Biochem. Sci.* **24,** 364–367.
62. Johnson, G. and Wu, T. T. (2001) Kabat database and its applications: future directions. *Nucleic Acids Res.* **29,** 205–206.
63. Nicholas, K. B., Nicholas, H. B. Jr, and Deerfield, D. W. II (1997) Genedoc: analysis and visualization of genetic variation. *EMBNEW.NEWS* **4,** 14.

64. Lefranc, M. P. (2001) IMGT, the international ImMunoGeneTics database. *Nucleic Acids Res.* **29,** 207–209.
65. Blundell, T. L., Sibanda, B. L., Sternberg, M.J.E., and Thornton, J. M. (1987) Knowledge-based prediction of protein structures and the design of novel molecules. *Nature* **326,** 347–352.
66. Greer, J. (1991) Comparative modeling of homologous proteins. *Methods Enzymol.* **202,** 239–252.
67. Padlan, E. A., Davies, D. R., Pecht, I., Givol, D., and Wright, C. (1977) Model-building studies of antigen-binding sites in the hapten-binding site of mopc-315. *Cold Spring Harbor Symp. Quant. Biol.* **41,** 627–637.
68. De la Paz, P., Sutton, B. J., Darsley, and Rees, A. R. (1986) Modeling of the combining sites of three anti-lysozyme monoclonal antibodies and of the complex between one of the antibodies and its epitope. *EMBO J.* **5,** 415–425.
69. Smith-Gill, S. J., Mainhart, C., Lavoie, T. B., Feldmann, R. J., Drohan, W., and Brooks, B. R. (1987) A three-dimensional model of an anti-lysozyme antibody. *J. Mol. Biol.* **194,** 713–724.
70. Stanford, J. M. and Wu, T. T. (1981) A predictive method for determining possible three-dimensional foldings of immunoglobulin backbones around antibody combining sites. *J. Theor. Biol.* **88,** 421.
71. Fine, R. M., Wang, H., Shenkin, P. S., Yarmush, D. L., and Levinthal, C. (1986) Predicting antibody hypervariable loop conformations. II: minimization and molecular dynamics studies of MCPC603 from many randomly generated loop conformations. *Proteins* **1,** 342–362.
72. Bruccoleri, R. E., Haber, E., and Novotny, J. (1988) Structure of antibody hypervariable loops reproduced by a conformational search algorithm. *Nature* **335,** 564–568.
73. Bendig, M. M. and Jones, S. T. (1996) Rodent to human antibodies by CDR grafting, in McCafferty, J., Hoogenboom, H. R., Chiswell, D. J. (eds) Antibody engineering—a practical approach. Oxford University Press, New York, NY,
74. Foote, J. and Winter, G. (1992) Antibody framework residues affecting the conformation of the hypervariable loops. *J. Mol. Biol.* **224,** 487–499.
75. Padlan, E. A. (1994) Anatomy of the antibody molecule. *Mol. Immunol.* **31,** 169–217.
76. Chothia, C., Novotny, J., Bruccoleri, R., and Karplus, M. (1985) Domain association in immunoglobulin molecules—the packing of variable domains. *J. Mol. Biol.* **186,** 651–663.
77. Marshall, R. D. (1972) Glycoproteins. *Annu. Rev. Biochem.* **41,** 673–702.
78. Gooley, A. A., Classon, B. J., Marschalek, R., and Williams, K. L. (1991) Glycosylation sites identified by detection of glycosylated amino acids released from Edman degradation: the identification of Xaa-Pro-Xaa-Xaa as a motif for Thr-O-glycosylation. *Biochem. Biophys. Res. Commun.* **178,** 1194–1201.
79. Pisano, A., Redmond, J. W., Williams, K. L., and Gooley, A. A. (1993) Glycosylation sites identified by solid-phase Edman degradation: O-linked glycosylation motifs on human glycophorin A. *Glycobiology* **3,** 429–435.

8

Selection of Human Antibodies from Phage Display Libraries

James D. Marks and Andrew Bradbury

1. Introduction

Phage display can be used to generate human antibodies to virtually any antigen. This requires a number of crucial steps that are common to all molecular diversity technologies: the creation of diversity, coupling of phenotype (antibody protein) to genotype (antibody gene), selection, amplification, and analysis (reviewed in **ref.** ***1–3***). Antibodies with affinities comparable to those obtained using traditional hybridoma technology can be selected from either large naïve antibody libraries or from immune phage antibody libraries. The affinity of initial isolates can be further increased, to levels unobtainable in the immune system, by using the selected antibodies as the basis for construction and selection of libraries in which the antibody gene is diversified (*see* Chapters 19 and 20). This approach is one of the few methods for generating fully human antibodies, especially to evolutionarily conserved or self proteins ***(4)***. Although in theory there are many different choices to be made when constructing a phage antibody library, most published work has followed a tried and tested formula. In particular, filamentous phage-based phagemid vectors have been used almost exclusively, and the phage minor coat protein pIII has been the display coat protein most frequently used ***(2,5)***. Libraries have been made using both the single-chain Fv (scFv) ***(6–10)*** and Fab format ***(11,12)***.

In Chapter 6, we described methods for capturing antibody heavy (V_H) and light (V_L)-chain variable region diversity using PCR, from either immune or non-immune donors. We also described methods for combining V_H and V_L genes to create repertoires of scFv antibody fragments and their cloning to create phage antibody libraries in which the scFv gene is fused to the phage minor coat protein pIII. In this chapter, we discuss selection and characterization of antigen-specific antibodies from phage antibody libraries.

From: *Methods in Molecular Biology, Vol. 248: Antibody Engineering: Methods and Protocols*
Edited by: B. K. C. Lo © Humana Press Inc., Totowa, NJ

Antigen-specific antibodies are isolated from phage antibody libraries by repetitive rounds of selection on antigen. Enrichment for specific antibodies can be 100–1,000 fold *(13)*, and thus two to five rounds of selection are necessary, depending on whether an immune or naïve antibody library is used and on the stringency of selections. A multitude of selection methods have been described, including selection on antigen adsorbed onto solid supports *(8)*, conjugated to columns or BIAcore sensor chips *(12,14)*, selection using biotinylated antigen *(15)*, peptides *(16)*, panning on fixed prokaryotic *(17)*, or mammalian *(18)* cells, tissue-culture cells *(19–21)*, fresh cells *(22,23)*, subtractive selection using sorting procedures such as fluorescence-activated cell sorting (FACS) *(24)*, and MACS *(20)*, enrichment on tissue sections or pieces of tissue *(25)*, selection for internalization (*26,27*; *see* Chapter 11) and, in principle, selection in living animals, as reported for peptide phage libraries *(28)*. More complex selection methods include "Pathfinder/Proximol" *(29)* and methods based on the restoration of phage infectivity via the antibody antigen interaction *(30,31)*. In general, selection methods can be divided into two broad classes: those which attempt to isolate antibodies against known antigens, and those that use phage antibodies as a research tool to target unknown antigens (e.g., against cell-surface antigens). This chapter will concentrate on selections using known antigens. Selection for internalizing antibodies binding cell-surface antigens are described in Chapter 11.

Phage antibody selection is optimal when purified antigen is available. Although relatively small quantities of antigen are required for selection, far greater amounts are typically needed for screening, for example by enzyme-linked immunosorbent assay (ELISA). In general, approx 500 μg–1 mg of antigen is enough to complete a generous series of selections and screenings. This process can be accomplished with as little as 100 μg of antigen, if the antigen is used at low concentrations and can be stored and reused.

The most common selection method involves coating plastic immunotubes with the antigen. This approach is described in **Subheading 3.1.** These tubes have a capacity of 650 ng antigen/cm^2, but have the potential problem that epitopes become partially denatured, leading to the selection of antibodies that do not recognize the native antigen. Although rare, this can be avoided by either first coating tubes with an antigen-specific antibody or by using soluble biotinylated antigen in solution *(15)* with capture of the antigen and bound phage with magnetic streptavidin or avidin beads. The latter approach is described in **Subheading 3.2.**

2. Materials

2.1. Selection of Phage Antibodies on Immobilized Antigen

1. 75 × 12 mm Nunc Maxisorp immunotubes 444202 (VWR Scientific).
2. Phosphate-buffered saline (PBS): 154 m*M* NaCl, 8.1 m*M* Na_2HPO_4, 1.9 m*M* NaH_2PO_4, pH 7.4.

3. 2% MPBS: 2% skimmed milk powder in PBS.
4. PBS/Tween: 0.1% Tween 20 in PBS.
5. 100 m*M* triethylamine.
6. 1 *M* Tris-HCl, pH 7.4.
7. Exponentially growing *E. coli* TG1 (Stratagene) or DH5αF′ (Invitrogen).
8. 2X TY media: dissolve 16 g bacto-tryptone, 10 g yeast, 5 g NaCl in 1 L of distilled water, and autoclave the solution.
9. 2X TY/amp/glu media: 2X TY media containing 100 μg/mL ampicillin and 2% glucose.
10. TYE/amp/glu plates: 100- and 150-mm TYE agar plates containing 100 μg/mL ampicillin and 2% glucose.
11. 50% glycerol, sterile-filtered.
12. Purified phage antibody library (prepared according to **Subheading 3.3.**).

2.2. Selection Using Soluble Biotinylated Antigen

1. Streptavidin magnetic beads (Dynal Biotech).
2. Avidin magnetic beads (MPG).
3. Magnetic 1.5-mL tube holder (Dynal Biotech).
4. PBS.
5. 2% MPBS.
6. Biotinylation kit (Pierce).
7. PBS/Tween.
8. 100 m*M* HCl.
9. 1 *M* Tris-HCl, pH 7.4.
10. Exponentially growing *E. coli* TG1.
11. 2X TY media (*see* **Subheading 2.1, item 8**).
12. 2X TY/amp/glu media (*see* **Subheading 2.1., item 9**).
13. TYE amp/glu plates (*see* **Subheading 2.1., item 10**).
14. Purified phage antibody library (prepared according to **Subheading 3.3.**).

2.3. Preparation of Phage Antibodies from Phage Library Glycerol Stocks and from Selection Rounds

1. 500-mL sterile polypropylene centrifuge bottles (*see* **Note 1**).
2. 2-L culture flasks.
3. RC5 centrifuge and GS3 rotor (Sorvall).
4. 2X TY/amp/glu media (*see* **Subheading 2.1., item 9**).
5. 100- and 150-mm TYE amp/glu agar plates (*see* **Subheading 2.1, item 10**).
6. 2X TY/amp/kan media: 2X TY media containing 100 μg/mL ampicillin and 25 μg/mL kanamycin.
7. TYE/kan/glu plates: 100 mm TYE agar plates containing 25 μg/mL kanamycin and 2% glucose.
8. Helper phage: VCSM13, R408 or M13K07 (Stratagene, New England Biolabs, or Amersham Biosciences; *see* **Note 2**).
9. PEG/NaCl solution: 20% (w/v) polyethylene glycol 6,000, 2.5 *M* NaCl.

10. PBS.
11. Library glycerol stock (prepared according to **Subheading 3.1.**).
12. Exponentially growing *E. coli* TG1.

2.4. Expression of Phage Antibodies in Microtiter Plates

1. Sterile 96-well round-bottom microtiter plates made for bacterial culture—e.g., Nunc 62162 (VWR Scientific).
2. 2X TY/amp/glu media (*see* **Subheading 2.1., item 9**).
3. 2X TY/amp/kan media (*see* **Subheading 2.3., item 6**).
4. 2X TY/amp/glu/gly media: 2X TY/amp/glu media containing 30% glycerol.
5. Colonies resulting from plating of selected libraries (prepared from **Subheadings 3.1.** or **3.2.**).

2.5. Phage Antibody ELISA

1. 96-well ELISA plates—e.g., Nunc Maxisorp 442404 (VWR Scientific).
2. 2% and 4% MPBS.
3. PBS/Tween.
4. HRP-conjugated mouse anti-phage monoclonal antibody (MAb) (Amersham Biosciences).
5. TMB liquid substrate system (Sigma T-8665).
6. Stop solution: 2 *N* H_2SO_4.
7. Phage preparation from **Subheading 3.4.**

2.6. Expression of Soluble ScFv Antibodies

1. Sterile 96-well round-bottom microtiter plates made for bacterial culture—e.g., Nunc 62162.
2. 2X TY/amp/0.1% glu media: 2X TY media containing 100 μg/mL ampicillin and 0.1% glucose.
3. 2X TY/amp/IPTG media: 2X TY media containing 100 μg/mL ampicillin and 3 m*M* IPTG isopropyl-β-D-galactopyranoside (IPTG).
4. Bacterial colonies harboring phage antibodies (*see* **Note 3**).

2.7. Soluble ScFv ELISA in Microtiter Plates

1. 96-well ELISA plates—e.g., Nunc Maxisorp 442404 (VWR Scientific).
2. 9E10 MAb (Santa Cruz Biotech) (*see* **Note 4**).
3. PBS.
4. PBS/Tween.
5. 2% and 4% MPBS.
6. HRP-conjugated anti-mouse immunoglobulins (Sigma).
7. TMB liquid substrate system.
8. Stop solution: 2 N H_2SO_4.
9. Antibody fragment-containing supernatants prepared from **Subheading 3.6.**

3. Methods

3.1. Selection of Phage Antibodies on Immobilized Antigen

1. Coat an immunotube with 2 mL of antigen (10–1,000 μg/mL) in PBS at room temperature overnight. Carbonate buffer can also be used if appropriate (*see* **Note 5**).
2. Wash the tube 3× with PBS and block the tube with 4 mL of 2% MPBS at 37°C for 1 h.
3. Wash the tube 3× with PBS, add 1 mL of 10^{12}–10^{13} phage antibodies (*see* Chapter 6 for phage library stock creation and **Subheading 3.3.** for phage preparation and phage titering) diluted in 1 mL of 2% MPBS, and incubate for 30 min on a rotating turntable followed by an additional 1.5 h standing (*see* **Note 6**).
4. Wash the tube once with PBS/Tween and twice with PBS by pouring buffer in and out of the tube (*see* **Note 7**).
5. Elute the bound phage by adding 1.0 mL of 100 m*M* triethylamine, capping the tube, and rotating it on a turntable for 8 min (*see* **Note 8**).
6. Transfer the eluant to a new tube and neutralize immediately by adding 0.5 mL of 1.0 *M* Tris-HCl, pH 7.4, and mixing.
7. Add one-half of the eluant (0.75 mL) to 10 mL of exponentially growing *E. coli* TG1 or DH5αF′ (A_{600} ~0.5) and incubate the culture without shaking at 37°C for 30 min.
8. Plate 1 μL and 0.1 μL of the culture on 100 mm TYE/amp/glu plates to titer the eluted phage (*see* **Note 9**). Plating these two amounts of culture will cover an out put titer of 10^4 –10^7 at an easily counted number of colonies. During later rounds of selection, it may be necessary to plate additional dilutions, as the titer tends to increase.
9. Centrifuge the remainder of the culture at 2,800*g* for 15 min, resuspend in 0.5 mL of 2X TY media, plate on two 150 mm TYE/amp/glu plates, and incubate at 37°C overnight.
10. The next day, add 3 mL of 2X TY/amp/glu media to each plate and scrape the bacteria from the plate with a bent glass rod. Make glycercol stocks by mixing 1.4 mL of bacteria and 0.6 mL of 50% glycerol. Save stock at –70°C.
11. Prepare phagemid particles for the next round of selection from the glycerol stocks by infection with helper phage as described in **Subheading 3.3.** It is only necessary to prepare phage from a 50-mL culture for subsequent rounds of selection.
12. Repeat the selection, starting with **step 1,** for a total of 2–4 rounds. Increase the stringency of washes after the first round by washing 20× in PBS/Tween and 20 times in PBS before phage elution (*see* **Note 10**). Increased stringency can also be provided by carrying out one or two 30-min washes with PBS and PBS/Tween (*see* **Notes 11** and **12**).

3.2. Selection Using Soluble Biotinylated Antigen

An alternative to selecting antibodies on immobilized antigen is to select the antibodies in solution. This solves problems related to antigens that change conformation when directly coated onto solid surfaces. It is also easier to select on the basis of binding affinity, because phage that display multiple copies of an antibody do not have a selective advantage over those that do not. It is also easier to limit the antigen concentration. In order to select phage antibodies in solution, antigen is biotinylated using any of the kits sold by immunochemical reagent companies. After incubating the library with biotinylated antigen, the antigen and bound phage antibodies are separated from unbound phage antibodies using either streptavidin or avidin magnetic beads. After suitable washing, phage antibodies are eluted with either acid or alkaline solutions.

1. Biotinylate the antigen using the kit as recommended by the manufacturer. If dithiothreitol (DTT) will be used for elution, use NHS-SS biotin for biotinylation.
2. Block a 1.5-mL microcentrifuge tube with 2% MPBS at room temperature for 1 h, then wash the tube once with PBS.
3. Block 100 µL of streptavidin-magnetic beads with 1 mL of 2% MPBS at room temperature for 1 h in a 1.5-mL microcentrifuge tube. After blocking, collect the beads by pulling to one side with the magnetic tube holder. Discard the buffer.
4. Pre-deplete the library of streptavidin binders by incubating 10^{12-13} phage (10^{13} for a first-round selection, 10^{12} for subsequent rounds) in 2% MPBS with 100 µL of streptavidin-magnetic beads for 1 h in a vol of 1 mL. Pull the beads to one side with the magnetic tube holder. Collect the phage antibodies and discard the beads (*see* **Note 13**).
5. Add the 1 mL of pre-depleted phage antibodies to the blocked tube and then add biotinylated antigen (100–500 nm). Incubate, rotating on a turntable, at room temperature for 1 h.
6. Add 100 µL of the blocked streptavidin-magnetic beads to the tube and incubate on the rotator at room temperature for 15 min. Place the tube in the magnetic rack for 30 s. The beads will migrate toward the magnet.
7. Aspirate the tubes, leaving the beads on the side of the microcentrifuge tube. This is most effectively done with a 200-µL pipet tip on a Pasteur pipet attached to a vacuum source. Wash the beads (1 mL per wash) 7× with PBS/Tween, followed by twice with 2% MPBS, then once with PBS. Transfer the beads after every second wash to a fresh 1.5-mL tube to facilitate efficient washing.
8. Elute the phage by adding 1 mL of 100 m*M* triethylamine, capping the tube, and rotating end-over-end for 8 min. Draw the beads to one side of the tube with the magnet and transfer the solution to an Eppendorf tube containing 500 µL of 1 *M* Tris-HCl, pH 7.4 (*see* **Note 14**).
9. Proceed as described in **Subheading 3.1., step 7** onward to prepare phage for the next round of selection. Depending on the titer of the eluted phage, the antigen concentration can be reduced 10-fold for the next round of selection.
10. After 2–4 rounds of selection, the success of selection can be determined by performing a ELISA for antigen binding using the polyclonal phage prepared from each round of selection (*see* **Subheading 3.5.**). When the polyclonal phage are positive for antigen binding, individual phage antibodies can be analyzed for antigen binding **(Subheadings 3.4–3.7.)**.

3.3. Preparation of Phage Antibodies from Phage Library Glycerol Stocks and from Selection Rounds

1. Inoculate an appropriate number of 2-L culture flasks containing 250 mL of 2X TY/amp/glu media with the library glycerol stock (*see* **Note 15**). The A_{600} of the culture media should be <0.05. Plate 1 µL from each culture flask to determine the size of the initial inoculum (*see* **Note 16**).
2. Grow the culture at 37°C with shaking (300 rpm) to an A_{600} of 0.5 (approx 1.5–2.5 h) (*see* **Note 17**).
3. Add helper phage at a helper phage: bacteria ratio of 10–20:1.
4. Incubate the culture at 37°C for 30 min, standing in a water bath with occasional mixing, then incubate at 37°C for 30 min with shaking. Remove 1 µL from each flask, dilute in 1 mL of 2X TY media and plate 1 and 10 µL on TYE/amp/glu and TYE/kan/glu plates to determine the efficiency of helper phage infection. The number of colonies on each plate should be similar, with the proportion of kanamycin-resistant colonies greater than 10% of the number of ampicillin-resistant colonies.
5. Centrifuge the bacterial cultures in 500-mL bottles in a GS3 rotor at 2,800*g*. Remove the supernatant.
6. Resuspend the bacterial pellet(s) in 2X TY/amp/kan media and distribute the suspended culture into 2-L flasks containing no more than 250 mL of media (*see* **Note 18**). Grow with shaking (300 rpm) overnight at 25–30°C.
7. Centrifuge the bacteria at 6,000*g* in 500-mL bottles in a GS3 rotor for 30 min.
8. Transfer the supernatant to new 500-mL bottles and precipitate the phage by adding 1/10–1/5 vol of PEG/NaCl solution and leave on ice for 1 h. Phage should be visible as a clouding of the supernatant.
9. Pellet the phage in 500-mL bottles by centrifuging in a GS3 rotor at 3,000*g* for 15 min at 4°C. Discard the supernatant. Centrifuge the "dry" pellet again for 30 s to bring down the last drops of supernatant, and remove the liquid. Resuspend the pellet in 1/10 vol of PBS.
10. Centrifuge the bottles at 6,000*g* for 15 min to pellet bacterial debris and transfer the supernatant to a new tube.
11. Repeat **steps 8–10** to further purify the phage, resuspending in a final vol that is 1/50 of the original culture volume. Phage prepared can be used immediately for selection, or stored at –80°C in aliquots of 10^{13} phage per aliquot (*see* **Note 19**).
12. Determine the titer of the phage preparation. The expected titer will be approx 10^{11}–10^{13} TU/mL. Therefore dilute the phage preparation 1,000,000-fold in 2X TY/amp/glu media. This will decrease the phage concentration to 10^5–10^7 TU/mL. Add 10 µL of the diluted phage to 1 mL of exponentially growing *E. coli* TG1, incubate standing for 30 min at 37°C. Plate 100 µL, 10 µL, 1 µL, and 0.1 µL of the *E. coli* on TYE amp/glu plates and incubate at 37°C overnight. The actual titer is obtained by counting the number of colonies and multiplying by the dilution factor.

3.4. Expression of Phage Antibodies in Microtiter Plates

The ability of individual phage antibodies to bind antigen is typically determined by producing phage antibodies from individual colonies in microtiter plates and then performing an ELISA (*see* also **Subheading 3.5.**). An alternative is to express soluble scFv or Fab antibody fragment into the bacterial supernatant and use native scFv or Fab to detect antigen binding by ELISA *(8)* (*see* **Subheadings 3.6.,** and **3.7.**). In many phage vectors, this can be done without subcloning, as the phage display vector includes an amber stop codon (TAG) between the scFv or Fab gene and gene III *(5)*. Native scFv or Fab expression can be achieved by infecting the phage into a non-suppressor strain of *E. coli.* In fact, this is not necessary, as even in suppressor strains of *E. coli,* such as TG1, suppression is only 50%, and adequate amounts of soluble scFv are produced upon the addition of IPTG. In vector systems without an amber stop codon, the scFv or Fab gene can be subcloned into an appropriate secretion vector *(32,33)*. This can usually be accomplished "en mass" by subcloning each of the selection round outputs into the secretion vector. Phage ELISA is usually more sensitive than scFv or Fab ELISA, because of the thousands of copies of the phage major coat protein pVIII that is used for detection by the secondary antibody. However, a percentage of phage antibodies that bind by phage ELISA will not bind as scFv or Fab, because of poor expression or low affinity. Thus we prefer to utilize scFv or Fab ELISA to identify initial binding phage antibodies. Once binding phage antibodies are identified, the number of unique antibodies can be determined by fingerprinting of the scFv or Fab genes and DNA sequencing *(8,34)*. ScFv or Fab expression can then be scaled up, and antibody fragment can be harvested and purified from the bacterial periplasm *(33)*, usually by affixing a hexahistidine tag at the C-terminus of the antibody fragment *(35)*.

1. Use sterile toothpicks to pick individual colonies into 96-well microtiter plate wells containing 100 µL of 2X TY/amp/glu (*see* **Note 20**). Grow with shaking (300 rpm) at 30°C overnight, preferably in a microtiter plate holder (*see* **Notes 21** and **22**).
2. Add 50 µL of 2X TY/amp/glu/gly to each well and store the plate at –70°C. This is the master plate.
3. Use a 96-well sterile transfer device or pipet to inoculate 2 µL per well from the master plate to a 96-well plate containing 150 µL of 2X TY/amp/glu per well. Grow these cultures to an A_{600} of approx 0.5 (around 2.5 h), at 37°C with shaking.
4. Add 50 µL of 2X TY/amp/glu media containing 2×10^9 pfu/mL helper phage (diluted from stock) to each well. The ratio of phage to bacterium should be approx 20:1. Incubate the plate at 37°C for 30 min (*see* **Note 23**).
5. Centrifuge the plate at 2,000*g* for 10 min and remove and discard the supernatant using a multichannel pipet or suction device.
6. Resuspend each bacterial pellet in 150 µL of 2X TY/amp/kan media. Glucose is omitted in this step, since expression of the antibody depends on leaky expression from the lacZ promoter. Incubate the plate shaking (300 rpm) overnight at 30°C.
7. The next day, spin the plate at 2,000*g* for 10 min and use 50 µL/well of the supernatant for phage ELISA (**Subheading 3.5.**).

3.5. Phage Antibody ELISA

1. Coat the wells of a microtiter plate overnight at 4°C with 100 µL/well of protein antigen; 10 µg/mL in PBS is standard, but sometimes higher concentrations or a different binding buffer (such as carbonate buffer) are required (*see* **Note 24**).
2. Discard the antigen solution and wash the wells twice with PBS (*see* **Note 25**).
3. Block the wells by adding 200 µL of 2% MPBS and incubate at 37°C for 2 h.
4. Wash the wells 3× with PBS.
5. Add 50 µL of 4% MPBS to all wells.
6. Add 50 µL of culture supernatant containing phage antibodies (*see* **Subheading 3.4.**) to the wells. Mix the solution by pipetting up and down and incubate at room temperature for 1 h.
7. Discard the solution, and wash wells 3× with PBS/Tween and 3× with PBS.
8. Add 100 µL/well HRP-conjugated mouse anti-phage MAb diluted 1:5,000 in 2% MPBS and incubate at room temperature for 1 h.
9. Discard the secondary antibody and wash wells 3× with PBS/Tween and 3× with PBS.
10. Add 100 µL of TMB substrate solution to each well and incubate at room temperature for 10–30 min in the dark. The blue color should appear within a few minutes.
11. Quench the reaction by adding 50–100 µL of stop solution to each well.
12. Read the plate at 450 nm.

3.6. Expression of Soluble ScFv Antibody

1. Prepare master plates as described in **Subheading 3.4., steps 1** and **2.**
2. Use a 96-well sterile transfer device or pipet to transfer 2 µL per well from the master plate to a sterile 96-well microtiter plate containing 100 µL/well of 2X TY/amp/0.1% glu media (*see* **Note 26**).
3. Incubate the plate at 37°C, shaking, until the A_{600nm} is approx 0.9 (usually 2–3 h).
4. Add 50 µL of 2X TY/amp/IPTG media to each well (final IPTG concentration of 1 m*M*). Continue shaking the plate at 30°C for an additional 16 h.
5. Centrifuge the plates at 2,000*g* for 15 min and use 50 µL of the supernatant for ELISA (**Subheading 3.7**).

3.7. Soluble ScFv ELISA in Microtiter Plates

1. Coat the wells of a microtiter plate overnight at 4°C with 100 µL/well of protein antigen; 10 µg/mL in PBS is standard, but sometimes higher concentrations or a different binding buffer (e.g., carbonate buffer) are required.
2. Discard the antigen solution and wash the wells twice with PBS.
3. Block each well by adding 200 µL of 2% MPBS and incubate at 37°C for 2 h.
4. Wash the wells 3× with PBS.
5. Add 50 µL of 4% MPBS to all wells.
6. Add 50 µL of culture supernatant containing soluble scFv (from **Subheading 3.6.**) to the wells. Mix the solutions by pipetting up and down and incubate the plate at room temperature for 1 h.

7. Discard the solution, and wash the wells 3× with PBS/Tween and 3× with PBS. Add 100 µL of 9E10 MAb (diluted 1:1,000 in 2% MPBS) to each well. Incubate the plate at room temperature for 1 h.
8. Discard the antibody, and wash wells 3× with PBS/Tween and 3× with PBS.
9. Add 100 µL of HRP-conjugated anti-mouse immunoglobulins (diluted 1:2,000 in 2% MPBS) to each well. Incubate the plate at room temperature for 1 h.
10. Discard the secondary antibody, and wash the wells 3× with PBS/Tween and 3× with PBS.
11. Add 100 µL of TMB substrate solution and incubate the plate at room temperature for 10–-30 min in the dark. The blue color should appear within a few minutes
12. Quench the reaction by adding 50–100 µL/well of stop solution.
13. Read the plate at 450 nm.

4. Notes

1. Keep a dedicated set of centrifuge bottles and culture flasks for library rescue to avoid contamination by previously isolated binding phage antibodies.
2. M13K07 is usually used, although no direct comparison has been made between the different helper phages.
3. This protocol requires that the scFv or Fab antibody gene must be under control of the lac promoter. For phage libraries in vectors such as pHEN1 ***(5)***, which have an amber stop codon between the antibody fragment gene and gene III, no subcloning is necessary. For other display systems, the antibody fragment gene would need to be subcloned into an appropriate secretion vector under control of the lac promoter ***(32)***.
4. ELISA using soluble scFv is similar in principle to that using phage antibodies, with the exception that detection of the soluble scFv involves the use of an anti-tag antibody with a secondary detection antibody (e.g., anti-mouse Fc antibody if the tag antibody is a mouse monoclonal). The anti-tag antibody described here is 9E10, which recognizes the myc tag and is incorporated into the pHEN vector ***(5)***. 9E10 is available commercially, or the hybridoma cell line can be obtained from American Type Culture Collection (ATCC).
5. We typically coat with 50–100 µg/mL of antigen. For some antigens, it may be possible to improve selection for higher-affinity antibodies by reducing the coating antigen concentration during subsequent rounds of selection.
6. For a first round selection, 10^{13} phage should be used, for subsequent rounds 10^{12} are sufficient.
7. Since there are relatively few copies of each individual phage antibody in large libraries, the first round of selection should be relatively non-stringent (only a few washes). In general, increasing the number of washes at this stage is counterproductive, as it reduces the number of phage eluted but does not increase the proportion of positive phage (I. Tomlinson, personal communication and **ref. *18***). After the first round of selection, the number of each individual phage antibody is far higher, and subsequent washes can be far more stringent.

8. Longer incubations with triethylamine may render the phage non-infectious ***(8)***. Alternative methods of elution include 1 mL of 100 m*M* HCl (10 min) ***(36)***, 1 mL of 0.2 *M* glycine, pH 2.5 ***(37)***, or 0.5 mL of trypsin (1 mg/mL in PBS for 10–60 min at 20–37°C) provided that there is a trypsin site (arginine or lysine) between antibody and gene III. Phage are resistant to trypsin for prolonged periods. Additional elution methods include: 0.1 *M* DTT when biotin is linked to antigen by a disulfide bridge ***(12)***, proteolytic cleavage of a site (e.g., using factor X or genenase) inserted between antibody and gene III ***(38)***, competition with excess antigen or antibodies to the antigen ***(39)***, or by adding bacteria. In a comparison of different methods using a number of antibodies recognizing a single antigen with widely varying affinities, the most effective method was shown to be 0.1 *M* HCl ***(40)***. Bacterial elution, in particular, was not found to be effective in eluting high-affinity phage.
9. The titer of the input and output phage can be used to calculate the recovery ratio (input/output ratio) which should increase each round of selection. In general, the expected first-round output is 10^4–10^7 phage particles. The phage titer usually increases between rounds of selection, and this indicates that selection is occurring. However, examples of successful selection with no change in output titer and failed selections with increases in titers have both been observed. In general, after the first round of selection, approx 1% of phage bind antigen. This increases to up to 100% after 2–4 rounds of selection. However, the number of different antibodies typically decreases with each round of selection.
10. This reduces the output to 10,000–500,000 phage, but with a proportion of positives that can be as high as 100% after the second round.
11. There are two ways to favor the selection of antibodies with higher affinity. One is to increase the stringency in the washing steps after the first round, and the other is to reduce the antigen concentration during subsequent rounds of selection (*see* **Subheading 3.2.**). Washing stringency can be increased either by increasing the number of washes or by a prolonged incubation (e.g., 5–30 min) in the washing buffer.
12. In theory, a selection involving two antibodies with differing affinities will favor the antibody with the higher affinity. In practice, display levels (a function of antibody folding efficiency, susceptibility to proteolysis, and tendency to aggregate) and toxicity to the bacterial host are two additional factors that may adversely affect selection efficiency. It is for this reason that if selections are continued for many rounds, diversity usually becomes reduced to a single clone, which sometimes may not even bind the antigen and has a selective advantage because of deletion ***(41)***. Antibodies that display better are usually selected for two reasons: first, a greater proportion of phage will have antibody on their surface—normally, only 1–10% of phagemid display an antibody fragment ***(42)***—and second, because a greater proportion of phage will display two copies of antibody, contributing to an avidity effect. It is clear that antibodies that are toxic for their bacterial hosts will be discriminated against in any selection process.
13. One problem with this method is that antibodies that recognize streptavidin are also isolated. This can be avoided by depleting the library of streptavidin binders prior

to selection, alternating between using avidin and streptavidin in selection rounds (these two biotin-binding proteins are very different in structure and do not crossreact), or by labeling the antigen with biotin disulfide N-hydroxy-succinimide ester (NHS-SS biotin) and eluting with DTT *(12)*.

14. If NHS-SS biotin was used to label the antigen, instead add 100 µL of 1 m*M* DTT, cap the tube, and rotate end-over-end for 5 min. Draw the beads to one side with the magnet, and transfer the solution to another Eppendorf tube.
15. In most phage display vectors, the expression of the antibody fragment is driven by the lacZ promoter. Although this is a leaky promoter and probably could be improved, it works quite well if a few guidelines are followed. Before phage are rescued, bacteria containing the library are grown to logarithmic phase in the presence of glucose. The glucose inhibits the activity of the lacZ promoter, and thus reduces the expression of both gIIIp and antibody. This has two important effects: the toxicity resulting from gIIIp and antibody is reduced, thus preventing library bias; and the inhibition of gIIIp expression permits bacterial pilus expression, allowing infection by helper phage. If glucose is omitted, the efficiency of infection drops dramatically, sharply reducing the effective size of the library.
16. Prior to beginning this protocol, calculate the number of bacteria per mL in the library stock by carrying out titrations on TYE/amp/glu plates. The number of bacteria for a particular A_{600} is lower for a library than for untransformed bacteria, perhaps because library bacteria containing plasmids are larger, or because dead bacteria are present. Once titrations have been related to the A_{600}, the absorbance can be used to calculate bacterial numbers. For phage preparation (library rescue), the size of the initial inoculum from the library stock should be 5× more bacteria than the library size. After inoculation into the culture media, the bacterial concentration should not exceed an A_{600} of 0.05. The use of 5× more bacteria than the library size ensures representation of all library members. An initial A_{600} of less than 0.05 for the culture ensures multiple doubling times prior to the addition of helper phage, increasing the efficiency of helper phage infection. The minimal culture volume for the rescue must be determined based on library size and a maximal initial A_{600} of 0.05. For example, a library containing 1.0×10^{10} members would require an initial inoculum of 5.0×10^{10} bacteria. Since an A_{600} of $1.0 = 1.0 \times 10^9$ bacteria, to keep the A_{600} of the initial culture <0.05, a volume of 1 L would be required. This culture volume should be equally distributed into 250 mL of 2X TY/amp/glu media in 2-L flasks.
17. A temperature of 37°C is essential for pilus expression, which is requisite for helper phage infection. It is important to avoid overgrowing the culture (A600 <0.5), or infection with helper phage will be inefficient because of loss of the pilus.
18. Use a volume at least twice the initial culture volume to allow bacterial growth. Glucose has been removed from the culture medium to allow leaky expression of the antibody-gIIIp fusion protein.
19. Phage can be further purified by filtration through a 0.45-µm sterile filter (Minisart NML; Sartorius). Although this is useful when long-term storage at 4°C is envis-

aged, it can lead to a significant loss of titer. The best selection results are obtained when fresh phage are used, since there is a slow proteolysis of the displayed antibody with time, even at –80°C. Proteolysis is significantly reduced by purifying phage using cesium chloride-gradient centrifugation. After such purification, displayed proteins are far more stable. The expected yield is 1–5 × 10^{11} phage/mL of original culture volume, although this may decrease by 50% after cesium chloride purification.

20. Controls should include wells containing i) wild-type *E. coli,* ii) *E. coli* expressing a phage construct that displays no antibody fragment, and iii) *E. coli* expressing a phage construct that displays an irrelevant antibody fragment. If available, positive controls that bind the antigen should be included.
21. To avoid evaporation, microtiter plates should be incubated in a closed box as far as possible from the incubator fan. Place a damp paper towel in the box.
22. The first time growth in microtiter plates is carried out, shaking conditions should be tested by placing bacteria in alternate wells and growing overnight. Conditions are suitable if growth occurs only in inoculated wells and not in adjacent uninoculated wells (contamination).
23. Keep the plate at 37°C to avoid loss of pilus expression. The presence of glucose inhibits gIIIp production and allows pilus expression.
24. If the amount of antigen is limited, it can be recovered and reused after coating. In this case, overnight incubation at 4°C is recommended.
25. Wash by submerging the plate into buffer, removing the air bubbles in the wells by agitation, and shaking the buffer out into the sink. To remove the last drops, bang the microtiter, plate upside down on a pile of paper towels.
26. The method described here relies on the amount of glucose (0.1%) present in the starting medium being low enough to be metabolized by the time the inducer (IPTG) is added. This avoids any centrifugation steps and reduces the risk of contamination.

References

1. Winter, G., Griffiths, A. D., Hawkins, R. E., and Hoogenboom, H. R. (1994) Making antibodies by phage display technology. *Annu. Rev. Immunol.* **12,** 433–455.
2. Marks, J. D., Hoogenboom, H. R., Griffiths, A. D., and Winter, G. (1992) Molecular evolution of proteins on filamentous phage: mimicking the strategy of the immune system. *J. Biol. Chem.* **267,** 16,007–16,010.
3. Marks, C. and Marks, J. D. (1996) Phage libraries: a new route to clinically useful antibodies. *N. Engl. J. Med.* **335,** 730–733.
4. Griffiths, A. D., Malmqvist, M., Marks, J. D., Bye, J. M., Embleton, M. J., and McCafferty, J. (1993) Human anti-self antibodies with high specificity from phage display libraries. *EMBO J.* **12,** 725–734.
5. Hoogenboom, H. R., Griffiths, A. D., Johnson, K. S., Chiswell, D. J., Hudson, P., and Winter, G. (1991) Multi-subunit proteins on the surface of filamentous phage: methodologies for displaying antibody (Fab) heavy and light chains. *Nucleic Acids Res.* **19,** 4133–4137.

6. Bird, R. E., Hardman, K. D., Jacobson, J. W., Johnson, S., Kaufman, B. M., Lee, S. M., et al. (1988) Single-chain antigen-binding proteins. *Science* **242,** 423–426.
7. Huston, J. S., Levinson, D., Mudgett, H. M., Tai, M. S., Novotny, J., Margolies, M. N., et al. (1988) Protein engineering of antibody binding sites: recovery of specific activity in an anti-digoxin single-chain Fv analogue produced in Escherichia coli. *Proc. Natl. Acad. Sci. USA* **85,** 5879–5883.
8. Marks, J. D., Hoogenboom, H. R., Bonnert, T. P., McCafferty, J., Griffiths, A. D., and Winter, G. (1991) By-passing immunization: Human antibodies from V-gene libraries displayed on phage. *J. Mol. Biol.* **222,** 581–597.
9. Sheets, M. D., Amersdorfer, P., Finnern, R., Sargent, P., Lindqvist, E., Schier, R., et al. (1988) Efficient construction of a large nonimmune phage antibody library: the production of high-affinity human single-chain antibodies to protein antigens. *Proc. Natl. Acad. Sci. USA* **95,** 6157–6162.
10. Vaughan, T. J., Williams, A. J., Pritchard, K., Osbourn, J. K., Pope, A. R., Earnshaw, J. C., et al. (1996) Human antibodies with sub-nanomolar affinities isolated from a large non-immunized phage display library. *Nat. Biotechnol.* **14,** 309–314.
11. de Haard, H. J., van Neer, N., Reurs, A., Hufton, S. E., Roovers, R. C., Henderikx, P., et al. (1999) A large non-immunized human Fab fragment phage library that permits rapid isolation and kinetic analysis of high affinity antibodies. *J. Biol. Chem.* **274,** 18,218–18,230.
12. Griffiths, A. D., Williams, S. C., Hartley, O., Tomlinson, I. M., Waterhouse, P., Crosby, W. L., et al. (1994) Isolation of high affinity human antibodies directly from large synthetic reperoires. *EMBO J.* **13,** 3245–3260.
13. McCafferty, J., Griffiths, A. D., Winter, G., and Chiswell, D. J. (1990) Phage antibodies: filamentous phage displaying antibody variable domains. *Nature* **348,** 552–554.
14. Malmborg, A. C., Duenas, M., Ohlin, M., Soderlind, E., and Borrebaeck, C. (1996) Selection of binders from phage displayed libraries using the BIAcore biosensor. *J. Immunol. Methods* **198,** 51–57.
15. Hawkins, R. E., Russell, S. J., and Winter, G. (1992) Selection of phage antibodies by binding afinity: mimicking affinity maturation. *J. Mol. Biol.* **226,** 889–896.
16. Persic, L., Horn, I. R., Rybak, S., Cattaneo, A., Hoogenboom, H. R., and Bradbury, A. (1999) Single-chain variable fragments selected on the 57–76 p21Ras neutralising epitope from phage libraries recognise the parental protein. *FEBS Lett.* **25,** 112–116.
17. Bradbury, A., Persic, L., Werge, T., and Cattaneo, A. (1993) From gene to antibody: the use of living columns to select phage antibodies. *Bio/Technology* **11,** 1565–1569.
18. Mutuberria, R., Hoogenboom, H. R., van der linden, E., de Bruine, A. P., and Roovers, R. C. (1999) Model systems to study the parameters determining the success of phage antibody selections on complex antigens. *J. Immunol. Methods* **231,** 65–81.
19. Cai, X. and Garen, A. (1995) Anti-melanoma antibodies from melanoma patients immunized with genetically modified autologous tumor cells: selection of specific

antibodies from single-chain Fv fusion phage libraries. *Proc. Natl. Acad. Sci. USA* **92,** 6537–6541.
20. Siegel, D. L., Chang, T. Y., Russell, S. L., and Bunya, V. Y. (1997) Isolation of cell surface-specific human monoclonal antibodies using phage display and magnetically-activated cell sorting: applications in immunohematology. *J. Immunol. Methods* **206,** 73–85.
21. Hoogenboom, H. R., Lutgerink, J. T., Pelsers, M. M., Rousch, M. J., Coote, J., Van Neer, N., et al. (1999) Selection-dominant and nonaccessible epitopes on cell-surface receptors revealed by cell-panning with a large phage antibody library. *Eur. J. Biochem.* **260,** 774–784.
22. Palmer, D. B., George, A. J., and Ritter, M. A. (1997) Selection of antibodies to cell surface determinants on mouse thymic epithelial cells using a phage display library. *Immunology* **91,** 473–478.
23. Figini, M., Obici, L., Mezzanzanica, D., Griffiths, A., Colnaghi, M. I., Winter, G., et al. (1998) Panning phage antibody libraries on cells: isolation of human Fab fragments against ovarian carcinoma using guided selection. *Cancer Res.* **58,** 991–996.
24. de Kruif, J., Terstappen, L., Boel, E., and Logtenberg, T. (1995) Rapid selection of cell subpopulation-specific human monoclonal antibodies from a synthetic phage antibody library. *Proc. Natl. Acad. Sci. USA* **92,** 3938–3942.
25. Van Ewijk, W., de Kruif, J., Germeraad, W. T., Berendes, P., Ropke, C., Platenburg, P. P., and Logtenberg, T. (1997) Subtractive isolation of phage-displayed single-chain antibodies to thymic stromal cells by using intact thymic fragments. *Proc. Natl. Acad. Sci. USA* **94,** 3903–3908.
26. Becerril, B., Poul, M.-A., and Marks, J. D. (1999) Towards selection of internalizing antibodies from phage libraries. *Biochem. Biophys. Res. Comm.* **255,** 386–393.
27. Poul, M.-A., Becerril, B., Nielsen, U. B., Morrison, P., and Marks, J. D. (2000) Selection of internalizing human antibodies from phage libraries. *J. Mol. Biol.* **301,** 1149–1161.
28. Pasqualini, R. and Ruoslahti, E. (1996) Organ targeting in vivo using phage display peptide libraries. *Nature* **380,** 364–366.
29. Osbourn, J. K., Derbyshire, E. J., Vaughan, T. J., Field, A. W., and Johnson, K. S. (1998) Pathfinder selection: in situ isolation of novel antibodies. *Immunotechnology* **3,** 293–302.
30. Malmborg, A. C., Soderlind, E., Frost, L., and Borrebaeck, C. A. (1997) Selective phage infection mediated by epitope expression on F pilus. *J. Mol. Biol.* **273,** 544–551.
31. Spada, S. and Pluckthun, A. (1997) Selectively infective phage (SIP) technology: a novel method for in vivo selection of interacting protein-ligand pairs. *Nat. Med.* **3,** 694–696.
32. Schier, R., Bye, J. M., Apell, G., McCall, A., Adams, G. P., Malmqvist, M., et al. (1996) Isolation of high affinity monomeric human anti-c-erbB-2 single chain Fv using affinity driven selection. *J. Mol. Biol.* **255,** 28–43.
33. Schier, R., McCall, A., Adams, G. P., Marshall, K., Yim, M., Merritt, H., et al. (1996) Isolation of picomolar affinity anti-c-erbB2 single-chain Fv by molecular

evolution of the complementarity determining regions in the centre of the antibody combining site. *J. Mol. Biol.* **263,** 551–567.
34. Gussow, D. and Clackson, T. (1989) Direct clone characterization from plaques and colonies by the polymerase chain reaction. *Nucleic Acids Res.* **17,** 4000.
35. Hochuli, E., Bannwarth, W., Dobeli, H., Gentz, R., and Stuber, D. (1988) Genetic approach to facilitate purification of recombinant proteins with a novel metal chelate adsorbent. *Bio/Technology* **6,** 1321–1325.
36. Roberts, B. L., Markland, W., Ley, A. C., Kent, R. B., White, D. W., Guterman, S. K., et al. (1992) Directed evolution of a protein: selection of potent neutrophil elastase inhibitors displayed on M13 fusion phage. *Proc. Natl. Acad. Sci. USA* **89,** 2429–2433.
37. Kang, A. S., Barbas, C. F., Janda, K. D., Benkovic, S. J., and Lerner, R. A. (1991) Linkage of recognition and replication functions by assembling combinatorial antibody Fab libraries along phage surfaces. *Proc. Natl. Acad. Sci. USA* **88,** 4363–4366.
38. Ward, R., Clark, M., Lees, J., and Hawkins, N. (1996) Retrieval of human antibodies from phage-display libraries using enzymatic cleavage. *J. Immnol. Methods* **189,** 73–82.
39. Clackson, T., Hoogenboom, H. R., Griffiths, A. D., and Winter, G. (1991) Making antibody fragments using phage display libraries. *Nature* **352,** 624–628.
40. Schier, R. S. and Marks, J. D. (1996) Efficient in vitro selection of phage antibodies using BIAcore guided selections. *Hum. Antib. hybrid.* **7,** 97–105.
41. de Bruin, R., Spelt, K., Mol, J., Koes, R., and Quattrocchio, F. (1999) Selection of high-affinity phage antibodies from phage display libraries. *Nat. Biotechnol.* **17,** 397–399.
42. Clackson, T. and Wells, J. A. (1994) In vitro selection from protein and peptide libraries. *Trends Biotechnol.* **12,** 173–184.

9

Production of Human Single-Chain Antibodies by Ribosome Display

Mingyue He, Neil Cooley, Alison Jackson, and Michael J. Taussig

1. Introduction

Antibodies are becoming increasingly important for both basic research and clinical applications. With the completion of genome sequencing, antibodies are required for the global analysis and detection of every encoded protein. In therapeutics, a number of antibodies have been used for disease treatment *(1)*.

Traditionally, monoclonal antibodies (MAbs) are produced through rodent immunization followed by hybridoma technology *(2)*. This method is laborious, and also poses difficulties in generating antibodies against self-antigens. Display technology provides an alternative to the hybridoma method for antibody production. By coupling the genotype and phenotype of individual proteins, the display method permits the selection of the genetic material (DNA or RNA) through the binding function of the MAb it encodes. In this in vitro manner, antibodies against self-antigens or those that are not normally immunogenic can also be produced *(3)*. Through repeated cycles of mutation and selection, it is also possible to develop antibodies with improved function. Several display methods have been demonstrated for antibody selection. They include cell-based systems, such as phage display (**ref.** *3*; *see* Chapter 8) and cell-surface display *(4,5)*, and cell-free methods, such as ribosome display *(6–8)* and mRNA display *(9)*.

Ribosome display uses a cell-free system to produce stable protein-ribosome-mRNA (PRM) complexes to provide the linkage of genotype and phenotype, allowing simultaneous selection of a desired protein together with its encoding mRNA. The mRNA is then recovered and amplified as DNA by reverse transcriptase-polymerase chain reaction (RT-PCR). Repeating the display cycle enriches the selected molecules, enabling rare species to be isolated. As a cell-free display system, ribosome display offers a number of advantages

From: *Methods in Molecular Biology, Vol. 248: Antibody Engineering: Methods and Protocols*
Edited by: B. K. C. Lo © Humana Press Inc., Totowa, NJ

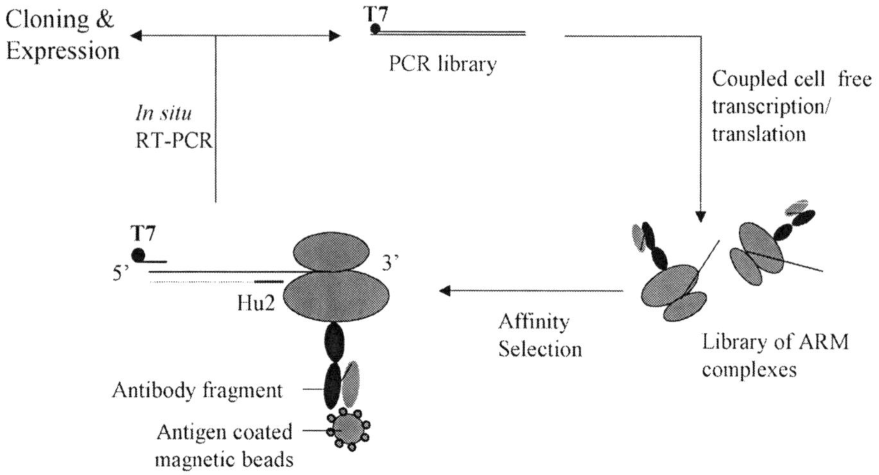

Fig. 1. ARM display cycle. T7: T7 promoter; ARM: antibody-ribosome-mRNA; RT-PCR: coupled reverse transcriptase-polymerase chain reaction. Hu2: a downstream primer for RT-PCR recovery.

over cell-based methods because it displays larger PCR libraries and effectively generates mutations for protein evolution in vitro ***(10)***. In addition, a cell-free system is capable of producing toxic, proteolytically sensitive and unstable proteins, or introducing modified amino acids (such as unnatural amino acids or chemically labeled amino acids) into proteins at desired positions during translation, expanding the range of proteins that may be displayed (*see* http://www.promega.com/techserv/tntbib.html).

Both prokaryotic and eukaryotic ribosome display systems have been developed for antibody generation *(7,8)*. This chapter describes the eukaryotic ribosome display method referred to as ARM (antibody-ribosome-mRNA) display *(8,11–12)*. A distinct feature in ARM display is the use of a novel *in situ* RT-PCR method to recover the genetic information after selection from intact ribosome complexes *(8)*, rendering this method more simple, rapid, and effective in protein selection. **Fig. 1** shows the ARM display cycle.

2. Materials

All the solutions, tubes, and tips used must be sterilized. Reagents should be nuclease-free. Precautions should be taken to avoid DNA contamination. It is recommended that solutions such as primers, RT-PCR buffer, washing buffer, and dNTPs should be stored at –20°C in aliquots.

2.1. Primers

The design of primers is based on de Haard et al. *(13)*, with some modifications for the construction of ribosome display libraries.

Table 1

Upstream primer	Sequence (5′→3′)
(1) $HuV_{H1b\text{-}7a}$/back	CAG(A/G)TGCAGCTGGTGCA(A/G)TCTGG
(2) HuV_{H1c}/back	(G/C)AGGTCCAGCTGGT(A/G)CAGTCTGG
(3) HuV_{H2b}/back	CAG(A/G)TCACCTTGAAGGAGTCTGG
(4) HuV_{H3b}/back	(G/C)AGGTGCAGCTGGTGGAGTCTGG
(5) HuV_{H3c}/back	GAGGTGCAGCTGGTGGAG(A/T)C(C/T)GG
(6) HuV_{H4b}/back	CAGGTGCAGCTACAGCAGTGGGG
(7) HuV_{H4c}/back	CAG(G/C)TGCAGCTGCAGGAGTC(G/C)GG
(8) HuV_{H5b}/back	GA(A/G)GTGCAGCTGGTGCAGTCTGG
(9) HuV_{H6a}/back	CAGGTACAGCTGCAGCAGTCAGG
Downstream primer	**Sequence (5′→3′)**
(1) HuIgM linker/for	GGA GAC GAG GGG GAA AAG GGT TGG

2.1.1. Primers for Generating V_H-linker Fragments

V_H-linker fragments that encode various Kabat V_H families ($V_{H1–7}$) are generated by PCR using one of the upstream primers in combination with the downstream primer (**Table 1**).

2.1.2. Primers for Generating κ Light Chain

κ chains of different Kabat families ($V_{\kappa 1–6}$) are produced using one of the upstream primers together with the downstream primer (**Table 2**). The underlined region indicates the sequence overlapping with HuIgM linker/for (*see* **Table 1**) for PCR assembly of κ chain with V_H-linker fragment to generate V_H/K format (**Fig. 3**). Underlined italics indicate the *Xba*I site for cloning.

2.1.3. Primers for Generating V_λ Fragments

V_λ fragments of different Kabat families ($V_{\lambda 1–9}$) are produced using one of the upstream primers together with the downstream primer (**Table 3**). The underlined region in the upstream primer indicates the sequence overlapping the HuIgM linker/for (*see* **Table 1**) for assembly with V_H-linker by PCR (**Fig. 3**). The underlined italics in the downstream primer indicate the overlapping sequence for assembly with constant region (C_κ) of the κ light chain (**Fig. 3**).

2.1.4. Primers for Generating Full-Length Construct

The full-length ribosome display construct is generated by PCR using an upstream primer, which contains a T7 promoter and a translation initiation signal (Kozak sequence), in combination with a downstream primer (**Table 4**).

Table 2

Upstream primer	Sequence (5′→3′)
(1) HuV$_{\kappa 1b}$/back	CTTTTCCCCCTCGTCTCCGACATCCAG(A/T)TGACCCAGTCTCC
(2) HuV$_{\kappa 2}$/back	CTTTTCCCCCTCGTCTCCGATGTTGTGATGACTCAGTCTCC
(3) HuV$_{\kappa 3b}$/back	CTTTTCCCCCTCGTCTCCGAAATTGTG(A/T)TGAC(A/G)CAGTCT CC
(4) HuV$_{\kappa 4b}$/back	CTTTTCCCCCTCGTCTCCGATATTGTGATGACCCACACTCC
(5) HuV$_{\kappa 5}$/back	CTTTTCCCCCTCGTCTCCGAAACGACACTCACGCAGTCTCC
(6) HuV$_{\kappa 6}$/back	CTTTTCCCCCTCGTCTCCGAAATTGTGCTGACTCAGTCTCC

Downstream primer	Sequence (5′→3′)
HuC$_{\kappa}$/for	G*CTCTAGA* ACACTCTCCCCTGTTGAAGCT

Table 3

Upstream primer	Sequence (5′→3′)
HuV$_{\lambda\ 1a}$/back	CTTTTCCCCCTCGTCTCCCAGTCTGTGCTGACTCAGCCACC
HuV$_{\lambda\ 1b}$/back	CTTTTCCCCCTCGTCTCCCAGTCTGTG(C/T)TGACGCAGCCGCC
HuV$_{\lambda\ 1c}$/back	CTTTTCCCCCTCGTCTCCCAGTCTGTCGTGACGCAGCCGCC
HuV$_{\lambda\ 2}$/back	CTTTTCCCCCTCGTCTCCCA(A/G)TCTGCCCTGACTCAGCCT
HuV$_{\lambda\ 3a}$/back	CTTTTCCCCCTCGTCTCCTCCTATG(A/T)GCTGACTCAGCCACC
HuV$_{\lambda\ 3b}$/back	CTTTTCCCCCTCGTCTCCTCTTCTGAGCTGACTCAGGACCC
HuV$_{\lambda\ 4}$/back	CTTTTCCCCCTCGTCTCCCACGTTATACTGACTCAACCGCC
HuV$_{\lambda\ 5}$/back	CTTTTCCCCCTCGTCTCCCAGGCTGTGCTGACTCAGCCGTC
HuV$_{\lambda\ 6}$/back	CTTTTCCCCCTCGTCTCCAATTTTATGCTGACTCAGCCCCA
HuV$_{\lambda\ 7-8}$/back	CTTTTCCCCCTCGTCTCCCAG(A/G)TCGTGGTGAC(C/T)CAGGAGCC
HuV$_{\lambda\ 9}$/back	CTTTTCCCCCTCGTCTCCC(A/T)GCCTGTGCTGACTCAGCC(A/C)CC

Downstream primer	Sequence (5′→3′)
HuJ$_{\lambda}$/for	*GACAGATGGTGCAGCCAC*ACCTA(A/G)(A/G)ACGGTGAGCTTGGTCCC

Table 4

Upstream primer	Sequence (5′→3′)
T7Ab/back	GCAGC*TAATACGACTCACTATAGGAACAGA***CCACC*<u>ATG</u>*** (C/G)AGGT(G/C)CA(G/C)*<u>CTCGAG</u>*(C/G)AGTCTGG

Downstream primer	Sequence (5′→3′)
HuC$_\kappa$/for	GC*<u>TCTAGA</u>* ACACTCTCCCCTGTTGAAGCT

The T7 promoter sequence is italicized. The Kozak sequence is indicated in bold. The initiation codon **ATG** is underlined. Underlined italics indicate restriction sites for cloning.

Table 5

Upstream primer	Sequence (5′→3′)
HuC$_\kappa$/back	ACTGTGGCTGCACCATCTG

Downstream primer	Sequence (5′→3′)
HuC$_\kappa$/for	GC*<u>TCTAGA</u>* ACACTCTCCCCTGTTGAAGCT

Underlined italics indicate a restriction site for cloning

2.1.5. Primers for Generating Constant Region (C$_\kappa$) of κ Light Chain

See **Table 5**.

2.1.6. Primers for in situ RT-PCR Recovery

Since the stalled ribosome occupies the 3′ end of the translated and selected mRNA, a downstream primer (Hu 2), which hybridizes at about 100 nt upstream of the 3′ end of the mRNA, is used to recover the genetic information by RT-PCR in combination with the upstream primer T7A1 (see **Table 6** and **Fig. 2**). As the use of Hu 2 produces a shortened DNA fragment in the first cycle, another primer Hu 3, which hybridizes upstream of Hu 2, is used in the second cycle. Similarly in the third cycle, a primer Hu 4 hybridizing upstream of Hu 3 is used **(Fig. 2)**. The DNA recovered in this way thus becomes progressively shorter in each cycle, but the full-length fragment can be regenerated in any cycle by overlapping PCR. The shortening only affects the constant domain of the κ light chain.

2.2. Molecular Biology Kits and Reagents

1. mRNA purification kit (Amersham Biosciences; Cat. #270-9255-01).

Table 6

Upstream primer	Sequence (5′→3′)
T7A1	GCAGC*TAATACGACTCACTATAGGAACAGA***CCACCATG**

Downstream primer	Sequence (5′→3′)
Hu 2	GCTCAGCGTCAGGGTGCTGCT
Hu 3	CTC TCCTGGGAGTTACC
Hu 4	GAAGACAGATGGTGCAGC

Italics indicate the T7 promoter. Kozak sequence and initiation codon ATG are in bold.

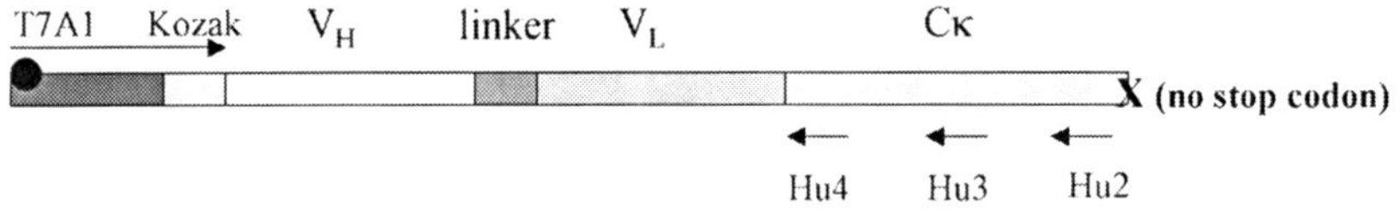

Fig. 2. A general antibody construct for ARM ribosome display. V_H: variable region of antibody heavy chain. V_L: variable region of antibody light chain. Cκ: constant region of human κ light chain. Hu2, Hu3, and Hu4 are downstream primers for *in situ* RT-PCR recovery.

2. Titan™ one tube RT-PCR system (Roche; Cat. #1888 382).
3. TOPO TA cloning kit (Invitrogen; Cat. #K4500-01).
4. Qiagen QIAEX II gel extraction kit (Qiagen; Cat. #20021).
5. QIAquick PCR purification kit (Qiagen; Cat. #28104).
6. TNT T7 quick for PCR DNA (Promega; Cat. #L5540).
7. Expand HiFi DNA polymerase (Expand™ high-fidelity PCR system: Roche; Cat. #1732 641).
8. Taq DNA polymerase (Qiagen; Cat. #201203).
9. AMV reverse transcriptase (Promega; Cat. #M5101).
10. 25 m*M* dNTPs: mix equal volumes of each 100 m*M* dNTP stock solution (Amersham Biosciences).
11. 100 m*M* DTT (Titan™ one tube RT-PCR system: *see* item 2).
12. Tosyl-activated Dynabeads M-280, 6.7 × 10^8/mL or 10 mg/mL (Dynal Biotech; Cat. #142.03).
13. Dynabeads M-280 streptavidin, 6.5 × 10^8/mL or 10 mg/mL (Dynal Biotech; Cat. #112.05/06).
14. RNase-free DNase I (Roche; Cat. #776 785).
15. Agarose.
16. 0.5-mL siliconized RNase-free microfuge tubes (Ambion; Cat. #12350).
17. Sterilized, diethylpyrocarbonate (DEPC)-treated distilled water (ddH_2O): autoclaved Milli-Q water containing 0.1% (v/v) DEPC.

2.3. Solutions

1. One-tube RT-PCR **Solution 1** (per 100 µL):

100 m*M* DTT (from Titan™ kit)	10 µL
10 m*M* dNTPs	4 µL
Upstream primer (16 µ*M*)	6 µL (*see* **Tables 1–5**)
Downstream primer (16 µ*M*)	6 µL (*see* **Tables 1–5**)
ddH_2O	74 µL

   Store at –20°C.
2. One-tube RT-PCR **Solution 2** (per 96 µL):

5X RT-PCR buffer (from Titan™ kit)	40 µL
ddH_2O	56 µL

   Store at –20°C.
3. Buffer A: 0.1 *M* sodium phosphate buffer, pH 7.4.
4. Buffer D: buffer A containing 0.1% bovine serum albumin (BSA).
5. Buffer E: 0.2 *M* Tris-HCl, pH 8.5, containing 0.1% BSA.
6. Phosphate-buffered saline (PBS): 1.63 m*M* Na_2HPO_4, 1.47 m*M* KH_2PO_4, 2.68 m*M* KCl, and 136 m*M* NaCl, pH 7.4.
7. EZ-link™ sulfo-NHS-LC-LC-biotin (Pierce; Cat. #21338). Solution is freshly made at a concentration of 1 mg/mL in water.
8. Antigen solution (0.5–1 mg/mL) in PBS.
9. 2X dilution buffer: PBS containing 10 m*M* magnesium acetate, stored at 4°C.
10. Washing buffer: PBS containing 0.1% Tween 20 and 5 m*M* magnesium acetate, stored at 4°C.
11. 10X DNase I digestion buffer: 400 m*M* Tris-HCl, pH 7.5, 60 m*M* $MgCl_2$, 100 m*M* NaCl, autoclaved and stored at 4°C.
12. 10% sodium azide.

3. Methods

3.1. Antibody Library Construction

Single-chain antibody libraries in the format of V_H/K and V_H/V_λ-C_κ (*see* **Note 1**) are constructed for ribosome display (**Fig. 3**). V_H/K is generated by linking the heavy-chain variable domain (V_H) to the complete κ chain. V_H/V_λ-C_κ is made by linking V_H to a fusion of variable region of λ chain (V_λ) and constant region of κ chain (C_κ). The heavy-chain "elbow" region, which is a continuation of the V_H domain, is used as the peptide linker to join heavy chains with light chains ***(8,11)***. The use of the "elbow" as a linker simplifies the process of library construction and avoids the introduction of nonhuman sequences. A T7 promoter and Kozak sequence are placed upstream to allow protein synthesis in the rabbit reticulocyte lysate. To generate stable ribosome complex, the stop codon is removed from the DNA using a primer lacking the stop codon ***(7,8)***. The C_κ domain provides a spacer region to allow an antibody fragment to be displayed on the surface of ribosome and to offer priming sites for RT-PCR recovery after selection.

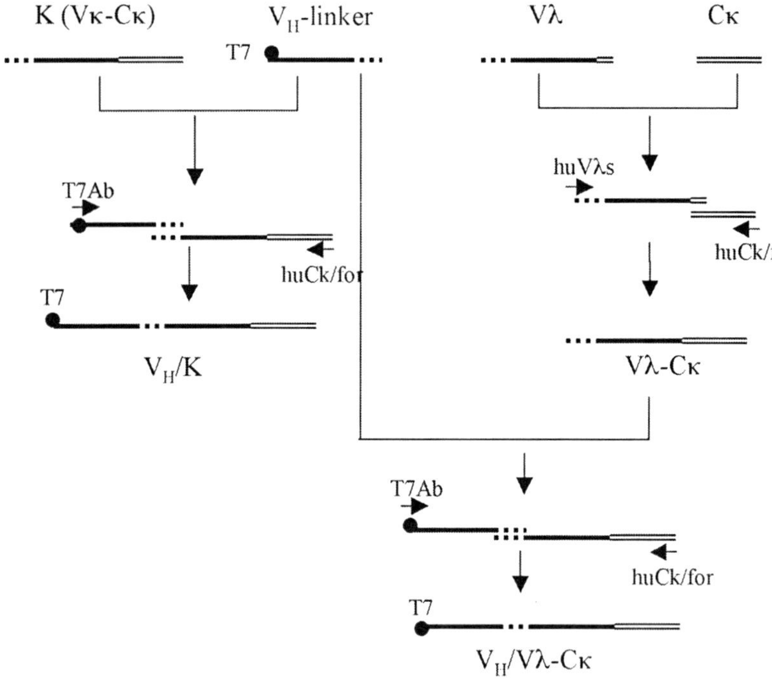

Fig. 3. A strategy for the construction of V_H/K and V_H/V_λ-C_κ PCR libraries from mRNA. The primer details are listed in **Subheading 2.1.**

1. Isolate total mRNA from human peripheral blood lymphocytes (PBLs) using the mRNA purification kit according to manufacturer's instructions.
2. Generate individual V_H-linker, κ, V_λ, and C_κ fragments by one-tube RT-PCR using the primers described under **Subheading 2.1.** Set up the one-tube RT-PCR mixture as follows:

Solution 1 (*see* **Subheading 2.3.**)	24 µL
Solution 2 (*see* **Subheading 2.3.**)	24 µL
Enzyme mix (from Titan™ kit)	1 µL (*see* **Note 2**)
mRNA	1 µL (10–50 ng).

A negative RT-PCR control (10 µL) is also set up without adding mRNA. Carry out RT-PCR thermal cycling: 1 cycle of 48°C for 45 min, followed by 94°C for 2 min; then 35 cycles of: 94°C for 30 s, 54°C for 1 min, 68°C for 2 min; finally, 1 cycle of 68°C for 7 min, then hold at 10°C.

3. Analyze 5–10 µL of the RT-PCR products by electrophoresis on a 1–2% agarose gel containing 0.5 µg/mL ethidium bromide. Purify individual DNA fragments of appropriate size (the expected size for V_H is ~400 bp, κ light chain should be ~700 bp, and V_λ is ~350 bp) using the QIAEX II gel extraction kit.
4. Assemble the purified DNA fragments using PCR (*see* **Fig. 3**):
   a. To generate the random assembly of different V_H and light chains, individual fragments generated from family-specific primers (*see* **Tables 1–3**) are mixed in

equal amounts to form pooled V_H-linker fragments, pooled V_λ fragments, and pooled κ chains.

b. V_H/K is generated by PCR assembly of the V_H-linker pool and κ chain pool through the overlapping sequence between the two fragments.

c. V_H/V_λ-C_κ is produced in two assembly steps: first, the V_λ pool is fused to C_κ to form V_λ-C_κ, then V_λ-C_κ is fused to the V_H-linker pool to produce V_H/V_λ-C_κ.

5. Set up the reactions as follows:

DNA fragment 1 (*see* **Fig. 3**)	5–25 ng
DNA fragment 2 (*see* **Fig. 3**)	5–25 ng
10X PCR buffer (from the Qiagen Taq kit)	2.5 µL
5X Q solution (from the Qiagen Taq kit)	5 µL
2.5 m*M* dNTPs	1 µL
Taq DNA polymerase	1 U
ddH_2O to final volume	25 µL.

Carry out 8 cycles of 94°C for 1 min; 54°C for 1 min; 72°C for 1 min.

6. Add 2 µL of the previous reaction to the following mixture for PCR amplification:

10X PCR buffer	5 µL
5X Q solution	10 µL
2.5 m*M* dNTPs	4 µL
16 µ*M* of upstream primer	1.5 µL
16 µ*M* of downstream primer	1.5 µL
Taq DNA polymerase	2.5 U
Distilled H_2O to final volume	48 µL.

Carry out 30 cycles of thermal cycling: 94°C for 30 s, 54°C for 1 min, 72°C for 1 min; finally, 1 cycle of 72°C for 7 min, then hold at 10°C.

7. Analyze the PCR products by agarose gel electrophoresis as described in **step 3.**
8. Confirm integrity of the assembly product by PCR mapping. Elute the PCR product from the gel using the QIAEX II gel extraction kit and then carry out PCR using combinations of primers, which allow diagnosis of product integrity (*see* **Note 3**).
9. Analyze diversity of the libraries by PCR cloning using the TOPO TA cloning kit, followed by DNA fingerprinting ***(14)*** and sequencing of individual clones (*see* **Note 4**).
10. Store the libraries for subsequent selections (*see* **Note 5**).

3.2. Preparation of Antigen-Coupled Magnetic Beads

Antigen-coupled Dynabeads are used to capture specific ribosome complexes. Protein antigens can be linked either directly onto tosyl-activated Dynabeads or indirectly onto streptavidin Dynabeads through biotinylation.

3.2.1. Coupling of Antigen to Tosyl-Activated Dynabeads

1. Wash Dynabeads 3× with 0.5 mL of buffer A using a magnetic particle concentrator (Dynal MPC). A convenient amount to prepare is 0.5–1 mg (50–100 µL) of beads.
2. Mix the beads with antigen in proportions of 20 µg antigen: 1 mg beads (100 µL) and vortex for 1 min. Incubate at 37°C for 16–24 h with slow rotation.

3. Remove the supernatant and wash the beads twice with 0.5 mL of buffer D at 4°C for 5 min, then treat with 0.5 mL of buffer E at 37°C for 4 h to block free tosyl groups.
4. Wash once with 0.5 mL of buffer D at 4°C for 5 min and resuspend in the original volume (100 μL, *see* **step 2**) in buffer D containing 0.02% sodium azide. The antigen-coupled beads are ready for use, and can be stored for 3–4 mo at 4°C.

3.2.2. Coupling of Biotinylated Proteins to Streptavidin Dynabeads

1. Mix antigen solution in PBS with sulfo-NHS-biotin solution (1 mg/mL) in proportions of 25 μg antigen: 1 μg sulfo-NHS-biotin. Incubate the mixture at room temperature for 30 min. Remove unreacted sulfo-NHS-biotin by dialysis against two 500-mL changes of PBS at 4°C overnight. The biotinylated protein can be stored at 4°C until use.
2. Wash 20 μL of streptavidin Dynabeads M-280 3× with buffer A as in **Subheading 3.2.1., step 1** and resuspend in 20 μL of PBS. Add 2 μg of biotinylated protein to the streptavidin Dynabeads M-280 in the proportions of 10 μg biotinylated protein: 1 μg beads. Incubate the mixture at room temperature for 30 min. Remove the supernatant and wash beads 3× with 50 μL of PBS. Finally, resuspend in the original volume (20 μL) of buffer D containing 0.02% sodium azide. Beads may be stored at 4°C for 3–4 mo.

3.3. ARM Ribosome Display

The PCR libraries are expressed in a coupled rabbit reticulocyte lysate (TNT) system to generate ARM complexes. The amount of library needed for display depends on the antibody affinity to be selected: antibodies with higher affinity are more sensitively recovered, and thus require less input DNA. More DNA is required for low-affinity antibodies. Repeated cycles can enrich rare species from a library. The effective display size is determined by the number of active ribosomes which, in the TNT mix, are present at approx 10^{14}/mL (supplier's information). As 1 μg of 1 kb DNA contains 9.1×10^{11} molecules, up to 5 μg of PCR library can be used by the procedure described here. This system can be scaled up for a larger library display (*see* **Note 6**).

1. Set up in vitro coupled transcription/translation to generate ribosome complexes. The following is a standard 50-μL mixture:

TNT T7 Quick for PCR DNA	40 μL
PCR DNA (from **Subheading 3.1., step 7**)	up to 5 μg
1 m*M* methionine (from TNT kit)	1 μL
100 m*M* magnesium acetate	0.5 μL (see **Note 7**)
ddH_2O	to 50 μL

   Incubate at 30°C for 60 min.
2. DNase I digestion to remove the input PCR fragment: to the mixture, add 40 U of DNase I together with 6 μL of 10X DNase I digestion buffer and ddH_2O to a final volume of 60 μL. Incubate the mixture at 30°C for an additional 15 min.

3. Dilute the mixture with 60 μL of cold 2X dilution buffer.
4. Add 1–2 μL of antigen-linked beads to the mixture and incubate at 4°C for 1–2 h with gentle shaking or vibration.
5. Wash the beads 5× with 100 μL of cold washing buffer, followed by two washes with 100 μL of cold ddH_2O.
6. Resuspend the selected beads in 10 μL of ddH_2O for RT-PCR.
7. Carry out the *in situ* RT-PCR recovery as follows:
   a. Set up the RT-PCR mixture:

Solution 1 (*see* **Subheading 2.3.**)	50 μL
Solution 2 (*see* **Subheading 2.3.**)	48 μL
Enzyme mix	2 μL (*see* **Note 2**)

   b. Dispense 10–20 μL of the previous mixture (*see* **Note 8**) into tubes for each individual sample and then add 2 μL of the selected beads. Carry out thermal cycling as follows: 1 cycle of 48°C for 45 min, followed by 94°C for 2 min; then 30–40 cycles of 94°C for 30 s, 54°C for 1 min, and 68°C for 2 min; finally, 1 cycle of 68°C for 7 min, then hold at 10°C.
8. Analyze the RT-PCR products by agarose electrophoresis as described in **step 3** (*see* **Notes 9** and **10**).
9. Recover the DNA fragments from the agarose gel using Qiagen gel extraction kit and re-amplify the DNA product by PCR for further manipulations such as repeating the display cycle or cloning and *E. coli* expression (*see* **Note 11**).

4. Notes

1. Although only the formats V_H/K and V_H/V_λ-C_κ are described in this chapter, the method is in principle equally applicable to other forms of single-chain antibodies, provided that a spacer domain is present at the C-terminus. The spacer should contain at least 30 amino acid residues to allow the V_H/V_L domain to be exposed on the ribosome surface.
2. The one tube RT-PCR can be carried out with comparable efficiency using AMV reverse transcriptase and Expand HiFi DNA polymerase in combination with the Titan™ RT-PCR buffer. For example, to a 50-μL RT-PCR reaction, 4–5 U (0.5 μL) of AMV reverse transcriptase and 2 U (0.5 μL) of Expand HiFi polymerase are added to the mixture.
3. PCR mapping is carried out by using combination of primers from different positions in the construct. If all PCR fragments show the expected size, it indicates that the correct construct has been obtained.
4. In general, library diversity is examined by analysis of individual clones using DNA fingerprinting first and then DNA sequencing. DNA fingerprinting is a simple and rapid way to analyze unique clones. Clones that show different restriction patterns should have different sequences. We usually sequence 50 clones to analyze the distribution of various gene families and lengths of CDR3.
5. The PCR libraries are usually stored in ddH_2O at –20°C for routine use. For long-term storage, ethanol-precipitate the DNA and store dry at –20°C.

6. The volume of TNT mixture used for ribosome display can be scaled up to 100 µL or down to 20 µL without significant reduction in recovery efficiency.
7. Magnesium acetate concentration added to TNT mixture during translation affects ARM generation and recovery. We have shown that antibodies can be more efficiently displayed and recovered with magnesium concentrations ranging from 0.5–2 m*M*.
8. PCR and RT-PCR can be performed in 10–100 µL according to applications. A negative control lacking a template should be always included in every experiment to evaluate DNA or mRNA contamination.
9. Nonspecific DNA recovery can occur for a number of reasons, including carryover of input DNA (despite DNase I digestion), nonspecific binding of ARMs or mRNA to beads through the washing procedure, or inadequate coating and blocking of the Dynabeads. PCR (RT-omitted) of the Dynabeads after selection can be used to detect contamination from input DNA. If necessary, further DNase I digestion on the Dynabeads may then be carried out to remove DNA completely. Increasing the number of washes, particularly using ddH_2O, can reduce background nonspecific binding of the complexes to beads.
10. Sometimes no RT-PCR product is detected after ribosome display cycle. This may merely indicate that the species being selected is rare and that further cycles are required. However, other factors can account for lack of recovery, including the efficiency of RT-PCR, DNA transcription/translation, RNA degradation or functionality of the nascent protein. The use of a standard internal control mRNA can monitor RT-PCR efficiency. It is possible to add RNase inhibitors to the washing buffer to prevent mRNA degradation when necessary.
11. The number of cycles required to detect enrichment depends on the nature of the antigen and library diversity. Enrichment usually refers to the relative increase of a target molecule to background after selection. It can be estimated by comparing the ratios of input DNA and recovered DNA in each cycle. Generally, 3–5 cycles should be sufficient to enrich a target from a library.

References

1. van Dijk, M. A. and van de Winkel, J. G. (2001) Human antibodies as next generation therapeutics. *Curr. Opin. Chem. Biol.* **5,** 368–374.
2. Kohler, G. and Milstein, C. (1975) Continuous cultures of fused cells secreting antibody-antigen interactions. *Nature* **256,** 495–497.
3. Winter, G., Griffiths, A. D., Hawkins, R. E., and Hoogenboom, H. R. (1994) Making antibodies by phage display technology. *Annu. Rev. Immunol.* **12,** 433–455.
4. Georgiou, G., Stathopoulos, C., Daugherty, P. S., Nayak, A. R., Iverson, B. L., and Curtiss, R. III (1997) Display of heterologous proteins on the surface of microorganisms: from the screening of combinatorial libraries to live recombinant vaccines. *Nat. Biotechnol.* **15,** 29–34.
5. Shusta, E. V., VanAntwerp, J., and Wittrup, K. D. (1999) Biosynthetic polypeptide libraries. *Curr. Opin. Biotechnol.* **10,** 117–122.
6. Mattheakis, L. C., Bhatt, R. R., and Dower, W. J. (1994) An *in vitro* polysome display system for identifying ligands from very large peptide libraries. *Proc. Natl. Acad. Sci. USA* **91,** 9022–9206.

7. Hanes, J. and Plükthun, A. (1997) *In vitro* selection and evolution of functional proteins by using ribosome display. *Proc. Natl. Acad. Sci. USA* **94,** 4937–4942.
8. He, M. and Taussig, M. J. (1997) Antibody-ribosome-mRNA (ARM) complexes as efficient selection particles for *in vitro* display and evolution of antibodies combining sites. *Nucleic Acids Res.* **24,** 5132–5134.
9. Roberts, R. W. and Szostak, J. W. (1997) RNA-peptide fusions for the *in vitro* selection of peptides and proteins. *Proc. Natl. Acad. Sci. USA* **94,** 12,297–12,302.
10. Hanes, J., Jermutus, L., Weber-Bornhauser, S., Bosshard, H. R., and Plükthun, A. (1998) Ribosome display efficiently selects and evolves high-affinity antibodies *in vitro* from immune libraries. *Proc. Natl. Acad. Sci. USA* **95,** 14,130–14,135.
11. He, M., Menges, M., Groves, M. A., Corps, E., Liu, H., Brüggemann, M., et al. (1999) Selection of a human anti-progesterone antibody fragment from a transgenic mouse library by ARM ribosome display. *J. Immunol. Methods* **231,** 105–117.
12. Taussig, M. J., Groves, M.A.T., Menges, M., Liu, H., and He, M. (1999) ARM complexes for *in vitro* display and evolution of antibody combining sites, in *Monoclonal Antibodies: A Practical Approach* (Shepherd, P. and Dean, C., eds.), Oxford University Press, pp. 91–109.
13. de Haard, H. J., van Neer, N., Reurs, A., Hufton, S. E., Roovers, R. C., Henderikx, P., et al. (1999) A large non-immunized human Fab fragment phage library that permits rapid isolation and kinetic analysis of high affinity antibodies. *J. Biol. Chem.* **274,** 18,218 18,230.
14. Marks, J. D., Hoogenboom, H. R., Bonnert, T. P., McCafferty, J., Griffiths, A. D., and Winter, G. (1991) By-passing immunization: Human antibodies from V-gene libraries displayed on phage. *J. Mol. Biol.* **222,** 581–597.

10

Production of Human Antibodies from Transgenic Mice

C. Geoffrey Davis, Xiao-Chi Jia, Xiao Feng, and Mary Haak-Frendscho

1. Introduction

Mice transgenic for human immunoglobulin genes offer the opportunity to investigators to derive fully human antibodies using well-established hybridoma technology. Several different strains of such mice are available, and can be obtained under certain conditions from various commercial entities *(1–3)*. The brand names of these mice include XenoMouse®, HuMAb® Mouse, TC™ mouse, and KM™ mouse. The leading transgenic mice all share certain properties. These include the following:

1. the endogenous murine heavy-chain locus has been inactivated by knocking out either all of the J_H genes or the C_μ gene *(4)*;
2. the endogenous murine kappa light-chain locus has been inactivated by knocking out the $J\kappa$ and/or the C_κ genes;
3. the endogenous murine lambda light-chain locus remains intact and functional;
4. the human heavy-chain transgene contains, at minimum, the majority of the human V_H genes, all the D genes, all the J_H genes, C_μ, C_δ, and at least one C_γ gene, together with regulatory elements; and
5. the human kappa light-chain transgene contains the majority of the human V_κ genes, all the J_κ genes, and the C_κ gene, together with regulatory elements.

The human transgenes have been shown to be fully compatible with the murine recombination machinery *(5–6)*. There is normal assembly of recombined V regions through the recombination of, in the case of the heavy-chain locus, human V_H, D, and J_H genes and, in the case of the light chain locus, human V_κ and J_κ genes. In both cases, N addition—i.e., the insertion of random nucleotides—occurs at the junctions. Yet, as is the case with humans, this

From: *Methods in Molecular Biology, Vol. 248: Antibody Engineering: Methods and Protocols*
Edited by: B. K. C. Lo © Humana Press Inc., Totowa, NJ

is more predominant in rearranged heavy-chain genes than in rearranged light-chain genes. Because it has access to multiple genetic elements and because it uses normal N addition, the recombination process in those human immunoglobulin transgenic mice with a majority of the human V-gene repertoire generates tremendous antibody diversity, comparable to that of humans.

Further, the human heavy-chain loci undergo normal class switching ***(1,6)***. Thus, in the course of B-cell development, surface immunoglobulin expression proceeds from co-expression of IgM and IgD to expression of a downstream heavy-chain class. The available selection of downstream heavy-chain classes depends upon the strain of mouse. The heavy-chain loci of the KM™ mice and TC™ mice have been introduced intact as minichromosomes. Thus, all human antibody classes and isotypes of heavy chains are available for class switching. However, it should be noted that mature IgM and IgA antibodies—in addition to heavy and light chains—also incorporate the J chain, which is encoded on a different chromosome and is of murine origin in all the transgenic mouse strains. IgA antibodies also can carry the secretory component, which also would be of murine origin in these transgenic mouse strains.

The XenoMouse® label actually applies to a collection of mouse strains. In all the XenoMouse® strains, the heavy-chain locus has been integrated into the mouse genome in the context of a yeast artificial chromosome (YAC) rather than as a minichromosome. These YACs have been engineered with the intention of constraining class switching from IgM to a single IgG isotype. Thus, there are individual XenoMouse® strains for the three IgG isotypes considered to be useful for therapeutics—IgG1, IgG2, and IgG4. The strain of transgenic mouse used influences the selection of screening reagents and the application of their antibodies.

With regard to light-chain composition, the kappa light-chain loci in all the available transgenic mouse strains are similar, and merit no special strain-specific considerations in screening. However, it is important to be aware that the murine lambda locus remains intact. As a result, the formation of human-mouse hybrid antibodies containing a human heavy chain and a murine lambda light chain can and does occur. The frequency of these hybrid antibodies is typically similar to that of $Ig\lambda^+$ B cells in normal mice, about 5%. The presence of such hybrids, although minor, mandates the incorporation of a screen that is specific for human kappa light chains to ensure the recovery of fully human antibodies.

Certain strains of human antibody transgenic mice, such as the XenoMouse®-KL strains, either in addition to or in place of a human kappa light-chain transgene, contain a human lambda light-chain construct. These mice also merit special consideration in the screening of their antibodies and, for this reason, a screen for human lambda light chains is included.

Investigators experienced in the generation of murine monoclonal antibodies (MAbs) will note that many of the protocols provided here, such as those for antigen preparation and for hybridoma fusion, can be equally well applied to either normal mice or human antibody transgenic mice. Our approach in writing this chapter has been to provide a start-to-finish manual containing all the steps necessary for successfully using transgenic mice to derive fully human antibodies. Although the techniques we have detailed were developed specifically in conjunction with the XenoMouse® technology, we have every reason to believe that they should be applicable to other strains of human antibody transgenic mice. Typically, the supplier will also provide standard protocols when shipping the mice. These protocols should be examined, and any variations from the techniques described in the following section should be carefully considered.

2. Materials

2.1. Preparation of Whole Cells for Immunization

1. Sterile phosphate-buffered saline (PBS): 8 g NaCl, 0.2 g KCl, 1.13 g Na_2HPO_4, 0.2 g KH_2PO_4, add distilled H_2O to dissolve, adjust pH to 7.2 with 1 N HCl; q.s. to 1 L with distilled H_2O.
2. Sterile microfuge tubes.

2.2. Immunization

1. PBS.
2. Complete Freund's adjuvant (Sigma; Cat. #F-5881).
3. Incomplete Freund's adjuvant (Sigma; Cat. #F-5506).
4. 70% ethanol.

2.3. Preparation of Lymphoid Cells

1. Dulbecco's Modified Eagle's Medium (DMEM) (4,500 mg/L glucose, 110 mg/L sodium pyruvate, no L-glutamine) (JRH Biosciences; Cat. #51444-79P), sterile-filtered.
2. Hybridoma medium: 15% fetal bovine serum (FBS) (qualified, heat-inactivated, hybridoma-tested) (Hyclone; Cat. #SH30070.03), 1% 200 m*M* L-glutamine (Sigma; Cat. #G2150), 1% 100X Non-essential amino acid solution for MEM (Sigma; Cat. #M 7145), 1% 100X Pen/Strep (10,000 U/mL penicillin + 10 mg/mL streptomycin) (Sigma; Cat. #P7539), 10 U/mL IL-6 (stock at 200,000 U/mL) (Boehringer Mannheim; Cat. #1299972), 1 vial/L OPI media supplement (oxaloacetate, pyruvate, bovine insulin) (Sigma; Cat. #O5003).
3. Hybridoma fusion medium (1/2 HA): 1 L of 50x HA (Sigma, Cat. #A9666), 1 L of Hybridoma medium.
4. Sterile, conical 50-mL centrifuge tubes.
5. ACK Lysis solution: 0.15 *M* NH_4Cl, 1.0 m*M* $KHCO_3$, 0.1 m*M* Na_2EDTA, sterile-filtered with 0.22-μm filter.
6. 70% ethanol.
7. Cell strainers, 70 μm (Falcon; Cat. #352350).
8. Pyrex brand tissue grinders (Fisher; Cat. #08-414-10C).
9. Water-jacketed incubator (Forma Scientific; 37°C, 10% CO_2).

2.4. Separation of B Cells

1. DMEM, sterile-filtered.
2. Mouse $CD90^+$ magnetic beads (Miltenyi Biotech; Cat. #491-01).
3. LS^+ column (Miltenyi Biotech; Cat. #424-01).

2.5. Cell Fusion

1. DMEM, sterile-filtered.
2. P3 or other myeloma cells—e.g., P3X63Ag8.653 (American Tissue Type Collection; Cat. #CRL-1580).
3. Hybridoma fusion medium (1/2 HA; *see* **Subheading 2.3.**).
4. PEG solution: 50% PEG 1,500 (Boehringer Mannheim; Cat. #783 641).
5. Sterile conical 15-mL centrifuge tubes.
6. Sterile 96-well tissue-culture plates (Costar; Cat. #3799).

2.6. Primary Screening for Antigen Binding

1. PBS.
2. Positive/negative control antibodies (antigen-specific).
3. Goat anti-hIgGFc horseradish peroxidase (HRP) (Caltag; Cat. #H10507).
4. Plate-coating buffer: 0.1 *M* carbonate buffer, pH 9.6 (8.4 g/L $NaHCO_3$).
5. Blocking buffer: 0.5% bovine serum albumin (BSA), 0.1% Tween 20, 0.01% Thimerosal in PBS.
6. 3,3′,5,5′-Tetramethylbenzidine (TMB) solution (BioFX Lab; Cat. #TMSK-0100-01).
7. TMB stop solution (BioFX Lab; Cat. #STPR-0100-01).
8. Washing buffer: 0.05% Tween 20 in PBS.
9. Hybridoma medium (*see* **Subheading 2.3.**).
10. 96-well enzyme-linked immunosorbent assay (ELISA) plates (Fisher; Cat. #12-565-136).
11. Sterile 24-well tissue-culture plates (Costar; Cat. #3526).
12. Streptavidin-coated plates (Fisher, Cat. #07-200-766 or #07-201-146).
13. ELISA plate reader, SPECTRAmax PLUS (Molecular Devices).

2.7. Secondary ELISA Screening

1. Items listed under **Subheading 2.6.**
2. Goat anti-hkappa-HRP (Southern Biotechnology; Cat. #2060-05).
3. Goat anti-hlambda-HRP (Southern Biotechnology; Cat. #2070-05).

2.8. Hybridoma Cloning

1. Hybridoma medium (**Subheading 2.3**).
2. Flat-bottomed 96-well tissue-culture plates (Costar; Cat. #3799).

3. Methods

3.1. Preparation of Whole Cells for Immunization

1. Harvest about 1×10^9 cells expressing the antigen of interest (*see* **Note 1**).
2. Suspend in serum-free buffer (e.g., sterile PBS) at 6×10^7 cells/mL.

3. Dispense 1-mL volumes into 18 sterile 1.5-mL microfuge tubes.
4. Spin down cell pellet and wash cells, by resuspension then centrifugation, 3× with sterile phosphate-buffered saline (PBS).
5. Pellet cells and remove final wash buffer.
6. Freeze cell pellets and store at –80°C.

3.2. Immunization

1. For whole-cell immunization, resuspend two cell pellets using 600 µL of PBS per tube (10^7 cells/100 µL). For soluble antigens, adjust protein concentration, if necessary, to 0.1 mg/mL. For the first immunization, emulsify antigen in Complete Freund's Adjuvant at a 1:1 (v:v) ratio according to manufacturer's directions.
2. For each mouse, inject approx 25 µL into the skin pad at each side of the base of the tail and 50 µL by the intraperitoneal route (*see* **Note 2**).
3. Repeat immunizations at intervals of 2 wk (*see* **Note 3**).

3.3. Preparation of Lymphoid Cells

1. On the day prior to the fusion, it is important to take several preparatory steps. Make sure that there is a clean, empty incubator for fusion plates. Wipe down surfaces with 70% ethanol. Empty water pan, clean with ethanol, and fill with 2–3 L of autoclaved, deionized H_2O. Make 1 L of 1/2 HA fusion medium. Filter 1 L of DMEM. Prepare two conical 50-mL tubes with ~15 mL of DMEM and leave in the refrigerator (this medium will be used to harvest the spleens and lymph nodes). Autoclave two tissue grinders in autoclave bags.
2. Tissue grinders are used to dissociate the lymphoid cells. In a sterile laminar flow hood, open the autoclave bag containing both parts of the tissue grinder, and place the glass tube part in a 50-mL tube rack. Transfer 15 mL of chilled DMEM (from **step 1**) into each glass tube. Place the pestle of each tissue grinder inside a 50-mL sterile centrifuge tube also in the tube rack.
3. Using the standard technique of cervical dislocation, sacrifice the five high-titer mice selected (**Subheading 3.2.**).
4. Harvest the spleens and pool them into 15 mL of chilled DMEM in the glass tube of one of the tissue grinders.
5. Harvest the popliteal and inguinal lymph nodes (one also can harvest the lumbar, thoracic, and ankle lymph nodes to be thorough) from the five mice and pool them with 15–20 mL of DMEM into the glass tube of the second tissue grinder.
6. To dissociate the lymphoid cells in the two tubes, insert the grinder into the glass tube and grind the tissue by moving the grinder up and down and twisting it. Continue grinding until the grinder can be pushed to the bottom of the tube and the color of the tissue is white. At that point, almost all of the cells are released from the tissue. Without taking out the grinder, pour the supernatant through a 70-µm cell strainer into a new 50-mL sterile centrifuge tube. Add DMEM to the cell suspension up to a total volume of 40–50 mL. Centrifuge at 800*g* for 5–7 min.
7. For the spleen-cell suspension only, it is necessary to lyse the red blood cells. This is accomplished by first removing the supernatant and then resuspending the cell

pellet gently but completely with 10 mL of ACK lysis solution. Next, add more ACK lysis solution up to a total volume of 30–40 mL. Let stand for 1–2 min before centrifuging at 800*g* for 5–7 min. Resuspend the cell pellet with 10 mL of DMEM. Add more DMEM up to a total volume of 40 mL. Filter spleen-cell suspension through a 70-μm cell-filter into a new 50-mL sterile centrifuge tube. Centrifuge at 800*g* for 5–7 min. For both preps, remove the supernatant and gently resuspend the cell pellet with 10 mL of DMEM. Add more DMEM up to a total volume of 40–50 mL. Centrifuge at 800*g* for 5–7 min. Remove the supernatant and resuspend in 10 mL of DMEM.

8. Count the cells (*see* **Note 4**).

3.4. Separation of B Cells

1. Centrifuge the cells in each tube at 800*g* for 5–7 min. Add 0.9 mL of DMEM per 10^8 lymphocytes to the cell pellet. Resuspend the cells gently but completely.
2. Add 100 μL of $CD90^+$ magnetic beads per 10^8 cells (the CD90+ beads remove T-cells, thereby enriching for B-cells). Mix gently and well.
3. Incubate the cells with the magnetic beads at 4°C for 15 min. During the incubation, prewash the LS^+ magnetic column with 3 mL of DMEM.
4. After the 15-min incubation, pipet the magnetically labeled cell suspension containing up to 10^8 positive cells (or up to 2×10^9 total cells) onto the LS^+ column. Allow the cell suspension to run through and collect the effluent as the CD90-negative fraction.
5. Wash the column with 3 × 3 mL of DMEM and collect the total effluent as the CD90-negative fraction (most of these cells are B cells).

3.5. Cell Fusion

1. Warm 1/2 HA fusion medium and filtered DMEM in a 37°C water bath. Put PEG solution in a 37°C incubator (*see* **Note 5**).
2. Harvest myeloma cells in a sterile centrifuge tube (either a 50-mL or a 250-mL tube, based on the number of cells harvested). Centrifuge at 800*g* for 5 min, remove supernatant, and resuspend in DMEM. Repeat wash and count cells. Keep in incubator until needed.
3. Combine P3 or other myeloma cells and B cell-enriched lymph nodes/spleen cells (from **Subheading 3.4.**) in a ratio of 1:5 (myeloma : lymph nodes/spleen cells) into a 50-mL conical tube. Mix well. Add more DMEM up to 40 mL if necessary (*see* **Note 6**).
4. Combine myeloma cells and lymph nodes/spleen cells in a ratio of 1:5 (myeloma:lymph nodes/spleen cells) into a 50-mL conical tube. Mix well. Add more DMEM up to 40 mL if necessary.
5. Centrifuge at 800*g* for 5–7 min. During centrifugation, bring the warmed 15-mL conical tube with 12 mL of DMEM in a beaker with 37°C water from the water bath into the hood. Also bring the PEG solution from the incubator into the hood.
6. After centrifuging, immediately remove supernatant. **Caution:** It is important to remove all supernatant from this cell pellet.

7. Using a timer, follow these fusion steps:
   a. Add 1 mL of PEG solution to cells over 1 min, while stirring slowly;
   b. Stir gently with the pipet for 1 additional min;
   c. Add 2 mL of DMEM over 2 min while stirring slowly;
   d. Add 8 mL of DMEM over 3 min while stirring slowly.
8. Centrifuge at room temperature, 400*g* for 5 min.
9. Remove supernatant. Add to the cell pellet the appropriate amount of 1/2 HA fusion medium based on the number of plates needed (18 mL for each fusion plate) (*see* **Note 7**).
10. *Gently* resuspend the cells and add 200 μL/well into 96-well plates or 50 μL/well in 384-well plates.

3.6. Primary Screening for Antigen Binding

1. Coat an appropriate number of ELISA plates with 50 μL/well of soluble antigen at 0.2–2 μg/mL in plate coating buffer or 100–300 ng/mL of biotinylated antigen if using a streptavidin-coated plate (*see* **Note 8**). Incubate at 4°C overnight or at 37°C for 2 h for soluble antigen or incubate at room temperature for 30 min for biotinylated antigen. Wash the plates 3× with washing buffer.
2. Add 200 μL/well of blocking buffer to each well and incubate at room temperature for 1 h. Wash the plates 3× with washing buffer.
3. Add 50 μL/well of hybridoma supernatant sample and positive and negative controls, and incubate at room temperature for 2 h (*see* **Note 9**). Wash the plates 3× with washing buffer.
4. Add 100 μL/well of secondary antibody goat anti-huIgGFc-HRP (dissolved diluted in blocking buffer or PBS). Incubate at room temperature for 1 h. Wash the plates 3× with washing buffer.
5. Add 100 μL/well of TMB solution. Allow color to develop for about 10 min (until negative control wells barely start to show color).
6. Add 50 μL/well of TMB stop solution and read on an ELISA plate reader at wavelength 450 nm.
7. Transfer cells from positive wells of 96-well fusion plates to fresh 24-well plates. Transfer the entire cellular contents from the well of a 96-well plate into a 24-well plate well with 1 mL of hybridoma medium and culture for approx 48 h.

3.7. Secondary ELISA Screening

1. Coat a number of ELISA plates equal to twice the number of culture plates with 50 μL/well of soluble or biotinylated antigen, using same conditions as that used for primary screening (*see* **Subheading 3.6.** and **Note 10**).
2. Wash the plates 3× with washing buffer. Add 200 μL/well of blocking buffer to each well and incubate at room temperature for 1 h. Wash the plates 3× with washing buffer.
3. Add 50 μL/well of hybridoma supernatant sample and positive and negative controls (*see* **Note 9**) to each of the two sets of plates and incubate at room temperature for 2 h. Wash the plates 3× with washing buffer.

4. Add 100 μL/well of secondary antibody goat anti-huIgGFc-HRP to one set of plates and goat anti-hkappa-HRP or goat anti-hlambda-HRP (depending on the strain of mice used for immunization) to the second set. Incubate at room temperature for 1 h. Wash the plates 3× with washing buffer.
5. Add 100 μL/well of TMB solution. Allow color to develop for about 10 min (until control wells barely start to show color).
6. Add 50 μL/well of TMB stop solution and read on an ELISA plate reader at wavelength 450 nm.

3.8. Hybridoma Cloning

1. Count cells from each line to be cloned (*see* **Note 11**). Based on the concentration of cells, calculate a serial dilution to make the following concentrations of cells in hybridoma media:
   a. 10 cells/well × 96 wells = 960 cells in 20 mL.
   b. 5 cells/well × 96 wells = 480 cells in 20 mL.
   c. 1 cell/well × 96 wells = 96 cells in 20 mL.
   d. 0.5 cells/well × 96 wells = 96 cells in 40 mL for 2 plates.
2. Plate at 200 μL/well in flat-bottom, 96-well tissue-culture plates. Generate a total of five plates of four different densities for each line cloned as follows:
   a. 10 cells/well × 1 plate.
   b. 5 cells/well × 1 plate.
   c. 1 cell/well × 1 plate.
   d. 0.5 cell/well × 2 plates.
3. The cloning plates are usually screened within 2 wk after cloning. Visually inspect plates under a microscope. Using a permanent marker, circle up to 20 wells in which only a single colony grows in each well, starting with the lowest density-wells, which are most likely to contain clonal populations. The high-density series is primarily plated for rescue, so that few or no clones are lost in this process.
4. Test the supernatants from the circled wells according to **Subheading 3.7.** to confirm antigen reactivity as well as the presence of human heavy and light chains.
5. Based on the screening results, select three clones from each line. Make frozen stocks using standard procedures.

4. Notes

1. Although soluble protein is typically preferable for immunization, whole cells expressing the antigen can also be used. In this latter case, it is preferable to use murine cells expressing 30,000 or more copies per cell of the target antigen. This protocol will provide sufficient cells for immunizing a single cohort of ten mice.
2. Injections are made at two sites, into the thick skin pad at the base of the tail (subcutaneous base-of-tail immunization) and into the peritoneal cavity (intraperitoneal immunization). Multiple subcutaneous injections near draining lymph nodes also have proven to be very productive, and are often amenable to an accelerated immunization schedule. For base-of-tail injections, the animal is restrained in a specially designed device that allows for exposure of the tail. The site of the injection is the

back of the mouse just above the junction with the tail. The area is swabbed with 70% ethanol and the needle is inserted at a very shallow angle to avoid penetrating muscle or other tissues. Note that each mouse receives 10^7 cells or 10 μg of soluble protein.

3. For all subsequent immunizations, use Incomplete Freund's adjuvant in the same proportions. Serum titers should be assayed (*see* **Subheading 3.6.**) approx 1 wk after the third immunization to determine whether additional immunizations are necessary prior to fusion. Once five mice have reached a sufficient titer, typically with an antigen-specific signal 3× over background at a 1:1,000 or greater dilution of serum, these mice may be selected for fusion. They should receive a boost immunization 4 d prior to the day scheduled for fusion. It is recommended that immunizations of the remaining mice be continued at 2-wk intervals so that they can be used for a subsequent fusion.
4. One should expect to recover 200–250 × 10^6 total lymphocytes from the lymph nodes of five immunized mice. There are approx 1–2 × 10^8 total lymphocytes recovered from each spleen.
5. Electro-cell fusion can be used in place of the PEG method and will typically result in markedly higher fusion efficiencies, but requires investment in additional instrumentation. Details of this procedure are supplied with shipments of XenoMouse® animals.
6. Other frequently used myeloma fusion partners are SP2/0 and NSO.
7. The number of plates depends on the number of lymph node/spleen cells: usually we plate 1 × 10^6 cells/plate, in 96- or 384-well plates. The low plating density is important because it frequently yields a single successful fusion per well. Thus we are removing any selective pressure from competing clones so that the greatest number of hybridomas is recovered for each fusion.
8. If the target is a membrane protein, it is preferable to use flow cytometric- or fluorometric microvolume assay technology (FMAT)-based screening employing a transfected cell line expressing 30,000 or more copies per cell and parallel screening on the negative control parental line.
9. Typically our positive controls are commercially available Abs, hybridoma supernatants generated in-house, or antigen-specific serum from immunized mice. Hybridoma medium constitutes the negative control solution added to the wells.
10. The purpose of this secondary screen is twofold, to confirm antigen reactivity and to confirm that the selected cell lines are expressing fully human antibodies. In this procedure, positive lines for the primary screen will be separately tested for the presence of both human heavy chains and human light chains. In addition, a third plate set may be made for confirming specificity. For example, if the immunogen had a carrier or expression tag, it will be necessary to screen all of the positive wells from the primary screen against the carrier or tag itself or by screening against an irrelevant protein with the same carrier or tag.
11. Hybridomas that have tested positive for antigen binding and the presence of human heavy and light chains should be cloned as quickly as possible in order to prevent loss from overgrowth of negative cells. The limiting dilution method is a

commonly used method for cloning and subcloning hybridoma cells. Note that cells should also be frozen down at this point using standard procedures.

References

1. Mendez, M. J., Green, L. L., Corvalan, C.R.F., Jia, X.-C., Maynard-Currie, C. E., Yang, X-d., et al. (1997) Functional transplant of megabase human immunoglobulin loci recapitulates human antibody response in mice. *Nat. Gen.* **15,** 146–156.
2. Fishwild, D. M., O'Donnell, S. L., Bengoechea, T., Hudson, D. V., Harding, F., Bernhard, S. L., et al. (1996) High-avidity human IgG kappa monoclonal antibodies from a novel strain of minilocus transgenic mice. *Nat. Biotechnol.* **14(7),** 845–851.
3. Tomizuka, K., Shinohara, T., Yoshida, H., Uejima, H., Ohguma, A., Tanaka, S., et al. (2000) Double trans-chromosomic mice: Maintenance of two individual human chromosome fragments containing Ig heavy and κ loci and expression of fully human antibodies. *Proc. Natl. Acad. Sci. USA* **97,** 722–727.
4. Jakobovits, A. J., Vergara, G. J., Kennedy, J. L., Hales, J. F., McGuinness, R. P., Casentini-Borocz, D. E., et al. (1993) Analysis of homozygous mutant chimeric mice: Deletion of the immunoglobulin heavy-chain joining region blocks B cell development and antibody production. *Proc. Natl. Acad. Sci. USA* **90,** 2551–2555.
5. Gallo, M. L., Ivanov, V. E., Jakobovits, A., and Davis, C. G. (2000) The human immunoglobulin loci introduced into mice: V (D) and J gene segment usage similar to that of adult humans. *Eur. J. Immunol.* **30,** 534–540.
6. Green, L. L. and Jakobovits, A. (1998) Regulation of B cell development by V gene complexity in mice reconstituted with human immunoglobulin YACs. *J. Exp. Med.* **188,** 483–495.

11

Selection of Internalizing Antibodies for Drug Delivery

James D. Marks

1. Introduction

Large non-immune (naïve) antibody gene diversity libraries displayed on filamentous phage have proven to be a reliable source of antibodies to any purified protein antigen *(1–3)*. In some instances, it is possible to directly select peptides and antibody fragments that bind cell-surface receptors from filamentous phage libraries by incubation of phage libraries with a target cell line *(4–8)*. This has led to a marked increase in the number of potential cell-targeting molecules. However, the isolation of cell type-specific antibodies from naïve libraries has been difficult because selections often result in crossreactive antibodies that bind to common cell-surface antigens *(9)*.

The ability of bacteriophage to undergo receptor-mediated endocytosis *(5,10)* indicates that phage libraries might be selected not only for cell binding, but also for internalization into mammalian or other cells. Such an approach would be particularly useful for generating ligands that could deliver drugs or toxins into a cell for therapeutic applications. Recently, a methodology has been developed *(11)* that allows isolation of specifically internalized phage and excludes phage that is merely bound to the cell surface. In the model system studied, phage could be selected on the basis of endocytosis, and enrichment ratios were significantly greater when phage were recovered from within the cell rather than from the cell surface. Enrichment ratios were also significantly higher when the phage were capable of crosslinking receptors, rather than merely binding. Crosslinking occurs when either bivalent antibody fragments such as diabodies are displayed in a phagemid system, or when scFv are displayed in a true phage vector. Thus, naïve antibody fragment libraries that are displayed using phage vector systems may prove more useful for the generation of internalizing antibodies than antibody fragments displayed using phagemid systems *(12,13)*.

From: *Methods in Molecular Biology, Vol. 248: Antibody Engineering: Methods and Protocols*
Edited by: B. K. C. Lo © Humana Press Inc., Totowa, NJ

This strategy of selection for internalization was employed to select scFv from a large naïve library *(2)* capable of undergoing endocytosis into breast tumor cells upon receptor binding *(14)*. Upon analysis of a large number of different antibodies isolated in this way, distinct cell-surface antigens recognized by the scFv were identified. Several of the scFv recognized the ErbB-2 growth-factor receptor, and another scFv bound the transferrin receptor. Interestingly, both the ErbB-2 and the transferrin receptors are rapidly internalized, and are specific markers for a number of cancers. Both the phage antibodies and the native scFv were rapidly endocytosed into cells that expressed the appropriate receptor. It is likely that selection on other cell types will identify cell-specific markers, because endocytosed receptors are more likely to be associated with specific cell functions (such as growth factor and nutrient transport receptors on cancer cells or Fc receptors on cells of the immune system).

Internalizing antibodies (or fragments of antibodies) are also required for many targeted therapies, such as targeted drugs (toxins, RNases, radioisotopes), targeted liposome therapy (e.g., for delivery of chemotherapeutics) or for targeted delivery of genes (especially of non-viral vectors). Not all antibodies that bind internalizing receptors are rapidly internalized. Rather, only antibodies to certain epitopes trigger internalization *(14)*. Recently, we have demonstrated that scFv antibody fragments selected for internalization using the protocols described here can be adapted for targeted drug delivery for cancer therapy *(15)*.

The strategy described here has aided us in isolating a number of cancer cell-specific antibodies. We have also used the protocol to isolate antibodies to the epidermal growth factor (EGF) receptor stably expressed on Chinese hamster ovary (CHO) cells *(16)*. All of the isolated antibodies have proven to be rapidly internalized into cells that express the target antigens, something that we did not observe with antibodies to the same receptors selected on recombinant antigen using traditional panning strategies *(14)*. The protocols used for this type of selection are outlined here.

2. Materials

2.1. Selection of Internalizing Antibodies from Phage Libraries

1. Appropriate mammalian cell-culture media.
2. Phage antibody library (*see* Chapters 6 and 8).
3. Glycine stripping buffer: 50 m*M* glycine, pH 2.5, 500 m*M* NaCl.
4. Phosphate-buffered saline (PBS): 154 m*M* NaCl, 8.1 m*M* Na_2HPO_4, 1.9 m*M* NaH_2PO_4, pH 7.4.
5. 1X trypsin/ethylenediaminetetraacetic acid (EDTA) solution (Sigma; Cat. #T3924).
6. 0.5 *M* Tris-HCl, pH 7.4.

7. 100 m*M* triethylamine.
8. Exponentially growing *E. coli* strain TG1 (Stratagene).
9. TYE/amp/glu plates: 100- and 150-mm TYE agar plates containing 100 µg/mL ampicillin and 2% glucose.
10. 2X TY media: dissolve 16 g of bacto-tryptone, 10 g of yeast extract, 5 g of NaCl in 1 L of distilled water; autoclave the solution.
11. 2X TY/amp/glu media: 2X TY media containing 100 µg/mL ampicillin and 2% glucose.
12. 50% glycerol, sterile-filtered.

2.2. Detection of Phage or ScFv Internalization by Fluorescence Microscopy

1. 15-mm round cover slips, sterile.
2. Sterile 6-well tissue-culture plates.
3. Appropriate mammalian cell-culture media.
4. Monoclonal phage preparations, sterile-filtered.
5. PBS.
6. Glycine stripping buffer (*see* **Subheading 2.1.**).
7. PBS containing 4% paraformaldehyde.
8. Cold 100% methanol.
9. PBS containing 1% bovine serum albumin (BSA).
10. Biotinylated anti-mouse Fc antibody (Amersham Biosciences).
11. Anti-myc antibody 9E10 (Santa Cruz).
12. Biotinylated anti-M13 antibody (Amersham Biosciences).
13. Streptavidin-Cy3 (Sigma; Cat. #S6402).
14. VECTOR shield containing DIAP (Vector Labs).

2.3. Detection of Phage Internalization by Green Fluorescent Protein (GFP) Reporter Gene Expression

1. Appropriate mammalian cell-culture media.
2. Monoclonal phage preparations, sterile-filtered.
3. PBS.
4. Six-well tissue-culture plates.
5. PBS containing 2% fetal calf serum (FCS).
6. 1X trypsin/EDTA solution.
7. Fluorescent microscope or flow cytometer.

3. Methods

3.1. Selection of Internalizing Antibodies from Phage Libraries

3.1.1. Phage Binding to Target Cells

1. Incubate subconfluent target cells plated in a 20-cm plate with 1 mL of the phage library (1×10^{13} pfu total phage) in 20 mL of cold cell-culture medium at 4°C for

1.5 h. The cell-culture media should be the same media that the cells were cultured in (*see* **Note 1**).
2. Wash gently 6× with 25 mL of PBS at room temperature.
3. Add 30 mL of prewarmed culture medium. Incubate at 37°C for 45 min to allow internalization.

3.1.2. Removal of Cell-Surface Bound Phage

1. Strip the cell surface of non-internalized phage by three 10-min washes with 20 mL glycine stripping buffer. Keep the glycine washes for phage titering (*see* **Note 2**).
2. Wash cells with 30 mL of PBS to neutralize.
3. Add 5 mL of trypsin/EDTA and incubate at 37°C until cells detach. Add 10 mL of PBS to the culture plate and transfer cells to a 15-mL centrifuge tube. Pellet cells by spinning at 500*g* for 5 min. Discard the supernatant.
4. Lyse the cell pellet by resuspending in 1 mL of 100 m*M* triethylamine and incubating at room temperature for 5 min. Neutralize with 1.5 mL of 0.5 *M* Tris-HCl, pH 7.4.

3.1.3. Re-infection of the Internalized Phage into E. coli

1. Add 1.22 mL of the eluted phage to 20 mL of exponentially growing *E. coli* strain TG1 (OD_{600} ~0.5). Store the remaining eluted phage at 4°C.
2. Incubate bacteria at 37°C for 30 min without shaking.
3. Titer TG1 infection by plating 2 µL and 20 µL onto 100 mm TYE/amp/glu plates (this is a 10^4 and 10^3 dilution, respectively).
4. Centrifuge the remaining bacteria solution at 1,700*g* for 10 min. Resuspend the bacterial pellet in 250 µL of 2X TY media and plate onto two 150-mm TYE/amp/glu plates. Incubate at 37°C overnight.
5. Add 3 mL of 2X TY/amp/glu media to each plate, and then scrape the bacteria from the plate with a bent glass rod. Make glycercol stocks by mixing 1.4 mL of bacteria and 0.6 mL of 50% glycerol (filtered). Save stock at –70°C.
6. Phage are then rescued for the next round of selection as described in Chapter 8.
7. After 2–4 rounds of selection, binding phage antibodies are identified by an appropriate screening technique, such as cell enzyme-linked immunosorbent assay (ELISA).

3.2. Detection of Phage or ScFv Internalization by Fluorescence Microscopy

3.2.1. Internalization into Cells Grown on Cover Slips

1. Place one sterile 15-mm round cover slip per well in a 6-well tissue-culture plate. Add 2 mL of appropriate tissue-culture media containing cells. Grow cells to 20–50% confluency.
2. Change media to remove dead cells and let the cells stand at 37°C for 10 min.
3. If *phage* is used for the internalization, add PEG concentrated and sterile-filtered phage to a final concentration of 10^{10} cfu/mL. For *scFv,* add purified and sterile-

filtered scFv to a final concentration of 20 μg/mL. Incubate at 37°C, 5% CO_2 atmosphere, for 2 h.

3.2.2. Stripping and Fixation of Cell Surface Bound ScFv/Phage

1. The cover slips are then washed 6× with PBS, followed by washing 3× for 10 min with glycine-stripping buffer.
2. After another wash with PBS, the cells are fixed on the coverslips with PBS containing 4% paraformaldehyde at room temperature for 5 min.
3. Wash the coverslips twice with 3 mL of PBS. The procedure can be interrupted at this point. The coverslips may be stored overnight in PBS.
4. Permeabilize the cells by incubating with cold 100% methanol for 10 min.
5. Wash the cells three times with PBS.

3.2.3. Antibody Staining for Internalized ScFv/Phage

1. Place cells in a new six-well plate and incubate with PBS containing 1% BSA for 30 min. Wash the cells once with PBS.
2. For phage staining: Add biotinylated anti-M13 antibody diluted 1/5,000 in PBS/1% BSA. For scFv staining: Add 9E10 antibody at 0.2 μg/mL in PBS/1% BSA. Incubate on a rocker at 4°C for 1 h.
3. Wash the cells ten times with 3 mL of cold PBS.
4. For phage staining: Add streptavidin-Cy3 diluted 1/1,000 in PBS/1% BSA and incubate on a rocker at 4°C for 30 min. For scFv staining: Add biotinylated anti-mouse Fc antibody diluted 1/200 in PBS/1% BSA. Incubate on a rocker at 4°C for 1 h. Wash ten times with cold PBS. Then add streptavidin-Cy3 diluted 1/1,000 in PBS/1% BSA and incubate on a rocker at 4°C for 30 min.
5. Wash the cells ten times with cold PBS.
6. Mount the coverslips onto microscope slides in 5 L of VECTOR shield containing DIAP. Slides can be stored at 4°C in the dark for up to a week. For long-term storage the slides may be kept at 20°C.

3.3 Detection of Phage Internalization by GFP Reporter-Gene Expression

Filamentous phage that displays internalizing antibody fragments can also be used to specifically target genes to mammalian cells expressing the specificity of the scFv *(17)*. When the gene targeted to the eukaryotic cell is a reporter gene, such as GFP, gene expression can be used as an assay for internalization of the scFv.

We have used two approaches to deliver reporter genes packaged in phage: in one approach, helper phages are used to infect *E. coli* containing a phagemid in which the reporter gene and eukaryotic promoter are cloned. Our lab has cloned the EGFP reporter gene and promoter into pHEN1, with SfiI/NotI cloning sites for scFv subcloning (the vector, pFROG, is available upon request from JDM). Phages recovered from the culture supernatant display an average of one scFv-pIII fusion protein, and 99% of them package the GFP reporter gene. In the other approach, the scFv gene is cloned into the fd phage genome (such as the vector fd-DOG1) for expression as a scFv-pIII fusion. fd-scFv phage are then used to infect *E. coli* containing a reporter phagemid vector (such as pcDNA3-GFP). Phages purified from the culture supernatant display multiple scFv-pIII fusion protein, and approx 50% package the reporter gene ***(17)***.

Many of the phage selected for internalization enter an intracellular trafficking pathway that ultimately leads to reporter gene expression. When GFP is used as a reporter gene, expression can be detected with as few as 2×10^7 cfu of phage and increases with increasing phage titer up to 10% of cells ***(17)***. The reporter gene expression from the internalized bacteriophage can be used as a rapid assay for internalization. The following protocol is used for the infection and expression of GFP reporter gene packaged in antibody-displaying phage.

3.3.1. Bacteriophage-Mediated Cell Infection

1. Phage preparations of scFv-phages containing a reporter gene are diluted at least 10-fold in complete cell-culture medium. Media and phage are filtered through a 0.45-μm filter.
2. Media containing the phage is added to 30–50% confluent cells grown in a 6-well tissue-culture plate.
3. After 48 h of incubation, the media is changed and cells are incubated for another 24–48 h.

3.3.2. Analysis for GFP Expression

1. **For fluorescence microscopy analysis:** media is aspirated, and cells are washed twice in PBS before analysis using standard protocols.

   For FACS analysis: cells are trypsinized with trypsin/EDTA and washed once in PBS containing 2% FCS and analyzed for GFP expression by FACS in the FL-1 channel using standard protocols.

4. Notes

1. It may be necessary to use a normal or related cell line in suspension to deplete for phage antibodies that bind common internalizing receptors. We have found this to be a critical step in eliminating crossreactive antibodies. The selecting cell line is grown adherent.
2. The stripping buffer may need to be adjusted for different cell types to ensure that washing does not lyse the cells. Parameters to investigate include pH and osmolality.

References

1. de Haard, H. J., van Neer, N., Reurs, A., Hufton, S. E., Roovers, R. C., Henderikx, P., et al. (1999) A large non-immunized human Fab fragment phage library that permits rapid isolation and kinetic analysis of high affinity antibodies. *J. Biol. Chem.* **274,** 18,218–18,230.

2. Sheets, M. D., Amersdorfer, P., Finnern, R., Sargent, P., Lindqvist, E., Schier, R., et al. (1998) Efficient construction of a large nonimmune phage antibody library: the production of high-affinity human single-chain antibodies to protein antigens. *Proc. Natl. Acad. Sci. USA* **95,** 6157–6162.
3. Vaughan, T. J., Williams, A. J., Pritchard, K., Osbourn, J. K., Pope, A. R., Earnshaw, J. C., et al. (1996) Human antibodies with sub-nanomolar affinities isolated from a large non-immunized phage display library. *Nat. Biotechmol.* **14,** 309–314.
4. Andersen, P. S., Stryhn, A., Hansen, B. E., Fugger, L., Engberg, J., and Buus, S. (1996) A recombinant antibody with the antigen-specific, major histocompatibility complex-restricted specificity of T cells. *Proc. Natl. Acad. Sci. USA* **93,** 1820–1824.
5. Barry, M. A., Dower, W. J., and Johnston, S. A. (1996) Toward cell-targeting gene therapy vectors: selection of cell-binding peptides from random peptide-presenting phage libraries. *Nat. Med.* **2,** 299–305.
6. Cai, X. and Garen, A. (1995) Anti-melanoma antibodies from melanoma patients immunized with genetically modified autologous tumor cells: selection of specific antibodies from single-chain Fv fusion phage libraries. *Proc. Natl. Acad. Sci. USA* **92,** 6537–6541.
7. de Kruif, J., Terstappen, L., Boel, E., and Logtenberg, T. (1995) Rapid selection of cell subpopulation-specific human monoclonal antibodies from a synthetic phage antibody library. *Proc. Natl. Acad. Sci. USA* **92,** 3938–3942.
8. Marks, J. D., Ouwehand, W. H., Bye, J. M., Finnern, R., Gorick, B. D., Voak, D., et al. (1993) Human antibody fragments specific for human blood group antigens from a phage display library. *Bio/Technology* **11,** 1145–1149.
9. Hoogenboom, H. R., Lutgerink, J. T., Pelsers, M. M., Rousch, M. J., Coote, J., Van Neer, N., et al. (1999) Selection-dominant and nonaccessible epitopes on cell-surface receptors revealed by cell-panning with a large phage antibody library. *Eur. J. Biochem.* **260,** 774–784.
10. Hart, S. L., Knight, A. M., Harbottle, R. P., Mistry, A., Hunger, H. D., Cutler, D. F., et al. (1994) Cell binding and internalization by filamentous phage displaying a cyclic Arg-Gly-Asp-containing peptide. *J. Biol. Chem.* **269,** 12,468–12,474.
11. Becerril, B., Poul, M. A., and Marks, J. D. (1999) Toward selection of internalizing antibodies from phage libraries. *Biochem. Biophys. Res. Comm.* **255,** 386–393.
12. Huie, M. A., Cheung, M.-C., Muench, M. O., Becerril, B., Kan, Y. W., and Marks, J. D. (2001) Antibodies to human fetal erythroid cells from a non-immune phage antibody library. *Proc. Natl. Acad. Sci. USA* **98,** 2682–2687.
13. O'Connell, D., Becerril, B., Roy-Burman, A., Daws, M., and Marks, J. D. (2002) Phage versus phagemid libraries for generation of human monoclonal antibodies. *J. Mol. Biol.* **321,** 49–56.
14. Poul, M.-A., Becerril, B., Nielsen, U. B., Morrison, P., and Marks, J. D. (2000) Selection of internalizing human antibodies from phage libraries. *J. Mol. Biol.* **301,** 1149–1161.
15. Nielsen, U. B., Kirpotin, D. B., Pickering, E. M., Hong, K., Park, J. W., Shalaby, R., et al. (2002) Therapeutic efficacy of anti-ErbB2 immunoliposomes targeted by a phage antibody selected for cellular internalization. *Biochim. Biophys. Acta.* **1591,** 109–118.

16. Heitner, T., Moor, A., Garrison, J. L., Hasan, T., and Marks, J. D. (2001) Selection of cell binding and internalzing epidermal growth factor receptor antibodies from a phage display library. *J. Immunol. Methods.* **248,** 17–30.
17. Poul, M. A. and Marks, J. D. (1999) Targeted gene delivery to mammalian cells by filamentous bacteriophage. *J. Mol. Biol.* **288,** 203–211.

12

Engineering Multivalent Antibody Fragments for In Vivo Targeting

Anna M. Wu

1. Introduction

Antibodies, especially in the genetically engineered form, provide a powerful and diverse set of reagents for the recognition of biological structures. Of key importance to those in the antibody field are advances in genomics and proteomics, fostered by technology such as gene-expression profiling, protein microarrays, and high-throughput validation assays, which are providing a multitude of new protein targets for investigation. In parallel, antibody technologies such as phage display (Chapter 8) and ribosome display (Chapter 9), are providing agents to specifically recognize these novel targets, with broad future applications in detection and intervention in fields as diverse as human health, agriculture, and the environment.

Once appropriate antigen/antibody pairs are identified, genetic engineering of antibodies allows full control over the format of the final protein produced. Development of recombinant antibodies for in vitro applications (e.g., assays) involves attention to universal factors such as specificity and affinity, as well as practical issues related to the final application such as conjugation/fusion to detectable or functional moieties and ease of production. Many additional issues come into play when developing antibodies for targeting applications in vivo. Biological aspects of the target antigen must be considered, including normal tissue expression, antigen density on the target tissue, and whether the antigen can be internalized or shed. Biological properties of the antibody must also be considered, including pharmacokinetics, route of clearance, and interaction with Fc receptors on immune cells or FcRB receptors in the endothelium or tissues. Protein engineering allows full control over characteristics such as affinity, valency, domain composition, flexibility, and orientation of the binding sites. In addition, for in vivo targeting applications, issues such as mol wt (including

From: *Methods in Molecular Biology, Vol. 248: Antibody Engineering: Methods and Protocols*
Edited by: B. K. C. Lo © Humana Press Inc., Totowa, NJ

whether the recombinant protein is above or below the threshhold for first-pass renal clearance), pI, and potential immunogenicity must be considered.

Furthermore, there is no "right" format for an engineered antibody that is intended for use in vivo. The ideal combination of properties will differ, depending on whether the antibody is intended for use in an unmodified form, whether it is conjugated to a radionuclide, toxin, drug, or other moiety for detection/intervention, or whether it is employed as a fusion protein. For example, if unmodified (naked; cold) antibodies are being developed for therapeutic applications, in many cases conventional intact antibodies are developed (following chimerization/humanization to reduce immunogenicity; *see* Chapter 7), since intact IgGs exhibit extended serum persistence. This allows dosing on an infrequent basis, which is convenient for patients and health care providers. On the other hand, if toxic agents are conjugated to antibodies, shorter half-lives may be more desirable to control normal tissue toxicity. For imaging using radiolabeled antibodies, rapid targeting and clearance are expected of tracers; for therapy using radiolabeled antibodies, the ability to deliver sufficient dose is foremost, but concerns about normal tissue toxicity remain critical. Ultimately, the final clinical application will dictate many of the properties that the ideal antibody must have. Fortunately, antibody engineers have been developing novel recombinant proteins for many of these applications, and considerable experience has been gained regarding their biochemical, biological, and targeting properties. This chapter reviews recent advances in developing mono-, bi-, tri-, and tetravalent engineered antibody fragments, with special consideration given to in vivo targeting properties of these molecules.

1.1. Single-Chain Variable Fragments

Single-chain variable fragments (scFv) have become the *de facto* building block for engineered antibody fragments. ScFvs are produced by assembling the genes that encode the heavy-chain variable region (V_H) and the light-chain variable region (V_L) of an antibody using sequences encoding a linker peptide to join the variable-region genes *(1,2)* (**Fig. 1**). As a result, the antigen-binding protein is expressed as a single polypeptide chain. Variable regions can generally be assembled in either order, although there are specific examples of antibodies in which one orientation has been more favorable than the other *(3)*. Although modeling studies suggest that linker peptides of 14–15 amino acid residues are long enough to span the distance between the N- and C-termini of the variable domains in an Fv, it is advisable to use longer linkers (18 amino acid residues or more) to favor folding of the polypeptides into monomeric scFv. Use of linkers of 3–12 amino acid residues in length results in diabody formation, and incorporation of linkers of intermediate length will frequently result in mixtures of monomeric and dimeric forms.

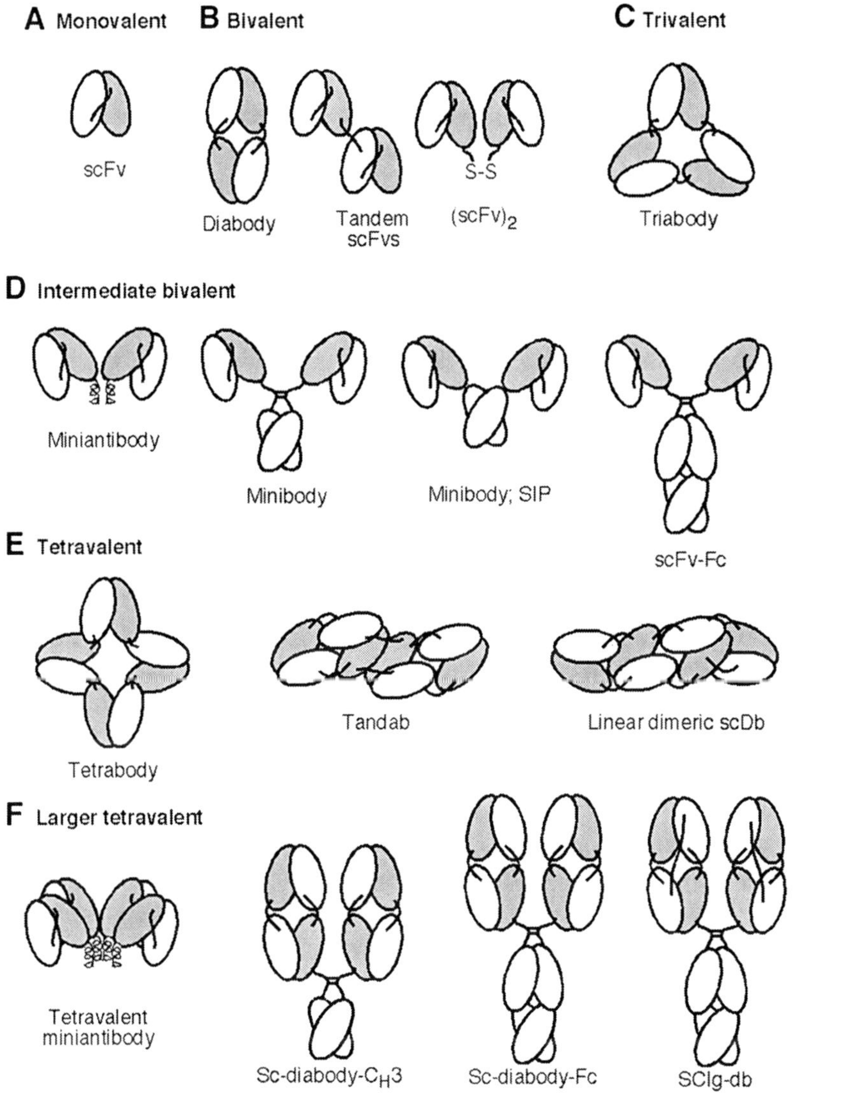

Fig. 1. Schematic diagram of many of the multivalent engineered antibodies described in the text. They are classified according to valency, and whether domains in addition to scFv units have been incorporated into each construct. In many cases, bispecific versions have been produced in addition to monospecific ones. In a growing number of examples, in vivo pharmacokinetic and targeting properties have been evaluated. This figure and the text are not all-inclusive, as new formats are constantly being developed and evaluated. Rather, this figure is intended to provide a general overview and a sense of the size and valency of representative forms. Please refer to the text for additional details.

The in vivo properties of radiolabeled forms of several different scFvs have been investigated in animal models, including in mice, rats, rabbits, and dogs (reviewed in **ref. *4***). Alpha- and beta-phase half-lives were short compared to larger fragments or intact antibodies, with terminal half-lives ranging from 1.5–4 h. Rapid blood clearance resulting from first-pass renal clearance of these ca. 25–30-kDa proteins results in low background activity and high contrast at early times following administration, provided that sufficient activity can be delivered to target tissue by the radiolabeled scFv. The cancer-targeting and imaging potential of several scFvs has been investigated for a variety of tumor-associated antigens (Tag-72, CEA, HER2, Tac, EpCAM, EGFR, and others; reviewed in **ref. *5***). Tumor localization is observed, often in the 1–5% injected dose per g (% ID/g) range, depending on the antibody, antigen, and specific model **(Fig. 2)**. ScFvs also offer the important advantage of facile penetration and distribution into tumors, in contrast to larger entities such as intact antibodies, which are often limited to perivascular distribution *(6)*.

Clinical evaluation of radiolabeled scFvs that are specific for targets in cancer has yielded results consistent with preclinical studies on the targeting potential of these antibody fragments. Begent and Chester and colleagues isolated an anti-carcinoembryonic antigen (CEA) scFv from a phage display library of murine scFvs. Appending a hexahistidine tag allowed purification by immobilized metal affinity chromatography (IMAC). The MFE-23 anti-CEA scFv was radiolabeled with I-123, a pure gamma emitter with a 13.2 h half-life, and imaging was evaluated in patients with CEA-positive malignancies *(7)*. Targeting was demonstrated to all known tumor deposits. Blood clearance was rapid, and α- and β-phase half lives were 0.42 and 5.3 h, respectively (ten patients). Subsequently the anti-CEA scFv was radiolabeled with I-125 and evaluated in 34 patients for radioimmunoguided surgery in conjunction with an intra-operative gamma probe. In this study, the β-phase half-life was longer (10.95 h) *(8)*. Separately, an I-123-radiolabeled anti-TAG-72 scFv based on the CC49 monoclonal antibody (MAb) was evaluated as a targeting and imaging agent in colorectal carcinoma patients by Larson and colleagues *(9)*. Rapid equilibration and clearance were observed with α- and β-half lives of 0.5 h and 10.5 h, respectively. Same-day imaging of primary tumors and metastatic lesions in the liver was possible, although tumor uptake was suboptimal.

1.2. ScFv Dimers and Diabodies

Several strategies have been employed to dimerize single-chain variable fragments to generate bivalent antigen-binding molecules (MW 55–60 kDa), and some of these variants have been evaluated for their targeting properties. Approaches have included production of tandem scFvs or covalent linkage of scFvs through disulfide bridging of cysteine residues appended to the C-terminus

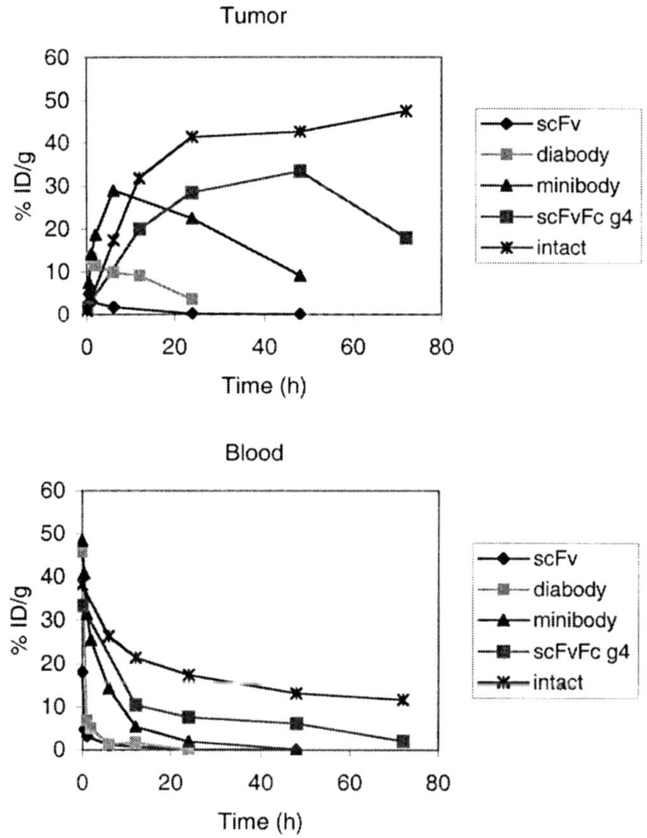

Fig. 2. Summary of the tumor-targeting and blood-clearance properties of a series of cognate engineered anti-CEA fragments based on the murine T84.66 antibody. Values are given as percent injected dose per g (% ID/g) of tumor tissue or blood. Radioiodinated fragments were evaluated in athymic mice bearing LS174T human colorectal carcinoma xenografts. Data are from references cited in **ref** ***(4)***, and unpublished studies (scFvFcγ4) of Drs. P. Yazaki, M. Sherman, and A. Wu.

through genetic engineering. By far the most straightforward approach has been the production of non-covalent dimers, also known as "diabodies." These are based on observations that certain scFvs have a tendency to spontaneously multimerize or aggregate ***(10)***. Subsequently, work by Winter's group and many other laboratories demonstrated that construction of scFvs with short (3–12 amino acid residues) linkers connecting the variable region genes strongly favored the generation of dimers or diabodies ***(11)***. X-ray structural analysis has confirmed that in these molecules, domain exchange has occurred, so that the V_L of each polypeptide has associated with the V_H from a separate polypeptide chain, resulting in a cross-paired structure *(12)*.

An initial comparison of $(scFv)_2$, generated using C-terminal cysteine tails for disulfide-bridge formation, showed superior targeting compared to the corresponding scFv or Fab fragment, in targeting of SK-OV-3 xenografts overexpressing c-erbB2 *(13)*. These still exhibited rapid clearance from the circulation in mouse models, but higher avidity resulting from bivalent binding sites has led to greatly improved retention of radioactivity in xenografts. Subsequently, targeting properties of radiolabeled cancer-specific diabodies has been extensively evaluated against a variety of targets *(14–17)*. In general, diabodies labeled with radioiodine or a variety of radiometals have been demonstrated to reach maximum tumor uptakes in the range of 10–15% ID/g at 1–2 h post injection, in murine tumor models *(5)* (**Fig. 2**).

1.3. Intermediate Molecular-Weight Bivalent Fragments

Larger, bivalent engineered antibody fragments are being developed to address a number of issues. ScFvs and diabodies, with mol wts below the threshhold for first-pass renal clearance, exhibit very rapid elimination from the circulation, with terminal half-lives in the range of 3–5 h. The limited blood activity (area under the curve) translates into short exposure of tumor or other target sites to circulating radiolabeled antibody, and places a limit on the maximum uptake of these species. Engineered fragments larger than approx 60 kDa should exhibit increased serum persistence. One common approach has been fusion of scFvs to protein domains that dimerize, an approach that also adds mass to the overall molecule. Examples include the use of self-assembling helices to produce "mini-antibodies" *(18)*, human IgG1 C_H3 domain to produce "minibodies" *(19)* or "SIPs" *(20)*, fos-jun leucine zippers *(21)*, kappa light-chain constant regions *(22)*, and a variety of other fusion partners. Several of these formats have been subjected to preclinical evaluation in murine tumor xenograft models. The anti-CEA minibody (scFv-C_H3 dimer, 80 kDa) was shown to achieve excellent tumor uptake in radioiodinated or radiometal labeled form, reaching 20–25% ID/g at 4–6 h post injection *(19)* (**Fig. 2**). Coupled with rapid blood clearance, excellent contrast was achieved, allowing imaging at early times using single-photon or positron emission tomography (PET) approaches *(19,23)*. This result has recently been confirmed using scFvs directed against the fibronectin ED-B domain, in which SIPs demonstrate substantially higher tumor uptake than the corresponding diabody. Mini-antibodies that are specific for HER2 also have shown improved tumor localization compared to the corresponding scFvs, although additional gains were made using tetravalent forms (*see* **Subheading 1.5.**).

Furthermore, fragments engineered to include the Fc portion of native antibodies will retain the extended half-lives characteristic of intact antibodies (days to weeks). This is because of the presence of a site at the junction between C_H2 and C_H3 that interacts with the FcRB receptor, which mediates serum persistence of intact antibodies ***(24)***. As an example, an scFv-Fc fusion protein developed from the anti-TAG-72 antibody CC49 has demonstrated high tumor localization coupled with a long serum half-life in murine models ***(25)***.

1.4. Triabodies

A systematic examination of linker lengths in scFv constructs has led to the observation that the use of extremely short or absent linkers between the variable regions induces formation of trimers, tetramers, and higher-order species ***(26–28)***. The precise length of linker required to induce trimer or tetramer formation varies, depending on the order of the variable regions (V_L-V_H vs V_H-V_L) and individual aspects of each specific antibody and its V regions. For example, NC10 scFvs assembled in the V_H-V_L orientation with 3–4 amino acid linkers formed diabodies and 0–2 amino acid, triabodies ***(29)***. In the opposite orientation, the trend was similar but the transition not so sharp, with a two-residue-linker yielding a mixture of diabodies, triabodies, and tetrabodies ***(30)***. Interestingly, for the HD37 anti-CD19 scFv (assembled V_H-V_L) the situation was more complex, with a 1-amino-acid linker yielding tetramers and zero yielding trimers ***(27)***. Pei and colleagues have determined the structure of a triabody assembled from non-cognate variable regions, confirming that domain exchange has occurred in these proteins ***(31)***. It will be of great interest to evaluate the biological properties of triabodies, which demonstrate high avidity in vitro, and should not be subject to rapid renal filtration in vivo as a result of their increased mass (ca. 80–90 kDa).

1.5. Tetravalent Antibody Fragments

Tetravalent forms of engineered antibody fragments have been generated by a variety of approaches. Continuing the approach outlined here, the laboratories of Hudson and Little have observed formation of scFv tetramers (tetrabodies; ca. 110–120 kDa) when linkers in the range of 0–1 amino acid were incorporated in the constructs. Again, the relationship between linker length and tetramer formation was a function of the specific antibody that was engineered, as well as order of assembly of the variable regions. As a result, considerable experimentation may be required in order to determine the configuration that gives the highest yield of tetramers. Furthermore, as one progresses to higher-order multimers, it is less likely that pure species will be obtained, and additional purification/size fractionation steps may be required to isolate the desired form. In addition, for all these constructs, the expression, purification, and storage conditions can affect the mix of multimers that are ultimately obtained. For example, many laboratories have observed that upon storage at high concentrations, previously pure species can rearrange and generate alternate forms, indicating that the V_H-V_L interactions that hold these multimers together can dissociate/reassociate, allowing re-equilibration of these non-covalently associated forms.

Several alternative approaches have been used to generate tetravalent species. Additional permutations based on scFvs joined using various combinations of linkers have been investigated. Although the tetrabodies are expected to be circularized molecules, linear forms such as tandem diabodies (tandabs) ***(32)***, tetravalent scFvs ***(33)***, and dimeric single-chain diabodies ***(34)*** of similar mol wt (ca. 110 kDa) have also been developed. One challenge in evaluating the biochemical properties of these multivalent molecules is determination of the valency of binding. An increase in functional affinity, caused by higher avidity, has been taken as evidence that all or most of the binding sites are active, although this will be difficult to determine precisely for monospecific agents.

Larger tetravalent constructs have been assembled using additional protein domains as multimerization modules or scaffolds. For example, the mini-antibody approach has been extended through the use of a short amphipathic helix that spontaneously tetramerizes ***(35)***. Gel-filtration analysis demonstrated efficient assembly into the expected 130-kDa tetramers and affinity was superior to bivalent miniantibodies. Alternatively, a short region of the *p53* protein that promotes tetramerization has been used as a self-assembling domain for multimerization of an anti-HER2 scFv ***(36)***. Substantially larger fusion proteins have been generated by fusion of diabodies to antibody Fc or C_H3 regions. A humanized CC49 single-chain diabody-Fc fusion protein exhibited the expected mol wt of 160 kDa and competed more effectively for binding to TAG-72 antigen than the corresponding parental murine antibody, humanized antibody, or scFv-Fc fusion protein ***(37)***. A bispecific single-chain diabody, binding to both CEA and β-galactosidase, was fused to either the human Ig1 Fc or just the C_H3 domain ***(38)***. The resultant proteins exhibited four functional binding sites with increased functional affinity. Finally, several groups have exploited the tetramerization properties of streptavidin to generate engineered antibody fragments that are not only multivalent, but also demonstrate very high avidity for biotin ***(39)***. This approach lends itself to multiple in vitro applications, but has demonstrated utility for in vivo targeting as the initial reagent forn pretargeted cancer therapy ***(40)***.

Of key interest is the evaluation of the performance of tetravalent engineered antibody fragments in vivo. The smallest of these, tandabs and tetravalent scFvs, have mol wts of approx 110 kDa, substantially above the threshhold for renal clearance. These constructs demonstrated higher functional avidity than the corresponding 55-kDa diabodies. Pharmacokinetic studies of radioiodinated bispecific tandabs in non-tumor-bearing mice indicated that the tandabs were more stable and persisted longer in the circulation compared to scFvs or diabodies; however, the terminal half-life was still quite short (1.5–2 h). The tumor-targeting and biodistribution properties of tetravalent CC49 scFvs have been determined for radioiodinated and radiometal-labeled forms ***(33,41)***. Tumor uptakes were highly favorable, reaching ca. 20% ID/g at 6 h post-administration for both radiolabels, and very good tumor-to-background ratios were achieved because of rapid blood clearance (elimination half-life, 5 h). The tumor-targeting properties of dimeric and tetrameric anti-HER2 mini-antibodies has also been evaluated in a murine xenograft model ***(36)***. Again, the tetravalent version proved superior to the divalent form, reaching higher tumor activity levels (43% ID/g at 24 h) and higher tumor-to-blood ratios. These results suggest a promising role for multivalent scFvs for cancer targeting and imaging. Finally, biological characterization of larger tetravalent constructs, such as the diabody-Fc fusions described here, should provide important alternatives. Fragments that incorporate intact Fc regions are expected to exhibit the prolonged serum persistence of intact antibodies, as demonstrated for scFv-Fc fusion proteins. These could be ideal molecules for applications in which tetravalent, mono-, or bi-specific reagents with long biological half-lives are desired.

1.6. Bispecific Reagents

One of the primary motivations for the development of tetravalent antibody constructs has been the desire to obtain bispecific *and* bivalent agents for immunotherapeutic applications. Bispecific antibodies can be used to link or recruit target cells (e.g., tumor cells) with effector cells (e.g., cytotoxic T-cells). Ideally, one would want bivalent binding to both targets for optimal engagement of the effector cells. All these formats readily lend themselves to production as bispecific reagents. The generation of recombinant, bispecific antibody fragments is discussed in greater detail in Chapter 13.

1.7. Summary

Investigators in the antibody field have been very creative in their approaches to designing and producing multivalent antibody fragments for targeting purposes. Nature has also cooperated in providing surprising new formats, including the cross-paired configurations found in diabodies, triabodies, and tetrabodies. What remains to be played out is a systematic evaluation of the performance of these various formats in biological systems, both preclinical, and ultimately in the clinical setting. These in vivo studies in turn will provide essential feedback to antibody engineers, and eventually contribute to a rational approach to the design of recombinant antibodies for specific final applications.

Above all, it is important that we have a diverse collection of multivalent antibody fragments from which to select the most appropriate form.

2. Materials

2.1. Construction of Genes that Encode Multivalent Engineered Antibody Fragments

1. Murine hybridoma cells that secrete antibody of interest.
2. TRIzol (Invitrogen).
3. Gene-Amp kit (Applied Biosystems).
4. AMV reverse transcriptase (Promega).
5. Synthetic oligonucleotide primers.
6. Thermal cycler (Applied Biosystems or other supplier).
7. Agarose gel electrophoresis apparatus.
8. SeaKem LE agarose (FMC).
9. Tris-acetate-EDTA gel electrophoresis buffer: 40 m*M* Tris-acetate, 1 m*M* EDTA, pH 8.0.
10. Restriction enzymes.
11. Manual DNA sequence analysis reagents/equipment or automated DNA sequencing facility.
12. QIAquick gel purification kit (Qiagen).
13. AmpliTaq DNA polymerase (Applied Biosystems)."

3. Methods

3.1. Construction of Genes that Encode Multivalent Engineered Antibody Fragments

3.1.1. Amplification of Variable-Region Genes from Hybridoma-Cell Lines

1. Prepare total RNA from 2×10^7 or more hybridoma cells expressing the monoclonal antibody (MAb) of interest, using a standard method such as guanidine thiocyanate/cesium chloride ultracentrifugation or using a product such as TRIzol, according to the manufacturer's instructions (*see* **Note 1**).
2. Set up separate reverse transcriptase-polymerase chain reaction (RT-PCR) reactions, one for the heavy-chain and one for the light-chain variable region. Each RT-PCR reaction should contain 5 µg of total RNA, 20 pmol of upstream (V-region consensus; e.g., *see* **ref. *42***) and downstream (constant region) primers, 2.5 m*M* deoxynucleotide 5′ triphosphates (dNTPs), 10X PCR buffer, and 0.1 m*M* dithiothreitol (DTT) in a 50-µL reaction. Commercially available kits such as Gene-Amp provide a convenient source of reagents.
3. Denature the reaction at 70°C for 10 min.
4. Add 10 U of AMV reverse transcriptase and allow cDNA synthesis to proceed at 37°C for 15 min.
5. Add 2 U of Taq polymerase polymerase and amplify in a thermal cycler programmed for 30 cycles at 1 min (94°C), 1 min (58°C), 2 min (72°C).

6. Gel-purify the PCR products, clone into the plasmid of interest, and confirm that the appropriate variable regions have been isolated by DNA sequence analysis (*see* **Note 2**).
7. Alternately, if variable regions have been obtained through methods such as phage display or ribosome display, sequenced variable-region genes will provide the templates for the next step.

3.1.2. Gene Assembly by Splice Overlap Extension PCR (SOE-PCR)

1. Fusion primers should be designed to extend across the planned junction of the two starting gene segments, with a six-amino-acid (18 nucleotide) overlap on each side. If linker sequences are to be added between the gene segments, the appropriate nucleic acid sequences are appended to the 5′ end of the primer. Again, these primers should be designed so that the fragments to be joined have an 18-nucleotide overlap (**Fig. 3**). Each gene segment is then amplified separately using a flanking (outside) primer and the appropriate overlap primer using standard PCR conditions, 25–30 cycles, to create modified gene segments that now include a short overlap. (*See* **Notes 3** and **4** regarding overall design issues).
2. Purify PCR products by agarose gel electrophoresis, and include a DNA standard of known concentration for quantification. Following ethidium bromide staining, estimate the amount of PCR product.
3. Recover the PCR fragment from the agarose using the QIAquick gel purification kit or a similar method, according to the manufacturer's instructions.
4. Set up the overlap extension to include: 100 ng of each overlap fragment, 2 μL of 250 μM dNTP mix, 5 μL of 10X PCR buffer, 0.5 U of Taq polymerase, and H_2O to a final volume of 50 μL.
5. Perform overlap extension in a thermal cycler programmed to run six cycles of 1 min (94°C), 2 min (37°C), and 4 min (72°C).
6. Transfer 10 μL of the previous reaction to a fresh tube and add: 1 μL of 250 μ*M* dNTP mix, 4 μL of 10X PCR buffer, 20 pmol each flanking (outside) forward and backward primers, 0.5 U of Taq polymerase, and H_2O to a final volume of 50 μL. Amplify 25–30 cycles using a standard PCR program.
7. Gel-purify fused PCR gene product and insert into a plasmid to be grown up for DNA sequence analysis and confirmation that the correct fusion has been produced. Transfer the engineered antibody gene to an expression vector.

3.1.3. Expression and Characterization of Multivalent Engineered Antibody Fragments

A detailed example of mammalian expression and purification of an engineered anti-CEA minibody is given in Chapter 15.

4. Notes

1. Many sources of antibody variable regions are available and are described in greater detail elsewhere in this volume. A key consideration is that each antibody has its own personality: variable regions vary. Each individual Fv (pair of light- and

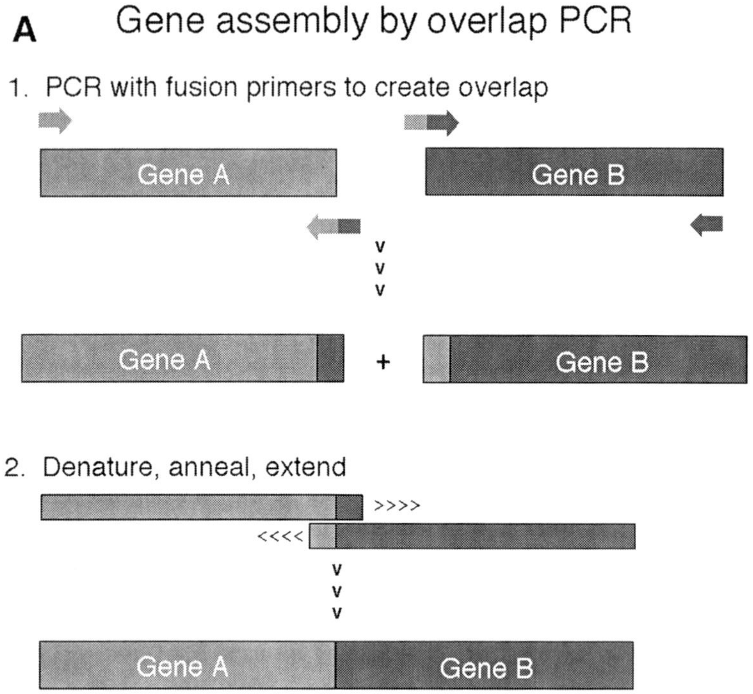

Fig. 3. Splice overlap extension method for fusing gene segments. Fusion primers are used in the PCR amplification of the individual gene segments, in order to create overlapping regions. The fragments are then denatured, annealed, extended, and reamplified using only the outermost primers to generate the fusion gene. (**A**) Direct fusion of two gene segments. (**B**) Addition of sequences encoding a linker peptide between two gene segments. Additional nucleotides are added to the 5′ ends of the overlap primers in order to add linker sequences (white/hatched/white in the final construct). At the same time, these primers are designed to generate the necessary overlap (hatched boxes) to allow the gene segments to be fused and reamplified. *(Figure continues)*

heavy-chain variable regions) will have its own physical properties such as pI, solubility, ability to fold, tendency to multimerize (for example, *see* **ref. *43***). Most antibody variable-region combinations can be successfully be converted into scFvs and higher-order forms; many work assembled in either orientation (V_H-V_L or V_L-V_H), but some antibody/antigen combinations may prefer one order of assembly over the other. The bottom line is that there is so much variation from antibody to antibody, that in order to ensure success it is best to embark on antibody engineering with with several candidates in mind.

2. When cloning antibody variable regions by RT-PCR, it is important to bear in mind that mutations can arise in any PCR. An error that occurs early in the amplification cycles can be propagated to many of the final clones. Thus, the DNA sequence must

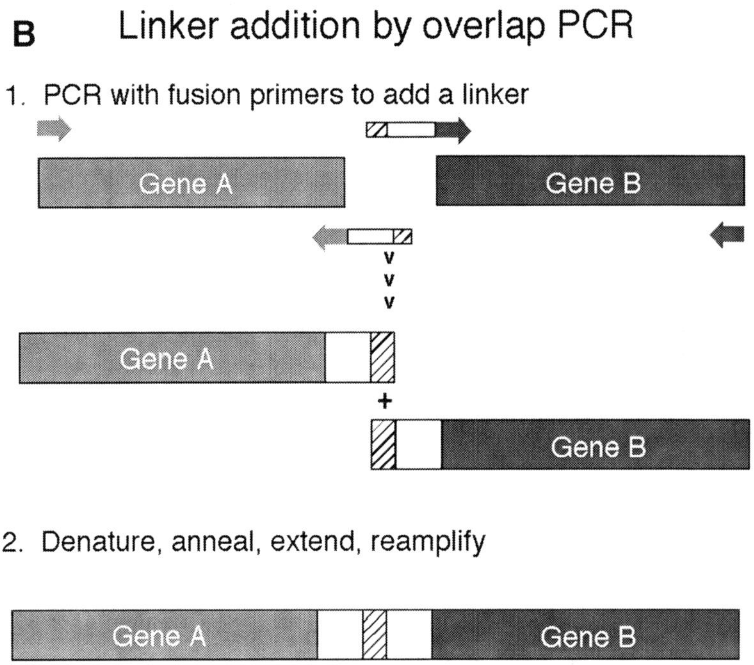

Fig. 3. *(Continued.)*

be determined from clones that have been generated in at least two independent RT-PCR reactions. Sequences should be checked to determine that endogenous kappa-chain transcripts have not been cloned. Comparison with databases such as Kabat (http://www.kabatdatabase.com; *see* also Chapter 2) will provide a useful indication as to whether the sequences cloned represent bona fide variable-region genes.

3. There seems to be considerable flexibility in linker design (*see* **ref. *44***). In general, linkers should be flexible and hydrophilic. The length of the linkers, as noted here, can be critical in determining the multimerization state. Furthermore, it is apparent that at this point, optimal linker length for generation of diabodies/triabodies/tetrabodies will probably still need to be determined empirically for any given antibody. Overly long linkers (in tandabs, covalent diabodies, or similar constructs) should be avoided, as they are often susceptible to proteolysis during production (especially if microbial expression is used) or when finally administered in vivo. As more knowledge and experience is gained, it should be possible to design linkers that are also non-immunogenic.
4. The various multivalent fragments described here should present the antigen-combining sites in a variety of spatial orientations and with differing flexibility. This is an area that will need further investigation. Ultimately, the optimal format must be complementary to the relevant epitope; however, it is presented on the target tissue. Furthermore, selection of the final configuration for a recombinant antibody frag-

ment will also depend on the intended function of the engineered antibody (e.g., imaging, drug delivery, bioactivity, or recruitment of effector cells).

References

1. Bird, R. E., Hardman, K. D., Jacobson, J. W., Johnson, S., Kaufman, B. M., Lee, S. M., et al. (1988) Single-chain antigen-binding proteins. *Science* **242,** 423–426.
2. Huston, J. S., Levinson, D., Mudgett-Hunter, M., Tai, M. S., Novotny, J., Margolies, M. N., et al. (1988) Protein engineering of antibody binding sites: recovery of specific activity in an anti-digoxin single-chain Fv analogue produced in Escherichia coli. *Proc. Natl. Acad. Sci. USA* **85,** 5879–5883.
3. Desplancq, D., King, D. J., Lawson, A.D.G., and Mountain, A. (1994) Multimerization behaviour of single chain Fv variants for the tumour-binding antibody B72.3. *Protein Eng.* **7,** 1027–1033.
4. Huston, J. S., George, A.J.T., Adams, G. P., Stafford, W. F., Jamar, F., Tsai, M.-S., et al. (1996) Single-chain Fv radioimmunotargeting. *Quarterly Journal of Nuclear Medicine* **40,** 320–323.
5. Wu, A. M. and Yazaki, P. J. (2000) Designer genes: recombinant antibody fragments for biological imaging. *Quarterly Journal of Nuclear Medicine* **44,** 268–283.
6. Yokota, T., Milenic, D. E., Whitlow, M., and Schlom, J. (1992) Rapid tumor penetration of a single-chain Fv and comparison with other immunoglobulin forms. *Cancer Res.* **52,** 3402–3408.
7. Begent, R.H.J., Verhaar, M. J., Chester, K. A., Casey, J. L., Green, A. J., Napier, M. P., et al. (1996) Clinical evidence of efficient tumor targeting based on single-chain Fv antibody selected from a combinatorial library. *Nat. Med.* **2,** 979–984.
8. Mayer, A., Tsiompanou, E., O'Malley, D., Boxer, G. M., Bhatia, J., Flynn, A. A., et al. (2000) Radioimmunoguided surgery in colorectal cancer using a genetically engineered anti-CEA single-chain Fv antibody. *Clin. Cancer Res.* **6,** 1711–1719.
9. Larson, S. M., El-Shirbiny, A. M., Divgi, C. R., Sgouros, G., Finn, R. D., Tschmelitsch, J., et al. (1997) Single chain antigen binding protein (sFv CC49)—First human studies in colorectal carcinoma metastatic to liver. *Cancer* **80** (Suppl.) 2458–2468.
10. Whitlow, M., Bell, B. A., Feng, S.-L., Filpula, D., Hardman, K. D., Hubert, S. L., et al. (1993) An improved linker for single-chain Fv with reduced aggregation and enhanced proteolytic stability. *Protein Eng.* **6,** 989–995.
11. Holliger, P., Prospero, T., and Winter, G. (1993) Diabodies: small bivalent and bispecific antibody fragments. *Proc. Natl. Acad. Sci. USA* **90,** 6444–6448.
12. Peresic, O., Webb, P. A., Holliger, P., Winter, G., and Williams, R. L. (1994) Crystal structure of a diabody, a bivalent antibody fragment. *Structure* **2,** 1217–1226.
13. Adams, G. P., McCartney, J. E., Tai, M.-S., Oppermann, H., Huston, J. S., Stafford, W. F., et al. (1993) Highly specific in vivo tumor targeting by monovalent and divalent forms of 741F8 anti-c-erbB-2 single-chain Fv. *Cancer Res.* **53,** 4026–4034.
14. Wu, A. M., Chen, W., Raubitschek, A. A., Williams, L. E., Fischer, R., Hu, S.-Z., et al. (1996) Tumor localization of anti-CEA single chain Fvs: Improved targeting by non-covalent dimers. *Immunotechnology* **2,** 21–36.

15. Wu, A. M., Williams, L. E., Zieran, L., Padma, A., Sherman, M. A., Bebb, G. G., et al. (1999) Anti-carcinoembryonic antigen (CEA) diabody for rapid tumor targeting and imaging. *Tumor Targeting* **4,** 47–58.
16. Adams, G. P., Schier, R., McCall, A. M., Crawford, R. S., Wolf, E. J., Weiner, L. M., and Marks, J. D. (1998) Prolonged in vivo tumour retention of a human diabody targeting the extracellular domain of human HER2/neu. *Br. J. Cancer* **77,** 1405–1412.
17. Viti, F., Tarli, L., Giovannoni, L., Zardi, L., and Neri, D. (1999) Increased binding affinity and valence of recombinant antibody fragments lead to improved targeting of tumoral angiogenesis. *Cancer Res.* **59,** 347–352.
18. Pack, P. and Pluckthun, A. (1992) Miniantibodies: use of amphipathic helices to produce functional, flexibly linked dimeric Fv fragments with avidity in Escherichia coli. *Biochemistry* **31,** 1579–1584.
19. Hu, S.-z., Shively, L., Raubitschek, A. A., Sherman, M., Williams, L. E., Wong, J.Y.C., et al. (1996) Minibody: a novel engineered anti-CEA antibody fragment (single-chain Fv-CH3) which exhibits rapid, high-level targeting of xenografts. *Cancer Res.* **56,** 3055–3061.
20. Li, E., Pedraza, A., Bestagno, M., Mancardi, S., Sanchez, R., and Burrone, O. (1997) Mammalian cell expression of dimeric small immune proteins (SIP). *Protein Eng.* **10,** 731–736.
21. De Kruif, J. and Logtenberg, T. (1996) Leucine zipper dimerized bivalent and bispecific scFv antibodies from a semi-synthetic antibody phage display library. *J. Biol. Chem.* **271,** 7630–7634.
22. McGregor, D. P., Molloy, P. E., Cunningham, C., and Harris, W. J. (1994) Spontaneous assembly of bivalent single chain antibody fragments in Escherichia coli. *Mol. Immunol.* **31,** 219–226.
23. Wu, A. M., Yazaki, P. J., Tsai, S., Nguyen, K., Anderson, A. L., McCarthy, D. W., et al. (2000) High-resolution microPET imaging of carcinoembryonic antigen-positive xenografts by using a copper-64-labeled engineered antibody fragment.PG. *Proc. Natl. Acad. Sci. USA* **97,** 8495–8500.
24. Ghetie, V. and Ward, E. S. (1997) FcRn: the MHC class I-related receptor that is more than an IgG transporter, in *Immunology Today,* Vol. 18, pp. 592–598.
25. Slavin-Chiorini, D. C., Kashmiri, S.V.S., Schlom, J., Calvo, B., Shu, L. M., Schott, M. E., et al. (1995) Biological properties of chimeric domain-deleted anticarcinoma immunoglobulins. *Cancer Res.* **55,** (Suppl.) 5957s–5967s.
26. Iliades, P., Kortt, A. A., and Hudson, P. J. (1997) Triabodies: single chain Fv fragments without a linker form trivalent trimers. *FEBS Lett.* **409,** 437–441.
27. Le Gall, F., Kipriyanov, S. M., Moldenhauer, G., and Little, M. (1999) Di-, tri- and tetrameric single chain Fv antibody fragments against human CD19: effect of valency on cell binding. *FEBS Lett.* **453,** 164–168.
28. Dolezal, O., Pearce, L. A., Lawrence, L. J., McCoy, A. J., Hudson, P. J., and Kortt, A. A., et al. (2000) ScFv multimers of the anti-neuraminidase antibody NC10: shortening of the linker in single-chain Fv fragment assembled in VL to VH orientation drives the formation of dimers, trimers, tetramers, and higher molecular mass multimers. *Protein Eng.* **13,** 565–574.

29. Atwell, J. L., Breheney, K. A., Lawrence, L. J., McCoy, A. J., Kortt, A. A., and Hudson, P. J. (1999) scFv multimers of the anti-neuraminidase antibody NC10: length of the linker between VH and VL domains dictates precisely the transition between diabodies and triabodies. *Protein Eng.* **12,** 597–604.
30. Hudson, P. J. and Kortt, A. A. (1999) High avidity scFv multimers: diabodies and triabodies. *J. Immunol. Methods.* **231,** *177–190.*
31. Pei, X. Y., Holliger, P., Murzin, A. G., and Williams, R. L. (1997) The 2.0-Åresolution crystal structure of a trimeric antibody fragment with noncognate VH-VL domain pairs shows are arrangement of VH CDR3. *Proc. Natl. Acad. Sci. USA* **94,** 9637–9642.
32. Kipriyanov, S. M., Moldenhauer, G., Schuhmacher, J., Cochlovius, B., Von der Lieth, C. W., Matys, E. R., et al. (1999) Bispecific tandem diabody for tumor therapy with improved antigen binding and pharmacokinetics. *J. Mol. Biol.* **293,** 41–56.
33. Goel, A., Colcher, D., Baranowska-Kortylewicz, J., Augustine, S., Booth, B. J., Pavlinkova, G., et al. (2000) Genetically engineered tetravalent single-chain Fv of the pancarcinoma monoclonal antibody CC49: improved biodistribution and potential for therapeutic application. *Cancer Res.* **60,** 6964–6971.
34. Singh, R. K., Joshi, V. K., Goel, R. K., Gambhir, S. S., and Acharya, S. B. (1996) Pharmacological actions of Pongamia pinnata seeds—a preliminary study. *Indian J. Exp. Biol.* **34,** 1204–1207.
35. Pack, P., Müller, K., Zahn, R., and Plükthun, A. (1995) Tetravalent miniantibodies with high avidity assembling in Escherichia coli. *J. Mol. Biol.* **246,** 28–34.
36. Willuda, J., Kubetzko, S., Waibel, R., Schubiger, P. A., Zangemeister-Wittke, U., and Pluckthun, A. (2001) Tumor targeting of mono-, di-, and tetravalent anti-p185(HER-2) miniantibodies multimerized by self-associating peptides. *J. Biol. Chem.* **276,** 14,385–14,392.
37. Santos, A. D., Kashmiri, S. V., Hand, P. H., Schlom, J., and Padlan, E. A. (1999) Generation and characterization of a single gene-encoded single-chain-tetravalent antitumor antibody. *Clin. Cancer Res.* **5,** 3118s–3123s.
38. Alt, M., Müller, R., and Kontermann, R. E. (1999) Novel tetravalent and bispecific IgG-like antibody molecules combining single-chain diabodies with the immunoglobulin gammal Fc or CH3 region. *FEBS Lett.* **454,** 90–94.
39. Dübel, S., Breitling, F., Kontermann, R., Schmidt, T., Skerra, A., and Little, M. (1995) Bifunctional and multimeric complexes of streptavidin fused to single chain antibodies (scFv). *J. Immunol. Methods* **178,** 201–209.
40. Goshorn, S., Sanderson, J., Axworthy, D., Lin, Y., Hylarides, M., and Schultz, J. (2001) Preclinical evaluation of a humanized NR-LU-10 antibody-streptavidin fusion protein for pretargeted cancer therapy. *Cancer Biother. Radiopharm.*
41. Goel, A., Baranowska-Kortylewicz, J., Hinrichs, S. H., Wisecarver, J., Pavlinkova, G., Augustine, S., et al. (2001) 99mTc-labeled divalent and tetravalent CC49 single-chain Fv's: novel imaging agents for rapid in vivo localization of human colon carcinoma. *J. Nucl. Med.* **42,** 1519–1527.
42. Dübel, S., Breitling, F., Fuchs, P., Zewe, M., Gotter, S., Welschof, M., et al. (1994) Isolation of IgG antibody Fv-DNA from various mouse and rat hybridoma cell

lines using the polymerase chain reaction with a simple set of primers. *J. Immunol. Methods* **175,** 89–95.
43. Wu, A. M., Tan, G. J., Sherman, M. A., Clarke, P., Olafsen, T., Forman, S. J., et al. (2001) Multimerization of a chimeric anti-CD20 single-chain Fv-Fc fusion protein is mediated through variable domain exchange. *Protein Eng.* **14,** 1025–1033.
44. Volkel, T., Korn, T., Bach, M., Muller, R., and Kontermann, R. E. (2001) Optimized linker sequences for the expression of monomeric and dimeric bispecific single-chain diabodies. *Protein Eng.* **14,** 815–823.

13

Production of Recombinant Bispecific Antibodies

Roland E. Kontermann, Tina Völkel, and Tina Korn

1. Introduction

Recombinant bispecific antibodies have great potential as diagnostic and therapeutic reagents *(1,2)*. Applications of recombinant bispecific antibodies include the recruitment of effector molecules (e.g., enzymes, complement components, immunoglobulins), effector cells (cytotoxic T-lymphocytes, natural killer [NK] cells, macrophages), or gene-therapeutic vectors (e.g., adenoviruses) to other molecules or target cells *(3–10)*. Through genetic engineering, molecules with the desired properties can be generated and produced in various prokaryotic or eukaryotic expression systems *(11)*. For example, by starting from antibody fragments isolated from human combinatorial libraries by phage display, completely human bispecific molecules can be generated.

A large number of different recombinant bispecific antibody formats have been developed during the past decade *(12)*. In this chapter, we will focus on the three formats that have been widely used for various applications. These formats are diabodies, single-chain diabodies (scDbs), and tandem scFvs (single-chain antibodies) (**Fig. 1**). All three formats are small, bispecific molecules with a mol wt of 50–60 kDa produced by assembly of the variable domains of two antibodies with different specificities. Their small size facilitates tissue penetration, but also results in a rapid clearance from circulation *(13)*. This must be considered when designing recombinant bispecific molecules for therapy. In order to increase the serum half-life of these small bispecific molecules, their size can be increased by fusion to additional protein domains, such as the immunoglobulin C_H3 or Fc part *(14,15)*. The resulting dimeric molecules possess molecular masses greater than 100 kDa, and exhibit a prolonged serum half-life.

Diabodies are heterodimers of two polypeptide chains of the structure V_HA-V_LB and V_HB-V_LA expressed in the same cell *(3)*. Functional antigen-binding sites are formed by crossover pairing of the variable light- and heavy-chain

From: *Methods in Molecular Biology, Vol. 248: Antibody Engineering: Methods and Protocols*
Edited by: B. K. C. Lo © Humana Press Inc., Totowa, NJ

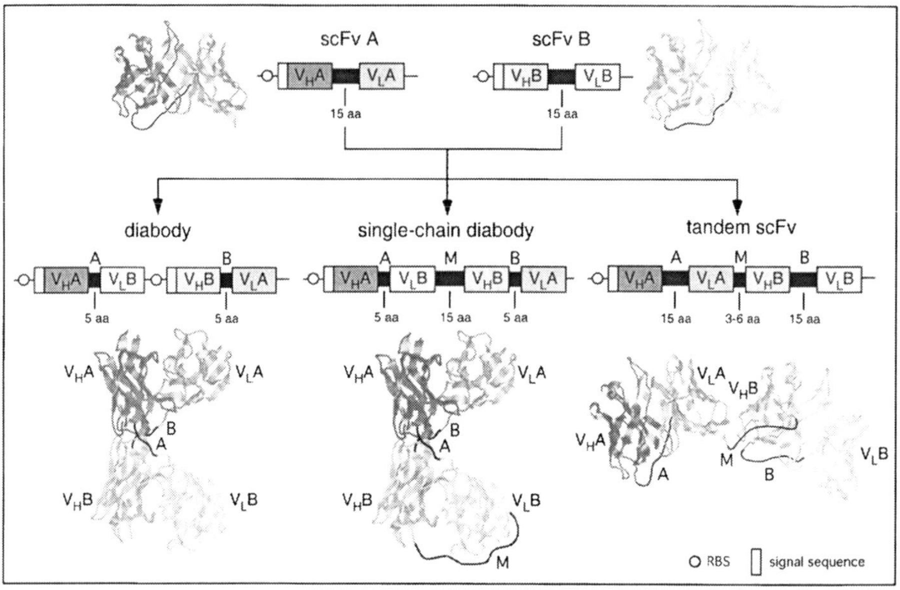

Fig. 1. Format and structure of recombinant bispecific antibodies: diabodies, single-chain diabodies, and tandem scFv.

domains of the two chains. They possess a rigid structure and can be expressed at high yields in bacteria. However, because of random association of the two polypeptide chains, nonfunctional homodimers are also formed, thus generating heterogenous populations of molecules. Pure populations of bispecific diabodies can be obtained by affinity chromatography.

In the scDb format, the two polypeptide chains used to generate a bispecific diabody are connected through an additional middle linker *(7)*. Consequently, all molecules with a molecular mass of 50–60 kDa are bispecific. ScDbs can be expressed in bacteria, although yields are usually lower than those obtained for diabodies. ScDb can also be expressed in eukaryotic cells and displayed on the surface of cells and bacteriophage ***(16,17)***. They can also be used to generate tetravalent, bispecific IgG-like molecules by simple fusion of the scDb to the Ig γ1 Fc or C_H3 region ***(14)***. By applying phage display, we have recently identified a panel of optimized non-repetitive linker sequences for the generation of scDbs ***(17)***. By varying the length of the three linkers of a scDb, it is also possible to produce bispecific scDb homodimers that possess two binding sites for each antigen and a mol wt of approx 110 kDa ***(13,17)***. These molecules have a much longer serum half-life because of their larger size.

The third format, tandem scFv (taFv), is produced by connecting two scFv molecules through a short middle linker ***(18–20)***. Although this format has a very flexible structure and is the most simple one to generate, difficulties in the expression of soluble taFvs in bacteria have been reported. Thus, refolding strategies or eukaryotic expression systems have been used for the production of soluble tandem scFvs ***(12,18,20,21)***. Recent studies have shown that it is possible to express taFvs in bacteria in a soluble and active form—e.g., by using appropriate linker sequences, as shown in this chapter.

2. Materials

2.1. Construction of Bispecific Diabodies

1. Bacterial expression vector pAB1 ***(4)***.
2. Thermocycler Robocycler (Stratagene).
3. Thermostable DNA polymerases and reaction buffers: Taq polymerase (e.g., from Amersham Biosciences), Pfu polymerase, and herculase (Stratagene).
4. Deoxynucleotide 5′ triphosphates (dNTPs) (Amersham Biosciences).
5. 1X TAE buffer: 40 m*M* Tris-acetate, 1 m*M* EDTA, pH 8.0.
6. 1% agarose gel: disolve 1 g of agarose in 1X TAE buffer. Boil in a microwave oven until completely melted, cool down to approx 45°C, and add 3 µL of ethidium bromide (10 mg/mL).
7. Restriction enzymes *Asc*I, *Bst*EII, *Not*I, *Sac*I, *Sfi*I, and restriction enzyme buffers (New England Biolabs).
8. Qiaquick gel extraction kit (Qiagen).
9. Primers for amplification of individual V_H and V_L domains (restriction sites are shown in italic, dots indicate bases that must be derived from individual sequences. For annealing, these regions should have a length of approx 20 bases):
   a. V_HABstFor: 5′-CGA GGA GAC *GGT GAC C*AG … …–3′ (reverse primer that anneals in the 3′-end of V_LA and introducing a *Bst*EII site in the region encoding framework 4);
   b. V_LABstSacBack: 5′-ATC CT*G GTC ACC* GTC TCC TCG GGC GGT GGC GGA TCC GAT ATC *GAG CTC* … …–3′ (anneals in the 5′-end of V_LA and introduces a *Bst*EII site, the GGGGS-linker encoding region and a *Sac*I site into V_LA);
   c. V_HBAscBack: 5′-TAA *GGC GCG CC*A ATG ATT ACG CCA AGC TTT CTA GAG-3′ (anneals in the 5′-untranslated region preceding the pelB signal sequence of scFv B and introduces an *Asc*I site; if using another vector, the sequence must be adjusted);
   d. V_HBSacFor: 5′-TCG *GAG CTC* GAT GTC CGA TCC GCC ACC GCC … …–3′ (reverse primer that anneals in the 3′-end of V_HB and introduces DNA encoding a five-amino acid GGGGS linker and a *Sac*I site);
   e. V_LBBstBack: 5′-GAT CT*G GTC ACC* GTC TCC TCA GGC GGT GGC GGA TCG … …–3′ (anneals in the 5′-end of V_LB and introduces a *Bst*EII site);
   f. V_LBAscFor: 5′-GCA T*GG CGC GCC* TTA TTA … … 3′ (reverse primer that anneals in the 3′-end of V_LB and introduces an *Asc*I site);
   g. LMB2: 5′-GTA AAA CGA CGG CCA GT-3′;
   h. LMB3: 5′-CAG GAA ACA GCT ATG ACC A-3′.

10. DNA ligase and buffer (Promega).
11. Luria-Bertani medium (LB): 5 g of NaCl, 10 g of bacto-tryptone, and 5 g of yeast extract per 1 L.
12. LB/amp/1% glucose medium: LB medium containing 100 μg/mL ampicillin and *1%* glucose.
13. DNA midiprep kit (Qiagen).

2.2. Construction of Bispecific scDbs

1. Primers for amplification of individual V_H and V_L domains:
   a. V_HABstFor, V_LABstSacBack, V_HBSacFor, and V_LBBstBack (*see* **Subheading 2.1.**);
   b. MV_HBAscBack: 5′-TAA G*GG CGC GCC* TCG GCT GGT AAT ACT AGT … …–3′ (anneals in the 5′-end of V_HB and introduces an *Asc*I site followed by DNA encoding the second part of middle linker M);
   c. V_LBMAscFor: 5′-GCA *GGC GCG CC*C AGC ATT ACT ATC ACT ACC … …–3′ (reverse primer that anneals in the 3′-end of V_LB and introduces DNA encoding the first part of middle linker M followed by an *Asc*I site);
2. Restriction enzymes *Asc*I, *Bst*EII, *Not*I, *Sac*I, *Sfi*I, and restriction enzyme buffers.
3. Bacterial expression vector pAB1 ***(4)***.
4. Thermocycler Robocycler.
5. Thermostable DNA polymerases and reaction buffers (*see* **Subheading 2.1.**).
6. dNTPs.
7. 1% agarose gel and 1X TAE buffer (*see* **Subheading 2.1.**).
8. DNA ligase and buffer.
9. DNA midiprep kit.

2.3. Construction of taFvs

1. Primers for amplification of individual scFv fragments:
   a. LMB3 and LMB2 (*see* **Subheading 2.1.**);
   b. V_LAXhoFor: 5′-GCA GGA TGC GGC CGC CCG TTT CAG *CTC GAG* CTT GGT GCC … …–3′ (reverse primer that anneals in the 3′-end of scFv A and introduces a *Xho*I site);
   c. V_HBXhoBack: 5′-TCC AAT *CTC GAG* CTG AAA CGG AGT ACT GAT GGT AAT ACT … …–3′ (anneals in the 5′-end of scFv B and introduces a *Xho*I site followed by DNA encoding the middle linker N).
2. Restriction enzymes *Not*I, *Sfi*I, and *Xho*I, and restriction enzyme buffers (New England Biolabs);
3. Bacterial expression vector pAB1 ***(4)***.
4. Thermocycler Robocycler.
5. Thermostable DNA polymerases and reaction buffers (*see* **Subheading 2.1.**).
6. dNTPs.
7. 1% agarose gel and 1X TAE buffer (*see* **Subheading 2.1.**).
8. DNA ligase and buffer.
9. DNA midiprep kit.

2.4. Transformation and Screening of Positive Clones

1. LB medium (*see* **Subheading 2.1., item 11**).
2. LB agar for plates: LB medium containing 15 g/L agar.
3. LB/amp/1% glucose agar plates: LB agar containing 100 µg/mL ampicillin and 1% glucose.
4. *E. coli* TG1 (Stratagene).
5. Ice-cold 100 m*M* $CaCl_2$.
6. Glycerol.
7. Screening primers: LMB2 and LMB3 (*see* **Subheading 2.1.**).
8. Thermocycler Robocycler.
9. Thermostable DNA polymerases and reaction buffers (*see* **Subheading 2.1.**).
10. dNTPs.
11. 1% agarose gel and 1X TAE buffer (*see* **Subheading 2.1.**).

2.5. Expression and Purification

1. LB medium (*see* **Subheading 2.1, item 11**).
2. LB/amp/1% glucose medium: LB medium containing 100 µg/mL ampicillin and 1% glucose.
3. LB/amp/0.1% glucose medium: LB medium containing 100 µg/mL ampicillin and 0.1% glucose.
4. Isopropyl-β-D-galactopyranoside (IPTG).
5. 2-L Erlenmeyer flasks.
6. Periplasmic extraction buffer (PPB): 30 m*M* Tris-HCl, pH 8.0, 1 m*M* EDTA, 20% sucrose.
7. Lysozyme (powder).
8. 1 *M* $MgSO_4$.
9. Dialysis tubing (10-kDa cutoff) and clamps.
10. Reagents for immobilized metal affinity chromatography (IMAC):
11. IMAC loading buffer: 50 m*M* sodium phosphate buffer, 500 m*M* NaCl, 20 m*M* imidazole, pH 7.5.
12. IMAC wash buffer: 50 m*M* sodium phosphate buffer, 500 m*M* NaCl, 35 m*M* imidazole, pH 7.5.
13. IMAC elution buffer: 50 m*M* sodium phosphate buffer, 500 m*M* NaCl, 100 m*M* imidazole, pH 7.5.
14. Ni-NTA resin (Qiagen).
15. Polypropylene columns (Biorad).
16. Bradford reagent (Biorad).
17. 96-well microtiter plate.
18. Phosphate-buffered saline (PBS), pH 7.5: 137 m*M* NaCl, 3 m*M* KCl, 8 m*M* Na_2HPO_4, 1.5 m*M* KH_2PO_4.
19. Quartz cuvet.
20. Standard sodium dodecyl polyacrylamide gel electrophoresis (SDS-PAGE), apparatus.

2.6. ELISA

1. 96-well flat-bottom flexible Falcon microtiter plates (Becton-Dickinson).
2. PBS.
3. 50 m*M* sodium carbonate buffer, pH 9.6.
4. 2% MPBS: PBS containing 2% (w/v) skimmed milk powder.
5. Horseradish peroxidase (HRP)-conjugated mouse anti-His-tag antibody (Santa Cruz Biotechnology).
6. Anti-Myc-tag antibody 9E10 (Genosys).
7. HRP-conjugated goat anti-mouse antibody (Dianova).
8. TMB substrate buffer: 100 m*M* sodium phosphate buffer, pH 6.0.
9. TMB substrate solution: add 100 µL of 10 mg/mL 3,3′-5,5′-tetramethylbenzidine (TMB) in dimethyl sulfoxide (DMSO) and 2 µL of 30% H_2O_2 per 10 mL of TMB substrate buffer.
10. Microtiter plate reader Spectromax 340 (MGW-Biotech).

2.7. Flow Cytometry (FACS)

1. 15-mL Polystyrol tubes (Nunc).
2. 6-mL Polystyrol tubes (Becton Dickinson Falcon).
3. Anti-His-tag antibody (Dianova).
4. Cy3-conjugated goat anti-mouse antibody (Dianova).
5. 1% fetal calf serum (FCS)/PBS (filtered).
6. PBS.
7. 3.7% formalin/PBS.
8. Fluorescence-activated cell sorting (FACS) Calibur (Becton Dickinson).
9. EDTA/PBS: 0.2 mg/mL EDTA in PBS.

3. Methods

3.1. Construction of Bispecific Diabodies

Bispecific diabodies in the V_H-V_L orientation are generated by expressing two polypeptide chains of the format V_HA-V_LB and V_HB-V_LA in the same cell. The genes that encode these chains are combined in a single expression plasmid (pAB1). Each coding sequence is preceded by a ribosome binding site (RBS) and the pelB signal sequence-encoding region (**Fig. 2**). For purification and detection, one chain also contains at its 3′-end a sequence for the Myc-tag and a hexahistidyl-tag. For the construction of a bispecific diabody, appropriate cloning sites must be introduced into binding site A. There is a *Bst*EII site in the 3′-region of the V_H gene and a *Sac*I site in the 5′-region of the V_L gene. The genes encoding the V_H and V_L domains of binding site B are then amplified by PCR, introducing a *Bst*EII and a *Asc*I site into V_LB and an *Asc*I site and a *Sac*I site into the V_HB domain, which also contains the vector-encoded RBS and signal sequence. The V_LB and V_HB fragments are then inserted into binding site A

```
               LMB3                      HinD III                           RBS
... CAGGAAACAGCTATGACCATGATTACGCCAAGCTTCTAGAGCATGCAAATTCTATTTCAAGGAGACAGTCATA

                                                                        Sfi I
ATG AAA TAC CTA TTG CCT ACG GCA GCC GCT GGA TTG TTA TTA CTC GCG GCC CAG CCG GCC
 M   K   Y   L   L   P   T   A   A   A   G   L   L   L   L   A   A   Q   P   A
 pelB signal sequence

                     Pst I   Sal I   Xho I                Asc I       Not I
ATG GCC CAG GTG CAG CTG CAG GTC GAC CTC GAG ATC AAA CGA GGC GCG CCA GCG GCC GCA
 M   A   Q   V   Q   L   Q   V   D   L   E   I   K   R   G   A   P   A   A   A

GAA CAA AAA CTC ATC TCA GAA GAG GAT CTG AAT GGG GCC GCA CAT CAC CAT CAT CAC CAT
 E   Q   K   L   I   S   E   E   D   L   N   G   A   A   H   H   H   H   H   H
 Myc-tag                                                 His-tag

       EcoR I        LMB2
TAA TAAGAATTCCACTGGCCGTCGTTTTAC ...
 *
```

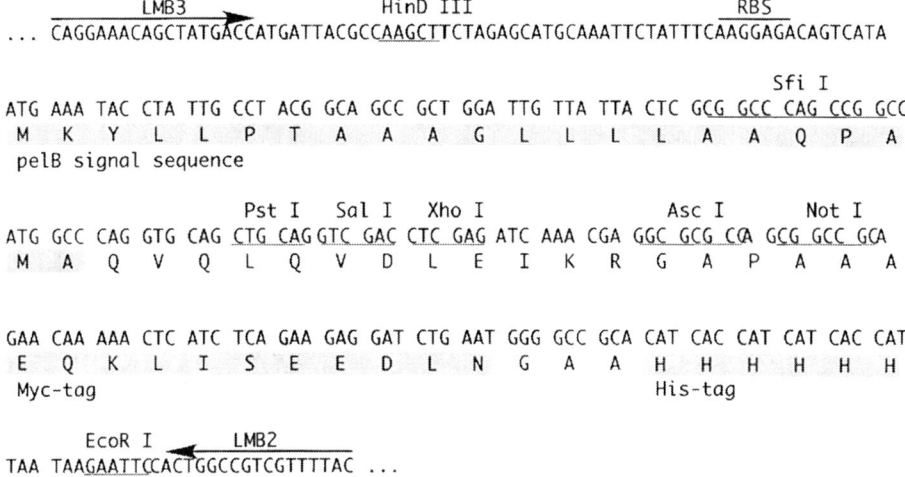

Fig. 2. Sequence of the cloning site of expression vector pAB1. The positions of screening primers LMB3 and LMB2 are indicated by arrows. The ribosome binding site (RBS), the pelB signal sequence, and the Myc- and His-tag are shown as boxes.

contained in the expression plasmid by a two-fragment ligation step (**Fig. 3**). As starting material, we use scFv fragments cloned in a bacterial expression vector (e.g., pAB1), which already contains a RBS and a signal sequence.

3.1.1. Binding Site A: Introduction of a BstEII and a SacI Site

1. PCR-amplify DNA encoding the V_H domain of the first antibody (A) with primers LMB3 and V_HABstFor introducing a *Bst*EII site at the C-terminal region (if not already present). For PCR, we routinely use Taq polymerase as well as a proof-reading polymerase (*Pfu,* herculase) and 25 cycles for amplification. PCRs are performed in a total volume of 50 µL containing 250 µ*M* dNTPs, 1.5 m*M* $MgCl_2$, 1 pmol of each primer, and 0.1 µg template DNA. Each cycle consists of 30 s at 94°C, 1 min at 50°C, and 1 min at 72°C.
2. PCR-amplify DNA encoding the V_L domain of the first antibody (A) with LMB2 and a V_LABstSacBack encoding the C-terminal region of the V_H domain (covering the *Bst*EII site), a five-residue linker A, and introducing a *Sac*I site at the N-terminal region of the V_L domain (*see* **Fig. 3**).
3. Analyze PCR products on a 1% agarose gel in 1X TAE buffer and isolate the fragments using the Qiaquick gel extraction kit.
4. Digest the V_HA fragment with *Sfi*I and *Bst*EII and the V_LA fragment with *Bst*EII and *Not*I.
5. Clone into bacterial expression vector pAB1 digested with *Sfi*I and *Not*I and transform into *E. coli* TG1 (*see* **Subheading 3.4.**). The insert encodes a bivalent diabody Db-A.

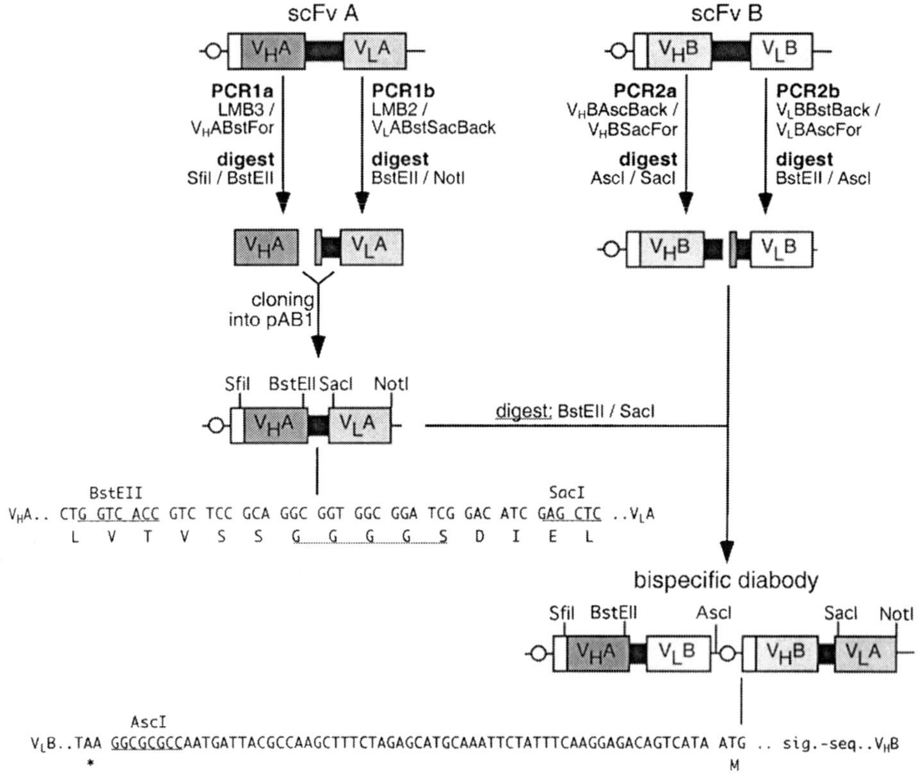

Fig. 3. Construction of a bispecific diabody (Db). In a first step, a *Bst*EII and a *Sac*I site are introduced into the variable fragments of binding site A. The variable fragments of binding site B are then amplified, introducing the short linkers A and B as well as an *Asc*I site in front of the untranslated region of the V_HB fragment. These fragments are then combined with binding site A to generate a bispecific diabody construct.

6. Prepare plasmid DNA (pAB1-Db-A) from a 50-mL overnight culture grown in LB/amp/1% glucose medium using a DNA plasmid midiprep kit.
7. Digest plasmid DNA with *Bst*EII and *Sac*I.

3.1.2. Binding Site B: Amplification and Insertion of Cloning Sites

1. PCR-amplify DNA encoding the V_H domain of the second antibody (B) with V_HBAscBack, annealing in the 5′-untranslated region preceding the pelB leading sequence and beginning with an *Asc*I site, and V_HBSacFor encoding a five-residue linker B and a *Sac*I site.
2. PCR-amplify DNA encoding the V_L domain of the second antibody (B) with V_LBAscFor, encoding the N-terminal half of the linker M beginning with an *Asc*I site and a backward primer encoding a five-residue linker A and a *Bst*EII site.

3. Analyze PCR products on a 1% agarose gel in 1X TAE buffer and isolate the fragments using the Qiaquick gel extraction kit.
4. Digest the V_HB fragment with *Asc*I and *Sac*I and the V_LB fragment with *Bst*EII and *Asc*I.

3.1.3. Combining DNA Encoding Binding Sites A and B

1. Clone digested fragments V_HB and V_LB into plasmid pAB1-Db-A digested with *Bst*EII and *Sac*I. Use a molar ratio of insert to vector of approx 3:1 in a 20-µL reaction and 1 U of DNA ligase. Incubate overnight at 15°C and transform into *E. coli* TG1 (*see* **Subheading 3.4.**). The insert encodes a bispecific diabody (**Fig. 3**).
2. Screen for positive clones by PCR with primers LMB3 and LMB2 as described in **Subheading 3.4.** Positive clones yield a PCR product of 1.6 kb.
3. Use positive clones for further analysis (**Subheadings 3.5–3.7**).

3.2. Construction of Bispecific ScDbs

To avoid non-functional diabody formation comprised of homodimers, the two polypeptide chains are fused together by a middle-linker M that generates a bispecific single-chain diabody. Linker M consists of the non-repetitive 15-amino acid sequence GSDSNA*GRAS*AGNTS *(17)*. The underlined amino acids (in **Fig. 4**) are encoded by a nucleotide sequence containing the *Asc*I restriction site. The coding sequence for the scDb is preceded by a RBS and the pelB signal sequence-encoding region. At the 3′-end of the gene encoding the antibody-fragment sequences coding for the Myc-tag and the hexahistidyl-tag are added, which are needed for purification and detection. The construction of bispecific scDbs is very similar to the construction of bispecific diabodies (**Subheading 3.1.**). Differences occur in the sequences of the oligonucleotides.

1. Introduce a *Bst*EII and a *Sac*I site into binding site A as described in **Subheading 3.1.1.**
2. PCR-amplify DNA encoding the V_H domain of the second antibody (B) with $MV_HBAscBack$, annealing at the 5′-end of V_HB, beginning with an *Asc*I site and encoding the C-terminal part of linker M, and $V_HBSacFor$ encoding a five-residue linker B and a *Sac*I site.
3. PCR-amplify DNA encoding the V_L domain of the second antibody (B) with V_LB-MAscFor, annealing at the 3′-end of V_LB, ending with an *Asc*I site and encoding the N-terminal part of linker M, and $V_LBBstBack$ encoding a five-residue linker A and a *Bst*EII site.
4. Analyze PCR products on a 1% agarose gel in 1X TAE buffer and isolate the fragments using the Qiaquick gel extraction kit.
5. Digest the V_HB fragment with *Asc*I and *Sac*I and the V_LB fragment with *Bst*EII and *Asc*I.
6. Combine DNA-encoding binding sites A and B as described in **Subheading 3.1.3.**

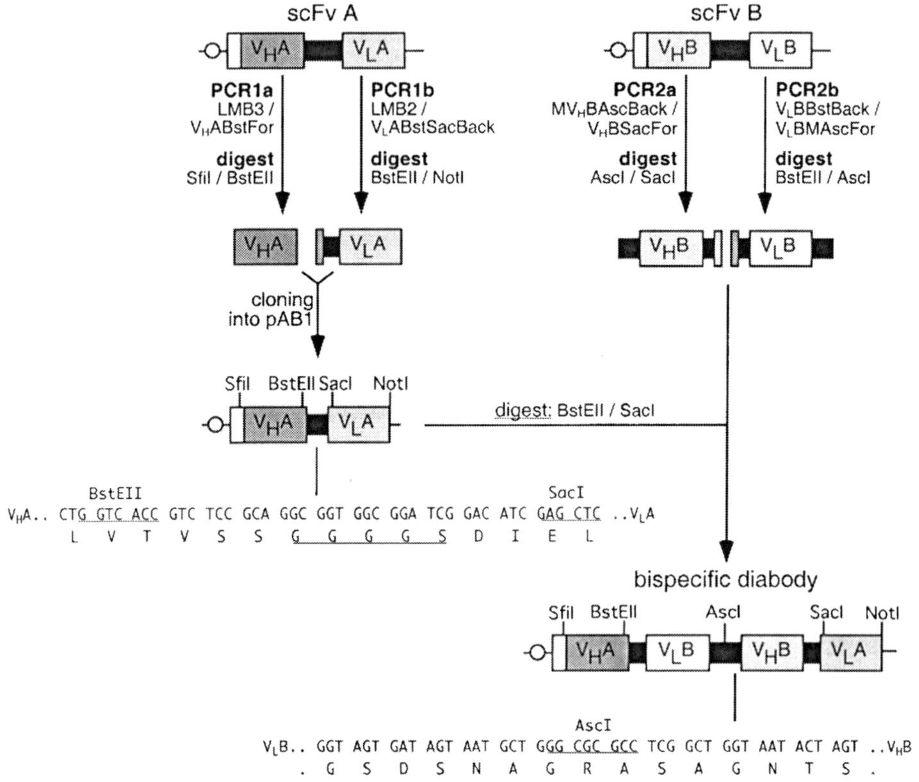

Fig. 4. Construction of a bispecific scDb. As for the construction of bispecific diabodies (**Subheading 3.1.**) a *Bst*EII and a *Sac*I site are introduced into the variable fragments of binding site A. The variable fragments of binding site B are then amplified, introducing the short linkers A and B as well as a linker M sequence containing an *Asc*I site in the middle. These fragments are then combined with binding site A to generate a bispecific scDb construct.

3.3. Generation of Bispecific taFv

In order to generate a bispecific taFv the polypeptide chains of two scFvs are fused together with a middle-linker M. The sequence of linker M was selected from a taFv library and consists of the following six amino acids: STDGNT. Similar to the scDb, the coding sequence of the taFv is preceded by a RBS and the pelB signal sequence-encoding region. At the C-terminus of the taFv sequence, a Myc-tag and a hexahistidyl-tag are added that enable the purification and detection of the antibody construct. For the construction of a taFv, appropriate cloning sites must be introduced into scFv A (binding

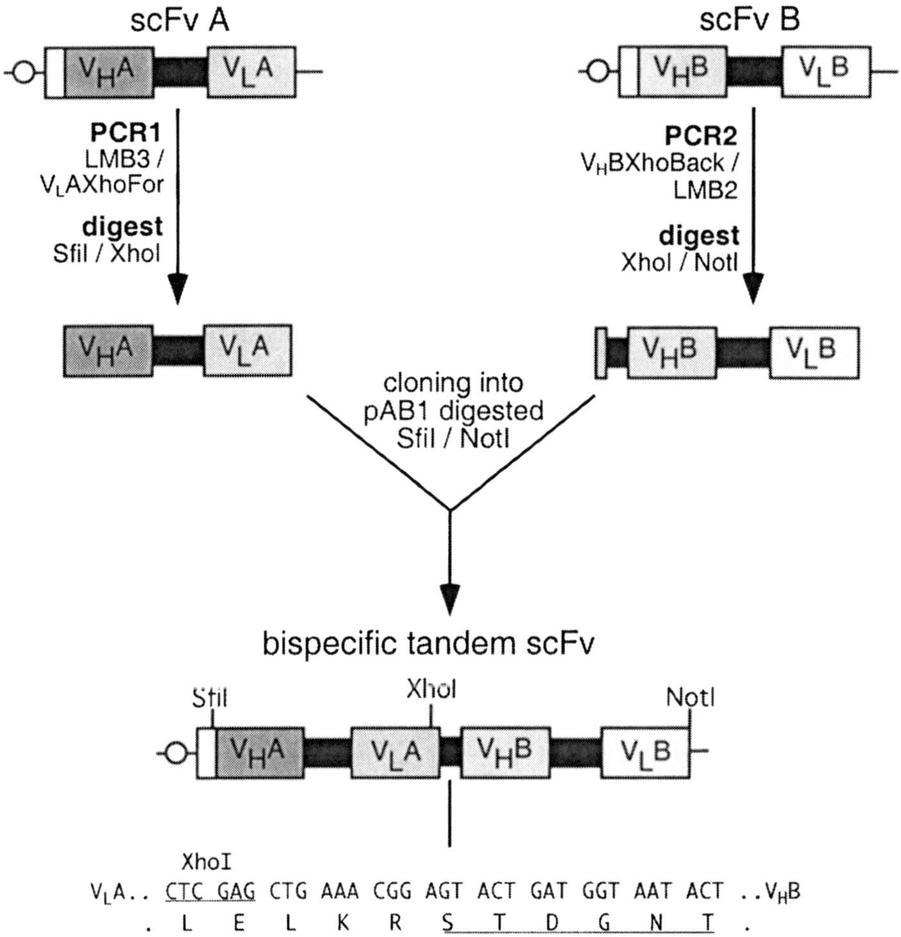

Fig. 5. Construction of a bispecific taFv. ScFv A is amplified to introduce a *Xho*I site at the 3′-end of V_LA and scFv B is PCR-amplified to introduce a *Xho*I site and a linker M-encoded sequence at the 5′-end of V_HB. These fragments are then combined to generate a bispecific taFv construct.

site A) and scFv B (binding site B). By PCR-amplification of scFv A a *Sfi*I site is introduced at the 5′-region and a *Xho*I site at the 3′-region. To connect the scFv B with the scFv A, a *Xho*I site and the linker M must be introduced at the 5′-region of scFv B. At the 3′-region of this scFv, a *Not*I site must be introduced. The two resulting fragments are then inserted in a bacterial expression vector (e.g., pAB1) by a two-fragment ligation step. In this case, the bacterial expression vector already contains the RBS and the pelB leader sequence **(Fig. 5)**.

3.3.1. Binding Site A: Introduction of a SfiI and a XhoI Site

1. PCR-amplify DNA encoding the scFv A with primers LMB3 and V_LAXhoFor for introducing a *Sfi*I site at the 5′-region and a *Xho*I site at the 3′-region. Perform PCR as described in **Subheading 3.1.1.**
2. Analyze PCR product on a 1% agarose gel in 1X TAE buffer and isolate the fragment using the Qiaquick gel extraction kit.
3. Digest the scFv A fragment with *Sfi*I and *Xho*I.

3.3.2. Binding Site B: Amplification, Insertion of Cloning Sites, and Linker M

1. PCR-amplify DNA encoding the scFv B with primers LMB2 and V_HBXhoBack annealing at the 5′-end of scFv B, beginning with a *Xho*I site and encoding the linker M.
2. Analyze PCR product on a 1% agarose gel in 1X TAE buffer and isolate the fragment using the Qiaquick gel extraction kit.
3. Digest the scFv B-fragment with *Xho*I and *Not*I.

3.3.3. Combining DNA Encoding ScFv A and ScFv B

1. Clone digested fragments scFv A and scFv B into bacterial expression vector digested with *Sfi*I/*Not*I (*see* **Subheading 3.1.3.**) and transform into *E. coli* TG1 (*see* **Subheading 3.4.**). The insert encodes a bispecific taFv **(Fig. 5)**.
2. Screen for positive clones by PCR with primers LMB3 and LMB2 as described in **Subheading 3.4.** Positive clones yield a PCR product of 1.6 kb.
3. Use positive clones for further analysis **(Subheadings 3.5 to 3.6.)**.

3.4. Transformation and Screening of Positive Clones

3.4.1. Preparation of Competent Bacteria

1. Grow an *E. coli* TG1 overnight culture in LB medium at 37°C.
2. Inoculate 50 mL of LB medium with 500 µL of the TG1 overnight culture and incubate shaking at 37°C until an OD_{600} of 0.4 is reached.
3. Centrifuge cells at 500*g* for 15 min.
4. Resuspend pellet in 20 mL of ice-cold 100 m*M* $CaCl_2$ and incubate on ice for 30 min.
5. Centrifuge cells at 500*g* for 15 min.
6. Resuspend cells in 2 mL of LB medium and incubate on ice for 2 h.
7. Add glycerol to a final concentration of 20%.
8. Use directly for transformation or store at –80°C.

3.4.2. Transformation

1. Add 10 µL of ligation reaction from **Subheadings 3.1., 3.2., or 3.3.** to 100 µL of competent cells.
2. Incubate on ice for 30 min.

3. Perform a heat shock by incubating cells for 60 s at 42°C in a water bath.
4. Immediately put cells on ice and incubate for 2–3 min.
5. Add 1 mL of LB and incubate shaking at 37°C for 45 min.
6. Centrifuge at 3,000*g* for 2 min and resuspend cell pellet in 100 µL of LB.
7. Plate onto a LB/amp/1% glucose plate and incubate at 37°C overnight.

3.4.3. Identification of Positive Clones by PCR Screening

1. Pick individual colonies with a toothpick and inoculate 20 µL of PCR reaction mixture (*see* **Subheading 3.1.1., step 1**) containing primers LMB3 and LMB2.
2. In order to obtain bacterial colonies, streak toothpick onto LB/amp/1% glucose plates. Incubate plates at 37°C overnight (this is your master plate).
3. Run PCR (94°C denaturation for 30 s, 50°C annealing for 1 min, 72°C extension for 1.5 min, 30 cycles).
4. Analyze PCR on a 1% agarose gel in 1X TAE buffer. Positive clones yield a PCR product of approx 1.6 kb.
5. Confirm correct sequence by DNA sequencing analysis.

3.5. Expression and Purification

1. Grow a 15-mL overnight culture of your antibody construct in LB/amp/1% glucose medium at 37°C.
2. Inoculate 1 L of LB/amp/0.1% glucose medium with 10 mL of overnight culture.
3. Incubate shaking at 37°C until an OD_{600} of 0.8 is reached (this takes approx 2.5 h).
4. Add IPTG to a final concentration of 1 m*M* and incubate shaking at 23°C for 3 h.
5. Pellet cells by centrifugation at 4,500*g* for 15 min.
6. Resuspend cells in 100 mL of PPB buffer.
7. Add lysozyme to a final concentration of 50 µg/mL.
8. Incubate on ice for 15 min.
9. Add $MgSO_4$ to a final concentration of 10 m*M*.
10. Centrifuge at 11,000*g* for 15 min.
11. Dialyze supernatant against IMAC loading buffer overnight at 4°C.
12. Centrifuge solution for 15 min at 11,000*g*.
13. Meanwhile, pour 1 mL of Ni-NTA resin into a chromatography column and equilibrate with IMAC loading buffer.
14. Load sample onto column by gravity.
15. Wash column with 10–20 mL of IMAC wash buffer.
16. Elute bound protein with IMAC elution buffer and collect 500-µL fractions.
17. Identify positive fractions by adding 10–20 µL of eluted fractions to 100 µL of Bradford reagent (in a microtiter plate). Positive fractions show a blue reaction.
18. Combine positive fractions and dialyze against 2–5 L of PBS overnight.
19. Measure OD_{280} of dialyzed sample.
20. Protein concentration can be determined using the calculated extinction coefficient (ε) of the antibody molecule (ε = number of tryptophans × 5,690 + number of tyrosines × 1,280). Molarity (M) of the protein solution is then calculated as:

OD_{280}/ε. Protein concentration can be calculated using the known mol wt of the antibody molecule (*see* **Notes 1** and **2**).

21. Check purity of the sample by SDS-PAGE. Single bands of 50–60 kDa should be observed for scDbs and taFvs, and two bands of 25–30 kDa should be seen for bispecific diabodies.
22. Store purified protein at –20°C.

3.6. ELISA

1. Coat a microtiter plate with your antigen at 1–10 µg/mL overnight in a suitable buffer (PBS or carbonate buffer, pH 9.6) at 4°C. Use one or more appropriate proteins as negative controls.
2. Block remaining binding sites with 2% MPBS at room temperature for 1–2 h.
3. Dilute antibody fragments in 2% MPBS and pipet 100 µL of the dilution per well. Incubate at room temperature for 1 h.
4. Wash plate 6× with PBS.
5. Dilute HRP-conjugated anti-His-tag antibody 1:1,000 in 2% MPBS and pipet 100 µL per well. Alternatively, dilute anti-Myc-tag antibody 9E10 to 10 µg/mL in 2% MPBS. Incubate at room temperature for 1 h.
6. Wash plate 6× with PBS.
7. If 9E10 has been used, dilute HRP-conjugated goat-anti-mouse-IgG antibody 1:1,000 in 2% MPBS and pipet 100 µL per well. Incubate at room temperature for 1 h.
8. Wash plate 6× with PBS.
9. Add 100 µL of TMB substrate solution per well and incubate until blue color has developed. Stop reaction by adding 50 µL of 1 *M* sulfuric acid. Read plate at 450 nm in a microtiter plate reader (*see* **Note 3**).

3.7. Flow Cytometry (FACS)

Use cells that are not incubated with antibodies, and incubated with an irrelevant antibody (isotype) as well as negative cells as controls.

1. Detach cells with 0.2 mg/mL EDTA/PBS (1×10^6 cells per sample).
2. Wash cells once with PBS and once with 1% FCS/PBS.
3. Incubate cells with 10–20 µg/mL antibody fragments in 1% FCS/PBS on ice for 30 min.
4. Wash cells once with 1% FCS/PBS.
5. Dilute anti-His-tag antibody 1:200 in 1% FCS/PBS. Pipet 100 µL to the cells and resuspend the cells carefully. Incubate on ice for 30 min.
6. Wash cells once with 1% FCS/PBS.
7. Dilute goat-anti-mouse-Cy3 antibody 1:200 in 1% FCS/PBS. Pipet 100 µL to the cells and resuspend the cells carefully. Incubate on ice for 30 min.
8. Wash cells once with 1% FCS/PBS and once with PBS (optional: fix cells with 3.7% formalin/PBS for 10 min, then wash cells twice with PBS).
9. Resuspend washed cells in 500 µL PBS and analyze by flow cytometry (e.g., in a FACS Calibur).

4. Notes

1. If only small amounts of antibody molecules are isolated from the periplasm, check to determine whether antibody molecules are secreted into the cell-culture supernatant.
2. Expression may not be possible in bacteria (especially with taFv). Use other expression systems such as *Pichia pastoris,* mammalian cells, or plants (*see* Chapters 15 and 17 and **ref. *11***).
3. It is essential to check for bispecific binding of the recombinant constructs in order to ensure that both binding sites are functional. If no binding is observed for one or even both binding sites, different formats should be tested. In addition, it may be necessary to switch the order of the binding sites.

References

1. Segal, D. M., Weiner, G. J., and Weiner, L. M. (1999) Bispecific antibodies in cancer therapy. *Curr. Opin. Immunol.* **11,** 558–562.
2. Withoff, S., Helfrich, W., de Leij, L. F., and Molema, G. (2001) Bispecific antibody therapy for the treatment of cancer. *Curr. Opin. Mol. Ther.* **3,** 353–362.
3. Holliger, P., Prospero, T. D., and Winter, G. (1993) "Diabodies", small bivalent and bispecific antibody fragments. *Proc. Natl. Acad. Sci. USA* **90,** 6444–6448.
4. Kontermann, R. E., Martineau, P., Cummings, C. E., Karpas, A., Allen, D., Derbyshire, E., et al. (1997) Enzyme immunoasssays using bispecific diabodies. *Immunotechnology* **3,** 137–144.
5. Kontermann, R. E., Wing, M. G., and Winter, G. (1997) Complement recruitment using bispecific diabodies. *Nat. Biotechnol.* **15,** 629–631.
6. Lffler, A., Kufer, P., Lutterbuse, R., Zettl, F., Daniel, P. T., Schwenkenbecker, J. M., et al. (2000) A recombinant bispecific single-chain antibody, CD19 × CD2, induces rapid and high lymphoma-directed cytotoxicity by unstimulated T lymphocytes. *Blood* **95,** 2098–2103.
7. Brüselbach, S., Korn, T., Vlkel, T., Mller, R., and Kontermann, R. E. (1999) Enzyme recruitment and tumor cell killing in vitro by a bispecific single-chain diabody. *Tumor Targeting* **4,** 115–123.
8. Cochlovius, B., Kipriyanov, S. M., Stassar, M. J., Schuhmacher, J., Benner, A., Moldenhauer, G., et al. (2000) Cure of Burkitt's lymphoma in severe combined immunodeficiency mice by T cells, tetravalent CD3 × CD19 tandem diabody, and CD28 costimulation. *Cancer Res.* **60,** 4336–4341.
9. Haisma, H. J., Grill, J., Curiel, D. T., Hoogeland, S., van Beusechem, V. W., Pinedo, H. M., et al. (2000) Targeting of adenoviral vectors through a bispecific single-chain antibody. *Cancer Gene Ther.* **7,** 901–904.
10. Nettelbeck, D. M., Miller, D. W., Jerome, J., Zuzarte, M., Watkins, S. J., Hawkins, R. E., et al. (2001) Retargeting of adenovirus to endothelial cells by a bispecific single-chain diabody directed against the adenovirus knob domain and human endoglin (CD105), *Mol. Ther.* **3,** 882–891.
11. Kontermann, R. E. and Dbel, S. (2001) Antibody Engineering. Lab Manual Series—Springer-Verlag, Heidelberg.

12. Kriangkum, J., Xu, B., Nagata, L. P., Fulton, R. E., and Suresh, M. R. (2001) Bispecific and bifunctional single chain recombinant antibodies. *Biomolecular Eng.* **18,** 31–40.
13. Kipriyanov, S. M., Moldenhauer, G., Schuhmacher, J., Cochlovius, B., von der Lieth, C. W., Matys, E. R., et al. (1999) Bispecific tandem diabody for tumor therapy with improved antigen binding and pharacokinetics. *J. Mol. Biol.* **293,** 41–56.
14. Alt, M., Mller, R., and Kontermann, R. E. (1999) Novel tetravalent and bispecific IgG-like antibody molecules combining single-chain diabodies with the immunoglobulin γ1 Fc or CH3 region. *FEBS Lett.* **454,** 90–94.
15. Park, S. S., Ryu, C. J., Kang, Y. J., Kashmiri, S.V.S., and Hong, H. J. (2000) Generation and characterization of a novel tetravalent bispecific antibody that binds to hepatitis B virus surface antigens. *Mol. Immunol.* **37,** 1123–1130.
16. Kontermann, R. E. and Mller, R. (1999) Intracellular and cell surface displayed single-chain diabodies. *J. Immunol. Methods* **226,** 179–188.
17. Vlkel, T., Korn, T., Bach, M., Mller, R., and Kontermann, R. E. (2001) Optimized linker sequences for the expression of monomeric and dimeric single-chain diabodies. *Protein Eng.* **14,** 815–823.
18. Mallender, W. D., Ferreira, S. T., Voss, E. W. Jr., and Coelho-Sampaio, T. (1994) Inter-active-site distance and solution dynamics of a bivalent-bispecific single-chain antibody molecule. *Biochemistry* **33,** 10100–10108.
19. Gruber, M., Schodin, B. A., Wilson, E. R., and Kranz, D. M. (1994) Efficient tumor cell lysis mediated by a bispecific single chain antibody expressed in Escherichia coli. *J. Immunol.* **152,** 5368–5374.
20. Hayden, M. S., Linsley, P. S., Gayle, M. A., Bajorath, J., Brady, W. A., Norris, N. A., et al. (1994) Single-chain mono- and bispecific antibody derivatives with novel biological properties and antitumor activity from a COS cell transcient expression system. *Ther. Immunol.* **1,** 3–15.
21. Fischer, R., Schumann, D., Zimmermann, S., Drosard, J., Sack, M., and Schillberg, S. (1999) Expression and characterization of bispecific single-chain Fv fragments produced in transgenic plants. *Eur. J. Biochem.* **262,** 810–816.

III

Antibody Expression and Optimization

14

Expression and Isolation of Recombinant Antibody Fragments in *E. coli*

Keith A. Charlton

1. Introduction

Central to the rapid growth in the study and application of recombinant antibodies in recent years is the ability to reliably and cost-effectively produce large quantities of functional protein. In contrast to glycosylated whole monoclonal antibodies (MAbs), functional derivatives (antibody fragments) carrying antigen (Ag)-binding activities, and comprising at minimum the antibody variable heavy- and light-chain domains, can be readily produced in bacterial cells *(1)*. The range of antibody (Ab) fragments that can be produced in this way includes $F(ab)_2$ and Fab′, which can also be generated by proteolytic digestion of MAb, and smaller variants such as Fv, scFv, and scAb (single-chain antibody). Even single domains that comprise only the heavy-chain or the light-chain variable region can display Ag-binding activity, although these are not discussed specifically in this chapter. The smallest derivatives that retain the Ag-binding specificity of parental MAbs are Fv fragments. However, as the two domains are held together only by non-covalent hydrophobic interactions, they are prone to dissociate. This can be overcome by chemical crosslinking or the introduction of inter-chain disulfide bonds, or more often by the inclusion of a short peptide linker into expression constructs to produce scFv. Further variants include the addition of a range of suitable (C-terminal) tags for the purification or detection of expressed protein, such as a human Cκ domain (scAb) *(2)*, c-myc, or hexahistidine.

Ab fragments have a number advantages over MAbs in clinical applications related to their small size. In particular, scFvs, at approx 1/6 the size of whole antibodies, are less immunogenic (when derived from nonhuman origins), have the potential for greater tissue penetration, and are more rapidly cleared from nonspecific tissues. They afford additional opportunities such as the covalent

From: *Methods in Molecular Biology, Vol. 248: Antibody Engineering: Methods and Protocols*
Edited by: B. K. C. Lo © Humana Press Inc., Totowa, NJ

attachment of drugs or radioactive isotopes for targeted therapy or imaging and can be engineered to exhibit dual specificities *(3)* (*see* Chapter 13). In most applications, for reasons of ease of transformation and manipulation, *E. coli* is the expression host of choice for laboratory research. Although other systems may produce higher yields of some antibodies, as discussed in Chapters 15–17, bacteria are unrivaled for the very rapid production and purification of protein for subsequent analysis *(4)*. The majority of expression vectors currently used are based on the pUC family *(5)*. These include the *ColE*1 replication origin to maintain high copy number under optimal growth conditions, such as 37°C. However, copy number can be reduced dramatically when cells are grown at suboptimal temperatures. Essential to efficient high expression of antibody fragments is tight control over transcription. The *lac* (and hybrid *tac*) promoter system is very popular because it is readily induced with isopropyl-1-β-D-thiogalactopyranoside (IPTG) and is repressed by glucose. The latter is of particular importance when expressing proteins that are toxic to the host such as exported antibody fragments, as excessive basal transcription can slow or arrest bacterial growth. Under these conditions, plasmids are not efficiently replicated, and can easily be lost. An in-depth discussion of the various vector systems available is beyond the scope of this chapter, and it will be assumed that readers have access to suitable expression vectors.

There are three locations from which bacterially expressed proteins can be isolated *(6)*—the cytoplasm, the periplasm, and the culture supernatant **(Fig. 1)**. *E. coli* usually secretes very few proteins into the extracellular medium. Recombinant proteins exported to the medium are least likely to be degraded by bacterial proteases, and purification is simplified by the lack of contaminating proteins. Conversely, the levels of protein found in culture supernatants are generally very low, and the volumes of material to be processed are very high. Antibody fragments directed to the periplasm are often found in culture supernatants, a process often referred to as "secretion into the medium." More accurately, the observations are usually the result of leakiness of the outer membrane under expression conditions, possibly because of the accumulation of insoluble aggregates within the periplasm or interference with the export of essential outer membrane components. Proteins expressed without an N-terminal signal sequence remain in the cytoplasm and form aggregates of insoluble folding intermediates or inclusion bodies. This can result in high protein yields, protection from proteolysis, and protection of the host cell against the toxic effects of export. The protein is produced in a non-functional form, and must be solubilized and re-folded using time-consuming and labor-intensive procedures. Final yields of biologically active material are often very low when using this approach.

A significant advance in the development of Ab engineering came about when it was demonstrated that Ab fragments could be exported to the bacterial

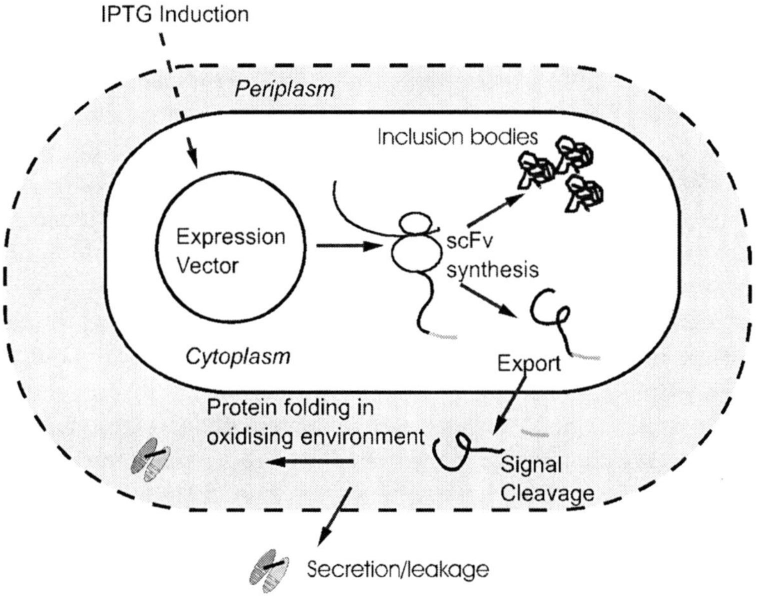

Fig. 1. Schematic representation of the localization of recombinant Ab fragments expressed in *E. coli* using different vectors and under different culture conditions.

periplasm by the inclusion of an N-terminal secretion signal sequence. Within the periplasmic space, Ab fragments are able to fold correctly, aided by the formation of intra-domain disulfide bonds made possible by the oxidative environment, and remain in a soluble form ***(7,8)***. The choice of bacterial signal sequence does not appear to be of major importance, as several have been used successfully, such as OmpA, PelB, PhoA, OmpF, β-lactamase, and many others ***(9–14)***. Upon transport into the periplasmic space, signal peptides are cleaved leaving the correct N-terminal sequence. High-level expression of recombinant proteins in *E. coli,* and subsequent export to the periplasm often result in the formation of insoluble aggregates ***(15)***. This not only reduces protein (functional) yield, but can also lead to growth arrest and cell lysis. The accumulation of insoluble material was once believed to reflect the solubility and stability of the Ab fragments involved. More recent research suggests that the solubility and stability of folding intermediates determine the extent of aggregation ***(16)***.

It is generally accepted that such effects can be reduced by induction and expression of Ab fragments (scFv, scAb, and Fab) at 20–25°C ***(16,17)***. This does not prevent aggregation completely, but does appear to increase the proportion of correctly folded functional protein produced. Significant improvements in the yields of soluble recombinant Ab fragments have also been obtained by identifying and overexpressing *E. coli* periplasmic chaperones such as Skp ***(18,19)*** and other proteins involved in modulating folding, such as the peptidylprolyl *cis,trans*-isomerase FkpA ***(20)***. Periplasmic aggregation can also be reduced by induction of cultures in growth medium supplemented with osmotic stress-inducing concentrations of non-metabolized compounds. These include 1 *M* sorbitol and 2.5 m*M* glycine betaine ***(3,21)***, or 0.4 *M* sucrose ***(22)***.

In a number of studies of individual recombinant Ab fragments, specific residues have been identified that contribute to poor protein folding. Some of these are exposed hydrophobic residues located at the interface between variable and constant domains ***(23)***, and thus are not problematic for Fab fragments, while others are located in both framework and CDR regions ***(24–26)***. In each case, mutation of the residue(s) resulted in improved solubility and decreased toxicity. A final consideration is that of codon usage. Many antibody fragments include codons that may not be effectively used in the host organism chosen for expression ***(27)***. Generally, with *E. coli,* there is little to be gained by engineering Ab fragment genes to include only those codons used most frequently. However, there may be advantages to precluding those codons used least frequently.

Recombinant Ab fragments expressed in the periplasm of *E. coli* can be readily released by lysozyme/EDTA treatment and mild cold osmotic shock. The crude Ab fragment is most conveniently purified from the periplasmic release fraction or culture supernatant by immobilized metal affinity chelate chromatography (IMAC) via a C-terminal poly-histidine tag incorporated into the expression vector. Detailed methods for purification of expressed recombinant Ab fragments are given in Chapter 22.

2. Materials

All media and buffers are prepared with ultrapure water (>18 MΩ/cm) and autoclaved. Antibiotics and glucose are added after cooling media to 50°C. Quantities given are for 1 L of medium.

2.1. Expression of Recombinant Antibody Fragments (for Subheadings 3.1., 3.2., and 3.4.)

1. *E. coli* strain XL1-Blue {*sup*E44 *hsd*R17 *rec*A1 *end*A1 *gyr*A46 *thi rel*A1 *lac*F′[*pro*AB + *lac*I^q *lac*ZΔM15 Tn10(*tet*^{r})]} (Stratagene).
2. *E. coli* strain TG1 {K12 D*(lac-pro) sup*E *thi hsd*D5/F′*tra*D36 *pro*A$^+$B$^+$ *lac*I^q *lac*ZΔM15} (Stratagene).
3. 2TY medium: 16 g Bacto-peptone, 5 g Bacto-yeast extract, 5 g NaCl, pH 7.5.
4. 2TY-amp/glu: 2TY medium supplemented with 100 µg/mL ampicillin and 2% (w/v) glucose.
5. 2TY-amp/suc: 2TY medium, 0.4 *M* sucrose (*see* **Note 1**), 100 µg/mL ampicillin.

6. TB (Terrific broth): 12 g Bacto-tryptone, 24 g Bacto-yeast extract, 4 mL glycerol dissolved in 800 mL milli-Q water, made up to 900 mL and autoclaved. Separately make up and sterilize a 100-mL solution containing 2.31 g KH_2PO_4 and 12.54 g K_2HPO_4, and add to the cooled broth.
7. TB-amp/glu: TB medium supplemented with 100 µg/mL ampicillin and 2% (w/v) glucose.
8. TB-amp/suc: TB medium, 0.4 *M* sucrose, 100 µg/mL ampicillin (*see* **Note 1**).
9. LB (Luria-Bertani) broth: 10 g Bacto-peptone, 5 g Bacto-yeast extract, 5 g NaCl, pH 7.5.
10. LB-amp/glu: LB medium supplemented with 100 µg/mL ampicillin and 2% (w/v) glucose.
11. 1 *M* IPTG stock solution (store at –20°C).

2.2. Isolation of Recombinant Antibody Fragments (for Subheadings 3.3. and 3.5.)

1. Amicon high-performance stirred ultrafiltration cell (Millipore).
2. Amicom YM 10 membrane with 10-kDa cut-off (Millipore).
3. ST buffer: 200 m*M* Tris-HCl pH 7.5, 20% (w/v) sucrose.
4. 1 *M* EDTA stock solution.
5. Hen egg lysozyme (Sigma-Aldrich; Cat. #L-6876).
6. Dialysis buffer: 50 m*M* Tris-HCl, 1 *M* NaCl, 10 m*M* imidazole, pH 7.0.
7. STE buffer: 20% sucrose, 100 m*M* Tris-HCl, pH 7.4, 10 m*M* EDTA.
8. TE buffer: 10 m*M* Tris HCl, pH 7.4, 1 m*M* EDTA.

3. Methods

A variety of methods for expressing and isolating recombinant Ab fragments in *E. coli* are available in the literature, which can lead to confusion among those who are not familiar with this area. The experience of our lab (and others) is that different protocols are best applied to different Ab fragment formulations. With this in mind, separate methods are given for scFv, scAb, and Fab fragments. The reader should also note that variations in the sequences of different antibodies and the use of different vectors will affect protein expression. It is therefore recommended that some small-scale experiments be carried out to optimize conditions for each clone.

3.1. Expression of scFv Fragments

1. Inoculate 10 mL of 2TY-amp/glu with a single fresh colony of XL1-Blue transformed with the scFv expression vector and grow at 37°C overnight.
2. Dilute the overnight culture 1:50 in fresh 2TY-amp/glu (*see* **Note 2**), and incubate at 37°C with vigorous shaking (200–250 rpm) to an optical density 600 nm (OD_{600}) of 0.6–0.8.
3. Pellet the cells by centrifugation at 3,000*g* for 20 min at 20°C.

4. Resuspend the cells in the same volume of 2TY-amp/suc (*see* **Note 1**) and incubate at 20–25°C for 1 h (*see* **Note 3**).
5. Add IPTG to a final concentration of 0.1 m*M* (*see* **Note 4**) and continue incubation at 20–25°C (*see* **Note 5**) overnight (16 h) with shaking (*see* **Notes 6, 7**).
6. Collect the cells by centrifugation at 6,000*g* for 30 min. Retain the culture supernatant and keep on ice (*see* **Note 8**).

3.2. Expression of scAb Fragments

1. Inoculate 10 mL of TB-amp/glu with a single fresh colony of XL1-Blue transformed with the scAb expression vector and grow overnight at 37°C.
2. Dilute the overnight culture 1:50 in fresh TB-amp/glu, and incubate at 37°C with vigorous shaking (200–250 rpm) for 7–8 h (*see* **Note 2**).
3. Pellet the cells by centrifugation at 3,000*g* for 20 min at 20°C and resuspend in fresh TB-amp/glu. Incubate overnight at 20–25°C with shaking (*see* **Note 9**).
4. Pellet the cells as in **step 3** and resuspend in the same volume of fresh TB-amp or TB-amp/suc (*see* **Note 8**). Incubate for 1 h at 20–25°C for cells to recover.
5. Induce the culture by addition of IPTG to a final concentration of 0.1 m*M* (*see* **Note 4**), and continue incubation for 4–16 h (*see* **Notes 6, 7**).
6. Collect the cells by centrifugation at 6,000*g* for 30 min. Retain the culture supernatant and keep on ice (*see* **Note 8**).

3.3. Isolation of Soluble scFv/scAb Fragments

1. Remove any remaining cells or cell debris from the retained culture supernatant by centrifuging at 20,000*g* for 45 min at 4°C.
2. Concentrate the solution using an Amicon ultrafiltration cell and YM 10 membrane (*see* **Note 11**).
3. Remove any remaining culture supernatant from cell pellets by pipetting or blotting with tissue paper.
4. Prepare fractionation buffer by adding 1 *M* EDTA to ice-cold ST buffer to a final concentration of 1 m*M*, and lysozyme to 500 µg/mL (*see* **Note 12**).
5. Resuspend the cell pellet in 10% culture volume ice-cold fractionation buffer (*see* **Note 13**) and shake gently on ice for 15 min (*see* **Note 14**).
6. Add an equal volume of ice-cold water and continue shaking gently on ice for an additional 15 min.
7. Pellet the spheroplast suspension by centrifugation at 20,000*g* for 30 min at 4°C, and carefully decant the supernatant containing released periplasmic proteins into a clean tube (*see* **Note 15**).
8. Pass the periplasmic release fraction through a 0.45-µm disposable filter.
9. The periplasmic release fraction can be combined with the concentrated culture supernatant **(step 2)**, and dialyzed against 2–3 changes 50 m*M* Tris-HCl, 1 *M* NaCl, 10 m*M* imidazole, pH 7.0 (*see* **Note 16**).
10. For a detailed description of immobilized metal affinity chromatography (IMAC) purification, *see* Chapter 22.

3.4. Expression of Fab Fragments

1. Inoculate 10 mL of LB-amp/glu with a single fresh colony of XL1-Blue or TG1 transformed with the Fab expression vector and grow overnight at 37°C (*see* **Note 17**).
2. Inoculate the overnight culture into 1 L of fresh LB-amp (*see* **Note 2**) and incubate with vigorous shaking at 37°C to OD_{600} 0.6–0.8.
3. Induce the culture by addition of IPTG to a final concentration of 0.1 m*M* (*see* **Note 4**), and continue incubation at 20–25°C for 16–24 h.
4. Collect the cells by centrifugation at 6,000*g* for 30 min at 4°C.

3.5. Isolation of Fab Fragments

1. Remove any remaining culture supernatant from cell pellets by pipetting or blotting with tissue paper.
2. Resuspend the cell pellet in 10% culture volume ice-cold STE buffer and incubate on ice with gentle shaking for 1 h (*see* **Notes 14, 18**).
3. Pellet the cells by centrifugation at 6,000*g* for 30 min at 4°C.
4. Resuspend the pellet in 10% culture volume ice-cold TE buffer and incubate on ice with gentle shaking for 15 min.
5. Pellet the cells by centrifugation at 6,000*g* for 30 min at 4°C and carefully decant the supernatant containing released periplasmic proteins into a clean tube (*see* **Note 15**).
6. Dialyze the periplasmic release fraction against 2–3 changes 50 m*M* Tris-HCl, 1 *M* NaCl, 10 m*M* imidazole, pH 7.0 (*see* **Note 16**).
7. For a detailed description of IMAC purification, *see* Chapter 22.

4. Notes

1. 2TY-amp/suc (and TB-amp/suc) are prepared by adding 137 g sucrose to 1 L sterile 2TY-amp (TB-amp) medium immediately before use. The inclusion of 0.4 *M* sucrose can significantly increase yields of some Ab fragments, particularly scFvs *(22)*.
2. Improved growth of *E. coli* (and protein yields) in shake-flask cultures are obtained when cells are grown in baffled flasks that increase oxygenation of the growth medium. Single 500-mL cultures can be grown in 2-L flasks or multiple 50-mL cultures in 250-mL flasks.
3. When the growth medium is replaced, the culture will enter lag phase. Incubation in fresh medium for 1 h allows the culture to begin exponential growth prior to induction.
4. This concentration of IPTG was found to be optimal when low-density cultures are induced (e.g., induction at OD_{600} 0.6–0.8). For very high cell densities (**Subheading 3.2.**), higher concentrations (1 m*M*) can be used. Because IPTG is an expensive reagent, it is recommended that small-scale expressions are carried out to optimize conditions for each vector/antibody used.
5. Induction of antibody fragments at suboptimal growth temperatures reduces the accumulation of insoluble partially folded aggregates in the periplasmic space, thus increasing yields of functional protein.
6. Antibody fragments are a very heterologous group of proteins, and factors such as folding efficiency and host-cell toxicity will vary considerably. Cell lysis causes a

significant proportion of the antibody to leak into the culture supernatant, and the release of proteases and nucleic acids that reduce protein yields and make subsequent isolation and purification more complex. It is advisable to generate growth curves for each antibody under inducing conditions to determine the time at which cell lysis begins, indicated by a reduction in culture OD_{600}.

7. Increasingly, antibody fragments are generated using phage display technology. The repeated expression of scFvs in TG1 cells as fusions to the phage minor coat protein, and export to the periplasm, drives selection to favor those antibodies that express at least reasonably well. In comparison, antibody fragments derived from hybridoma cell lines exhibit a much wider range of expression yields, and can be very toxic to *E. coli.*
8. The inclusion of sucrose in growth medium results in an increase in the amount of protein localizing in the culture supernatant, particularly when prolonged induction times are used. In the case of scFvs, it may also promote the formation of dimeric antibody fragments (diabodies), in which the V_H of one molecule associates with the V_L of a second, and *vice versa* ***(3)***.
9. We have found that the use of very rich (TB) medium and growth of cultures to high cell density before induction gives high yields of scAb fragments over short induction times (≤4 h). Prolonged induction can lead to cell lysis.
10. Although the addition of sucrose to the growth medium can significantly increase yields of scFvs, we find improvements in yield of scAbs to be very variable. It is advisable to determine its effect for each different scAb.
11. Any other protein concentrator with a mW cutoff of 10–12 kDa may be used.
12. A high concentration of Tris-HCl together with EDTA serves to destabilize the bacterial outer membrane and facilitate release of periplasmic proteins. Lysozyme cleaves the peptidoglycan matrix that makes up the cell wall, releasing recombinant protein held behind or within the matrix ***(29)***.
13. When expressing scAbs at high cell density (**Subheading 3.2.**), it may be necessary to use 20% culture volume fractionation buffer because of the large cell pellet.
14. The cell pellet is best resuspended using a 10 mL Gilson with the end of the tip cutoff to increase the hole size or a disposable pipet and hand pump.
15. The cell pellet will be quite loose after fractionation and osmotic shock. It may be necessary to extend the spin time to firm up the pellet.
16. The addition of 10 m*M* imidazole to the dialysis buffer helps to prevent nonspecific binding of *E. coli* proteins to the Ni^{2+} resin during IMAC purification. It does not inhibit binding of recombinant proteins with a hexa-histidine tag. Other buffers can be substituted, as suggested by the nickel column supplier used.
17. Fab fragments can be successfully expressed in either XL1-Blue or TG1 cells. TG1 cells are reported to give yields as much as 10× greater than those observed with the same Fab fragment expressed in XL1-Blue ***(28)***. The improved yields result in part from the faster growth rate of TG1 and resulting higher cell densities, but they also express higher levels of soluble protein (Fab) per g wet cell weight.
18. Incubation in excess of 1 h can lead to leakage of the Fab fragments into the supernatant.

References

1. Huston, J. S., Levinson, D., Mudgett-Hunter, M., Tei, M., Novotny, J., Margolies, M. N., et al. (1988) Protein engineering of antibody binding sites: recovery of specific activity in an anti-digoxin single-chain Fv analogue produced in *Escherichia coli. Biochemistry* **85,** 5879–5883.
2. McGregor, D. P., Molloy, P. E., Cunningham, C., and Harris, W. J. (1994) Spontaneous assembly of bivalent single chain antibody fragments in *Escherichia coli. Mol. Immunol.* **31(3),** 219–226.
3. Kipriyanov, S. M., Moldenhauer, G., Schuhmacher, J., Cochlovius, B., Von der Lieth, C., Matys, E. R., et al. (1999) Bispecific tandem diabody for tumor therapy with improved antigen binding and pharmacokinetics. *J. Mol. Biol.* **293,** 41–56.
4. Better, M. and Horowitz, A. H. (1989) Expression of engineered antibodies and antibody fragments in microorganisms, in Langone, John J. (Ed) *Methods in Enzymology* vol. 178, Academic Press, San Diego, California, pp. 476–496.
5. Yannish-Perron, C., Vieira, J., and Messing, J. (1985) Improved M13 phage cloning vectors and host strains—nucleotide sequences of the M13mp18 and pUC19 vectors. *Gene* **33,** 103–119.
6. Ge, L., Knappik, A., Pack, P., Freund, C., and Plükthun, A. (1995) Expressing antibodies in *Escherichhia coli,* in *Antibody Engineering* (Borrebaeck, C.A.K., ed.), Oxford University Press, Inc., New York, NY, pp. 229–266.
7. Skerra, A. and Plükthun, A. (1998) Assembly of a functional immunoglobulin Fv fragment in *Escherichia coli. Science* **240,** 1038–1041.
8. Better, M., Chang, C. P., Robinson, R. R., and Horwitz, A. H. (1988) *Escherichia coli* secretion of an active chimeric antibody fragment. *Science* **240,** 1041–1043.
9. Skerra, A., Pfitzinger, I., and Plükthun, A. (1991) The functional expression of antibody Fv fragments in *Escherichia coli* and a generally applicable purification technique. *Bio/Technology* **9,** 273–278.
10. Dueñs, M., Váquez, J., Ayala, M., Söderlind, E., Ohlin, M., Péez, L., et al. (1994) Intra- and Extracellular expression of an scFv antibody fragment in *E. coli:* Effect of bacterial strains and pathway engineering using GroES/L Chaperonins. *Biotechniques* **16(3),** 476–477 and 480–483.
11. Somerville, J.E.J., Goshorn, S. C., Fell, H. P., and Darveau, R. P. (1994) Bacterial aspects associated with the expression of a single-chain antibody fragment in *E. coli. Appl. Microbiol. Biotechnol.* **42(4),** 595–603.
12. Denèe, P., Kovarik, S., Ciora, T., Gosselet, N., Béichou, J.-C., Latta, M., et al. (1989) Heterologous protein export in *Escherichia coli:* influence of bacterial signal peptides on the export of human interleukin 1β. *Gene* **85,** 499–510.
13. Hoffman, C. F. and Wright, A. (1985) Fusions of secreted proteins to alkaline phosphatase: an approach for studying protein secretion. *Proc. Natl. Acad. Sci. USA* **82,** 5107–5111.
14. Kadonaga, J. T., Gautier, A. E., Straus, D. R., Charles, A. D., Edge, M. D., and Knowles, J. R. (1984) The role of the β-lactamase signal sequence in the secretion of proteins by *Escherichia coli. J. Biol. Chem.* **259,** 2149–2154.

15. Kipriyanov, S. M., Düel, S., Breitling, F., Kontermann, R. E., and Little, M. (1994) Recombinant single-chain Fv fragments carrying C-terminal cysteine residues: production of bivalent and biotinylated miniantibodies. *Mol. Immunol.* **31(14),** 1047–1058.
16. Plükthun, A. (1993) Antibodies from *Escherichia coli,* in *Handbook of Experimental Pharmacology,* vol. 113: *The Pharmacology of Monoclonal Antibodies,* (Rosenberg, M. and Moore, G. P., eds.), Springer-Verlag, Berlin, Heidelberg, pp. 269–315.
17. Cabilly, S. (1989) Growth at sub-optimal temperatures allows the production of functional, antigen-binding Fab fragments in *Escherichia coli. Gene* **85(2),** 553–557.
18. Hayhurst, A. and Harris, W. J. (1999) *Escherichia coli* Skp chaperone co-expression improves solubility and phage display of single chain antibody fragments. *Protein Expr. Purif.* **15,** 336–343.
19. Bothmann, H. and Plükthun, A. (1998) Selection for a periplasmic factor improving phage display and functional periplasmic expression. *Nat. Biotechnol.* **16,** 377–380.
20. Bothmann, H. and Plükthun, A. (2000) The Periplasmic *Escherichia coli* Peptidylprolyl *cis,trans*-isomerase FkpA. 1. Increased fucntional expression of antibody fragments with and without *cis*-prolines. *J. Biol. Chem.* **275(22),** 17,100–17,105.
21. Blackwell, J. R. and Horgan, R. (1991) A novel strategy for production of highly expressed recombinant protein in an active form. *FEBS Lett.* **295,** 10–12.
22. Kipriyanov, S. M., Moldenhauer, G., and Little, M. (1997) High level production of soluble single chain antibodies in small-scale *Escherichia coli* cultures. *J. Immunol. Methods* **200,** 69–77.
23. Nieba, L., Honegger, A., Krebber, C., and Pluckthun, A. (1997) Disrupting the hydrophobic patches at the antibody variable/constant domain interface: improved *in vivo* folding and physical characterization of an engineered scFv fragment. *Protein Eng.* **10(4),** 435–444.
24. Knappik, A. and Pluckthun, A. (1995) Engineered turns of a recombinant antibody improve its in vivo folding. *Protein Eng.* **8(1),** 81–89.
25. Forsberg, G., Forsgren, M., Jaki, M., Norin, M., Sterky, C., Enhorning, A., et al. (1997) Identification of framework residues in a secreted recombinant antibody fragment that control production level and localization in *Escherichia coli. J. Biol. Chem.* **272(19),** 12,430–12,436.
26. Kipriyanov, S. M., Moldenhauer, G., Martin, A.C.R., Kupriyanova, O. A., and Little, M. (1997) Two amino acid mutations in an anti-human CD3 single chain Fv antibody fragment that affect the yield on bacterial secretion but not the affinity. *Protein Eng.* **10(4),** 445–453.
27. Wada, K., Wada, Y., Ishibashi, F., Gojobori, T., and Ikemura, T. (1992) Codon usage tabulated from the GenBank genetic sequence data. *Nucleic Acids Res.* **20,** 2111–2118.
28. Raffaï R., Vukmirica, J., Weisgraber, K. H., Rassart, E., Innerarity, T. I., and Milne, R. (1999) Bacterial expression and purification of the Fab fragment of a monoclonal antibody specific for the low density lipoprotein receptor-binding site of human apolipoprotein E. *Protein Expr. Purif.* **16,** 84–90.
29. French, C., Keshavarz-Moore, E., and Ward, J. M. (1996) Development of a simple method for the recovery of recombinant proteins from the *Escherichia coli* periplasm. *Enzyme Microb. Technol.* **19,** 332–338.

15

Expression of Recombinant Antibodies in Mammalian Cell Lines

Paul J. Yazaki and Anna M. Wu

1. Introduction

1.1. Recombinant Antibody Expression Systems

Since the advent of hybridoma technology, mammalian-cell culture has been employed for the expression and high-level production of monoclonal antibodies (MAbs). Recent adaptations in recombinant technology have developed the use of numerous prokaryotic and eukaryotic systems for the expression of heterologous molecules. The major systems used for MAb expression have been reviewed and compared *(1,2)*, and a number of the key methodologies are detailed within this series. Prokaryotic expression systems offer the potential of high production yields at a substantial reduced cost of goods (*see* Chapter 14). However, because of the complexity of the protein-folding pathway, bacterial expression has been limited to small antibody fragments, and in some cases may require refolding to produce a biologically active product. Another limitation is that recombinant MAbs expressed in bacteria are aglycosylated and can result in the reduction or loss of biological effector functions *(3)*. For large, multidomain molecules such as full-length MAbs or complex recombinant antibody fragments, eukaryotic systems such as mammalian or yeast expression have been utilized. Currently, nine USFDA-licensed Mabs for in vivo human use have been expressed from mammalian-cell culture for commercial production.

Ultimately, the key consideration in choosing a host organism is dictated by the intended purpose of the MAb product. For example, initial research studies may require only limited amounts of material to evaluate biological activity and specificity, whereas products intended for in vivo clinical use must conform to very stringent criteria. Some of the factors to be anticipated for clinical development include biological activity, specificity, molecular structure, scala-

From: *Methods in Molecular Biology, Vol. 248: Antibody Engineering: Methods and Protocols*
Edited by: B. K. C. Lo © Humana Press Inc., Totowa, NJ

bility in production, quality and purity of product, lack of toxicity and immunogenicity, and cost.

1.2. Mammalian Expression

For mammalian expression, there are two modes of expression—transient and stable expression. Although stable expression is required for long-term production, transient expression can be a valuable tool. The host-cell line most commonly used is the COS-1 monkey cell, which transformed by an origin defective SV40 mutant, allows high copy-number replication of plasmids that contain an SV40 origin of replication *(4)*. The high transcription of the plasmid DNA, although toxic to the cells, offers the potential for rapid expression of new recombinant molecules. Yields from transient expression have been reported to be 0.1–4 µg/mL *(5)*.

Stable expression requires the integration of the complete gene of interest into an actively transcribed region of the host chromosome. Transfection is normally achieved by chemical or electrical disruption of the host-cell membrane *(6)*. Because of the low efficiency of this random integration, a selection process is required that uses cytotoxic drugs or limiting essential growth components. The most commonly used selection method at the research scale is the NEO^R/G418 system. For commercial MAb production, the predominant expression/selection systems have been the dihydrofolate reductase (DHFR) system *(7)* and the glutamine synthetase (GS) system *(8)*.

In the DHFR system, mammalian expression vectors are designed to express the DHFR gene and antibody gene of interest. The plasmid is transfected into host DHFR-negative cells and transformants are selected for growth in the presence of the cytotoxic drug, methotrexate, an inhibitor of DHFR *(9)*. Growth in the presence of increasing concentrations of methotrexate results in significant DHFR gene amplification and, as a result of the co-integration of the gene of interest, high-level MAb expression *(10)*.

The GS system bases the selection process on the biosynthesis pathway of the amino acid glutamine, from the substrates glutamate and ammonia. Some mammalian host-cell lines such as myeloma NS0 cells are deficient in GS, and other myeloma Sp2/0 or non-lymphoid Chinese hamster ovary (CHO) cells have low endogenous levels of GS. The GS expression vector, pEE12, has been designed to contain a GS minigene along with the antibody gene of interest, allowing selection by growth in glutamine-free cell-culture media *(8)*. For those cell lines with low levels of endogenous GS, the GS inhibitor, methionine sulfoximine, allows selection, and at increasing concentrations, gene amplification can be achieved. An advantage of the GS system using NS0 cells is that cytotoxic drugs are not required for selection, allowing selective pressure to be maintained throughout production. In addition, in high-capacity bioreactors, the utilization of ammonia reduces the toxic levels that accumulate in the culture. The design of the pEE12/pEE6 expression vectors enables the dual expression of the immunoglobulin light and heavy chain. The GS system, which includes vectors and cell lines, is available through a license with Lonza Biologics, Slough, UK.

The selection of mammalian host cell will determine the recombinant MAb's glycosylation pattern, and a choice must be made as to culture them as adherent or suspension cells. The choice of the appropriate cell line has been the focus of several comprehensive reviews comparing Sp2/0, NS0 ***(11)*** or CHO cells ***(5,12)***.

In a detailed side-by-side comparison, the CHO and NS0 cells showed identical protein secretion rates for a humanized MAb, ranging from 26–31 $\mu g/10^6$ cells/24 h ***(13)***. Because of the rapid advances of the biotechnology sector, MAb secretion levels have exceeded 2 g/L ***(14)***. Recently, this area of industrial large-scale MAb production has been well-reviewed ***(15,16)***.

Additional factors that may need to be considered include promoter strength, genetic stability, productivity without selective pressure, and scale-up to desired levels of production. However, the most important factor that dictates expression levels and gene stability is the site of integration into the host genome. Since transfection from these systems is a random process, the success of isolating a stable high-producing clone is dependent on screening a large number of transformants.

1.3. Anti-CEA Recombinant Antibodies

In this chapter, we describe the production of a series of recombinant MAbs and antibody fragments against carcinoembryonic antigen (CEA) ***(17,18)*** that are in clinical development (*see* Chapter 12). This cognate family of anti-CEA recombinant molecules has the same variable regions, providing an approach to modify the targeting and pharmacokinetics properties of antibodies in the delivery of a radioisotope payload to the tumor site for the imaging and therapy of cancer ***(19,20)***.

The basis for this recombinant work was the highly specific anti-CEA T84.66 murine MAb ***(21,22)***. Recombinant antibody fragments utilizing the single-chain Fv format (scFv) provide antigen recognition in a single polypeptide which, encoded by a single gene, greatly facilitates expression in a heterologous system. Using the scFv as a building block, larger engineered molecules have been generated that include diabodies (scFv non-covalent dimer) ***(23)***, minibodies (scFv-CH3 covalent dimer) ***(24)***, and scFvFc ***(25)***. The diabody and the minibody molecules were initially expressed in bacteria, but the minibody displayed proteolysis and a 10-fold lower expression compared to mammalian expression ***(24)***. In addition to these recombinant antibody fragments, several versions of full-

Table 1
Anti-CEA Activity Levels of T84.66 Recombinant Antibodies

Recombinant antibody construct	Anti-CEA activity
T84.66 diabody	25 μg/mL
T84.66 minibody	50 μg/mL
T84.66 scFvFc	80 μg/mL
T84.66 chimeric MAb	7 μg/mL
T84.66 humanized MAb	25 μg/mL

length humanized MAbs were expressed. Protocols have been developed to establish a uniform platform technology for a wide range of molecules, including the diabody, with the intent of production for pilot clinical trials. **Table 1** provides a list of anti-CEA recombinant antibody molecules from our laboratory and their level of anti-CEA activity produced using the GS expression/selection system.

To provide specific examples, protocols are described for the production of the T84.66 minibody. Expansion of high-producing transfectomas into small to medium-scale hollow-fiber bioreactors can provide sufficient material for pilot clinical trials, in which relatively small quantities of protein are required *(26)*. Also included in this chapter is a general purification method that utilizes mild chromatography and elution conditions and brief protocols for biochemical characterization. These protocols have been used for a variety of diabodies and minibodies, and are assembled as an example of a successful path for production of recombinant MAbs. However, individual MAbs for different applications may perform better in alternate expression systems, host cells, or production approaches, and all of these should be considered.

2. Materials

2.1. Transient Expression

1. COS-1 cells (American Tissue Type Collection; Cat. #CRL-1650).
2. DME media, glutamine-free (JRH Biosciences; Cat. #51435).
3. Fetal bovine serum (FBS), heat-inactivated (Omega Scientific; Cat. #FB-02).
4. L-glutamine (Irvine Scientific; Cat. #9317).
5. Phosphate-buffered saline (PBS) (Irvine Scientific; Cat. #9240).
6. Trypsin ethylenediaminetetraacetic acid (EDTA) (Irvine Scientific; Cat. #9341).
7. Trypan blue (Sigma; Cat. #T8154).
8. LipoTAXI mammalian transfection kit (Stratagene; Cat. #204110).
9. pEE12 control plasmid, GS Expression system™ (Lonza Biologics).
10. Experimental plasmid, pEE12 minibody plasmid *(26)*.
11. Protein A Sepharose 4 Fast Flow (Amersham Biosciences; Cat. #17-0974-01).
12. rProtein L™-agarose (Affitech; Cat. #201-2).

13. 4X sodium dodecyl sulfate (SDS) gel loading buffer: 0.2 *M* Tris-HCl, pH 6.8, 8% sodium dodecyl sulfate, 0.4% bromophenol blue, 40% glycerol [add 0.4 *M* dithiothreitol (DTT) if running reduced samples].
14. Hemocytometer (VWR-Scientific Products; Cat. #15170-172).
15. 0.2-µm PES syringe filters, 25 mm (Whatman; Cat. #6A96-2502).
16. Disposable sterile pipets and cultureware (Falcon).
17. 5% CO_2 humidified incubator.
18. Light microscope.

2.2. Stable Expression

1. Murine myeloma NS0 cells (Lonza Biologics).
2. FBS, heat-inactivated.
3. NS0 non-selective media: DME media supplemented with 2 m*M* L-glutamine, 10% heat-inactivated fetal bovine serum (FBS) and 1X GS supplement (final concentration).
4. PBS.
5. Trypan blue.
6. pEE12 control plasmid.
7. pEE12 minibody plasmid.
8. *Sal*I restriction enzyme and reaction buffer (New England Biolabs; Cat. #R0138S).
9. Micropure-EZ™ enzyme remover (Millipore; Cat. #42529).
10. Disposable sterile pipets and cultureware.
11. Hemocytometer.
12. Gene Pulser II™ electroporation unit and 5-mm gap electroporation cuvet, Cat. #165-2088 (Bio-Rad) or;
13. Multiporator™ electroporation unit, 2-mm gap electroporation cuvet, Cat. #E4307000593 and hypo-osmolar electroporation buffer, Cat. #4308070.501 (Eppendorf).
14. 10% CO_2 humidified incubator. If not available, 5% CO_2 can be substituted.

2.3. Cell Culture

1. Trypan blue.
2. FBS, dialyzed and heat-inactivated (HyClone; Cat. # SH30079.03).
3. NS0 selective media: glutamine-free DME media supplemented with 10% dialyzed, heat-inactivated FBS and 1X GS supplement (final concentration).
4. Hemocytometer.
5. Flat-bottomed 6-, 24-, and 96-well sterile plates (Costar).
6. Disposable sterile pipets and cultureware.
7. Bottom viewing mirror (Dynatech).
8. Dimethyl sulfoxide Hybri-max® (DMSO) (Sigma; Cat. #D2650).

2.4. Enzyme-linked immunosorbent assay (ELISA)

1. Coating buffer: 0.2 *M* sodium bicarbonate buffer, pH 9.6, 0.02% sodium azide (Sigma).

2. PBST: PBS + 0.05% Tween 20 (Bio-Rad).
3. Blocking solution: 5% bovine serum albumin (BSA) (Sigma; Cat. #A9647) in PBST.
4. Goat anti-human IgG Fc alkaline phosphatase antibody (Jackson ImmunoResearch; Cat. #109-055-098).
5. Diethanolamine buffer: 10% (v/v) diethanolamine, pH 9.8 (J.T. Baker; Cat. #9227-01), 0.02% sodium azide (Sigma).
6. Alkaline phosphatase substrate: one Sigma 104® phosphatase substrate tablet (Sigma; Cat. #104-40T) in 40 mL of diethanolamine buffer.
7. 96-well ELISA plates.
8. Bio-Rad 450 plate reader (Bio-Rad).

2.5. Hollow-Fiber Bioreactor Production

1. Cell Pharm® 100 bioreactor (Biovest International).
2. Hollow-fiber Flowpath 10-kDa MW exclusion (Biovest International; Cat. #22010663).
3. Autoharvester and Flowpath (Biovest International; Cat. #12340637).
4. NS0 selective media (*see* **Subheading 2.3.**).

2.6. Purification of Anti-CEA Minibody

All solutions are made with nanopurified water or water for injection.

1. AG1-X8 strong anion resin (Bio-Rad; Cat. #140-1441).
2. PBS.
3. Macro-Prep ceramic hydroxyapatite (HA) resin, Type I (Bio-Rad).
4. HA column equilibration buffer: 0.05 *M* 2-[N-morpholino]ethanesulfonic acid sodium salt (MES), pH 6.5 (Sigma; Cat. #M2933).
5. HA column elution buffer: 0.04 *M* $K_2HPO_43H_2O$ (Mallinkrodt Baker; Cat. #7088), 0.05 *M* MES, pH 6.5.
6. Source 15Q resin (Amersham Biosciences; Cat. #17-0947-20).
7. 15Q column equilibration buffer: 0.05 *M* N- [2 hydroxyethyl] piperazine-N′- [2-ethanesulfonic acid] (HEPES), pH 7.4 (Fluka; Cat. #54457).
8. 15Q column elution buffer: 0.2 *M* NaCl (Fluka), 0.05 *M* HEPES, pH 7.4.
9. 0.2-µm PES vacuum filter unit, 1 L (Corning; Cat. #431098).
10. AP2, Cat. #WAT027501 and AP1, Cat. #WAT021901 column bodies (Waters).
11. Centriprep 30 concentrator (Millipore; Cat. #4306).
12. 0.2-µm syringe filter (Whatman; Cat. #6780-0402).

2.7. Biochemical Characterization

1. Ready-Gels, 10% polyacrylamide (Bio-Rad Laboratories; Cat. #161-1155).
2. Electrophoresis buffer: 10X Tris/glycine/SDS buffer (Bio-Rad; Cat. #161-0732).
3. Kaleidoscope prestained standards (Bio-Rad; Cat. #161-0324).
4. Bio-Safe Coomassie stain (Bio-Rad; Cat. #161-0786).
5. Mini Trans-Blot™ cell (Bio-Rad; Cat. #170-3930).

6. Transfer buffer: 10X Tris/glycine buffer (Bio-Rad; Cat. #161-0734).
7. Nitrocellulose membrane (Bio-Rad; Cat. #162-0115).
8. Alkaline phosphatase buffer: 0.1 *M* Tris-base, 0.1 *M* NaCl, 0.005 *M* $MgCl_2$, pH 9.6.
9. Development buffer: 33 μL of 5-bromo-4-chloro-3-indoyl phosphate (BCIP) and 66 μL of nitro blue tetrazolium (NBT) (Promega; Cat. #S3771) in 10 mL of alkaline phosphatase buffer.
10. Superdex 75 HR 10/30 column (Amersham Biosciences; Cat. #17-1047-01).
11. PBS.
12. Gel-filtration chromatography standards (Bio-Rad; Cat. #151-1901).
13. Bio-Rad protein assay kit I (Bio-Rad; Cat. #500-0001).
14. rProtein L™-agarose (Affitech; Cat. #201-2).
15. Protein L equilibration buffer: 0.05 *M* sodium citrate, pH 7.4.
16. Protein L wash buffer: 0.02 *M* sodium citrate, 0.02 *M* sodium phosphate, pH 7.4, 0.5 *M* NaCl.
17. 0.1 *M* sodium citrate, not pH-adjusted (Fluka; Cat. #71404).
18. 0.1 *M* citric acid, not pH-adjusted (Fluka; Cat. #27487).
19. EZ-Link™ Sulfo-NHS-LC-biotinylation kit (Pierce Chemical; Cat. #21430ZZ).
20. BIAcore 1000, biosensor SA chips, reagents and BIAevaluation 3.0 software (BIAcore).

3. Methods

3.1. Transient Expression in COS-1 Cells

1. Maintain COS-1 host cells by growth to near confluence in a T-75 flask containing DME media, 10% heat-inactivated FBS, and 2 m*M* L-glutamine (final concentration). Harvest cells by removing media, wash with PBS, add minimal amount of trypsin/EDTA to cover monolayer, watch under microscope until cells start to round up, tap side, and add 10 mL of complete media. Resuspend cells, and determine cell count by trypan blue exclusion using a hemocytometer, and replate at desired density.
2. Twenty-four hours prior to transfection, prepare 100-mm tissue-culture dishes, two plates per experimental DNA (*see* **Note 1**) or vector DNA. Into each plate, inoculate 8×10^5 cells in log phase growth with 5 mL of media.
3. Place in humidified 37°C and 5% CO_2 incubator overnight. Cells should be 60–80% confluent at the time of transfection.
4. On the day of transfection, transfer 900 μL of sterile serum-free antibiotic-free DME media to sterile 6-well culture plate, one well per transfection reaction.
5. Add 50 μL of LipoTaxi transfection reagent per well. Tap side to mix.
6. Add 7 μg of control plasmid or 10 μg of experimental plasmid DNA to respective wells.
7. Mix gently and incubate at room temperature for 15–30 min.
8. Remove culture media from 100-mm dishes by aspiration.
9. Add 1.5 mL of serum-free DME media to each transfection reaction. Transfer the entire mixture (2.5 mL) dropwise onto the cell monolayer while swirling.

10. Incubate for 4–6 h, then add 2.5 mL of DME media/20% heat-inactivated FBS; incubate overnight.
11. On the next day, replace media in plates with 5 mL of fresh media.
12. Incubate for 72 h, recover supernatant, pass through a 0.22-µm syringe filter, and store at 4°C.
13. Determine molecular size of antibody product in culture supernatant by Western blot analysis (*see* **Subheading 3.7.**).
14. Low expression of some constructs may require affinity resin precipitation. Take 1 mL of culture supernatant and 2–5 µL of appropriate resin; incubate overnight at 4°C in a microfuge tube. For antibodies or recombinant fragments that contain a Fc portion, use Protein A Sepharose, which binds to the IgG Fc domain. For scFv recombinant fragments use rProtein L™-agarose, which binds human or murine kappa light chains (*see* **Note 2**). After incubation, spin down resin in a microfuge, discard supernatant, wash with PBS, and resuspend in SDS gel-loading buffer. Proceed with sodium dodecyl sulfate polyacrylamide gel electrophoresis (SDS-PAGE) (**Subheading 3.7.**); prior to loading sample, spin down resin and apply supernatant in gel well.

3.2. *Stable Expression in NS0 Cells* (see Note 1)

1. Grow host NS0 cells in non-selective media and expand in a T-150 flask. Harvest cells in log phase by centrifugation, wash in PBS, and count viable cells by trypan blue exclusion using a hemocytometer.
2. Linearize 100 µg each of pEE12 minibody plasmid and a pEE12 control, using the restriction enzyme *Sal*I. Spin-filter using a Micropure-EZ™ Enzyme Remover to remove BSA and restriction enzyme from sample. Ethanol-precipitate, resuspend the DNA in sterile water, and quantify the amount of cut DNA by gel electrophoresis.
3. Electroporate a 5-mm gap cuvet containing 800 µL of PBS, 40 µg of DNA, and 10^7 NS0 cells using two pulses at 1,500 V, 3 µF from a Gene Pulser II™. Alternatively, electroporate a 2-mm gap cuvet containing 400 µL of hypo-osmolar electroporation buffer, 10 µg of DNA, and 2×10^6 cells using one pulse at 200 V, 100 µs from a Multiporator™ (*see* **Note 3**).
4. Place on ice for 5 min, then carefully pipet cells into a T-75 flask containing 10 mL of prewarmed nonselective media, and place into a humidified 37°C, 10% CO_2 incubator.

3.3. *Cell Culture*

1. After 72 h, determine the number of viable cells by trypan blue exclusion using a hemocytometer.
2. Dilute the transfected cells from each cuvet to a density of 5×10^4 cell/mL with selective media (glutamine-free). Pipet 10^3–10^4 cells per well of a 96-well plate, adding selective media to a total volume of 200 µL. A typical experiment would consist of five plates of 5×10^3 cells/well, ten plates of 10^3 cells/well and five plates of 10^4 cells/well (*see* **Note 4**).
3. Return plates to the incubator and leave undisturbed for 3 wk.

4. Using a bottom viewing mirror, mark wells containing single colonies.
5. When lightly confluent, determine antibody expression by removal of 100 µL of supernatant for an ELISA (**Subheading 3.4.**) and re-feed the clones. Score as either positive or negative.
6. When confluent, expand 96 of the positive wells into 24-well plates. One day prior to transfer, add 100 µL of fresh selective media, and the next day transfer to 24-well plates (2 mL). Re-feed the 96-well plates as a backup.
7. When the 24-well plates are confluent, assay by ELISA (**Subheading 3.4.**) for antibody quantitation.
8. Transfer the 24 highest-producing clones to 6-well plates (in 3 mL of selective media), and repeat quantitation assay. Expand the 6–12 best stable producers into T-25 flasks (10 mL), and repeat quantitation assay.
9. At the first available opportunity, freeze a limited number of cell-culture stocks in 20% heat-inactivated, dialyzed FBS/10% DMSO at –80°C.
10. Select three clones, subclone by limiting dilution, and adapt to 2% heat-inactivated, dialyzed FBS—or, if possible, serum-free conditions, and freeze stocks in liquid nitrogen.

3.4. ELISA

1. Coat 96-well plates with antigen at 100 µL/well—e.g., 2 µg/mL N-A3 recombinant CEA fragment in coating buffer, overnight at 4°C.
2. Wash plates 3× for 5 min with PBST, tap dry, and block with 150 µL/well of blocking solution at room temperature for 1 h.
3. Serially dilute culture supernatant (e.g., 1:50, 1:100, 1:200, 1:400, 1:800 1:1,600) and an antibody standard (e.g., antibody fragment 2 µg/mL). Add 100 µL of each dilution to a well, incubate at 37°C for 1 h, and wash 3× with PBST.
4. Add 100 µL/well of secondary antibody conjugated to alkaline phosphatase and incubate at 37°C for 1 h. The human C_H3 domain is detected by using the goat anti-human IgG Fc-AP antibody at a 1:20,000 dilution. Wash plates 3× in PBST, tap dry.
5. Develop color by adding 100 µL/well of alkaline phosphatase substrate, wait 15–30 min and read absorbance at 405 nm using a Biorad 450-plate reader. Determine antibody concentration based on serial dilutions of the antibody standard.

3.5. Hollow-Fiber Bioreactor Production

1. Set up a Cell Pharm® 100 bioreactor equipped with a single 1.5-sq. ft. hollow-fiber Flowpath (10 kDa MW exclusion) according to manufacturer's instructions. Use selective media supplemented with 2% FBS on the intracapillary space (ICS) and the extracapillary space (ECS). Monitor pH, glucose, lactate, and ammonia levels every other day. Make adjustments to the incoming O_2 and CO_2 levels to maintain the pH at 7.0–7.2. The ICS feed rate should be approx 2 L/wk and the recirculation rate should bc 130 mL/min.
2. Inoculate 10^8 cells into the ECS.
3. Connect an Autoharvester to the ECS and initially collect two harvests of 5 mL per day, and gradually increase to 12 harvests of 20 mL total per day (*see* **Note 5**).

3.6. Purification of Anti-CEA Minibody (see Note 6)

3.6.1. Processing of Bioreactor Harvest

1. Add to the bioreactor harvest a 50% slurry of AG1-X8/PBS to a final concentration of 5% (v/v). Under physiological conditions, most antibodies are electronegative, whereas the cell-culture contaminants (e.g., cell debris, DNA, and phenol red) are electropositive and will bind to the strong anion AG 1-X8 resin (*see* **Notes 7** and **8**).
2. Slowly rotate overnight at 4°C. Pour mixture into a 0.2-µm PES vacuum filter unit, allow the resin to settle, and apply vacuum.
3. Store material at 4°C until purification.

3.6.2. Ceramic Hydroxyapatite (HA) Chromatography

1. Dialyze the supernatant against ten column volumes of HA column equilibration buffer, three changes, 24 h at 4°C.
2. Prepare HA column, pour HA resin into AP2 column body (2 cm id × 10 cm h), equilibrate with HA column equilibration buffer, flow rate 5 mL/min.
3. Apply the supernatant (2 mL/min), and wash until the absorbance at 280 nm returns to baseline.
4. Elute with a linear gradient from 100% (v/v) HA equilibration buffer to 100% (v/v) HA elution buffer over 20 column volumes; collect 10-mL fractions.
5. Analyze fractions by Coomassie-stained SDS-PAGE under non-reducing conditions (*see* **Subheading 3.7.**). Pool minibody-enriched fractions and dilute with four volumes of 15Q column equilibration buffer.

3.6.3. Anion Exchange Chromatography

1. Prepare a Source 15Q column in AP1 column body (1 cm id × 10 cm h) and equilibrate with 15Q column equilibration buffer, flow rate 2.5 mL/min.
2. Apply the peak containing minibody from the HA column to the 15Q column and wash until the absorbance at 280 nm returns to baseline.
3. Elute with a linear gradient from 100% (v/v) 15Q column equilibration buffer to 100% (v/v) 15Q elution buffer over 20 column volumes.
4. Analyze fractions by SDS-PAGE (*see* **Subheading 3.7.**), pool purified minibody fractions and dialyze against PBS overnight at 4°C.
5. Concentrate the purified minibody to 8–20 mg/mL using a Centriprep-30 concentrator.
6. Under aseptic conditions, filter the sample using a 0.2-µm syringe filter.

3.7. Biochemical Characterization

3.7.1. SDS-PAGE

1. Electrophorese aliquot of samples and mol wt standards under non-reducing conditions on SDS-PAGE "Ready Gels" with electrophoresis buffer under established conditions *(27)*.
2. Detect proteins with Biosafe™ Coomassie stain according to manufacturer's instructions.

3.7.2. Western Blot Analysis

1. Fill a mini Trans-Blot™ cell with transfer buffer, and electrotransfer proteins from a duplicate gel to a nitrocellulose membrane according to established techniques *(28)*.
2. Use anti-human Fc AP antibody in 1:10,000 dilution as the reporting antibody.
3. Develop blot in dark with development buffer.

3.7.3. Size-Exclusion Chromatography

1. Pre-equilibrate a size exclusion Superdex 75 column using PBS as the mobile phase (0.5 mL/min), and standardize with gel-filtration chromatography standards.
2. Determine molecular size of minibody and analyze for the presence of aggregation.

3.7.4. Protein Concentration Determination

1. Determine the protein concentration by the Bio-Rad protein assay kit I using the supplied bovine gamma globulin as a standard.

*3.7.5. Protein L Antibody Quantitation (see **Note 2**)*

1. Prepare rProtein L™-agarose column in AP1 column body (1 cm id × 2.5 cm h) and equilibrate with Protein L equilibration buffer at 1 mL/min.
2. Inject 100 µL of sample, wash with ten column volumes of Protein L wash buffer.
3. Elute with linear gradient from 100% 0.1 *M* sodium citrate to 100% 0.1 *M* citric acid (15 column volumes). Monitor absorbance at 280 nm and calculate the area of the minibody peak. Generate a standard curve using a previously purified lot of minibody.

3.7.6. Affinity Analysis by BIAcore (see Chapter 23)

1. Biotinylate antigen using the EZ-Link™ Sulfo-NHS-LC-Biotin kit (Pierce).
2. Immobilize to SA biosensor chip that results in a R_{max} of 250 response units.
3. Perform kinetic analysis on concentrations of minibody (6.25–500 nm), regenerating the antigen surface between runs.
4. Calculate affinity using the bivalent model of BIAevaluation 3.0 software.

4. Notes

1. The T84.66 minibody gene was inserted into the pEE12 expression vector *(26)*, which allowed transient expression in COS-1 cells and stable expression in NSO cells These two transfections can be done sequentially, but normally because the DNA is available from the same plasmid preparation, the operations are started in parallel.
2. Recombinant Protein L is specific for human and murine kappa light chain. It does not bind bovine immunoglobulins that are present in FBS.
3. Electroporation is the most common method for introduction of foreign genes into myeloma cells, and the majority of our work involving the expression of anti-CEA minibodies used the BioRad Gene Pulser II™. Subsequently, the Eppendorf Multiporator™ was tested as an alternative, using buffers and conditions recommended

by the manufacturer. Although both instruments provided acceptable results, we have continued our work with the Multiporator™. The Multiporator™ requires 10-fold fewer cells, less DNA, and the hypo-osmotic buffer in the protocol allowed use of gentler electroporation conditions.

4. The process of selecting a stable, high-producing clone is very time-consuming; thus, it is important to initially screen a large number of clones (250–500). However, at each round of selection only the 10–20% top producers are expanded. Although production levels differed according to the antibody construct, the minibody production levels obtained in T-flasks were 20–50 μg/mL.
5. Our own production in the small Cell Pharm® 100 hollow-fiber bioreactor usually resulted in a three- to fivefold higher level of activity compared to T-flasks. Pilot runs produced between 100 and 300 mg of minibody in the harvested supernatant over a period of 2–4 mo. Scale-up production in a larger hollow-fiber bioreactor (Cell Pharm® 2000) resulted in gram quantities of crude minibody, ample for purification for pilot clinical trials.
6. Chromatography conditions have been optimized for the anti-CEA minibody [isoelectric point (pI), approx 4.8]*(26)*, and conditions may need to be modified depending on the pI and chromatographic properties of the protein of interest.
7. The clarification steps (AG1-X8) greatly enhanced the protein purification process. Reduction of cellular debris, DNA, and removal of the pH indicator dye—phenol red—have contributed to higher resolution and improved reproducibility of the chromatography steps.
8. An excellent general source of information on antibody purification can be found in the book *Purification Tools for Monoclonal Antibodies* ***(29)***, and detailed protocols can also be found in Chapter 22.

Acknowledgments

We are grateful to Chia-wei Cheung, Louise Shively, and Cheryl Clark for their expert technical assistance in developing the protocols outlined here. We also thank the many participants in the City of Hope Radioimmunotherapy Program for their ongoing support and contributions. Funding was provided by National Institutes of Health grants CA 43904 and CA 33572.

References

1. Verma, R., Boleti, E., and George, A.J.T. (1998) Antibody engineering: comparison of bacterial, yeast, insect and mammalian expression systems. *J. Immunol. Methods* **216,** 165–181.
2. Geisse, S. and Kocher, H. P. (1999) Protein expression in mammalian and insect cells systems. *Methods Enzymol.* **306,** 19–41.
3. Wright, A. and Morrison, S. (1997) Effect of glycosylation on antibody function: implications for genetic engineering. *Trends Biotechnol.* **15,** 26–32.
4. Gluzman, Y. (1981) SV40-transformed simian cells support the replication of early SV40 mutants. *Cell* **23,** 175–182.

5. Trill, J. J., Shatzman, A. R., and Ganquly, S. (1995) Production of monoclonal antibodies in COS and CHO cells. *Curr. Opin. Biotechnol.* **6,** 553–560.
6. Chu, G., Hayakawa, H., and Berg, P. (1987) Electroporation for the efficient transfection of mammalian cells with DNA. *Nucleic Acids Res.* **15,** 1311–1326.
7. Kaufman, R. J. (1990) Selection and coamplification of heterologous genes in mammalian cells. *Methods Enzymol.* **185,** 537–566.
8. Bebbington, C., Renner, G., Thomson, S., King, D., Abrams, D., and Yarranton, G. T. (1992) High-level expression of a recombinant antibody from myeloma cells using a glutamine synthetase gene as an amplifiable selection marker. *Biotechnology* **10,** 169–175.
9. Urlaub, G. and Chasin, L. A. (1980) Isolation of Chinese hamster ovary cell mutants deficient in dihydrofolate reductase activity. *Proc. Natl. Acad. Sci. USA* **77,** 4216–4220.
10. Page, M. J., Sydenham, M.A. (1991) High level expression of a humanized monoclonal antibody CAMPATH-1H in Chinese hamster ovary cells. *Biotechnology* **9,** 64–68.
11. Yoo, E. M., Chintalacharuvu, K. R., Penichet, M. L., and Morrison, S. L. (2002) Myeloma expression systems. *J. Immunol. Methods* **261,** 1–20.
12. Bebbington, C. R. (1991) Expression of antibody genes in nonlymphoid mammalian cells. *Methods: A Companion to Methods in Enzymology* **2,** 136–145.
13. Peakman, T. C., Worden, J., Harris, R. H., Cooper, H., Tite, J., Page, M. J., et al. (1994) Comparison of expression of a humanized monoclonal antibody in mouse NS0 myeloma cells and Chinese Hamster Ovary cells. *Hum. Antibod. Hybrid.* **5(1 and 2),** 65–74.
14. Zhou, W., Chen, C., Buckland, B., and Aunins, J. (1997) Fed-batch culture of recombinant NS0 myeloma cells with high monoclonal antibody production. *Biotechnol. Bioeng.* **55,** 783–792.
15. Chadd, H. E. and Chamow, S. M. (2001) Therapeutic antibody expression technology. *Curr. Opin. Biotechnol.* **12,** 188–194.
16. Chu, L. and Robinson, D. K. (2001) Industrial choices for protein production by large-scale. *Curr. Opin. Biotechnol.* **2,** 180–187.
17. Shively, J. E. and Beatty, J. D. (1985) CEA-related antigens: molecular biology and clinical significance. *Crit. Rev. Oncol. Hematol.* **2,** 355–399.
18. Hammarstrom, S. (1999) The carcinoembryonic antigen (CEA) family: structures, suggested functions and expression in normal and malignant tissues. *Sem. in Cancer Biol.* **9,** 67–81.
19. Wu, A. M. and Yazaki, P. J. (2000) Designer genes: recombinant antibody fragments for biological imaging. *Quarterly Journal of Nuclear Medicine* **44,** 268–283.
20. Goldenberg, D. M. (2002) Targeted therapy of cancer with radiolabeled antibodies. *Journal of Nuclear Medicine* **43,** 693–713.
21. Wagener, C., Yang, Y.H.J., Crawford, F. G., and Shively, J. E. (1983) Monoclonal antibodies for carcinoembryonic antigen and related antigens as a model system: a systematic approach for the determination of epitope specificities of monoclonal antibodies. *J. Immunol.* **130,** 2308–2315.

22. Neumaier, M., Fenger, U., and Wagener, C. (1985) Monoclonal antibodies for carcinoembryonic antigen (CEA) as a model system: identification of two novel CEA-related antigens in meconium and colorectal carcinoma tissue by Western blots and differential immunoaffinity chromatography. *J. Immunol.* **135,** 3604–3609.
23. Wu, A. M., Williams, L. E., Zieran, L., Padma, A., Sherman, M., Bebb, G. G., et al. (1999) Anti-carcinoembryonic antigen (CEA) diabody for rapid tumor targeting and imaging. *Tumor Targeting* **4,** 1–12.
24. Hu, S.-Z., Shively, L., Raubitschek, A. A., Sherman, M., Williams, L. E., Wong, J.Y.C., et al. (1996) Minibody: A novel engineered anti-carcinoembryonic antigen antibody fragment (single-chain Fv-C_H3) which exhibits rapid high-level targeting of xenografts. *Cancer Res.* **56,** 3055–3061.
25. Xu, X., Clarke, P., Szalai, G., Shively, J. E., Williams, L. E., Shyr, Y., et al. (2000) Targeting and therapy of carcinoembryonic antigen-expressing tumors in transgenic mice with an antibody-interleukin 2 fusion protein. *Cancer Res.* **60,** 4475–4484.
26. Yazaki, P. J., Shively, L., Clark, C., Cheung, C.-W., Le, W., Szpikowska, B., et al. (2001) Mammalian expression and hollow fiber bioreactor production of recombinant anti-CEA diabody and minibody for clinical applications. *J. Immunol. Methods* **253,** 195–208.
27. Laemmli, U. (1970) Cleavage of structural protein during the assembly of the head of bacteriophage T4. *Nature* **227,** 680–685.
28. Towbin, H., Staehelin, T., and Gordon, J. (1979) Electrophoretic transfer of proteins from polyacrylamide gels to nitrocellulose sheets: procedures and some applications. *Proc. Natl. Acad. Sci. USA* **76,** 4350–4354.
29. Gagnon, P. (1996) *Purification tools for monoclonal antibodies.* Validated Biosystems, Tuscon, AZ.

16

Human Antibody Production Using Insect-Cell Expression Systems

Mary C. Guttieri and Mifang Liang

1. Introduction

Insect-cell expression systems provide a reliable and effective means for generating biologically active human antibodies. Two methods are available for this purpose. Most notably, the baculovirus expression system is a well-established tool for rapidly producing functional immunoglobulins (Ig) and can generate antibody yields that rival or even surpass those achieved by other eukaryotic expression systems *(1–5)*. Alternatively, a recently described transfection method allows the continuous expression of antibodies from stably transformed insect cells *(6)*. Both methods offer an advantage over bacterial and yeast expression systems, as proteins generated in insect cells are correctly folded and processed *(7)*. Although all of the authentic protein modifications (e.g., glycosylation) provided by mammalian expression systems are not evident for insect cell-expressed antibodies, these differences do not affect important Ig activities, including complement lysis activity and antibody-dependent cell-mediated cytotoxicity *(8)*. When compared to mammalian-based systems, the methods required for generating and maintaining baculovirus recombinants and stable insect-cell transformants are simple and cost-effective, requiring incubation and storage without the support of carbon dioxide and liquid nitrogen, respectively. Furthermore, insect-cell expression systems provide a safer platform for producing therapeutic proteins because of the limited host range of baculoviruses and the absence of harmful factors (e.g., retroviral elements) potentially associated with mammalian cells.

1.1. Insect Cells for Expression of Foreign Proteins

Lepidopteran cell lines are the primary hosts for insect cell-mediated expression of foreign proteins, with the most commonly used cell lines derived

From: *Methods in Molecular Biology, Vol. 248: Antibody Engineering: Methods and Protocols*
Edited by: B. K. C. Lo © Humana Press Inc., Totowa, NJ

from the fall armyworm, *Spodoptera frugiperda* (S*f*), or from the cabbage looper, *Trichoplusia ni* (TN). Both S*f* and TN cell lines are adapted for growth in serum-free medium, reducing the costs incurred by cell culture reagents. Furthermore, these cell lines are susceptible to infection by the prototype baculovirus *Autographa californica* nuclear polyhedrosis virus (AcNPV). Cell line S*f*9, a subclone of ovarian cell line IPBL-S*f*-21-AE ***(9–10)***, and TN cell line BTI-TN-5B1-4, established from TN egg-cell homogenates ***(11)***, are commonly used for baculovirus expression of antibodies ***(1,12)***, whereas the latter is the only cell line to date that has been used to generate antibody-secreting insect-cell transformants ***(6)***.

1.2. Human Antibody Production by Baculovirus Expression

AcNPV is the primary baculovirus strain used for foreign gene expression. The virions are comprised of enveloped, rod-shaped nucleocapsids that possess circular, double-stranded DNA genomes of approx 131 kilobase pairs (kbp) ***(13)***. The viral life cycle is characterized by a temporal cascade of events that results in the generation of two virion phenotypes, budded virus and occluded virus ***(3,14)***. After entry into permissive cells by either fusion or endocytosis, virions are uncoated in the nucleus and viral DNA is replicated. This early phase of infection is composed of two successive events designated as immediate early and delayed early gene expression. Baculovirus immediate early genes are transcribed by the host RNA polymerase, resulting in the production of proteins required for delayed early gene expression. After or concomitant with viral DNA replication, the late phase of the viral life cycle is initiated. At this stage, baculovirus structural proteins, encoded by viral late genes, encapsidate viral DNA, subsequently producing progeny virus. Nucleocapsids are detected as early as 8 h postinfection (h p.i.), and by 10–12 h p.i., enveloped progeny virus are released from infected cells by budding through the plasma membrane. Budded viral particles can infect additional cells and are the vehicles by which virus is transmitted within cell cultures. At a very late stage of infection (>18 h p.i.), enveloped nucleocapsids are occluded in the nucleus with crystalline matrices comprised of a 29-kDa protein designated polyhedrin. Hyper-expressed late genes that are transcribed at dramatically high levels during this stage encode polyhedrin and another viral protein, p10. These late genes are not essential for baculovirus production in cell culture, as occluded virus functions in the spread of virus between insects rather than cell-to-cell transmission.

The hyper-expression of baculovirus late genes, reflected by the powerful polyhedrin and p10 promoters, and the fact that polyhedrin is not essential for productive infection in vitro make baculoviruses desirable vectors for foreign gene expression. Since the viral genome is too large for insertion of heterologous genes by in vitro ligation methods, homologous recombination in vivo is required. The foreign gene must first be cloned into a transfer vector so that it is under the control of an appropriate baculovirus promoter (e.g., polyhedrin) and is flanked by sequences homologous to the viral genome. The modified vector is then co-transfected with viral DNA into insect cells after which recombinant viruses that contain the foreign gene are generated and selected. Most commercial transfer vectors contain multiple cloning regions flanked by AcNPV DNA sequences that are homologous to the polyhedrin locus. Recombination between these vectors and viral DNA results in the production of non-occluded virus, thereby providing a basis for selection. Several strategies have been developed to improve the efficiency of recombination and subsequent selection of recombinant viruses. One strategy employs linearized AcNPV DNA that has been modified to contain a lethal deletion within a gene, open reading frame (ORF) 1629, encoding a capsid protein ***(15)***. Recombination with a transfer vector that contains the DNA deleted from the viral genome will restore ORF 1629 and thereby allow for the production of viable virus. In this way, the background of wild-type virus is dramatically decreased.

To generate baculovirus recombinants capable of expressing human antibodies, a dual-expression transfer vector, such as pAcUW51 (PharMingen), is required. Heavy- and light-chain antibody genes, including appropriate eukaryotic signal sequences, are respectively cloned, downstream of AcNPV late promoters (e.g., polyhedrin or p10) that are present in opposite orientation within the polyhedrin locus (**Fig. 1**). In this manner, the modified vector can be used to engineer a baculovirus recombinant that efficiently expresses both antibody genes. Upon infection of insect cells with such a viral recombinant, functional Ig is generated and secreted from the cells into the culture medium, where it accumulates in copious amounts and can be collected and then purified for further analysis. Because the baculovirus recombinant buds from infected cells, cell-culture supernatant can be collected, stored, and used as a source of virus for another round of infection and antibody production.

Antibody genes for insect-cell expression can be obtained from various sources such as hybridoma cell lines. Researchers often rely on phage display methods to generate large combinatorial libraries of antigen-specific single-chain Fv (scFv) or Fab antibody fragments (*see* **Note 1**) ***(16)***. Although this technology is a powerful tool, the fragments generated in this manner lack certain functions associated with the Fc domain of the antibody that are potentially necessary for therapeutic purposes. Therefore, methods for converting human scFv and Fab fragments into full-length antibodies are required. In this regard, a set of baculovirus expression cassette vectors was engineered to generate full-length antibody genes using phage display-selected human antibody fragments and to subsequently produce baculovirus recombinants that are capable of

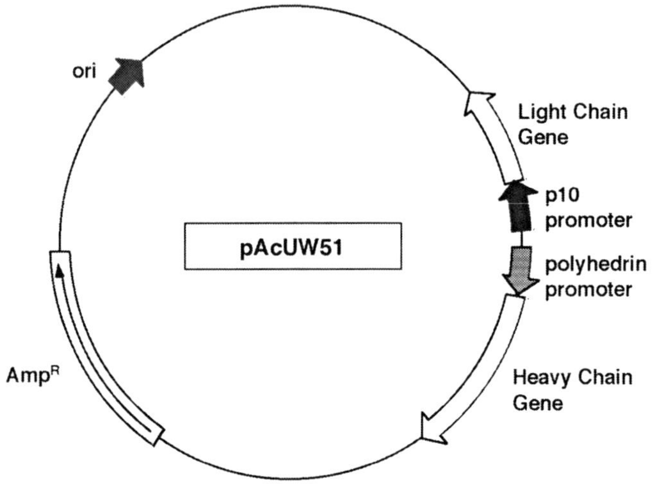

Fig. 1. Transfer vector for generation of baculovirus recombinants expressing human immunoglobulin. Heavy- and light-chain antibody genes are cloned into transfer vector pAcUW51, under control of the baculovirus late promoters polyhedrin and p10, which are situated in opposite orientation. The origin of replication (ori) and ampicillin resistance gene (Amp$^®$) are indicated.

expressing complete human IgG **(Fig. 2)** *(12)*. Transfer vector pAcUW51 was used as a backbone plasmid for constructing this cassette vector system, thereby providing baculovirus polyhedrin and p10 promoters in opposite orientation, an ampicillin resistance gene, an SV40 terminator, and the f1 origin of phage DNA replication. Cassette vectors pAc-K-CH3 or pAC-L-CH3 are respectively used to clone genes that code for scFv that possess kappa (κ) or lambda (λ) light-chain variable regions **(Fig. 2A)**. To clone genes encoding Fab fragments with κ or λ complete light chains, vectors pAc-K-Fc or pAC-L-Fc are used, respectively **(Fig. 2B)**. The light-chain insertion region in each of these vectors contains a κ or λ eukaryotic signal sequence proceeded by specific cloning sites. Vectors pAc-K-CH3 and pAC-L-CH3 also contain, respectively, the coding sequences for the complete κ and λ constant regions. The heavy-chain insertion region in each vector includes an authentic IgG signal sequence from IgG1 subgroup VHIII followed by specific cloning sites and the coding region for human IgG1 constant domains CH1, CH2, and CH3 (in vectors pAc-K-CH3 and pAC-L-CH3) or for human IgG1 constant domains CH2 and CH3 (in vectors pAc-K-Fc and pAC-L-Fc). Variable heavy-chain scFv genes or heavy-chain (Fd region) Fab genes are inserted downstream of the polyhedrin promoter using either *Xho*I and *Nhe*I (in vectors pAc-K-CH3 or pAC-L-CH3) or *Xho*I and *Spe*I restriction sites (in vectors pAc-K-Fc or pAC-

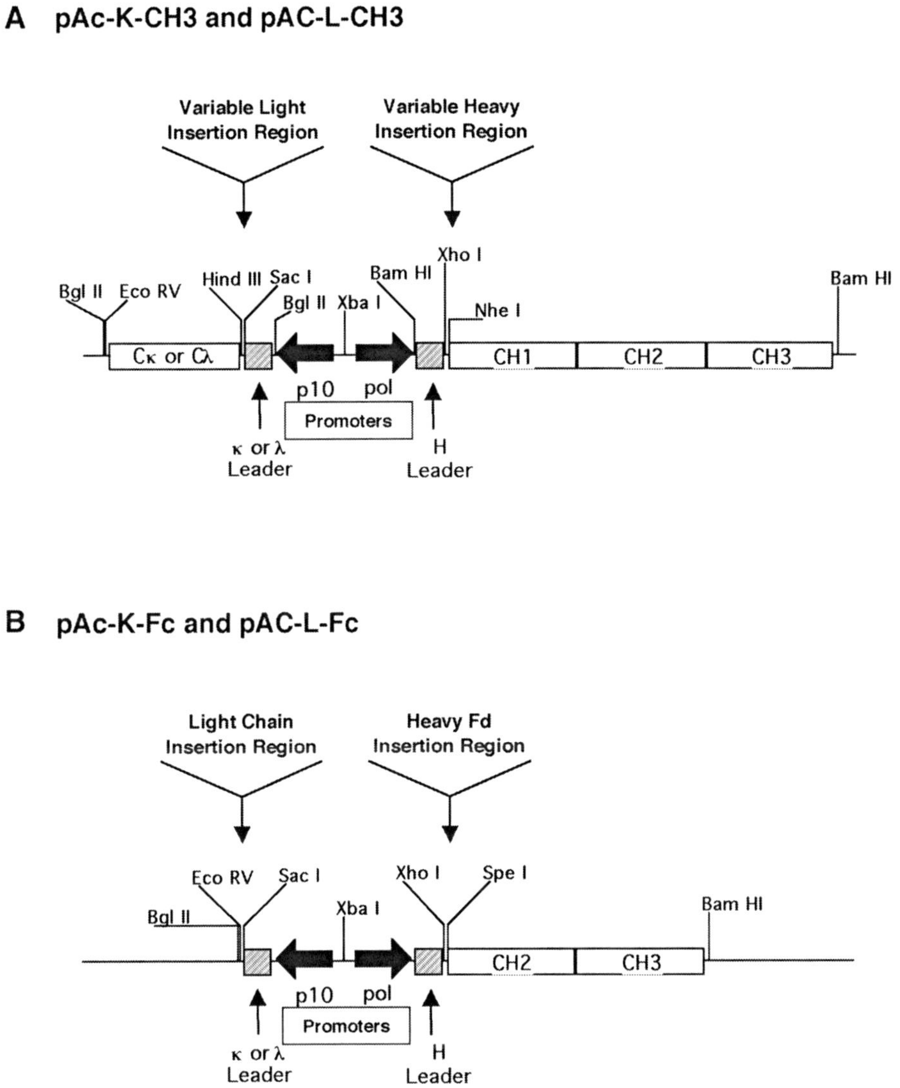

Fig. 2. Baculovirus transfer vector cassettes used for construction and subsequent expression of full-length human IgG genes by inserting coding sequences for phage display-selected antibody fragments *(12)*. **(A)** Vectors used to clone coding regions for light- and heavy-chain variable domains of scFv. **(B)** Vectors used to clone coding regions for light chains and heavy-chain Fd domains of Fabs. Abbreviations: C(x), constant human immunoglobulin regions; H, κ, or λ leader region coding for the respective human antibody signal peptide.

L-Fc), respectively, whereas κ or λ variable light-chain scFv genes or light-chain Fab genes are cloned, respectively, into *Sac*I and *Hind*III (in vectors pAc-K-CH3 or pAC-L-CH3) or *Sac*I and *Eco*RV sites (in vectors pAc-K-Fc or pAC-L-Fc) under control of the p10 promoter. This system was designed so that Fab genes derived from the pComb3 phagemid vector system ***(17)*** can be cloned into vector pAc-K-Fc or pAC-L-Fc without PCR amplification, whereas other genes may need to be amplified by using specific primers as outlined in **Subheading 3.2.2.** Recombinant baculoviruses that are generated by using this cassette system express full-length human light and heavy chains, which are correctly processed and assembled into functional antibodies that are subsequently secreted from infected cells.

1.3. Human Antibody Production in Stably Transformed Insect Cells

Although high yields of foreign protein can be generated by baculovirus expression, there are limitations associated with a system mediated by viral infection. Because baculovirus infection ultimately results in death of the host cell, recombinant protein expression is transient. In addition, the conventional viral late promoters driving expression are active at a stage of the infection cycle when normal cell function is impaired, and studies suggest that the infected host's secretory pathway can become compromised, further limiting productive expression ***(18)***. The potential obstacles associated with processing proteins in a damaged host can be averted by using the baculovirus immediate early gene promoter, IE1, for expression. Promoter IE1 is recognized by insect-cell RNA polymerase II at a very early stage of the infection cycle, which therefore favors the correct and efficient processing of proteins encoded by genes that are regulated by this promoter ***(14)***. This factor prompted the development of baculovirus expression vector systems that rely on the IE1 promoter ***(19)***. As an added benefit to using an IE1-based system, protein recovery can be accomplished sooner than permitted by conventional baculovirus expression systems.

Despite the advantages afforded by early expression of foreign genes, traditional baculovirus systems are preferred because of the high yields of expressed proteins afforded by the powerful viral late promoters. However, a continuous and consistent supply of expressed protein is often desirable, and in this instance, an IE1-mediated expression system is beneficial. Based upon the premise that the host insect-cell RNA polymerase recognizes the IE1 promoter, a transfer vector expression system was engineered for inserting foreign genes into the insect cell genome under control of the IE1 promoter, thereby providing constitutive expression of recombinant proteins in stably transformed insect cells ***(19–20)***. The commercially available pIE1-3, 4 transfer vector

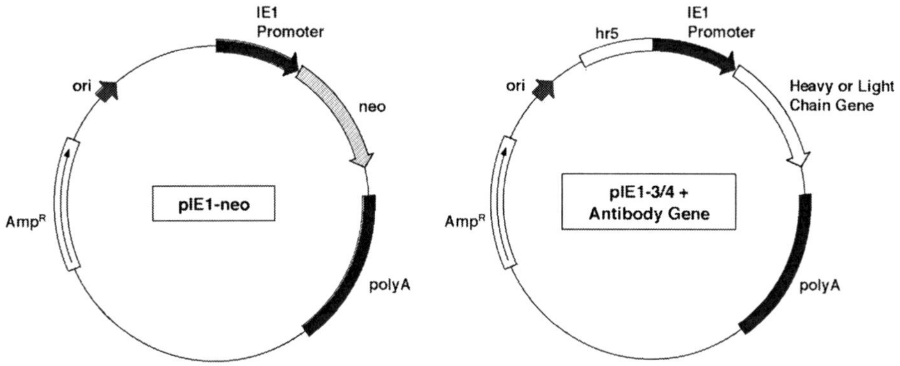

Fig. 3. Transfer vectors for integration of human light- and heavy-chain antibody genes into insect-cell genomes. Heavy- and light-chain genes are separately cloned downstream of the baculovirus early gene promoter IE1 in transfer vector pIE1-3 or pIE1-4. Insect cells are cotransfected with these constructs and selection plasmid pIE1-neo to generate stably transformed insect-cell lines that express human immunoglobulin. The origin of replication (ori), ampicillin resistance gene (Amp®), gene conferring resistance to the antibiotic G418 sulfate (neo), baculovirus enhancer element hr5, and polyadenylation sequences (polyA) are indicated.

series (Novagen) is used for this purpose. In addition to an ampicillin resistance gene and the baculovirus enhancer element hr5, the pIE1-3 and pIE1-4 vectors each contain the baculovirus immediate early IE1 promoter followed by a multiple cloning region and polyadenylation sequences. The orientation of the multiple cloning sites is the only difference between the two vectors. The gene of interest, including the ATG start codon, is cloned downstream of the IE1 promoter in either pIE1-3 or pIE1-4, after which insect cells are cotransfected with the modified vector and selection plasmid pIE1-neo (Novagen), which confers resistance to the antibiotic G418. Stably transformed insect cells that constitutively express the foreign gene are propagated initially under selective pressure with G418, and clonal cell lines are established by selecting and amplifying individual colonies. As recently demonstrated, this expression system can be used to generate stably transformed insect-cell lines that constitutively express biologically active human antibodies *(6)*. Full-length, heavy- and light-chain antibody genes are cloned separately into either pIE1-3 or pIE1-4, and insect cells are cotransfected with these constructs as well as pIE1-neo **(Fig. 3)**. After propagation under selective pressure, an immunological assay, described in **Subheading 3.3.4.3.**, is then used to screen and select transformed insect cells that express Ig. Although the yield of Ig generated in this manner is lower (approx 0.06 μg of IgG/mL) *(6)* than the yield obtained by recombinant baculovirus infection (approx 9 μg of IgG/mL) *(1)*, the convenience afforded by a continuous and consistent supply of Ig makes the IE1-mediated stable transformation expression system a valuable tool for antibody production. Recently, a cassette vector system using this technology was engineered for transformed insect-cell expression of full-length antibodies using genes that code for antibody fragments ***(21)***.

2. Materials

2.1. Insect-Cell Lines and Cell-Culture Maintenance

1. Insect cells: *Sf9* cells (Invitrogen; Cat. #114965-015) and TN High Five™ (H5) cells (BTI-TN-5B1-4) (Invitrogen; Cat. B855-02).
2. Insect-cell culture media: *Sf9* and TN H5 cells are respectively maintained under serum-free growth conditions using Sf-900 II SFM (Invitrogen; Cat. #10902-096) and Excell-405 medium (JRH Biosciences; Cat. #14405-1000M), both supplemented with 50 μg/mL of gentamicin sulfate (Sigma); 1X antibiotic-antimycotic solution (Invitrogen; Cat. #15240-096) yielding 100 U/mL of penicillin, 100 μg/mL of streptomycin, and 0.25 μg/mL of amphotericin B; and 1 m*M* L-glutamine (Invitrogen) (complete medium), unless specified otherwise. TC-100 (Invitrogen; Cat. #11600-061), TNM-FH (PharMingen; cat. #21227M), and IPL-41 (Invitrogen; Cat. #11405-081) insect-cell culture media, which are not modulated for serum-free growth, can also be used by adding 5% (v/v) heat-inactivated fetal bovine serum (FBS) (Invitrogen) and the previously mentioned supplements (FBS-complete medium). To propagate cells from a frozen seed, FBS-complete TC-100 or TNM-FH medium is routinely used. All media are stored at 4°C, protected from light.
3. Cell-culture flasks: 25 cm^2 (T25) flasks; 75 cm^2 (T75) flasks; and 250 mL Erlenmeyer flasks (Corning Costar).
4. Additional cell-culture supplies and reagents: cell storage buffer [10% (v/v) dimethyl sulfoxide (DMS), 90% (v/v) FBS]; cryogenic vials (Corning Costar); and, to determine cell concentration, a hemocytometer and Trypan Blue stain (Invitrogen).
5. Cell-culture incubator (27°C) and, for suspension cultures, a shaker incubator such as the Innova 4330 digital incubator shaker (New Brunswick Scientific).
6. Biological safety cabinet such as the BioGARD hood (The Baker Company) or suitable sterile workbench.

2.2. Preparation of Recombinant Baculoviruses

1. Insect-cell lines and cell-culture media: *Sf9* and TN H5 insect cells (Invitrogen); complete Sf-900 II SFM and Excell-405 media as well as FBS-complete TC-100 medium, unless specified otherwise (*see* **Subheading 2.1.**).
2. Bacteria: *Escherichia coli (E. coli)* DH5α competent cells (Invitrogen).
3. Baculovirus transfer vectors: pAcUW51 (PharMingen; Cat. #21205P); pAc-K-CH3, pAc-K-Fc, pAC-L-CH3, and pAC-L-Fc ***(12)***.
4. PCR reagents: ELONGASE amplification system, including ELONGASE enzyme mix, buffers; and 10 m*M* deoxynucleotide 5′ triphosphate (dNTP) mix (Invitrogen; Cat. #10481-018).

5. Restriction endonucleases and reaction buffers: *Bam*HI, *Bgl*II, *Eco*RV, *Hind*III, *Nhe*I, *Sac*I, *Spe*I, and *Xho*I (Invitrogen or New England Biolabs).
6. Cloning vector pEG5F (Promega).
7. Cloning reagents: T4 DNA ligase (New England Biolabs) and 10X ligase buffer [1X buffer contains 50 m*M* Tris-HCl, pH 7.5, 10 m*M* $MgCl_2$, 10 m*M* dithiothreitol (DTT), 1 m*M* ATP, 25 µg/mL of bovine serum albumin (BSA)]; Luria-Bertani (LB) medium (10 g of tryptone, 5 g of yeast extract, 5 g of NaCl, 1 mL of 1 *N* NaOH per L); LB medium containing 100 µg/mL ampicillin (Invitrogen); LB agar plates (LB medium supplemented with 15 g of agar per L) containing 100 µg/mL ampicillin; and S.O.C. medium (Invitrogen; Cat. #15544-034).
8. Plasmid DNA isolation and gel extraction reagents: QIAprep spin miniprep kit (Qiagen; Cat. #27104); Plasmid midi kit (Qiagen; Cat. #12143); and QIAquick gel extraction kit (Qiagen; Cat. #28704).
9. Baculovirus transfection reagents: BaculoGold™ baculovirus DNA (PharMingen; Cat. #21100D) and transfection buffer A & B set (PharMingen; Cat. #21483A).
10. Baculovirus plaque assay reagents: AgarPlaque Plus™ agarose (PharMingen; Cat. #21403A); BaculoGold™ protein-free insect medium (PharMingen; Cat. #21228M); and Neutral Red solution (Invitrogen; cat. #15330-079).
11. Immunofluorescence assay (IFA) reagents and supplies: 10- or 12-well microscope slides (Corning Costar); phosphate-buffered saline (PBS), pH 7.3 (10X stock solution contains 80 g of NaCl, 2 g of KCl, 11.5 g of Na_2HPO_4 $7H_2O$, 2 g of KH_2PO_4 per L); acetone (Sigma), Coplin jar or suitable container for fixing slides; fluorescein isothiocyanate isomer I (FITC)-conjugated, anti-human Fc- or Fab-specific antibody (Sigma); and a fluorescence microscope.
12. Enzyme-linked immunosorbent assay (ELISA) reagents and supplies: 0.1 *M* $NaHCO_3$, pH 8.6 (coating buffer); PBS and PBS with 3% (w/v) skim milk; unconjugated anti-human Fc- or Fab-specific antibody (Sigma); horseradish peroxidase (HRP)-conjugated, anti-human antibody (Sigma); TMB (3,3′,5,5′-Tetramethylbenzidine) peroxidase substrate (Kirkegaard and Perry; Cat. #50-76-05) and TMB stop solution (Kirkegaard and Perry; Cat. #50-85-05); flat-bottomed 96-well polyvinylchloride ELISA plates (Nunc); and ELISA equipment, including an automated ELISA microplate reader.
13. Antibody affinity purification: 0.45 µm low-protein binding filters (Corning Costar); protein G-sepharose affinity column (Amersham Biosciences); start buffer: 20 m*M* sodium phosphate, pH 7.0; elution buffer: 0.1 *M* glycine-HCl, pH 2.7; 1 *M* Tris-HCl, pH 9.0; and a suitable pump or syringe.
14. SDS-polyacrylamide gel electrophoresis (PAGE): NuPAGE Novex 10% Bis-Tris gel (Invitrogen; cat. #NP0301); NuPAGE LDS 4X sample buffer (Invitrogen; cat. #NP0007); NuPAGE antioxidant (Invitrogen; Cat. #NP0005); NuPAGE 10X sample-reducing agent (Invitrogen; cat. #NP0004); NuPAGE MOPS SDS 20x running buffer for Bis-Tris gels (Invitrogen; Cat. #NP0001); Coomassie brilliant blue R staining buffer [per 200 mL: 0.5 g of Coomassie brilliant R, 45% (v/v) methanol, 10% (v/v) glacial acetic acid, in H_2O]; destain buffer [per 200 mL: 45% (v/v) methanol, 10% (v/v) glacial acetic acid, in H_2O]; BenchMark™ prestained protein

ladder (Invtirogen; Cat. #10748-010); BSA protein standard (Pierce; Cat. #23209); and appropriate gel apparatus.

15. Cell-culture supplies: T25 and T75 flasks, 500-mL Erlenmeyer flasks, 60-mm cell-culture dish, and 6-well and 24-well cell-culture plates (Corning Costar).

2.3. Preparation of Stably Transformed Insect Cells

1. Insect-cell line and cell-culture media: TN H5 insect cells; complete Excell-405 and FBS-complete TNM-FH media, unless specified otherwise (*see* **Subheading 2.1.**).
2. Bacteria: *E. coli* DH5α competent cells.
3. Insect-cell expression plasmids: pIE1-3 (Novagen; Cat. #69090-3) or pIE1-4 (Novagen; cat. #69091-3) and pIE1-neo (Novagen; Cat. #70171-3).
4. PCR reagents: *see* **Subheading 2.2., item 4.**
5. Restriction endonucleases and reaction buffers: *Bam*HI, *Bgl*II, *Bsa*BI, *Dsa*I, *Not*I, *Nru*I, *Pme*I, and/or *Sac*II (Invitrogen or New England Biolabs).
6. Cloning reagents: *see* **Subheading 2.2., item 7.**
7. Plasmid DNA isolation and gel extraction reagents: *see* **Subheading 2.2., item 8.**
8. Insect-cell transfection reagents: Lipofectin reagent (Invitrogen; Cat. #18292-011) and G418 sulfate (Geneticin selective antibiotic) (Invitrogen; Cat. #10131-035).
9. ELISA reagents and supplies: goat anti-human Fab-specific antibody and goat HRP-conjugated, anti-human Fc-specific antibody (Sigma); PBS, pH 7.4; PBS with 0.05% Tween-20 (PBS-T); PBS-T with 3% skim milk (PBS-TS); TMB peroxidase substrate and TMB stop solution; flat-bottomed 96-well IEA/RIA ELISA plates (Corning Costar; Cat. #3590); and ELISA equipment, including an automated ELISA microplate reader.
10. Antibody affinity purification: 0.45-μm low-protein binding filters (Corning Costar), Centricon-10 filters (Amicon), monoclonal antibody (MAb) Trap GII antibody affinity purification kit (Amersham Biosciences; Cat. #17-1128-01), and a suitable pump or syringe.
11. SDS-polyacrylamide gel electrophoresis reagents: *see* **Subheading 2.2., item 14.**
12. Cell-culture supplies: 6-, 12-, 24-, and 96-well cell-culture plates as well as T25, T75, and 150 cm^2 (T150) cell-culture flasks.

3. Methods

3.1. Insect-Cell Culture

The cell-culture methods outlined in **Subheadings 3.1.1 to 3.1.4.** apply to *Sf*9 and TN H5 insect cells (*see* **Note 2**).

3.1.1. Culturing Insect Cells from a Frozen Stock

1. Rapidly thaw a vial of frozen insect cells by placing the vial in a 37°C water bath, with gentle agitation (*see* **Note 3**).
2. When cells are almost thawed, remove the vial from the water bath and transfer 1 mL of the cell suspension into a 15-mL centrifugation tube that contains 7 mL of FBS-complete TC-100 or TNM-FH medium prewarmed to room temperature (*see* **Note 4**).

3. Centrifuge the cell suspension at 230*g* at room temperature for 5–8 min (*see* **Note 5**).
4. After carefully discarding the supernatant, gently resuspend the cells in 5 mL of room temperature FBS-complete TC-100 or TNM-FH medium, and transfer the cell suspension to a T25 flask.
5. Incubate the flask at 27°C overnight, allowing the cells to adhere to the flask.
6. Carefully remove the medium from the flask and replace with 5 mL of fresh FBS-complete TC-100 or TNM-FH medium prewarmed to room temperature.
7. Continue to incubate the flask at 27°C until a confluent monolayer of cells has formed (*see* **Note 6**). Proceed with cell-culture maintenance as outlined in **Subheading 3.1.2.**

3.1.2. Cell-Culture Maintenance

*Sf*9 and TN H5 insect cells can be maintained either as monolayers in flasks or as suspension cultures in Erlenmeyer flasks. The latter format is used when large numbers of cells are required for amplifying baculovirus stocks or for generating high levels of foreign proteins. Serum-containing or serum-free medium can be used to maintain cells as either adherent or suspension cultures. Serum-free medium is preferred when analyzing and purifying antibodies secreted from baculovirus-infected or stably transformed insect cells. Conversely, the presence of serum is beneficial for adherent and suspension cultures because it aids, respectively, in the attachment of cells to the flask and reduces the stress on cells from hydrodynamic forces. The following method describes the maintenance of insect cells grown as monolayers in T25 (or T75) flasks with FBS-complete TC-100 or TNM-FH medium since, as described in **Subheading 3.1.1.,** this is the format used to initiate cultures from frozen cells. Insect-cell monolayers should be passaged when they become confluent (*see* **Notes 6** and **7**).

1. Warm FBS-complete TC-100 or TNM-FH medium to room temperature.
2. Transfer 4 mL (or 12 mL) of the prewarmed medium into each of three T25 (or T75) flasks.
3. Carefully remove and discard the medium from the T25 (or T75) flask containing the insect-cell monolayer so that 3 mL (or 9 mL) of medium remain.
4. Dislodge the cells from the surface of the flask by drawing up the remaining medium into a 10-mL pipet then gently dispensing the medium across the monolayer, using a side-to-side sweeping motion. Repeat this process, working from the bottom to the top of the flask, until the cells have detached.
5. Transfer 1 mL (or 3 mL) of the cell suspension into each of the T25 (or T75) flasks containing the prewarmed medium.
6. Distribute the cells evenly across the surface by gently rocking each of the flasks.
7. Incubate the flasks at 27°C until confluent monolayers are obtained, after which the cells are again passaged as described in **steps 1–6.**

3.1.3. Adaptation of Insect-Cell Cultures to Serum-free Medium

The above method describes the propagation of insect cells in serum-containing medium. However, it is often desirable to culture cells in serum-free medium. If the cell culture was initiated in the presence of serum, it is necessary to gradually adapt the cells to growth in serum-free conditions.

1. Allow insect cells to reach 50% confluency in a T25 flask (*see* **Note 7**).
2. Warm serum-free medium to room temperature, using complete Sf-900 II SFM or Excell-405 medium for *Sf9* or TN H5 insect cells, respectively.
3. Carefully remove approx 1 mL (25%) of serum-containing medium from the flask, and replace with the same volume of prewarmed serum-free medium.
4. Incubate the flask at 27°C until the cells reach confluency.
5. Dislodge the cells from the flask as in **Subheading 3.1.2., step 4,** using the entire volume of medium in the flask.
6. Transfer approx 2.5 mL (50%) of the cell suspension to a T25 flask that contains 2.5 mL complete medium comprised of 50% serum-containing medium and 50% serum-free medium prewarmed to room temperature.
7. Allow the cells to reach confluency at 27°C, then passage as in **steps 5** and **6,** except using complete medium comprised of 25% serum-containing medium and 75% serum-free medium.
8. Repeat **step 7** except using 100% complete serum-free medium, and proceed with cell-culture maintenance using this medium as outlined in **Subheading 3.1.2.**

3.1.4. Preparation and Storage of Frozen Insect Cells

1. Transfer insect cells from flasks into 15-mL (or 50-mL) centrifuge tubes when they are 98% viable and in the logarithmic phase of growth.
2. Pellet the cells by centrifugation at 230*g* for 10 min at 4°C.
3. Carefully remove all of the supernatant and gently resuspend the cells in cell storage buffer to a final density of 1×10^7 cells/mL.
4. Aliquot the cell suspension into cryogenic vials and incubate the vials at –70°C overnight.
5. Rapidly transfer the vials to liquid nitrogen for long-term storage.

3.2. Expression of Human Antibodies by Recombinant Baculoviruses

3.2.1. Selection of Baculovirus Transfer Vector

To generate recombinant baculoviruses that express full-length human antibodies, a transfer vector is required. Selection of this vector is determined by the source of antibody genes. As described in **Subheading 1.2.,** genes that code for full-length heavy and light chains are cloned into a dual-expression vector, such as pAcUW51 (**Fig. 1**). Alternatively, an appropriate baculovirus cassette vector *(12)* is required for expression of full-length human IgG using antibody genes that are derived from an scFv or Fab prokaryotic expression system (*see* **Note 1**). Cassette vector pAc-K-CH3 or pAC-L-CH3 is used to clone genes that code for scFv with κ or λ light chains, respectively **(Fig. 2A)**. Conversely, genes that encode Fab fragments with κ or λ light chains are cloned into cassette vector pAc-K-Fc or pAC-L-Fc, respectively **(Fig. 2B)**. Once the transfer vector is selected, the cloning strategy can be determined. In this regard, it is often necessary to amplify antibody genes by polymerase chain reaction (PCR) to generate necessary cloning sites, and for this purpose, specific oligonucleotide primers are required as described in **Subheading 3.2.2.**

3.2.2. Primer Design

Primers for PCR amplification of human antibody genes must be designed according to the sequences of the heavy- and light-chain genes and the transfer vector to be used (e.g., vector pAcUW51 or cassette vector pAc-K-Fc). Several issues must be considered in this regard, including the origin of the genes (e.g., genes derived from scFv expression systems), the presence of internal restriction sites, and the availability of start and stop codons. For expression, heavy- and light-chain antibody genes must possess eukaryotic signal sequences. Human antibody genes derived from eukaryotic expression systems have the appropriate leader sequences; however, antibody genes that are obtained from prokaryotic systems (e.g., phage display-derived scFv or Fab genes) must be cloned into transfer vectors that provide eukaryotic signal sequences (e.g., pAc-K-CH3 or pAc-K-Fc). Most importantly, antibody genes must be cloned in frame within the appropriate insertion region of the selected transfer vector (*see* **Note 8**). The baculovirus expression cassette vector system described in **Subheading 1.2.** allows for human Fab genes derived from the pComb3 phagemid vector system *(17)* to be rapidly cloned into expression vector pAc-K-Fc or pAC-L-Fc **(Fig. 2B)** without using PCR. To clone human antibody genes derived from other expression systems into one of the cassette plasmids (e.g., pAc-K-CH3) or into other transfer vectors (e.g., pAcUW51), it is often necessary to amplify the genes by PCR, and for this purpose, two primer sets are required. Each set consists of 5′ (forward) and 3′ (reverse) primers for amplifying the gene that codes for the heavy or light chain of the specific antibody or antibody fragment. When designing primer sets, the primer sequences must include the appropriate restriction sites required for cloning into the selected transfer vector, and amplification must allow for inclusion of start and stop codons if the transfer vector does not provide them. With regard to cloning genes that code for human antibody fragments, **Table 1** summarizes primers used for PCR amplification of coding sequences for the variable regions of heavy and light chains, heavy-chain (Fd) regions, or complete light chains. Genes encoding heavy-chain variable regions are amplified by using one of the

Table 1
Primers for Cloning Human Antibody Genes into Baculovirus IgG Expression Vectors

Human IgG variable-chain 5′ primers	
VH3a	5′-GAG GTG CAG CTC GAG GAG TCT GGG-3′*
VH1f	5′-CAG GTG CAG CTG CTC GAG TCT GGG-3′*
VH3f	5′-GAG GTG CAG CTG CTC GAG TCT GGG-3′*
VH4f	5′-CAG GTG CAG CTG CTC GAG TCG GG-3′*
VH6a	5′-CAG GTA CAG CTG CTC GAG TCA GG-3′
Human Kappa variable-chain 5′ primers*	
VK1a	5′-GAC ATC GAG CTC ACC CAG TCT CCA-3′
VK3a	5′-GAA ATT GAG CTC ACG CAG TCT CCA-3′
VK2a	5′-GAT ATT GAG CTC ACT CAG TCT CCA-3′
Human Lambda variable-chain 5′ primers	
VL1	5′-AAT TTT GAG GAG CTC CAG CCC CAC-3′
VL2	5′-TCT GCC GAG CTC CAG CCT GCC TCC GTG-3′*
VL3	5′-TCT GTG GAG CTC CAG CCG CCC TCA GTG-3′*
VL4	5′-TCT GAA GAG CTC CAG GAC CCT GTT GTG TCT GTG-3′*
VL5	5′-CAG TCT GTG GAG CTC CAG CCG CCC-3′
VL6	5′-5′-CAG ACT GAG GAG CTC CAG GAG CCC-3′*
Human IgG (Fd) heavy-chain 3′ primers (SpeI)*	
CG1z	5′-GCA TGT ACT AGT TTT GTC ACA AGA TTT GGG-3′
Human IgG variable-chain 3′ primer	
NheI-HR	5′- TGG GCC CTT GGT GCT AGC TGA GGA GAC GGT GACC-3′
Human Lambda variable-chain 3′ primer	
HindIII-LR	5′- GAC GGT AAG CTT GGT CCC TCC-3′
Human Kappa variable-chain 3′ primer	
HindIII-KR1	5′-CAG TTC GTT TGA TTT CAA GCT TGG TCCC-3′
HindIII-KR2	5′-CAG TTC GTC TGA TCT CAA GCT TGG TCCC-3′
Human Lambda-chain 3′ primer	
EcoRV-CL	5′-CCG GAT ATC TAG AAC TAT GAA CAT TCT GTA GG-3′
Human Kappa-chain 3′ primer	
EcoRV-CK	5′- CCG GAT ATC TAG AAC TAA CAC TCT CCC CTG TTGA-3′

* Primers identical to the pComb3 system primers *(22)*.

human IgG variable-chain 5′ primers and the human IgG variable-chain 3′ primer, whereas one of the 5′ primers and the human IgG (Fd) heavy-chain 3′ primer are used for amplifying genes that code for heavy-chain (Fd) regions. Similarly, genes that encode variable regions of κ or λ light chains are amplified with one of the human κ or λ variable-chain 5′ primers and one of the human κ or λ variable-chain 3′ primers. When amplifying genes that code for full-length κ or λ light chains of Fabs, one of the human κ or λ variable-chain 5′ primers and the human κ or λ chain 3′ primer are used (*see* **Note 9**). Once the cloning strategy is determined and primers, if required, are engineered, heavy- and light-chain antibody genes are inserted into the baculovirus transfer vector as outlined in **Subheading 3.2.3.**

3.2.3. PCR Amplification and Cloning into the Baculovirus Transfer Vector

The following protocol describes the method used to clone genes for human antibodies into transfer vector pAcUW51 or genes for human antibody fragments into cassette vectors pAc-K-CH3, pAC-L-CH3, pAc-K-Fc, or pAC-L-Fc *(12)*. The protocol for PCR amplification, if necessary, is outlined in **steps 1** and **2.**

1. For each PCR mixture, add 0.1 μg of DNA (e.g., plasmid containing light-chain Fab gene), 60 pmol of each 5′ (forward) and 3′ (reverse) primer, 10 μL of Buffer A, 10 μL of Buffer B, 1 μL of ELONGASE enzyme mix, and H_2O to a final volume of 100 μL.
2. Amplify full-length light- and heavy-chain genes by 35 repeated PCR cycles at 94°C for 1 min, 54°C for 1 min, and 72°C for 3 min. For amplifying light- and heavy-chain scFv or Fab genes, carry out 25 repeated PCR cycles at 94°C for 50 s, 54°C for 50 s, and 72°C for 1 min (*see* **Note 10**). Each is followed by a final incubation at 72°C for 10 min and is then cooled to 4°C.
3. If working with a pComb3-derived human Fab, double-digest pComb3 phagemid DNA containing the human Fab gene with *Sac*I and *Xba*I, to excise the light-chain gene, and with *Xho*I and *Spe*I, to excise the heavy-chain (Fd) gene.
4. Subject PCR products or digested phagemid DNA to agarose gel electrophoresis, and purify the appropriate DNA bands by using a QIAquick gel extraction kit.
5. Subclone the double-digested, purified phagemid DNA coding for the Fab light chain (*Sac*I and *Xba*I restriction fragment) into transfer vector pEG5F and then excise the gene from pEG5F by double digestion with *Sac*I and *Eco*RV. Purify the restriction fragment by gel electrophoresis as in **step 4** (*see* **Note 11**).
6. Digest the purified PCR product of DNA coding for the light chain of the full-length human antibody and transfer vector pAcUW51 with *Bgl*II *or* the purified PCR product of DNA coding for the light chain of the human Fab (not derived from phagemid pComb3) and vector pAc-K-Fc or pAC-L-Fc with *Sac*I and *Eco*RV *or* the purified PCR product of DNA coding for the light-chain variable region of the human scFv and vector pAc-K-CH3 or pAC-L-CH3 with *Sac*I and *Hind*III (*see* **Note 12**).
7. Ligate *Bgl*II-digested light chain DNA to digested vector pAcUW51 *or* *Sac*I and *Eco*RV double-digested light-chain Fab DNA (obtained by PCR amplification or by excision from vector pEG5F, *see* **steps 3–5**) to digested vector pAc-K-Fc or pAC-L-Fc *or* *Sac*I and *Hind*III double-digested, light-chain scFv DNA to vector

pAc-K-CH3 or pAC-L-CH3 by establishing a DNA molar ratio of 4:1 (insert:vector), using 1 U of T4 DNA ligase and incubating at 16°C overnight.

8. Transform *E. coli* DH5α competent cells with the ligated plasmid DNA by heat shock at 42°C for 45 s.
9. Plate the transformation mixture on LB agar plates containing 100 μg/mL ampicillin, and incubate the plate at 37°C overnight.
10. Pick transformants into LB medium containing 100 μg/mL ampicillin, incubate at 37°C overnight, extract DNA from the cultures by using a QIAprep spin miniprep kit, and conduct restriction endonuclease analysis to identify clones that contain the correct light-chain DNA insert.
11. Digest the purified PCR product of DNA coding for the heavy chain of the full-length human antibody and vector pAcUW51 containing the light-chain insert with *Bam*HI *or* the purified PCR product of DNA coding for the heavy-chain (Fd) region of the human Fab (not derived from phagemid pComb3) and vector pAc-K-Fc or pAC-L-Fc containing the light-chain insert with *Xho*I and *Spe*I *or* the purified PCR product of DNA coding for the heavy-chain variable region of the human scFv and vector pAc-K-CH3 or pAC-L-CH3 containing the light-chain insert with *Xho*I and *Nhe*I.
12. Ligate *Bam*HI-digested heavy-chain DNA to digested vector pAcUW51 containing the light-chain insert *or Xho*I and *Spe*I double-digested, heavy-chain (Fd) DNA derived from Fab genes (obtained by PCR amplification or by excision from vector pComb3, *see* **steps 3** and **4**) to digested vector pAc-K-Fc or pAC-L-Fc containing the light-chain insert *or Xho*I and *Nhe*I double-digested, heavy-chain DNA derived from scFv genes to vector pAc-K-CH3 or pAC-L-CH3 containing the light-chain insert as described in **step 7.**
13. Transform bacteria with the ligated plasmid DNA, analyze clones for the presence of the heavy-chain DNA insert as outlined in **steps 8–10,** and proceed with preparation of the recombinant baculovirus as described in **Subheading 3.2.4.**

3.2.4. Generation of Recombinant Baculoviruses

1. Prepare highly purified transfer vector DNA containing both the light- and heavy-chain DNA inserts by using a Plasmid midi kit and resuspend the purified DNA in ultra-pure H_2O at a final concentration of 0.5 μg/μL (*see* **Notes 13** and **14**).
2. Seed a 60-mm cell-culture plate with 5 mL of FBS-complete TC-100 medium containing a total of 2×10^6 *Sf*9 cells, and incubate at 27°C for 20–30 min to allow the cells to adhere to the plate (*see* **Notes 15** and **16**).
3. Combine 0.5 μg of BaculoGold™ baculovirus DNA and 2–5 μg of transfer vector DNA containing the light- and heavy-chain DNA inserts in a microfuge tube, mix, and incubate the mixture at room temperature for 5 min, then add 1 mL of transfection buffer B.
4. While the DNA mixture is incubating, carefully wash the *Sf*9 cells attached to the cell culture plate twice, each time with 2 mL of TC-100 medium *without serum and antibiotics.* After the second wash, add 1 mL of transfection buffer A so that the buffer completely covers all of the cells on the plate.

5. Transfer the entire DNA mixture in transfection buffer B to the plate by slowly adding the mixture in a dropwise manner, periodically rocking the plate gently to ensure even distribution and sufficient mixing (*see* **Notes 17** and **18**).
6. Incubate the plate at 27°C for 4 h.
7. Warm complete Sf-900 II SFM medium to room temperature.
8. Wash the plate once with 4 mL of room temperature complete Sf-900 II SFM medium and then incubate the cells for 4–5 d at 27°C in 3.5 mL of this medium. Check the cells periodically for signs of infection (*see* **Note 19**).
9. Remove the medium from the plate, and store at 4°C in a polypropylene tube (*see* **Note 20**). This stock is designated passage zero (P0).
10. Analyze the transfected cells and cell-culture medium for recombinant baculovirus expression of human antibodies by immunofluorescence staining and ELISA, respectively, as outlined in **Subheadings 3.2.5.** and **3.2.6.**

*3.2.5. Detection of Recombinant Baculovirus-Expressed Human Antibodies in Transfected Sf9 Insect Cells by Immunofluorescence (see **Note 21**)*

1. Carefully wash transfected *Sf9* cells on the plate from **Subheading 3.2.4., step 9,** twice with PBS (*see* **Note 22**).
2. Gently dislodge the cells from the surface of the plate by pipetting with 3.0–3.5 mL of PBS.
3. Transfer the cell suspension dropwise to the wells of multiple well slides, and air-dry the slides at room temperature.
4. Place the dried slides in a Coplin jar or suitable container; add acetone, covering the entire surface of the slides; incubate at room temperature for 10 min; remove the slides from the container and air-dry.
5. Add 25–30 μL of FITC-conjugated anti-human Fab- or Fc-specific antibody (diluted according to the manufacturer's specifications) to each of the wells on the dried slides containing fixed cells, and incubate the slides at 37°C for 30 min.
6. Wash the slides 3 × with PBS, and observe the transfected cells by immunofluorescence microscopy.

*3.2.6. Detection of Recombinant Baculovirus-Expressed Human Antibodies in Cell-Culture Medium from Transfected Sf9 Insect Cells by ELISA (see **Note 23**)*

1. Dilute anti-human Fc or Fab-specific antibody to a concentration of 5–10 μg/mL in coating buffer (0.1 *M* $NaHCO_3$, pH 8.6), and add 100 μL of the diluted antibody to each well of a flat-bottomed 96-well polyvinylchloride ELISA plate.
2. Incubate the plate at 37°C for 1 h.
3. Wash the wells 3× with PBS, and to each well, add 100 μL of transfected *Sf9* cell-culture medium from **Subheading 3.2.4., step 9,** diluted 1:1 with PBS containing 3% skim milk (*see* **Note 24**).
4. Incubate the plate at 37°C for 1 h.

5. Wash the wells 3× with PBS, and to each well, add 100 μL of HRP-conjugated anti-human antibody diluted according to the manufacturer's specifications.
6. Incubate the plate at 37°C for 1 h.
7. Wash the wells 5× with PBS, add TMB substrate solution, incubate at room temperature for 15 min, add TMB stop solution, and measure the O.D. at 450 nm.

3.2.7. Plaque Purification, Titration, and Amplification of Recombinant Baculoviruses

After confirming antibody expression in transfected *Sf*9 insect cells as outlined in **Subheadings 3.2.5.** and **3.2.6,** recombinant baculoviruses must be plaque-purified to maintain stability and to generate high-titer viral stocks that can then be used to infect large numbers of cells for producing high yields of antibody. The plaque assay can also be used to determine the recombinant viral titer.

3.2.7.1. Titration and Purification of Recombinant Baculoviruses by Plaque Assay

1. Seed each well of a 6-well cell-culture plate with 3 mL of complete Sf-900 II SFM or FBS-complete TC-100 medium containing a total of 2×10^6 *Sf*9 insect cells, and incubate at 27°C for 1 h to allow the cells to attach (*see* **Note 25**).
2. Serially dilute (from 10^{-2} to 10^{-7}) the P0 recombinant viral stock from **Subheading 3.2.4., step 9,** with complete Sf-900 II SFM medium.
3. Carefully remove the medium from the wells of the plate, and add 200 μL of each virus dilution to the respective well.
4. Gently distribute the virus solution over the cell surface, and incubate the plate at 27°C for 1 h, with occasional rocking (*see* **Note 26**).
5. While the plate is incubating, prepare a 2% agarose solution using AgarPlaque Plus™ agarose and protein-free insect medium, heat this solution in a microwave for approx 30 s to 1 min to dissolve the agarose (solution should begin to boil), and approx 15 min before using it for the plaque assay, place the solution in a 40°C water bath to cool.
6. After incubating the plate, dilute the agarose solution 1:1 with complete Sf-900 II SFM medium, prewarmed to room temperature, and pipet 3 mL of the solution into each well of the plate by tilting the plate and carefully dispensing the solution down the side of the well.
7. Allow the agarose to solidify on a level surface and then incubate the plate at 27°C for approx 5 d (*see* **Note 26**).
8. After incubating the plate, prepare an agarose solution as described in **steps 5** and **6,** except include 2% Neutral Red solution.
9. Add 3 mL of the Neutral Red-containing agarose solution to each well of the plate, allow the agarose to solidify, and incubate the plate at 27°C for 12–24 h.
10. Observe and count plaques under natural or white light, identifying the plaques by marking the bottom side of the plaque assay plate.
11. Determine the viral titer by using the following formula: plaque-forming units (pfu)/mL = plaque number × highest dilution × 5.

12. Seed each well of a 24-well cell-culture plate with 1 mL of complete Sf-900 II SFM medium containing approx 4 × 10^5 *Sf9* insect cells, and incubate at 27°C for 30 min to allow the cells to attach.
13. Carefully pick a plaque by penetrating the agarose with a Pasteur pipet and drawing up with a pipet bulb.
14. Transfer the agarose plug to one of the *Sf9* insect cell-containing wells of the culture plate.
15. Repeat **steps 13** and **14** for at least five more plaques.
16. Incubate the plate at 27°C until infection is visibly apparent (*see* **Note 19**).
17. Collect the infected cell-culture supernatant from the well, and store at 4°C (*see* **Notes 20** and **27**). This plaque-purified viral stock is designated P1 and is amplified as outlined in **Subheading 3.2.7.2.**

3.2.7.2. Virus Amplification

The following describes the method used to generate high-titer viral stocks that can then be utilized to infect large numbers of cells for the purpose of generating high yields of antibody.

1. Seed a T25 flask with 2 × 10^6 *Sf9* insect cells in 5 mL of complete Sf-900 II SFM medium (*see* **Note 28**), and incubate the flask at 27°C for 1 h.
2. Replace the medium in the flask with the same volume of fresh complete Sf-900 II SFM medium prewarmed to room temperature, and add 50 µL of the P1 viral stock obtained from **Subheading 3.2.7.1., step 17.**
3. Incubate the flask at 27°C for 5–7 d or until the cells are completely lysed.
4. Transfer the infected cell-culture medium from the flask into a 15-mL polypropylene centrifuge tube, and centrifuge the tube at 1,000*g* at 4°C for 10 min.
5. Collect the supernatant from the tube, and store at 4°C in a polypropylene tube (*see* **Note 20**). This viral stock is designated P2.
6. Analyze the infected cells and cell-culture supernatant for recombinant baculovirus expression of human antibodies by immunofluorescence staining and ELISA, respectively, as outlined in **Subheadings 3.2.5.** and **3.2.6.**
7. Repeat **steps 1–6** using a T75 flask seeded with 5 × 10^6 *Sf9* cells and infected with 200–300 µL of the P2 viral stock from **step 5.**
8. Repeat **steps 1–6** using 2.5 mL of P2 viral stock and a 250-mL suspension *Sf9* cell culture containing 1 × 10^6 cells/mL. For this infection, incubate the flask for 7–10 d in an incubator shaker at 27°C, with shaking at 80 rpm. Upon collection of the infected cell culture medium, increase the centrifugation time to 15 min.
9. Determine the titer of the recombinant baculovirus stock obtained in **step 8** by plaque assay as outlined in **Subheading 3.2.7.1., steps 1–11,** and then proceed with high-yield antibody production as described in **Subheading 3.2.8.** (*see* **Note 29**).

3.2.8. Recombinant Baculovirus Infection Protocol for Efficient Human Antibody Expression

1. Seed a T75 flask with 7.5 mL of complete Excell-405 medium containing 3 × 10^7 TN H5 insect cells (*see* **Note 2**), and incubate the flask at 27°C for 1 h to allow the cells to attach.

2. To the flask, add recombinant baculovirus from the high-titer stock obtained in **Subheading 3.2.7.2., step 8,** at a multiplicity of infection (MOI) of 10^{-3} (*see* **Note 30**) (usually 2 mL of virus per T75 flask).
3. Incubate the flask at room temperature for 1 h, with occasional rocking.
4. Add 7.5 mL of complete Excell-405 medium prewarmed to room temperature, and incubate the flask at 27°C until cell viability is at 30–40%, observing the cells for cytopathic effects (*see* **Notes 19** and **31**).
5. Transfer the infected cell-culture medium to a 15-mL polypropylene centrifuge tube, centrifuge at 1,000*g* at 4°C for 15 min, filter the supernatant through a 0.45-µm low-protein binding filter, and store at 4°C in a polypropylene tube until ready to process for antibody affinity purification as outlined in **Subheading 3.2.9.**

3.2.9. Affinity Purification of Baculovirus-Expressed Human Antibodies

1. Warm the filtered supernatant collected in **Subheading 3.2.8., step 5,** to room temperature.
2. Prewarm the following at room temperature: a 1-mL pre-packaged protein G-sepharose affinity column, start buffer (20 m*M* sodium phosphate, pH 7.0), elution buffer (0.1 *M* glycine-HCl, pH 2.7), and 1 *M* Tris-HCl, pH 9.0.
3. Equilibrate the column with at least three column volumes of start buffer.
4. Using a syringe or pump, apply the supernatant from **step 1** to the column and then wash the column with five column volumes of start buffer.
5. Elute antibody from the column with three column volumes of elution buffer, collecting 1 mL fractions off the column (*see* **Note 32**).
6. Quickly add approx 25–30 µL of 1 *M* Tris-HCl, pH 9.0, to each fraction to adjust the pH to 7–8.
7. Determine the antibody concentration in each fraction by using the following formula and spectrophotometry **(1)**:

$$\text{Ig concentration mg/mL} = (1.55 \times A_{280}) - (0.76 \times A_{260})$$

Antibody concentration can also be determined by SDS-polyacrylamide gel electrophoresis (SDS-PAGE) as outlined in **Subheading 3.4.** (*see* **Note 33**).

3.3. Expression of Human Antibodies in Stably Transformed Insect Cells

3.3.1. Selection of Transfer Vector

The guidelines for selection of a transfer vector used for producing antibody-secreting stably transformed insect cells are similar to the parameters for choosing vectors used for generating Ig-expressing baculovirus recombinants, as described in **Subheading 3.2.1.** Unlike the vast array of baculovirus transfer vectors suitable for this purpose, only one expression vector system, series pIE1-3-4, has been used to date to establish human antibody-secreting insect cell lines. Although a transformed insect-cell cassette vector system was recently reported on ***(21)***, commercial vectors are not yet available for expression of full-length human IgG using antibody genes derived from scFv or Fab prokaryotic systems, and as a result, complete antibody genes are a prerequisite for this mode of insect-cell expression, as described here. With the pIE1-3-4 system, light- and heavy-chain antibody genes are cloned into separate plasmids, and for this purpose, vector pIE1-3 and/or vector pIE1-4 can be utilized **(Fig. 3)**. These vectors differ only in the orientation of restriction enzyme sites within the multiple cloning regions of the plasmids. Therefore, vector selection is primarily based upon restriction endonuclease requirements, which are dictated by the presence or absence of internal restriction sites within the antibody genes. As a result, the coding sequences of heavy- and light-chain genes should be known to select the appropriate transfer vector and, as outlined in **Subheading 3.3.2.**, to subsequently establish a suitable cloning strategy.

3.3.2. Primer Design and PCR Amplification

Full-length human heavy- and light-chain antibody genes can be cloned directly into respective transfer vectors pIE1-3 and/or pIE1-4 by endonuclease digestion if restriction enzyme sites are available and if this cloning strategy provides for inclusion of start and stop codons as well as eukaryotic signal sequences. Often, this is not possible, and as a result, the antibody genes must be amplified by PCR to permit cloning. Specific oligonucleotide primers are designed for this purpose according to several parameters, including the sequences of the antibody genes, available cloning sites in the vector, and requirements for various expression elements (e.g., stop codons or eukaryotic signal sequences). As outlined in **Subheading 3.2.2.**, two primer sets are required, with each set comprised of 5′ (forward) and 3′ (reverse primer) primers used for amplification of either the full-length heavy- or light-chain gene. Appropriate restriction enzyme sites must be incorporated into the primer sequences, and amplification must result in the inclusion of important genetic elements (e.g., stop codons). Once the primers are engineered, PCR amplification of full-length human antibody genes is carried out as described in **Subheading 3.2.3., steps 1** and **2.**

3.3.3. Cloning into Transfer Vectors pIE1-3 and pIE1-4

This section describes the method used to clone human heavy- and light-chain antibody genes into transfer vector pIE1-3 and/or vector pIE1-4. The multiple cloning region of each plasmid is comprised of *Bam*HI, *Bgl*II, *Bsa*BI, *Dsa*I, *Not*I, *Nru*I, *Pme*I, and *Sac*II restriction enzyme sites, and as heavy- and light-chain genes are cloned separately with this system, either vector or both can be used. As described in **Subheadings 3.3.1.** and **3.3.2.**, the sequences of the antibody genes ultimately determine the cloning strategy utilized.

1. If the heavy- and light-chain genes are amplified by PCR as outlined in **Subheading 3.3.2.**, subject the PCR products to agarose gel electrophoresis, purify the

appropriate DNA bands by using a QIAquick gel extraction kit, and digest the purified products with the appropriate restriction enzymes (e.g., *Bam*HI or *Bgl*II) for cloning into the selected transfer vector(s).
2. If PCR amplification is not necessary, obtain DNA restriction enzyme fragments that contain the complete heavy- or light-chain gene by using enzymes (e.g., *Bam*HI or *Bgl*II) that permit cloning into the selected transfer vector (*see* **Note 34**).
3. Digest transfer vector pIE1-3 and/or pIE1-4 with the appropriate restriction enzymes (e.g., *Bam*HI or *Bgl*II) that will allow insertion of the antibody genes obtained in **step 1** or **2.**
4. Using 1 U of T4 DNA ligase and a DNA molar ratio of 4:1 (insert:vector), ligate digested light-chain DNA to digested vector pIE1-3 or pIE1-4 overnight at 16°C. Repeat for heavy-chain DNA so that two recombinant plamsids are generated, one with the heavy-chain gene and the other with the light-chain gene.
5. Transform *E. coli* DH5α competent cells with each ligated plasmid DNA preparation by heat shock at 42°C for 45 s.
6. Plate the transformation mixtures on LB agar plates containing 100 μg/mL ampicillin and incubate the plates at 37°C overnight.
7. Pick transformants into LB medium containing 100 μg/mL ampicillin, incubate at 37°C overnight, extract DNA from the cultures by using a QIAprep spin miniprep kit, and conduct restriction enzyme analysis to identify clones that contain the correct heavy- or light-chain DNA insert.

3.3.4. Generation of Stably Transformed Insect-Cell Lines

3.3.4.1. Transfection

1. Prepare highly purified transfer vector DNA containing the light-chain gene and vector DNA containing the heavy-chain gene by using a Plasmid midi kit, and resuspend each in ultra pure H_2O at a final concentration of 1.0 μg/μL (*see* **Notes 13** and **14**).
2. Seed a well of a 12-well cell-culture plate with 1×10^6 TN H5 insect cells in 1 mL of Excell-405 serum-free medium *without antibiotics,* and incubate the plate at 27°C for 1 h to allow the cells to attach to the well (*see* **Note 35**).
3. In a microfuge tube, mix 1 μg of selection plasmid pIE1-neo, 3 μg of recombinant vector pIE1-3 or pIE1-4 containing the light-chain gene, and 3 μg of recombinant vector pIE1-3 or pIE1-4 containing the heavy-chain gene in a total volume of 7 μL.
4. Mix 2 μL of Lipofectin reagent with 1 μL of ultra-pure H_2O, add the diluted reagent to the tube containing the pIE1-neo and recombinant plasmid DNA mixture, mix gently, and incubate the suspension at room temperature for 15 min.
5. Carefully add the DNA-Lipofectin suspension to the well containing cells in a dropwise manner, and incubate the plate at 27°C overnight (*see* **Note 36**).
6. Add 1 mL of FBS-complete TNM-FH medium to the well, and incubate at 27°C overnight (*see* **Note 36**).
7. Gently dislodge the cells from the well by using a 5-mL pipet to remove the medium in the well and gently disperse it across the monolayer.

8. Transfer the cell suspension to a 15-mL polypropylene tube, and bring the volume of the suspension up to 10 mL with FBS-complete TNM-FH medium prewarmed to room temperature.
9. Add the antibiotic G418 sulfate (Geneticin) to the cell suspension to a final concentration of 1 mg/mL.
10. Distribute the cell suspension into the wells of a 96-well cell-culture plate at a volume of 100 µL/well, and incubate the plate at 27°C for 1 wk (*see* **Notes 36** and **37**).
11. Carefully remove the medium from each well, and replace it with 100 µL of fresh FBS-complete TNM-FH medium prewarmed to room temperature and containing G418 at a concentration of 1 mg/mL (*see* **Note 38**).
12. Incubate the plate at 27°C for 1 wk (*see* **Notes 36** and **37**).
13. Carefully remove the medium from each well, and replace it with 100 µL of fresh FBS-complete TNM-FH medium prewarmed to room temperature but *not* containing G418 (*see* **Note 39**).
14. Incubate the plate at 27°C for 1 wk, while observing the wells for cell growth. Mark the bottom of any well in which growth of only a single colony of cells is apparent (*see* **Note 40**).
15. Repeat **steps 13** and **14.**
16. When wells containing single colonies exhibit confluent monolayers, the cells are expanded as described in **Subheading 3.3.4.2.** (*see* **Note 41**).

3.3.4.2. Clonal Expansion

1. Carefully remove the medium from a well containing a transformed cell monolayer derived from a single colony identified in **Subheading 3.3.4.1., step 14.**
2. Gently pipet 200 µL of fresh FBS-complete TNM-FH medium, prewarmed to room temperature, across the monolayer to detach the cells from the well surface.
3. Using a wide-bore pipet tip (*see* **Note 42**), transfer 150 µL of the cell suspension to a well of a 24-well cell-culture plate containing 450 µL of room temperature complete Excell-405 medium.
4. Incubate the plate at 27°C until a confluent monolayer forms, replacing the medium with 0.5 mL of fresh complete Excell-405 every 4 d to ensure efficient growth (*see* **Notes 36** and **41**).
5. When a confluent monolayer is visible, analyze the transformed cell line for expression of human antibodies by ELISA of the cell-culture medium in the well as outlined in **Subheading 3.3.4.3.**

3.3.4.3. Rapid Detection of Human Antibodies in Cell-Culture Medium from Transformed TN H5 Insect Cells by ELISA

1. Dilute goat anti-human Fab-specific antibody 1:1,600 in PBS, pH 7.4; and add 100 µL to each well of a flat-bottomed 96-well IEA/RIA ELISA plate.
2. Incubate the plate at 4°C overnight.
3. Wash the wells once with PBS containing 0.05% Tween-20 (PBS-T), add 100 µL of PBS-T containing 3% skim milk (PBS-TS) to each well, and incubate the plate at room temperature for 1 h.

4. Wash the wells 3× with PBS-T, and to each well add 100 µL of transformed TN H5 cell-culture medium from **Subheading 3.3.4.2., step 5,** diluted 1:1 with PBS-TS (*see* **Note 43**).
5. Incubate the plate at 37°C for 1 h.
6. Wash the wells 4× with PBS-T, and to each well add 100 µL of goat HRP-conjugated, anti-human Fc-specific antibody diluted 1:1,600 in PBS-TS.
7. Incubate the plate at 37°C for 1 h.
8. Wash the wells 4× with PBS-T, and to each well, add TMB peroxidase substrate, incubate at room temperature for 15 min, add TMB stop solution, and measure the O.D. at 450 nm.
9. Cell lines that exhibit high titers of human antibodies by ELISA of cell-culture medium are further expanded as described in **Subheading 3.3.4.4.** (*see* **Note 44**).

3.3.4.4. Expansion of Stably Transformed TN H5 Insect Cell Lines

1. Gently dislodge the cells from the well selected in **Subheading 3.3.4.3., step 9,** by using a wide-bore pipet tip (*see* **Note 42**) to remove the medium in the well and gently disperse it across the monolayer.
2. Transfer the cell suspension to the well of a 6-well cell-culture plate containing 2 mL of complete Excell-405 medium prewarmed to room temperature, and incubate the plate at 27°C until a confluent monolayer forms.
3. When confluent, gently dislodge the cells from the well by using a 5-mL pipet to remove the medium in the well and gently disperse it across the monolayer.
4. Transfer the cell suspension to a T25 flask containing 3 mL of complete Excell-405 medium prewarmed to room temperature.
5. Incubate the flask at 27°C until a confluent monolayer forms, and confirm antibody expression by ELISA as outlined in **Subheading 3.3.4.3.**
6. Continue expanding the cell line by dislodging the cells from the T25 flask and transferring the cell suspension to a T75 flask containing 10 mL of complete Excell-405 medium prewarmed to room temperature. Incubate the flask at 27°C.
7. The stably transformed insect-cell line is maintained as described in **Subheading 3.1.2.** using complete Excell-405 medium.
8. Freeze the cell line for storage at a concentration of $\geq 5 \times 10^5$ cells/mL as outlined in **Subheading 3.1.4.**

3.3.5. Protocol for Efficient Expression of Human Antibodies in Stably Transformed Insect Cells

1. Seed a T150 flask with 1×10^7 stably transformed TN H5 insect cells in 30 mL of complete Excell-405 medium, and incubate the flask at 27°C for 72 h.
2. Transfer the cell-culture medium to a 50-mL polypropylene centrifugation tube, centrifuge at 1,000*g* at 4°C for 15 min, and filter the supernatant through a 0.45-µm low-protein binding filter.
3. Confirm antibody expression by ELISA as outlined in **Subheading 3.3.4.3.**
4. Store the filtered cell-culture supernatant at 4°C in a polypropylene tube until ready to process for antibody affinity purification as described in **Subheading 3.3.6.**

3.3.6. Affinity Purification of Insect Cell-Expressed Human Antibodies

1. Warm the filtered supernatant collected in **Subheading 3.3.5., step 2** to room temperature.
2. Prewarm to room temperature the components from a MABTrap GII antibody affinity purification kit including a 1-mL HiTrap protein G column, binding buffer, elution buffer, and neutralizing buffer.
3. Dilute the binding and elution buffers 1:10 with ultra-pure H_2O.
4. Wash the column with 5 mL of ultra-pure H_2O.
5. Equilibrate the column with 3 mL of diluted binding buffer.
6. Using a syringe or pump, apply the supernatant from **step 1** to the column and then wash the column with 5 mL of binding buffer.
7. Elute antibody from the column with 5 mL of elution buffer, collecting 1-mL fractions off the column into tubes containing 100 μL of neutralizing buffer.
8. Determine the antibody concentration by spectrophotometry as described in **Subheading 3.2.9., step 7,** and by SDS-PAGE as outlined in **Subheading 3.4.** (*see* **Note 45**).

3.4. SDS-PAGE of Human Antibodies Expressed by Baculovirus Recombinants and in Stably Transformed Insect Cells

SDS-PAGE can be used to examine the integrity of light and heavy chains of human antibodies expressed by baculovirus recombinants and in stably transformed insect cells. Furthermore, antibody concentration can be determined by this procedure.

1. Dilute BSA stock to 1 mg/mL, 0.5 mg/mL, 0.25 mg/mL, and 0.125 mg/mL in ultra-pure H_2O.
2. Add 10 μL of each BSA dilution to separate tubes, and mix each with 5 μL of NuPAGE LDS sample buffer, 2 μL of NuPAGE sample reducing agent, and 3 μL of ultra-pure H_2O.
3. Serially dilute affinity-purified antibodies obtained in **Subheading 3.2.9.** or **3.3.6.,** and mix 10 μL of each dilution with sample buffer, reducing agent, and ultra-pure H_2O as described in **step 2.**
4. Boil the BSA and antibody samples for 5 min, spin briefly in a microfuge, and immediately load wells of a NuPAGE Novex 10% Bis-Tris gel. Include BenchMark pre-stained protein ladder in one well as a mol wt marker.
5. Run the gel at ≤ 200 V until the dye-front reaches the bottom of the gel (approx 35–50 min) using NuPAGE MOPS SDS running buffer and NuPAGE antioxidant.
6. Place the gel in Coomassie brilliant blue R staining buffer [per 200 mL: 0.5 g of Coomassie brilliant R, 45% (v/v) methanol, 10% (v/v) glacial acetic acid, in H_2O] at room temperature overnight, with gentle rocking.
7. Transfer the gel to destain buffer [per 200 mL: 45% (v/v) methanol, 10% (v/v) glacial acetic acid, in H_2O] and incubate at room temperature, with gentle rocking, until protein bands become visible. A standard result is shown in **Fig. 4.**
8. Compare antibody concentrations to the Coomassie blue-stained protein standards.

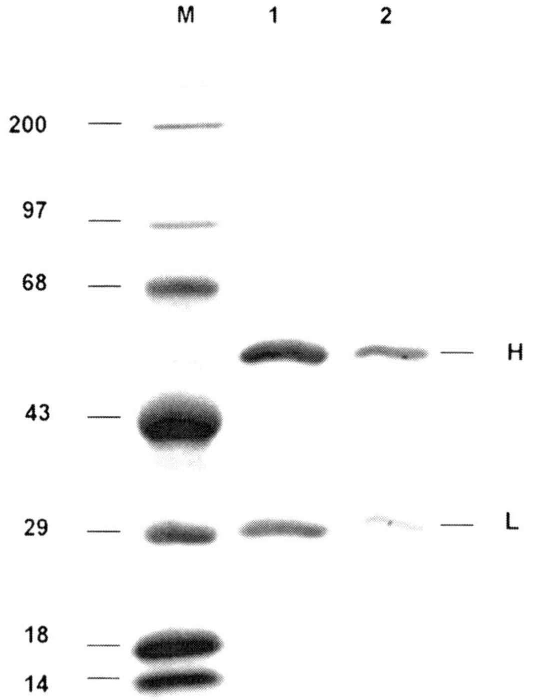

Fig. 4. Polyacrylamide gel electrophoresis and Coomassie blue staining of human Puumala virus-specific antibodies that were affinity-purified from cell-culture supernatants of hybridoma cells (1) or stably transformed insect cells (2). The heavy (H) and light (L) chains are indicated. The sizes of mol wt markers (M) are shown to the left of the panel. (Reprinted from *Journal of Immunological Methods,* 246; Guttieri M. C., Bookwalter C., and Schmaljohn C., "Expression of a human, neutralizing monoclonal antibody specific to Puumala virus G2-protein in stably-transformed insect cells," pp. 97–108, 2000, with permission from Elsevier Science.)

4. Notes

1. Fab antibody fragments are comprised of a single full-length light chain and the variable region as well as constant domain CH1 of a single heavy chain (Fd region), whereas scFv fragments consist of the variable regions of a single light- and single heavy chain joined by a peptide linker.
2. *Sf9* insect cells are routinely used for generating recombinant baculoviruses and for amplifying virus, whereas TN H5 insect cells are used for producing high levels of baculovirus-expressed antibody and for generating antibody-secreting insect cell transformants.
3. Once frozen insect cells have thawed, do not leave the cells in the 37°C water bath, as cell death will result.

4. Insect cells can be thawed in insect cell-culture medium with serum (e.g., TC-100) or in serum-free medium (e.g., Sf-900 II SFM). However, the former medium is preferred because serum assists in the attachment of insect cells to the flask. Once the cell culture is established, the cells can be adapted for growth in serum-free conditions as outlined in **Subheading 3.1.3.**
5. Cell damage can result if insect cells are subjected to centrifugation speeds that exceed 250*g*.
6. Confluency is described as the presence of a single layer of cells over the entire area of attachment in the flask.
7. Typically, insect cells grown as monolayers or in suspension double approximately every 24 h. When adherent cultures reach confluency, cells are diluted 1:5 upon passaging. Suspension cultures should be passaged when they reach 1×10^6 cells/mL, and at this time, the cells should be diluted to 5×10^5 cells/mL.
8. When cloning scFv or Fab coding sequences in frame into a baculovirus transfer cassette vector, the cloning site must match the correct amino acid positions determined from the first start codon of the mRNA.
9. Stop codons are provided in both the light- and heavy-chain cassettes of vectors pAc-K- CH3, pAC-L-CH3 but only in the heavy-chain cassettes of vectors pAc-K-Fc and pAC-L- Fc. Therefore, the sequence of the 3′ primer used for amplifying genes encoding full-length κ or λ light chains of Fabs must include a stop codon when cloning into baculovirus transfer cassette vectors pAc-K-Fc or pAC-L-Fc, respectively.
10. PCR amplification is not required to clone light- and heavy-chain Fab genes derived from the pComb3 phagemid expression system into baculovirus cassette vector pAc-K-Fc or pAC-L-Fc.
11. Although PCR amplification is not required to clone genes that code for light chains of pComb3-derived Fabs into baculovirus cassette vector pAc-K-Fc or pAC-L-Fc, the light- chain gene must first be subcloned into vector pEG5F (Promega) to add the *Eco*RV 3′ cloning site.
12. The light-chain gene is generally cloned first into a baculovirus transfer vector, although this may vary depending on the cloning strategy, which is partly determined by the presence of internal restriction sites within the light- and heavy-chain genes.
13. To increase transfection efficiency, purify plasmid DNA from a bacterial culture that does not exceed an O.D. $_{600\ nm}$ of 1.5.
14. Before transfecting insect cells, it is imperative that the integrity of the heavy- and light-chain coding sequences within the transfer vectors be confirmed by DNA sequence analysis. In this manner, any mutations within the ORFs of the genes can be identified.
15. *Sf*9 cells should be evenly distributed across the surface of the plate before transfecting with baculovirus DNA. If the cells do not adhere within 30 min, they are not suitable for transfection. Furthermore, the density of cells on the plate should not exceed 60–70%.
16. Several plates of *Sf*9 cells should be seeded to establish multiple transfections with a specific baculovirus transfer vector and to establish wild-type viral DNA as well as mock transfection controls.

17. During co-transfection with baculovirus DNA, a calcium phosphate/DNA precipitate will form that turns the medium in the plate slightly turbid.
18. Keep transfection buffer B on ice to ensure high transfection efficiency.
19. Baculovirus-infected insect cells may exhibit enlarged nuclei when compared to uninfected cells. The appearance of cytopathic effects is delayed in recombinant baculovirus-infected *Sf*9 cells as compared to cells infected with wild-type virus. Furthermore, occlusion bodies will not be visible in recombinant baculovirus infections.
20. Baculovirus stocks can be stored at 4°C for at least 6 mo or at –70°C for long-term storage.
21. The immunofluorescence assay can be used to detect antibody expression in baculovirus-infected insect cells as well as transfected cells.
22. When carrying out immunofluorescence staining of transfected or infected *Sf*9 cells, the cells must be washed with PBS to remove serum that can inhibit fixation of cells to the slides. Also, include a negative control by using *Sf*9 cells that are mock-transfected (or -infected).
23. ELISA can be used to detect antibody expression in baculovirus-infected insect cells as well as transfected cells.
24. As a negative control, add 100 µL of mock-transfected (or -infected) cell-culture medium to a well(s) when using ELISA to detect baculovirus-expressed antibodies.
25. For efficient plaque production, *Sf*9 insect cells should be at a density of approx 70% in each well of the plaque assay plate, with 98% of the cells viable.
26. Plaque assay plates should be incubated in a moist chamber (e.g., a box containing a moist tissue) to prevent evaporation.
27. For plaque purification of recombinant baculoviruses, observe the cells in the well for the presence of polyhedra and, if visible, repeat plaque purification to ensure that a clonal stock of recombinant virus is obtained.
28. Because antibody expression is confirmed after each round of virus amplification, recombinant baculovirus-infected *Sf*9 cells are cultured in serum-free medium.
29. Amplification typically results in a viral titer of approx 1×10^8 pfu/mL.
30. Using the baculovirus titer obtained by plaque assay (described **in Subheading 3.2.7.1., steps 1–11**), MOI can be calculated as follows:

$$\text{MOI (pfu/cell)} = \frac{\text{Viral titer (pfu/mL)} \times \text{mL of inoculum}}{\text{Number of total cells to be infected}}$$

From this, the volume of virus inoculum can be calculated as follows:

$$\text{Volume of inoculum (mL)} = \frac{\text{MOI} \times \text{total number of cells to be infected}}{\text{Viral titer (pfu/mL)}}$$

31. Lower yields of antibody will be obtained from baculovirus-infected cell cultures if cell viabilities are lower than 30% or higher than 40%.
32. Antibody can also be affinity-purified from baculovirus-infected cell-culture supernatants by using the MAb Trap GII antibody affinity purification kit (Amersham Biosciences) as described in **Subheading 3.3.6.**

33. Baculovirus-expressed Ig is secreted into the cell-culture medium at levels that reach 10–20 mg/L.
34. Light-chain or heavy-chain antibody genes can be excised from other vectors for cloning into transfer plasmids pIE1–3 or pIE1–4 if the appropriate restriction enzyme sites are available (e.g., full-length antibody genes can be sublconed from baculovirus transfer vector pAcUW51 or cassette vectors pAc-K-CH3, pAc-K-Fc, pAC-L-CH3, and pAC-L-Fc into transfer vectors pIE1–3 or pIE1–4).
35. Several wells should be seeded with TN H5 insect cells to establish multiple transfections with pIE1-based recombinant vectors and to establish negative as well as mock transfection controls.
36. Cell-culture plates containing transfected TN H5 insect cells should be incubated in a moist chamber (e.g., a box containing a moist tissue) to prevent evaporation.
37. Transfected TN H5 insect cells that do not acquire the gene for G418 resistance by recombination with selection plasmid pIE1-neo will not survive when grown in medium containing the antibiotic G418. After transfection, vast cell debris will therefore be apparent while the transfected cells are grown under selective pressure; however, this debris will not inhibit the growth of cells that have acquired antibiotic resistance.
38. When growing newly transfected TN H5 insect cells in the presence of the antibiotic G418, it is potentially beneficial for cell growth to use FBS-complete TNM-FH medium that is diluted 1:1 with FBS-complete TNM-FH *conditioned* medium derived from the supernatant of a wild type TN H5 insect-cell culture. To do so, the conditioned medium (supernatant) must first be clarified by centrifugation. Any residual cells remaining in the supernatant after clarification will not survive when grown in the presence of the G418 antibiotic. Conditioned medium should *not* be used when transfected cells are no longer maintained under selective pressure.
39. As indicated by previous studies, selective pressure is not necessary for continuous and stable antibody expression ***(6)***.
40. If it is not possible to identify wells with single colonies of TN H5 cells after transfection with pIE1-based plasmids as a result of high cell concentrations within the wells, the cells should be gently resuspended in complete TNM-FH medium prewarmed to room temperature, further diluted in this medium, distributed among several wells of a 96-well cell-culture plate, incubated at 27°C, and processed as described in **Subheading 3.3.4.1., steps 14–16.**
41. Multiple copies of antibody genes can integrate into TN H5 insect-cell genomes upon transfection with pIE1-based plasmids ***(6)***. Because the mechanism and location of integration are not known, this process can interfere with normal cell function, thereby resulting in changes in morphology or growth rate. Transfected cells usually require more time to reach a confluent monolayer than wild-type cells.
42. When transferring small volumes of cells with a pipet tip, use a wide-bore tip to reduce cell damage. This pipet tip can be created by snipping off the end and autoclaving the modified tip before use.
43. As a negative control, add 100 µL of wild-type cell-culture medium to a well(s) when using ELISA to detect antibodies expressed by transformed TN H5 insect cells.

44. In addition to their ability to express antibodies, transformed TN H5 cell lines should be selected for expansion based upon their stable growth in culture (*see* **Note 41**).
45. Antibodies expressed in stably transformed insect cells are secreted into the cell-culture medium at levels that reach 0.06 μg Ig/10^6 cells.

Acknowledgments

The authors greatly appreciate the assistance of Dr. Anita McElroy and Dr. Jay Hooper for critical review of the manuscript.

References

1. Liang, M., Guttieri, M., Lundkvist, A., and Schmaljohn, C. (1997) Baculovirus expression of a human G2-specific, neutralizing IgG monoclonal antibody to Puumala virus. *Virology* **235,** 252–260.
2. Nesbit, M., Fu, Z. F., McDonald-Smith, J., Steplewski, Z., and Curtis, P. J. (1992) Production of a functional monoclonal antibody recognizing human colorectal carcinoma cells from a baculovirus expression system. *J. Immunol. Method* **151,** 201–208.
3. Potter, K. N., Yucheng, L. I., and Capra, J. D. (1993) Antibody production in the baculovirus expression system. *Intern. Rev. Immunol.* **10,** 103–112.
4. Poul, M. A., Cerutti, M., Chaabihi, H., Devauchelle, G., Kaczorek, M., and Lefranc, M. P. (1995) Design of cassette baculovirus vectors for the production of therapeutic antibodies in insect cells. *Immunotechnology* **1,** 189–196.
5. Poul, M. A., Cerutti, M., Chaabihi, H., Ticchioni, M., Deramoudt, F. X., Bernard, A., et al. (1995) Cassette baculovirus vectors for the production of chimeric, humanized, or human antibodies in insect cells. *Eur. J. Immunol.* **25,** 2005–2009.
6. Guttieri, M. C., Bookwalter, C., and Schmaljohn, C. (2000) Expression of a human, neutralizing monoclonal antibody specific to Puumala virus G2-protein in stably-transformed insect cells. *J. Immunol. Methods* **246,** 97–108.
7. Verma, R., Boleti, E., and George, A.J.T. (1998) Antibody engineering: comparison of bacterial, yeast, insect and mammalian expression systems. *J. Immunol. Methods* **216,** 165–181.
8. Boyd, P. N., Lines, A. C., and Patel, A. K. (1995) The effect of the removal of sialic acid, galactose, and total carbohydrate on the functional activity of Campath-1H. *Mol. Immunol.* **32,** 1311–1318.
9. O'Reilly, D. R., Miller, L. K., and Luckow, V. A. (eds.) (1992) *Baculovirus Expression Vectors. A Laboratory Manual.* W. H. Freeman & Co., New York, NY.
10. Vaughn, J. L., Goodwin, R. H., Thompkins, G. J., and McCawley, P. (1977) The establishment of two insect cell lines from the insect *Spodoptera frugiperda* (Lepidoptera: Noctuidae). *In Vitro* **13,** 213–217.
11. Wickham, T. J., Davis, T., Granados, R. R., Shuler, M. L., and Wood, H. A. (1992) Screening of insect cell lines for the production of recombinant proteins and infectious virus in the baculovirus expression system. *Biotechnol. Prog.* **8,** 391–395.

12. Liang, M., Dubel, S., Li, D., Queitsch, I., Li, W., and Bautz, E.K.F. (2001) Baculovirus expression cassette vectors for rapid production of complete human IgG from phage display selected antibody fragments. *J. Immunol. Methods* **247,** 119–130.
13. Friesen, P. D. and Miller, L. K. (2001) Insect viruses, in *Fields Virology, 4th Edition,* (Knipe, D. M. and Howley, P. M., eds.), Lippincott Williams and Wilkins, Philadelphia, PA, pp. 599–628.
14. Blissard, G. W. and Rohrmann, G. F. (1990) Baculovirus diversity and molecular biology. *Annu. Rev. Entomol.* **35,** 127–155.
15. Kitts, P. A. and Possee, R. D. (1993) A method for producing recombinant baculovirus expression vectors at high frequency. *Biotechniques* **14,** 810–817.
16. Burton, D. R. (1991) Human and mouse monoclonal antibodies by repertoire cloning. *Trends Biotechnol.* **9,** 169–175.
17. Barbas, C. F., Kang, A. S., Lerner, R. A., and Benkovic, S. J. (1991) Assembly of combinatorial antibody libraries on phage surfaces: The gene III site. *Proc. Natl. Acad. Sci. USA* **88,** 7978–7982.
18. Jarvis, D. L. and Summers, M. D. (1989) Glycosylation and secretion of human tissue plasminogen activator in recombinant baculovirus-infected insect cells. *Mol. Cell Biol.* **9,** 214–223.
19. Jarvis, D. L., Weinkauf, C., and Guarino, L. A. (1996) Immediate-early baculovirus vectors for foreign gene expression in transformed or infected insect cells. *Protein Expr. Purif.* **8,** 191–203.
20. Jarvis, D. L., Fleming, J.G.W., Kovacs, G. R., Summers, M. D., and Guarino, L. A. (1990) Use of early baculovirus promoters for continuous expression and efficient processing of foreign gene products in stably transformed lepidopteran cells. *Biotechnology* **8,** 950–955.
21. Guttieri, M. C., Sinha, T., Bookwalter, C., Liang, M., and Schmaljohn, C. S. (2003) Cassette vectors for conversion of Fab fragments into full-length human IgG_1 monoclonal antibodies by expression in stably transformed insect cells. *Hybridoma and Hybridomics* **22**, 135–145.
22. Kang, A. S., Burton, D., and Lerner, R. A. (1991) Combinatorial immunoglobulin libraries in phage λ, in *Methods: A Companion to Methods in Enzymology, Volume 2, New Techniques in Antibody Generation* (Burton, D. and Lerner, R., eds.), Academic Press, San Diego, CA, pp. 111–118.

17

Antibody Production in Transgenic Plants

**Eva Stoger, Stefan Schillberg, Richard M. Twyman,
Rainer Fischer, and Paul Christou**

1. Introduction

Antibodies bind with great affinity and specificity to their target antigens, allowing them to be exploited in research, medicine, agriculture, and industry *(1–3)*. It is estimated that more than 1,000 antibody-based pharmaceuticals are currently in development, and about 200 of these are already undergoing clinical evaluation in humans. Such widespread use of antibodies would benefit from a safe, convenient, and cost-effective system for large-scale production.

Traditionally, monoclonal antibodies (MAbs) have been produced using murine hybridoma cell lines *(4)*. However, advances in recombinant DNA techniques, improvements in gene transfer procedures and the development of phage display technology now open the way for the production of high-affinity recombinant antibodies in diverse expression systems *(5–8)*. These technological advances also allow the expression of humanized antibodies and antibody derivatives [such as single-chain Fv (scFv) fragments and antibody fusion proteins], thus broadening the range of potential applications.

The best-established systems for recombinant antibody production are mammalian cell lines for full-length immunoglobulins (*see* Chapter 15) and bacterial cells for antibody fragments (*see* Chapter 14). Mammalian cells, particularly Chinese hamster ovary (CHO) and murine NS0 myeloma cells, are favored because they efficiently assemble complex multi-subunit proteins (such as immunoglobulins) and are believed to synthesize glycans similar to those found on human glycoproteins *(9)*. However, the culture of mammalian cells is expensive in terms of equipment, media, and the need for skilled personnel, and there is limited opportunity for scale-up. There is also concern that mammalian cell lines could harbor human pathogens such as viruses and prions, or oncogenic DNA sequences. Furthermore, there is now evidence that the

From: *Methods in Molecular Biology, Vol. 248: Antibody Engineering: Methods and Protocols*
Edited by: B. K. C. Lo © Humana Press Inc., Totowa, NJ

rodent cell lines discussed above do not produce the same glycan patterns as human cells *(10)*. Similar reservations apply to the use of transgenic animals *(11)*. Bacterial cells are used predominantly for the production of scFvs because these are simple, aglycosylated antibody derivatives (e.g., *see* **ref.** ***12***). Full-length, glycosylated antibodies cannot be produced using this expression system because bacteria do not add glycan chains to proteins. Furthermore, the folding and assembly of complex proteins is inefficient in bacteria, resulting in the formation of insoluble complexes known as inclusion bodies. Another disadvantage of bacteria is their production of endotoxins, which are co-purified with the antibody.

Transgenic plants have a number of distinct advantages over culture-based systems for antibody production *(13–15)*. The main advantage is the anticipated cost savings, reflecting the large amount of biomass that can be produced in a short time with no need for specialized equipment or expensive media. Sowing, growing, and harvesting can be carried out using traditional agricultural practices and unskilled labor. Moreover, scale-up can be achieved rapidly and inexpensively, simply by cultivating more land. Heterologous proteins accumulate to high levels in plant cells, and plant-derived antibodies are virtually indistinguishable from those produced by hybridoma cells. Protein synthesis, secretion, and folding, as well as post-translational modifications such as signal peptide cleavage, disulfide-bond formation and the initial stages of glycosylation, are very similar in plant and animal cells. Only minor differences in glycan structure have been identified, such as the absence of terminal sialic acid residues and the presence of the plant-specific linkages α1,3-fucose, and β1–2 xylose *(16)*. Regardless of these differences, studies using mice administered a recombinant IgG isolated from plants showed that the recombinant antibody was not immunogenic *(17)*. There are no data from human studies as yet, because the only plant-derived antibodies at the clinical trials stage have been designed for topical application. However, in anticipation of possible immunogenicity, plants have been modified to express human galactosyltransferases in order to make the glycan chains of recombinant proteins more similar to those found in humans *(18)*.

A variety of different plants can be used as expression hosts, and we have chosen to describe three popular systems—tobacco, rice and wheat—with which we have the most experience. Tobacco is advantageous because of the immense amount of biomass produced (tobacco can be cropped several times a year) and the availability of simple transformation protocols. Cereals produce less biomass, but have other advantages. For example, antibodies can be expressed in the seeds, and remain stable at ambient temperatures for years. This is desirable if the transgenic material must be stored or transported a long distance before processing (tobacco leaves must be dried or chilled for this purpose). Cereals also lack the toxic alkaloids that are present in tobacco, which must be removed during downstream processing. Wheat has the lowest producer price of all the major crops. Several recent publications discuss the relative merits of different plant-based production systems for recombinant proteins *(14,19–21)*.

Despite the anticipated economic advantages of transgenic plants for antibody expression, the majority of the production costs (>90%) arise from downstream processing *(19,22)*. In this chapter, we outline a simple procedure for the isolation and purification of full-length antibodies from tobacco, based on their affinity for protein A. This involves the homogenization of transgenic material, which causes the release of proteases and oxidizing agents (e.g., phenols or tannins) from subcellular compartments. Furthermore, all cell debris must be removed from the crude extract before further processing. To circumvent problems caused by these initial steps, we have developed a cross-flow filtration and fast-flow affinity chromatography protocol which results in purification to >95% *(23)*. Although protein A binds some V_H domains, the purification of recombinant antibody derivatives such as scFv fragments requires alternative procedures because such interactions are not universal. For example, scFvs can be purified by immobilized metal affinity chromatography (IMAC; *see* Chapter 22) using a C-terminal $(His)_6$ affinity tag attached to the recombinant protein *(24,25)*. However, although tags are useful for detection and purification, they may be undesirable in antibodies that are destined for clinical use. A case-by-case decision for each recombinant antibody is necessary and the intended application is often a major factor in the choice of the most appropriate antibody type and expression host *(14)*.

2. Materials

2.1. Plant Material

1. Tobacco (*Nicotiana tabacum* L.): the leaf disk transformation protocol for tobacco (**Subheading 3.1.**) is optimized for cultivar Petit Havana SR1, but other cultivars can also be used. Leaf disks can be obtained from plants grown in the field, but those from sterile-grown plants are preferred because of reduced risk of contamination and faster growth. The preparation of sterile axenic shoots is described in **Subheading 3.1.1.**
2. Rice (*Oryza sativa* L.): the transformation protocol for rice (**Subheading 3.2.**) has been used successfully with many different varieties *(26)*. Typically, rice plants should be maintained in a climate-controlled growth chamber at 28/24°C day/night temperature with 70%/85% humidity with a 14-h photoperiod until fully acclimated. The plants are then transferred to the greenhouse (24–28°C, 70% humidity), and seeds can be collected.
3. Wheat (*Triticum aestivum* L.): the transformation protocol for wheat (**Subheading 3.3.**) is optimized for the variety Bobwhite *(27)*. Plants should be maintained

in a greenhouse or growth room at 15/12°C day/night temperature with a 10-h photoperiod during the first 40 days, followed by maintenance at 21/18°C day/night temperature with a 16-h photoperiod thereafter. Seeds can be collected after another 8 wk.

2.2. Laboratory Equipment and Reagents

1. Controlled environment room (required for rice plants only).
2. Plant culture room with illumination system.
3. Laminar flow hood.
4. Steel punches, 5-mm diameter (for tobacco transformation).
5. Stereomicroscope with light (for rice and wheat transformation).
6. Fine forceps (two pairs) and scalpel.
7. 30°C/37°C incubator.
8. Whatman 3MM filter paper (Whatman).
9. Rotary shaker.
10. Petri dishes.
11. Culture pots and tubes.
12. Parafilm.
13. Sterile, double-distilled water.
14. 70% (v/v) ethanol.
15. 20% and 50% (w/v) sodium hypochlorite solutions.
16. Tween 20.

2.3. Growth Media and Additives

2.3.1. Basic Components

1. MS (Murashige and Skoog) salts with minimal organics (**ref.** *28*; *see* **Note 1**).
2. CC salts with minimal organics (**ref.** *29; see* **Notes 1** and **2**).
3. Sucrose (tissue-culture grade).
4. Agar (for tobacco) or type I agarose (for rice and wheat). Agarose can also be used for tobacco, but is more expensive than agar.
5. Mannitol (Sigma).
6. Sorbitol (Sigma).

2.3.2. Additives and Stock Solutions (see **Note 3**)

1. Kanamycin sulfate (Sigma). Stock solution: 100 mg/mL in H_2O (*see* **Note 4**).
2. Hygromycin B (Sigma). Stock solution: 100 mg/mL in H_2O (*see* **Note 4**).
3. Phosphinothricin (PPT) supplied as glufosinate-ammonium powder (Sigma). Stock solution: 10 mg/mL in H_2O (*see* **Note 4**).
4. Cefotaxime sodium (Duchefa). Stock solution: 200 mg/mL in H_2O.
5. 2,4-Dichlorophenoxyacetic acid (2,4-D) (Sigma). Stock solution: 2 mg/mL in dimethyl sulfoxide (DMSO).
6. Kinetin (Sigma). Stock solution: 1 mg/mL in H_2O.
7. 6-Benzylaminopurine (6-BAP) (Sigma). Stock solution: 1 mg/mL in DMSO.

Table 1
MS and CC Basic Salts and Minimal Organics *(29,30)*

Ingredients	MS	CC
Macroelements (mg/L)		
Make up as a 10X stock in sterile distilled water		
$CaCl_2.2H_20$	440	588
KH_2PO_4	170	136
KNO_3	1,900	1,212
$MgSO_4.7H_2O$	370	247
NH_4NO_3	–	64
Microelements (mg/L)		
Make up as a 100X stock in sterile distilled water		
$CoCl_2.6H_2O$	0.025	0.025
$CuSO_4.5H_2O$	0.025	0.025
H_3BO_3	6.2	3.1
KI	0.83	0.83
$MnSO_4.4H_2O$	22.3	11.15
$Na_2MoO_4.2H_2O$	0.25	0.25
$ZnSO_4.7H_2O$	8.6	5.76
Iron (mg/L)		
Make up as 100X stock in sterile distilled water		
$FeSO_4.7H_2O$	27.8	27.8
$Na_2EDTA.2H_2O$	37.3	37.3
Organic components (mg/L)		
Make up as 100X stock in sterile distilled water		
Filter-sterilize.		
Thiamine HCl	0.1	8.5
Pyridoxine HCl	0.5	1
Glycine	2	2
Myo-inositol	100	90
Nicotinic acid	0.5	6
Casein hydroxylate	1,000	1,000

All stocks should be stored at 4°C. The macroelements, microelements, and iron can be added before autoclaving but the organic compounds should be added after cooling.

8. α-Naphthaleneacetic acid (NAA) (Sigma). Stock solution: 1 mg/mL in H_2O.
9. Zeatin (Sigma). Stock solution: 10 mg/mL in 50% (v/v) ethanol.
10. Betabactyl (Glaxo-SmithKline). Stock solution: ticarcillin:clavulanic acid (25:1).

2.3.3. Media Composition

All media should be prepared immediately before use with ultrapure water (>18 MΩ cm) and autoclaved or filter-sterilized (0.2 μm). Heat-sensitive components (selective agents, phytohormones) must be added after cooling to below 50°C.

2.3.3.1. Tobacco Transformation

1. Basic MS medium for tobacco cultivation (tMS). Make up MS salts and minimal organics as shown in **Table 1** or from a ready-mixed powder (*see* **Note 1**). Add 20 g of sucrose and 8 g of agar per L of medium. Before adding the agar, adjust pH to 5.8 with 1 *N* NaOH (*see* **Note 5**).
2. tMSHCK: tMS plus 0.1 mL of NAA stock, 1 mL of 6-BAP stock, 1 mL of kanamycin sulfate stock, and 2 mL of cefotaxime stock (*see* **Subheading 2.3.2.**).
3. tMSCK: tMS plus 1 mL of kanamycin sulfate stock and 2 mL of cefotaxime stock (*see* **Subheading 2.3.2.**).
4. KCB water: distilled water plus 1 mL of kanamycin-sulfate stock, 1 mL of cefotaxime stock, and 1 mL of Betabactyl stock per L (*see* **Subheading 2.3.2.**).

2.3.3.2. Rice Transformation

1. Basic CC medium for rice cultivation (*see* **Note 6**). Make up CC salts and minimal organics as shown in **Table 1** or from a ready-mixed powder (*see* **Note 1**). Add 20 g of sucrose and 6 g of type I agarose per L of medium. Before adding the agarose, adjust pH to 6.0 with 1 *N* NaOH.
2. CCH (CC with phytohormones): CC medium plus 2 mg/L of 2,4-D.
3. CCHO (osmoticum medium): CCH plus 0.2 *M* mannitol and 0.2 *M* sorbitol.
4. CCHS (selection medium): CCH plus 50 mg/L hygromycin B.
5. CCR (regeneration medium): CC basic salts and minimal organics with 6% type I agarose, **containing only 1% (w/v) sucrose and no phytohormones,** supplemented with 50 mg/mL hygromycin B.
6. rMSR (rice rooting medium): MS basic salts and minimal organics with 6% type I agarose, **containing only 1% (w/v) sucrose and no phytohormones,** supplemented with 50 mg/L hygromycin B.

2.3.3.3. Wheat Transformation

1. Basic MS medium for wheat cultivation (wMS). Make up MS salts and minimal organics as shown in **Table 1** or from a ready-mixed powder (*see* **Note 1**). Add 30 g of sucrose and 6 g of type I agarose per L of medium. Before adding the agarose, adjust pH to 6.0 with 1 *N* NaOH.
2. wMSH (wMS with phytohormones): wMS plus 2 mg/L 2,4-D.
3. wMSHO (osmoticum medium): wMSH plus 0.2 *M* mannitol and 0.2 *M* sorbitol.
4. wMSS (selection medium): wMS plus 10 mg/L Zeatin and 2–4 mg/L PPT.
5. wMSR (regeneration medium): **half-strength** MS salts, also with **half the normal concentration of sucrose** (15 g/L) but the normal amount of agarose, supplemented with 2–4 mg/L PPT.

2.4. Transformation

2.4.1. Agrobacterium-Mediated Transformation of Tobacco

1. A suitable *Agrobacterium tumefaciens* strain transformed with the binary expression vector containing the antibody transgene (**refs. *30,31***; *see* **Note 7**).
2. LB medium: 10 g of bacto-tryptone, 5 g of bacto-yeast extract and 10 g of NaCl per L of water; adjust pH to 7.0 with 5 *N* NaOH and sterilize by autoclaving for 20 min.
3. Whatman 3MM filters.

2.4.2. Particle Bombardment (Rice and Wheat)

1. Stock solution of transforming plasmid DNA carrying antibody transgene (*see* **Notes 7–9**).
2. Stock solution of plasmid DNA carrying selectable marker gene (*see* **Note 10**).
3. Bio-Rad particle accelerator gun and accessories (*see* **Note 11**).
4. Gold particles (Bio-Rad) (*see* **Note 12**).
5. 100 m*M* spermidine.
6. 30% PEG (mW 8,000).
7. 100% ethanol.
8. TE buffer: 10 m*M* Tris-HCl, pH 8.0, 1 m*M* EDTA.
9. 2.5 *M* $CaCl_2$.
10. Sonicator.

2.5. Affinity Purification of Antibodies

2.5.1. Equipment

1. Benchtop cross-flow filtration system.
2. Low-pressure chromatography system.

2.5.2. Chemicals and Buffers

All chemicals are analytical-grade or molecular biology-grade, unless otherwise specified. Aqueous solutions must be prepared with deionized water. Chromatography buffers are filter-sterilized (0.2 μm) and degassed before use.

1. Extraction buffer: 200 m*M* Tris-HCl, pH 7.5, 5 m*M* EDTA, 0.1 m*M* dithiothreitol (DTT), 0.1% (v/v) Tween 20, freshly prepared before use (*see* **Note 13**).
2. Phosphate-buffered saline (PBS): 137 m*M* NaCl, 2.7 m*M* KCl, 8.1 m*M* Na_2HPO_4, 1.5 m*M* KH_2PO_4, pH 7.4.
3. Binding buffer: PBS supplemented to 500 m*M* NaCl.
4. Elution buffer 1: 100 m*M* sodium citrate, pH 5.0.
5. Elution buffer 2: 100 m*M* sodium citrate, pH 3.0.

2.5.3. Chromatography Media

1. Protein A Hyper D (Invitrogen).
2. Sephacryl S300 HR (Amersham Biosciences).

3. Methods

3.1. Transformation of Tobacco Plants

This is a simple protocol for gene transfer to tobacco leaf disks using *Agrobacterium tumefaciens* **(30)**. The manipulation of tobacco plants, leaf disks, and callus should be carried out using sterilized tools and equipment in a laminar flow hood to avoid microbial contamination. We recommend that all glassware is cleaned with double-distilled water (no detergents), covered with aluminum foil (*see* **Note 14**), and sterilized at 180°C for 4 h.

3.1.1. Preparation of Sterile Axenic Shoot Cultures

1. Wash tobacco seeds briefly with sterile H_2O in a Petri dish.
2. Surface sterilize seeds in 70% (v/v) ethanol for 5 min.
3. Wash seeds once with sterile H_2O.
4. Transfer seeds to Petri dishes containing tMS medium. Place ten seeds, evenly spaced, in each dish.
5. Seal Petri dishes with Parafilm (*see* **Note 15**).
6. Incubate dishes at 24°C and 3,000 lux with a 16-h photoperiod. Seeds will germinate after approx 3–5 d.
7. As soon as the seedlings touch the lids of the Petri dishes (2–3 wk), transfer them to culture pots containing tMS medium. Place three seedlings in each pot and continue cultivation until leaf disks are required.

3.1.2. Preparation of Leaf Disks

1. After 3–4 wk of axenic shoot growth in the culture pots, remove fully developed, healthy looking leaves with a scalpel and forceps, place them (one at a time) in a glass Petri dish lined with autoclaved Whatman 3MM filter paper, and cut out leaf disks using a sterile punch or scalpel.

3.1.3. Infection with Agrobacterium tumefaciens and Selection of Transgenic Tissue

1. Two days before transformation, take a single colony or 10 µL of glycerol stock of the *Agrobacterium* strain containing the recombinant binary vector.
2. Transfer to 10 mL of LB medium containing 50 mg/L kanamycin and incubate at 28°C overnight on an orbital shaker at 250 rpm.
3. Transfer 100 µL of the culture to fresh medium with no antibiotics. Incubate at 28°C overnight on an orbital shaker at 250 rpm.
4. Adjust the OD_{600} of the *Agrobacterium* suspension to 1.0 with liquid MS medium. Return culture to the orbital shaker at 28°C for 2 h.
5. Transfer the leaf disks to a beaker containing 50–100 mL of *Agrobacterium* suspension. Ensure good contact between bacteria and leaf disks. Incubate at room temperature for 30 min.

6. Transfer the leaf disks onto sterile, pre-wetted Whatman filters in Petri dishes, seal with Parafilm, and incubate at 26–28°C in the dark for 2 d.
7. Wash disks with KCB water and transfer them onto tMSHCK dishes. Cultivate at 25°C, 3,000 lux with 16-h illumination from now on. After approx 2 wk, callus tissue will begin to form. The development of shoots will require at least another 2 wk (*see* **Note 16**).

3.1.4. Regeneration of Transgenic Plants

1. After leaf disk transformation, whole plants can be regenerated for analysis, antibody purification or breeding (*see* **Note 17**).
2. When internodes have formed on shoots, dissect the shoots from the callus and transfer them to fresh tMSHCK dishes. Cut off shoots right next to the callus, but without co-transferring callus tissue. Continue cultivation at 25°C as previously described. Transgenic shoots will survive, whereas wild-type shoots will bleach.
3. As soon as the shoots touch the lids of the dishes, transfer them to tMSCK culture pots (three plants per pot) for root induction. Cultivation conditions are unchanged. Roots should form within 2 wk, or accidentally co-transferred callus tissue may have to be removed.
4. In order to obtain seeds, cultivate the transgenic plants for another 3–4 wk before transferring into soil (*see* **Note 18**).

3.2. Transformation of Rice Plants

As discussed for tobacco, the manipulation of seeds, explants, and callus should be carried out using sterilized tools and equipment in a laminar flow hood to avoid microbial contamination. We recommend that all glassware should be cleaned with double-distilled water (no detergents), covered with aluminum foil (*see* **Note 14**) and sterilized at 180°C for 4 h.

3.2.1. Preparation of Rice Explants for Bombardment

1. Collect seeds from greenhouse-grown rice plants. Immature embryos are required for the following protocol (*see* **Note 19**), but an alternative procedure can be used for mature seed-derived callus (*see* **Note 20**).
2. Wash seeds in 500 mL of tap water plus 0.5 mL of Tween 20.
3. Transfer work to the laminar flow hood. Soak the seeds for 1 min in 70% (v/v) ethanol and rinse in distilled water.
4. Immerse seeds in 500 mL of 50% (w/v) sodium hypochlorite containing 0.5 mL of Tween 20 and agitate for 5 min.
5. Rinse seeds at least 3× in sterile distilled water.
6. Dehusk seeds under a stereomicroscope using two pairs of fine forceps. Dip each dehusked seed briefly in the 50% (w/v) sodium hypochlorite solution and immediately transfer to sterile distilled water.
7. Pour off the water and sterilize for 10 min in 50% (w/v) sodium hypochlorite (*see* **Note 21**).

8. Rinse the seeds 3× in sterile distilled water and excise the immature embryos using fine forceps. Place the embryos scutellum-uppermost on CCH medium.
9. Incubate the plates in darkness at 24°C for 2 d.
10. Four hours before bombardment, transfer embryos to CCHO medium (*see* **Note 22**).

3.2.2. Preparation of DNA-Coated Gold Particles for Bombardment

1. Prepare high-quality plasmid DNA at a concentration of 10 mg/mL (*see* **Note 8**).
2. Mix 2.5 mg of gold particles with 5 µg of the selectable marker plasmid and 10 µg of the plasmid carrying the antibody transgene. Add TE buffer to 100 µL. Vortex for 30 s.
3. Add 100 µL of 100 m*M* spermidine to protect the DNA during the precipitation process. Vortex for 30 s.
4. Add 100 µL of 30% PEG (MW 8,000) and vortex for 30 s.
5. Slowly add 100 µL of 2.5 *M* $CaCl_2$ with continuous vortexing. When all the $CaCl_2$ has been added, vortex for an additional 10 min.
6. Centrifuge at 12,000*g* for 30 s and discard supernatant.
7. Wash with 200 µL of 100% ethanol and centrifuge at 12,000*g* for 30 s.
8. Add 100 µL of 100% ethanol and sonicate briefly to break up clumps of particles.
9. Spot 5–10 µL of the suspension onto the center of a carrier disk and allow to air dry.

3.2.3. Bombardment Using Bio-Rad Helium Gun (PDS 1000/He)

1. Follow manufacturer's guidelines for loading the carrier disc into the barrel of the helium gun.
2. Place plant material in the targeting chamber.
3. Bombard plant material at 900–1,300 psi (*see* **Note 23**).
4. Bombard embryos twice with a 4-h interval between bombardments (maintain the embryos in darkness at 24°C on CCHO during the interval).

3.2.4. Selection and Regeneration of Transgenic Rice Plants

1. Transfer bombarded embryos to CCH medium. Incubate in darkness at 24°C for 2 d.
2. For the induction of transformed embryogenic callus, transfer embryos to CCHS medium and incubate in darkness at 24°C for 2 wk.
3. Subculture the callus recovered under selective conditions at 2-wk intervals on fresh CCHS medium in darkness at 24°C (*see* **Note 24**).
4. At the end of the third round of selection, transfer explants to CCR medium for regeneration. Incubate plates at 24°C under low light (100 µE) for 10 d and stronger light (130 µE) for the subsequent 10 d, with an 18-h photoperiod.
5. For rooting, transfer regenerating callus to rMSR medium under the stronger light conditions previously described.
6. When transgenic rice plantlets are approx 10 cm high, they can be transferred to soil (*see* **Note 25**).

7. Maintain transgenic rice plants in a growth chamber at 28°C day/24°C night temperature, 70% day/85% night humidity and a 14-h photoperiod for 3 mo. Then transfer to greenhouse conditions (24–28°C, 70% humidity, sunlight, and approx 14-h photoperiod).

3.3. Transformation of Wheat

3.3.1. Preparation of Wheat Embryos

1. Dehusk wheat seeds and transfer to 20% (w/v) sodium hypochlorite for 10 min under a laminar flow hood.
2. Rinse the seeds at least 3× with distilled water.
3. Excise embryos and place, scutellum-uppermost, on plates containing wMSH medium.
4. Leave plates in darkness at 24°C.
5. At least 4 h prior to bombardment, transfer embryos to wMSHO medium (*see* **Note 22**).

3.3.2. Particle Bombardment

1. Prepare DNA-coated gold particles as described in **Subheading 3.2.2.**
2. Spot 5–10 µL of the suspension onto the center of a carrier disk and allow to air dry.
3. Follow manufacturer's guidelines for loading carrier disc into the barrel of the helium gun.
4. Place wMSHO dish containing excised wheat embryos in the targeting chamber.
5. Bombard explants once at 900–1,300 psi (*see* **Note 23**).

3.3.3. Regeneration of Transgenic Wheat Plants

1. For callus induction, transfer bombarded wheat embryos to wMSH medium and incubate in darkness at 24°C for 2 wk.
2. Transfer callus to wMSS (selection medium) and maintain under stronger light conditions (130 µE, 18-h photoperiod) at 24°C.
3. Transfer transgenic callus onto half-strength MS medium (wMSR) for shoot regeneration. Maintain under strong light conditions (130 µE), 18-h photoperiod, 24°C. After two rounds of selection (2 wk each), transfer elongated shoots to tissue-culture tubes containing the same medium.
4. When transgenic wheat plantlets are approx 10 cm high, they can be transferred to soil (*see* **Note 26**).
5. Maintain transgenic wheat plants at 15/12°C day/night temperature with a 10-h photoperiod during the first 40 days, followed by maintenance at 21/18°C day/night temperature with a 16-h photoperiod thereafter.

3.4. Isolation of Full-Size Antibodies by Affinity Chromatography

The protocol described here has been optimized for tobacco cultivar Petit Havana SR1 (*see* **Note 13**). It uses cross-flow filtration for clarification and fast-flow affinity chromatography for rapid processing *(23)*. The combination of these techniques significantly speeds up processing and minimizes product loss during initial purification (*see* **Note 27**).

1. Homogenize 0.5 g of tobacco leaf tissue in 1 mL of protein extraction buffer in a precooled mortar.
2. Centrifuge the homogenate at 14,000*g* at 4°C for 10 min.
3. Transfer the clear supernatant to a new sample tube and store the plant extract at 4°C for further analysis.
4. Filter through a 0.2-µm membrane in cross-flow filtration system (*see* **Note 28**).
5. Meanwhile, equilibrate a prepacked Protein A Hyper D column (20-mL matrix) with five column volumes of PBS (*see* **Note 29**).
6. Pass filtrate through the column at ~300 cm/h using a low-pressure chromatography system.
7. Wash column with five column volumes of binding buffer.
8. Wash with elution buffer 1 until the baseline is reached at A_{280}.
9. Elute bound antibodies with three column volumes of elution buffer 2.
10. Pool antibody-containing fractions and immediately adjust to pH 7.2 with 1 *M* Tris-HCl, pH 8.0.
11. Immediately re-equilibrate the affinity matrix with five column volumes of PBS.
12. Pass antibody-containing fractions through Sephacryl S300 HR gel-filtration column (3.4 × 80 cm) for further purification, removal of aggregates (*see* **Note 30**) and transfer of the purified product into a suitable buffer for final processing and storage.
13. Pool antibody-containing fractions, filter-sterilize, and use for characterization—e.g., by Western blot, ELISA, surface plasmon resonance (SPR) (*see* **Chapter 23**)—or store aliquots at –20°C (*see* **Note 31**).

4. Notes

1. The ingredients for these basic media are listed in **Table 1,** although ready-mixed powders can be purchased from many suppliers (e.g., Gibco-BRL, Flow Labs, Sigma).
2. CC medium is required only for the transformation of immature *indica* rice embryos.
3. Some chemicals (e.g., kanamycin sulfate) may require warming to 37°C and/or prolonged stirring before they dissolve. Before use, briefly vortex all stock solutions.
4. The choice of selective agent depends on the selectable marker used for transformation. Kanamycin is used for tobacco transformation if the selectable marker is *npt*II or *aph*II, as these genes encode phosphotransferase enzymes that neutralize all aminoglycoside antibiotics. Hygromycin is preferred in rice because kanamycin is ineffective. Hygromycin B is used with the marker *hpt* (from *Klebsiella* spp., encoding hygromycin phosphotransferase). Although hygromycin selection can also be used with wheat, we have found that PPT selection kills non-transformed tissue more efficiently in the tissue culture protocol described here. PPT selection is used with the markers *bar* or *pat,* from *Streptomyces hygroscopicus,* both of which encode phosphinothricin acetyltransferase.

5. The use of (expensive) glucose as a carbon source, although recommended by some authors, showed no significant influence on plant growth or transformation efficiency in our hands.
6. CC medium is used for the culture of immature *indica* rice embryos ***(26)***. However, if mature seed-derived callus is used as the explant, MS medium is used instead ***(32,33)***. The basic culture medium, rMSH, is MS salts plus 30 g/L sucrose, 6 g/L type I agarose, and 2.5 mg/L 2,4-D. An osmoticum medium (rMSHO) is used prior to bombardment, and this is rMSH plus 0.4 *M* mannitol and 0.4 *M* sorbitol. A selection medium (rMSHS) is required to isolate transgenic callus and this is rMSH plus 30 mg/L hygromycin B. The regeneration medium (rMSHR) is basic MS salts and 6 g/L type I agarose plus 0.5 mg/L NAA, 2 mg/L 6-BAP, and 30 mg/L hygromycin B, with 3 g/L maltose used instead of sucrose. The rooting medium is similar to rMSR as defined in **Subheading 2.3.2.2.** (basic MS salts and 6 g/L type I agarose plus 1 g/L sucrose, no phytohormones), but contains only 30 mg/L hygromycin B. For the transformation protocol using mature seed-derived callus, *see* **Note 20.**
7. The binary vector (*Agrobacterium* transformation) or transforming plasmid (particle bombardment) contains the antibody transgene within a plant expression construct. The construct is designed for high-level expression and is optimized to facilitate protein accumulation. Several general principles of construct design have become clear during our work on antibody production, but it should be kept in mind that these principles have been developed for tobacco, rice, and wheat, so modification may be necessary for successful antibody production in alternative hosts. A strong and constitutive promoter is useful for antibody expression. The CaMV 35S promoter is favored for dicots, and the maize ubiquitin-1 (*ubi*-1) promoter is more active in monocots. A polyadenylation site is essential, and is included in all plant expression vectors. Protein synthesis can be maximized by using a construct in which the Kozak sequence has been optimized for plants and native 5′- and 3′-untranslated regions are replaced with plant translational enhancers, such as the tobacco mosaic virus omega sequence. In some cases it may also be necessary to modify the coding region of the transgene to match the codon preference of the host plant. The most important factor is protein targeting. Full-length antibodies need to be targeted to the secretory pathway in order to fold and assemble properly and to undergo correct glycosylation. Indeed, antibodies of all classes generally accumulate to higher levels when targeted to the secretory pathway, although there are several examples of high-level accumulation of scFvs in the cytosol. This is probably because of the behavior of different antibody sequences, which may affect stability. Further increases in yield can be obtained by retaining the antibody in the secretory pathway. The chemical environment of the ER lumen is well-suited to antibody accumulation. The highest yields are generally achieved by including in the expression construct a C-terminal KDEL sequence that causes proteins to be retrieved from the Golgi apparatus and returned to the endoplasmic reticulum (ER) in the manner of a resident ER protein. The factors affecting antibody expression in plants have been reviewed in detail elsewhere ***(14,21,34,35)***.

8. Only high-quality supercoiled plasmid DNA should be used. We find that QIAGEN Maxiprep plasmid kits are satisfactory for this purpose. The eluted plasmid DNA should be ethanol-precipitated and redissolved in TE buffer to a concentration of 10 mg/mL. Check the concentration and purity using a UV spectrophotometer.
9. As an alternative to plasmid DNA, the expression construct can be isolated from the plasmid by restriction digestion and used for transformation. The same total amount of DNA is used. We find that this modification can reduce the transgene copy number and simplify the structure of the transgenic locus in regenerated plants, reducing the likelihood of transgene silencing *(36)*.
10. The selectable marker gene may be included on the same plasmid as the antibody transgene, but this is unnecessary. Co-transformation with two plasmids results in co-integration at the same locus in most plants *(37)*. Co-integration also occurs when linear minimal DNA fragments are used—e.g., the transgene and marker gene as separate fragments each bracketed by a promoter and polyadenylation site (**ref.** ***38***; *see* **Note 9**).
11. We use a Bio-Rad pneumatic helium gun, model PDS 1000/He. When considering investing in or hiring a particle gun, the manufacturers will advise on setting up and operating procedures. Bottled helium gas, carrier sheets, rupture disks, and stopping sheets are available from the same supplier.
12. Tungsten particles are less expensive than gold ones, but we find gold particles more efficient for transformation, possibly because tungsten particles are less uniform in size and are more reactive than gold, resulting in more damage to the target tissue (reviewed in **ref.** ***38***). The size of the particles also influences transformation efficiency. Larger particles are generally faster and penetrate further, but cause more damage. For rice and wheat transformation, we recommend particles 0.7–0.9 µm in diameter.
13. The extraction buffer has been optimized for *Nicotiana tabacum* cv. Petit Havana SR1. It prevents oxidation and reduces degradation and inactivation of recombinant antibodies in crude plant-cell extracts. Other tobacco varieties or plant species may require different buffer compositions.
14. The use of expensive covers for the culture vessels is not necessary. Aluminum foil is easily co-sterilized, and provides a satisfactory safety level against contamination.
15. Axenic shoot cultures and callus grow slightly better without Parafilm because there is better oxygen transfer, but the risk of contamination is increased.
16. In some cases, shoot induction is better if NAA concentration is reduced or the hormone is completely omitted. This should be tested in parallel.
17. The propagation of transformed callus results in non-homogeneous cell populations with variable levels of antibody production because of position effects and varying transgene copy numbers *(39)*. Regenerated plants are regarded as offspring derived from a single cell. Transgenic plants can be self-fertilized to produce homozygous offspring in order to increase production levels. Furthermore, crossing independent high producers results in double transgenic plants with further improved yields of recombinant protein unless the transgene is affected by epigenetic silencing. Seeds of these plants can be stored indefinitely.

18. Seeds of *Nicotiana tabacum* cv. Petit Havana SR1 can be obtained after a relatively short regeneration time (4 mo). Other varieties may need more than 7 mo.
19. The optimal age of immature rice embryos for transformation is 12–15 d. The maturity of the embryos can be estimated by pressing the seeds between the fingers. Immature seeds leave a milky liquid residue on the fingers, whereas mature seeds do not.
20. Callus derived from mature seeds can also be used as an explant for transformation ***(32,33)***. In this case, the seeds are tougher and must be dehusked with sandpaper. Sterilize as described for immature seeds, but incubate dehusked seeds in 50% (w/v) sodium hypochlorite for 15 min with agitation before rinsing and drying. Excise the mature embryos with forceps and place, scutellum-uppermost, onto MS medium supplemented with 2.5 mg/L 2,4-D for 5 d. Transfer to rMSHO osmoticum medium (*see* **Note 6**) and bombard twice, as discussed for immature embryos. After bombardment, transfer to rMSH medium (*see* **Note 6**) plus 30 mg/L hygromycin B. Incubate in darkness for 3 wk at 24°C. Take embryogenic callus produced within the scutellar region of the mature embryos and subculture on fresh selection medium in darkness for 3 wk at 24°C. Repeat subculture at 3-wk intervals, discarding dead callus at each subculture. After three rounds of selection, transfer surviving embryos to rMSHR medium (*see* **Note 6**) and maintain as for immature embryos. Regenerated plantlets should be transferred to modified rMSR rooting medium (MS medium supplemented with 6 g/L type I agarose, 1 g/L sucrose, and 30 mg/L hygromycin B; *see* **Note 6**) and maintained under stronger light (130 μE, 18-h photoperiod) at 24°C. When transgenic rice plantlets are approx 10 cm high, they can be transferred to soil in the controlled environment room and then to the greenhouse, as discussed in **Subheading 2.2.**
21. Handle the dehusked seeds very carefully from this stage onward and use very gentle agitation.
22. Osmoticum treatment has been shown to improve the efficiency of particle bombardment. The exact basis is unclear. It may remove water droplets from the seed surface that deflect the metal particles, or it may change the physiology of the cell, making it more amenable to transformation ***(40)***.
23. The optimal pressure and distance between the gun and the target must be determined empirically for each type of explant. For optimization, a transient expression assay is useful. Bombard the explant with a DNA construct comprising the visible marker *gusA* driven by a strong and constitutive promoter such as maize *ubi-1* (for cereals) or CaMV 35S (for dicots). Optimal transformation occurs when many small spots of GUS activity are detected 1–2 d after bombardment. If there are large sectors of GUS-positive tissue, this usually indicates that the particle velocity is too high and the tissue is dead. Reduce the pressure or increase the distance to the target and try again.
24. Dead callus turns black. Discard all the dead callus at every subculture because this prevents the proliferation of escapes (non-transformed cells that survive under selection). Escapes may inhibit the growth of transformed callus.

25. We use a 4:2:1 mix of topsoil, sand, and pouzzelane (EuroPouzzelane, Lattes, France).
26. We use John Innes No. 2 compost.
27. The only major "contamination product" found in this protocol is a fragment of the immunoglobulin heavy chain. The addition of protease inhibitors (other than EDTA) and antioxidants such as polyvinylpolypyrrolidone (PVPP) during the first downstream processing step is unnecessary because it does not prevent this cleavage of the heavy chain. Furthermore, we do not recommend the use of PVPP because it clogs the filter membranes during clarification.
28. The use of cross-flow filtration is highly recommended for clarification because conventional membrane filters (even with large cross-sectional areas) are quickly clogged by plant-cell debris even if the crude extract is centrifuged at high speed.
29. This step is optimized for murine IgG2a/b. The purification of IgG1 may require different binding buffers. Please refer to the manufacturer's instructions.
30. The "cleaved" heavy-chain product (*see* **Note 27**) usually associates with the affinity-purified full-size antibodies and cannot be removed by gel-filtration chromatography. If necessary, it can be removed by size fractionation procedures such as preparative gel electrophoresis.
31. Store aliquots of purified recombinant antibodies at 4°C for subsequent characterization. Avoid repeated freeze/thaw cycles, as these will cause protein degradation and loss of activity.

References

1. Chadd, H. E. and Chamow, S. M. (2001) Therapeutic antibody expression technology. *Curr. Opin. Biotechnol.* **12,** 188–194.
2. Schillberg, S., Zimmermann, S., Zhang, M.-Y., and Fischer, R. (2001) Antibody-based resistance to plant pathogens. *Transgenic Res.* **10,** 1–12.
3. Gavilondo, J. V. and Larrick, J. W. (2000) Antibody production technology in the millennium. *Biotechniques* **29,** 128–145.
4. Kĥler, G. and Milstein, C. (1975) Continuous cultures of fused cells secreting antibody of predefined specificity. *Nature* **256,** 495–497.
5. Kipriyanov, S. M. and Little, M. (1999) Generation of recombinant antibodies. *Mol. Biotechnol.* **12,** 173–201.
6. Green, L. (1999) Antibody engineering via genetic engineering of the mouse: xenomouse strains are a vehicle for the facile generation of therapeutic human monoclonal antibodies. *J. Immunol. Methods* **231,** 11–23.
7. Griffiths, A. and Duncan, A. (1998) Strategies for selection of antibodies by phage display. *Curr. Opin. Biotechnol.* **9,** 102–108.
8. Sidhu, S. S. (2000) Phage display in pharmaceutical biotechnology. *Curr. Opin. Biotechnol.* **11,** 610–616.
9. Chu, L. and Robinson, D. K. (2001) Industrial choices for protein production by large-scale cell culture. *Curr. Opin. Biotechnol.* **12,** 180–187.
10. Raju, T. S., Briggs, J., Borge, S. M., and Jones, A. J. S. (2000) Species-specific variation in glycosylation of IgG: evidence for the species-specific sialylation and

branch-specific galactosylation and importance for engineering recombinant glycoprotein therapeutics. *Glycobiolgy* **10,** 477–486.
11. Houdebaine, L. M. (2000) Transgenic animal bioreactors. *Transgenic Res.* **9,** 305–320.
12. Sáchez, L., Ayala, M., Freyre, F., Pedroso, I., Bell, H., Falcó, V., et al. (1999) High cytoplasmic expression in *E. coli,* purification and *in vitro* refolding of a single chain Fv antibody fragment against the hepatitis B surface antigen. *J. Biotechnol.* **72,** 13–20.
13. Giddings, G. (2001) Transgenic plants as protein factories. *Curr. Opin. Biotechnol.* **12,** 450–454.
14. Stoger, E., Sack, M., Fischer, R., and Christou, P. (2002) Plantibodies: applications, advantages and bottlenecks. *Curr. Opin. Biotechnol.* **13,** 161–166.
15. Daniell, H., Streatfield, S. J., and Wycoff, K. (2001) Medical molecular farming: production of antibodies, biopharmaceutical and edible vaccines in plants. *Trends Plant Sci.* **6,** 219–226.
16. Cabanes-Macheteau, M., Fitchette-Laine, A. C., Loutelier-Bourhis, C., Lange, C., Vine, N., Ma, J., et al. (1999) N-Glycosylation of a mouse IgG expressed in transgenic tobacco plants. *Glycobiology* **9,** 365–372.
17. Chargelegue, D., Vine, N., van Dolleweerd, C., Drake, P. M., and Ma, J. (2000) A murine monoclonal antibody produced in transgenic plants with plant-specific glycans is not immunogenic in mice. *Transgenic Res.* **9,** 187–194.
18. Bakker, H., Bardor, M., Molthoff, J. W., Gomord, V., Elbers, I., Stevens, L. H., et al. (2001) Galactose-extended glycans of antibodies produced by transgenic plants. *Proc. Natl. Acad. Sci. USA* **98,** 2899–2904.
19. Kusnadi, A. R., Nikolov, Z. L., and Howard, J. A. (1997) Production of recombinant proteins in transgenic plants: practical considerations. *Biotechnol. Bioeng.* **56,** 473–484.
20. Stoger, E., Sack, M., Perrin, Y., Vaquero, C., Torres, E., Twyman, R. M., et al. (2002) Practical considerations for pharmaceutical antibody production in different crop systems. *Mol. Breeding* **9,** 149–158.
21. Schillberg, S., Fischer, R., and Emans, N. (2003) Molecular farming of recombinant antibodies in plants. *Cell. Mol. Life Sci.* **60,** 433–445.
22. Evangelista, R. L., Kusnadi, A. R., Howard, J. A., and Nikolov, Z. L. (1998) Process and economic evaluation of the extraction and purification of recombinant β-glucuronidase from transgenic corn. *Biotechnol. Prog.* **14,** 607–614.
23. Fulton, S. P. (1994) Large-scale processing of macromolecules. *Curr. Opin. Biotechnol.* **5,** 201–205.
24. Ford, C. F., Suominen, I., and Glatz, C. E. (1991) Fusion tails for the recovery and purification of recombinant proteins. *Protein Expr. Purif.* **2,** 95–107.
25. Nygren, P. A., Stahl, S., and Uhlen, M. (1994) Engineering proteins to facilitate bioprocessing. *Trends Biotechnol.* **12,** 184–188.
26. Christou, P., Ford, T., and Kofron, M. (1991) Production of transgenic rice (*Oryza sativa* L.) plants from agronomically-important *indica* and *japonica* varieties via electric discharge particle acceleration of exogenous DNA into immature zygotic embryos. *Bio/Technol.* **9,**957–962.

27. Altpeter, F., Vasil, V., Srivastava, V., Stoger, E., and Vasil, I. K. (1996) Accelerated production of transgenic wheat (*Triticum aestivum* L.) plants. *Plant Cell Rep.* **16,** 12–17.
28. Murashige, T. and Skoog, F. (1962) A revised medium for rapid growth and bioassays with tobacco tissue cultures. *Physiol. Plant.* **15,** 473–497.
29. Potrykus, I., Harms, C. T., and Lorz, H. (1979) Callus formation from cell culture protoplasts of corn (*Zea mays* L.). *Theor. Appl. Genet.* **54,** 209–214.
30. Horsch, R. B., Fry, J. E., Hoffman, N. L., Eicholtz, D., Rogers, S. G., and Fraley, R. T. (1985) A simple and general method for transferring genes into plants. *Science* **227,** 1229–1231.
31. Voss, A., Niersbach, M., Hain, R., Hirsch, H. J., Liao, Y. C., Kreuzaler, F., et al. (1995) Reduced virus infectivity in N. tabacum secreting a TMV-specific full-size antibody. *Mol. Breed.* **1,** 39–50.
32. Sudhakar, D., Duc, L. T., Bong, B. B., Tinjuangjun, P., Maqbool, S. B., Valdez, M., et al. (1998) An efficient rice transformation system utilizing mature seed- derived explants and a portable, inexpensive particle bombardment device *Transgenic Res.* **7,** 289–294.
33. Valdez, M., Cabera-Ponce, J. L., Sudhakar, D., Herrera-Estrella, L., and Christou, P. (1998) Transgenic Central American, West African and Asian elite rice varieties resulting from particle bombardment of foreign DNA into mature seed-derived explants utilizing three different bombardment devices. *Ann. Bot.* **82,** 795–801.
34. Fischer, R. and Emans, N. (2000) Molecular farming of pharmaceutical proteins. *Transgenic Res.* **9,** 279–299.
35. Schillberg, S., Emans, N., and Fischer, R. (2002) Antibody molecular farming in plants and plant cells. *Phytochem. Rev.* **1,** 45–54.
36. Fu, X., Duc, L. T., Fontana, S., Bong, B. B., Tinjuangjun, P., Sudhakar, D., et al. (2000) Linear transgene constructs lacking vector backbone sequences generate low-copy-number transgenic plants with simple integration patterns. *Transgenic Res.* **9,** 11–19.
37. Christou, P. and Swain, W. F. (1990) Cotransformation frequencies of foreign genes in soybean cell cultures. *Theor. Appl. Genet.* **90,** 97–104.
38. Twyman, R. M. and Christou, P. Plant transformation technology—particle bombardment, in *Handbook of Plant Biotechnology* (Christou, P., ed.), John Wiley & Sons, NY (in press).
39. Twyman, R. M., Stoger, E., Kohli, A., and Christou, P. (2002) Foreign DNA: integration and expression in transgenic plants, in *Genetic Engineering: Principles and Practice, Volume 24* (Setlow, J. K., ed.), Plenum Press, NY.
40. Vain, P., McMullen, M. D., and Finer, J. J. (1993) Osmotic treatment enhances particle bombardment-mediated transient and stable transformation of maize. *Plant Cell Rep.* **12,** 84–88.

18

Directed Mutagenesis of Antibody Variable Domains

Kevin Brady and Benny K. C. Lo

1. Introduction

The ability to mutate antibodies at the genetic level is a powerful tool in the antibody engineer's toolbox. Specific properties of these molecules, such as stability, specificity, and in some cases catalysis, can be thoroughly studied using mutagenesis techniques. Harnessed to other techniques such as affinity measurements (*see* **Chapters 23–25**), catalytic assays or biophysical characterization, mutagenesis can reveal the exact atomic mechanisms involved. Mutagenesis also provides the engineer with the ability to incorporate improved functionality into the antigen-binding site.

Two methods are described here, which should allow the reader to perform site-directed mutagenesis of antibody variable-domain genes (V_L, V_H, and single-chain Fv [scFv]). The choice of methods is left to the reader, as each have their own advantages and disadvantages. This decision is often dictated by the sequence itself. Although the first method requires the use of a commercial kit, a number of non-proprietary methods are available *(1)*. Both techniques allow the introduction of one or more mutations in a single experiment. Insertions and deletions are also possible, thus enabling the user to engineer tags and leader sequences in or out of the DNA.

2. Materials

2.1. Oligonucleotide-Directed Mutagenesis

1. Plasmid vector harboring the antibody gene to be mutated.
2. Transformer™ site-directed mutagenesis kit (Clontech), encompassing 10X annealing buffer, 10X synthesis buffer, T4 DNA polymerase, T4 DNA ligase, and *E. coli* BMH71-18 *mutS* strain.
3. Restriction enzymes and reaction buffers (New England Biolabs).
4. Plasmid preparation kit—e.g., Qiaprep spin miniprep kit (Qiagen).

From: *Methods in Molecular Biology, Vol. 248: Antibody Engineering: Methods and Protocols*
Edited by: B. K. C. Lo © Humana Press Inc., Totowa, NJ

5. Luria Broth (LB): 10 g of tryptone, 5 g of yeast extract, and 10 g of NaCl made up to 1 L in deionized water; autoclave at 121°C for 15 min.
6. Antibiotic for resistance provided by vector.
7. LB/antibiotic plates: 10 g of tryptone, 5 g of yeast extract, 10 g of NaCl, and 15 g of agar made up to 1 L in deionized water; autoclave at 121°C for 15 min, let cool to 50°C, and add appropriate antibiotic. Pour 20 mL into a 10-cm Petri dish; once cool, store at 4°C for up to 4 wk.
8. Oligonucleotide primers for mutagenesis (40 pmol/µL), high-performance liquid chromatography (HPLC)-purified (*see* **Note 1**).
9. TSS solution: 8.5 mL of LB, 0.5 mL of dimethyl sulfoxide (DMSO), 1 g of polyethylene glycol 8,000, 1 mL of 0.5 *M* $MgCl_2$. Sterile-filter through a 0.22-µm filter, and store at 4°C for up to 2 wk.
10. Chemically competent *E. coli* strains TOP10F′ (Invitrogen) or Novablue (Novagen).
11. Double-distilled water, sterile-filtered through a 0.22-µm filter.
12. NuSieve 3:1 agarose (FMC Bioproducts) and appropriate gel electrophoresis equipment.
13. 1X TBE buffer: 0.09 *M* Tris-borate, 2 m*M* ethylenediaminetetraacetic acid (EDTA) in deionized water.
14. Ethidium bromide solution, 10 mg/mL (Sigma).

2.2. Overlap Extension Polymerase Chain Reaction (PCR) Mutagenesis

1. Amplification primers A and D, and mutagenic primers B and C (all at 40 pmol/µL; *see* **Fig. 2**), (HPLC)-purified.
2. Thermal cycling PCR block.
3. Taq DNA polymerase [e.g., AmpliTaq DNA polymerase (Applied Biosystems)] or proof-reading DNA polymerase [e.g., Platinum *Pfx* DNA polymerase (Invitrogen)], with supplied PCR buffers.
4. 10 m*M* deoxyadenosine 5′ triphosphate (dATP), deoxythymidine 5′ triphosphate (dTTP), deoxycytidine 5′ triphosphate (dCTP) and deoxyguanosine 5′ triphosphate (dGTP) (Sigma). Mix 62.5 µL of each dNTP, and add 750 µL of water. This yields 2.5 m*M* dNTPs.
5. Mineral oil (Sigma).
6. PCR purification kit [e.g., Wizard PCR preps purification system (Promega)].
7. NuSieve 3:1 agarose and appropriate gel electrophoresis equipment.
8. 1X TBE buffer (*see* **Subheading 2.1.**).
9. Ethidium bromide solution, 10 mg/mL.
10. 0.5-mL Eppendorf tubes.
11. Double-distilled water, sterile-filtered through a 0.22-µm filter.

3. Methods

3.1. Oligonucleotide-Directed Mutagenesis

This technique requires the use of a mutagenic primer that anneals to the wild-type DNA, along with a specific selection strategy for the mutant (*see* **Note 1**). No subcloning is required, as circular (and hopefully mutated) DNA is

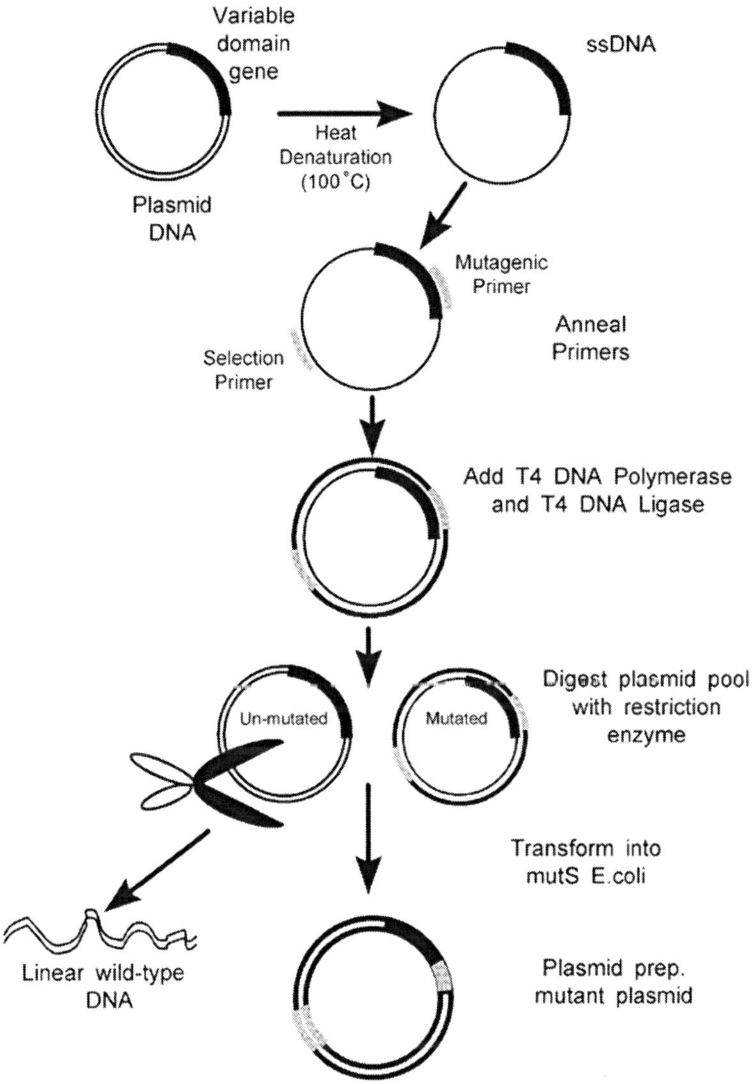

Fig. 1. Oligonucleotide directed mutagenesis protocol summary.

retained throughout. There are a number of kits available on the market, and all of these have the same theme. The Transformer™ site-directed mutagenesis kit, which is discussed here, is based on the methodology of Deng and Nickoloff *(2)*. The protocol is summarized in **Fig. 1.**

1. In a 0.5-mL Eppendorf tube, place 2 µL of 10X annealing buffer, 1 µL of plasmid DNA (100 ng/µL; *see* **Note 2**), 2 µL of selection primer, and 2 µL of mutagenic primer. Make up to 20 µL with water. Vortex, then centrifuge briefly.

2. Incubate at 100°C on a thermal cycler block for 3 min. Chill immediately on ice for 5 min.
3. Briefly centrifuge mixture and add 3 µL of 10X synthesis buffer, 4 U (1 µL) of T4 DNA polymerase, 5 U (1 µL) of T4 DNA ligase, and 5 µL of water. Mix by gentle pipetting a few times, centrifuge briefly, and incubate at 37°C for 2 h.
4. Stop reaction by heating to 70°C for 5 min. Leave tube to cool to room temperature.
5. Add 1 µL of the chosen restriction enzyme (*see* **Notes 3** and **4**). Incubate at the required temperature for 2 h.
6. Mix 100 µL of ice-cold chemically competent BMH71-18 *mutS E. coli* (*see* **Notes 5** and **6**) with 5 µL of the reaction mixture in a 20-mL Universal tube. Incubate on ice for 20 min, gently swirling the mixture every 5 min.
7. Transfer to a 42°C water bath for 1 min. Immediately add 1 mL of prewarmed LB media. Incubate at 37°C for 1 h with shaking at 220 rpm.
8. After 1 h, add 4 mL of LB containing the required antibiotic (**Note:** in addition to the antibiotic for the maintenance of the antibody plasmid tetracycline is also added to a final concentration of 50 µg/mL for the maintenance of BMH71-18 *mutsS E. coli*). Incubate overnight at 37°C with shaking at 220 rpm.
9. Perform plasmid isolation using a plasmid preparation kit according to the manufacturer's instructions.
10. Set up a second restriction digest using 2.5 µL of plasmid DNA from **step 9,** 2 µL of 10X restriction digest buffer, and 20 U (1 µL) of the required restriction enzyme. Make up to 20 µL with water. Incubate at 37°C for 2 h, then add an additional 1 µL of enzyme to the mixture. Incubate for an additional hour.
11. Transform chemically competent *E. coli* (TOP10F′ or Novablue) with 5 µL of the restriction digest mixture according to the manufacturer's instructions.
12. Spread 50 µL of the transformation suspension onto LB/antibiotic plates, and incubate at 37°C overnight. Inoculate several colonies separately into 5 mL of LB/antibiotic and grow overnight at 37°C with shaking at 220 rpm. Perform plasmid isolation from these individual cultures and confirm mutagenesis by DNA sequencing and restriction digest (*see* **Note 7**).

3.2. Overlap Extension PCR Mutagenesis

This is a more efficient methodology *(3)* for generating mutants than the previous method, because only the mutated DNA is handled. However, subcloning into the target plasmid vector is required. **Fig. 2** summarizes the strategy involved (*see* **Notes 8** and **9**).

1. Use standard PCR conditions for each reaction as defined here. In a PCR tube add:
   1 µL of template DNA(100 ng/µL; *see* **Note 10**),
   1 µL of each primer (40 pmol/µL; *see* **Note 11**),
   4 µL of 2.5 m*M* dNTPs,
   10 µL of 10X PCR buffer with 15 m*M* $MgCl_2$, and H_2O to 99 µL.

   Overlay mixture with 40 µL of mineral oil (omit if using thermal cyclers with a heated lid).

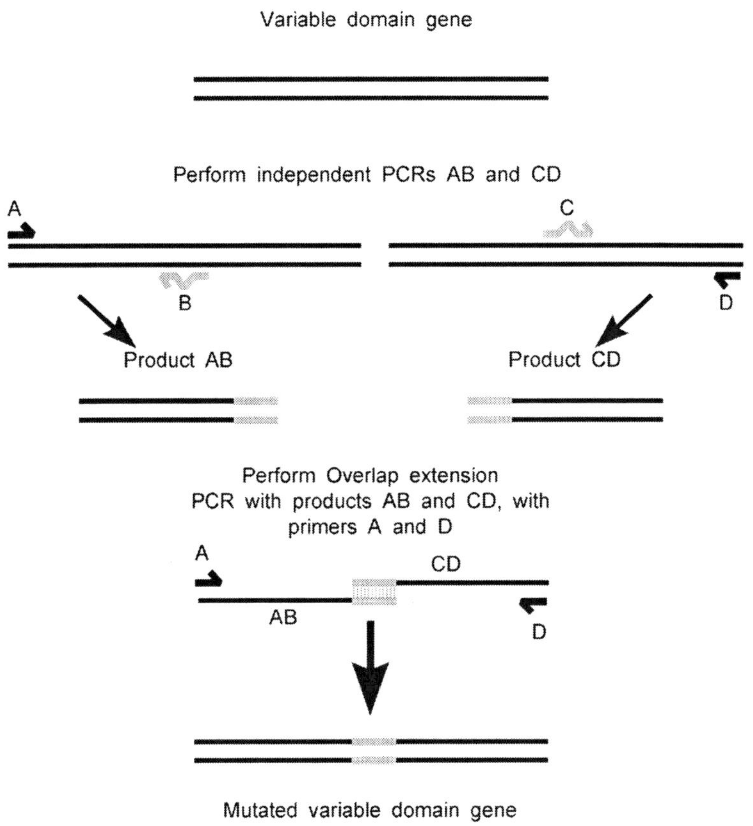

Fig. 2. Overlap extension PCR mutagenesis protocol summary.

2. Prepare reactions AB and CD according to **step 1.**
3. Heat the reactions to 94°C for 5 min (hot start) then add 1 U (1 μL) of Taq DNA polymerase. Proceed with 20 cycles of 94°C (1 min), 55°C (1 min), and 72°C (1 min) (*see* **Notes 12** and **13**). Incubate at 72°C for 10 min for final extension.
4. Visualize PCR products by running 5 μL of each reaction on a 1% agarose gel in 1X TBE buffer with ethidium bromide solution added to 0.2 μg/mL. Optimize PCR and repeat if necessary (*see* **Notes 10–12**).
5. Clean up PCR products with the Wizard PCR Preps DNA purification system according to the manufacturer"s instructions.
6. Run overlap extension PCR, by adding approximately equal amounts of products AB and CD (evaluate by eye from agarose gel, usually 1 μL of each) without primers A and D. Run for five cycles, then add primers A and D to complete PCR for 20 cycles.
7. Check gene reconstruction on a 1% agarose gel.
8. Digest with appropriate restriction enzymes and ligate into desired pre-digested plasmid.
9. Verify mutation by DNA sequencing.

4. Notes

1. The technique requires the user to design two primers (*see* **Fig. 1**). The mutagenic primer will incorporate the desired mutation into the gene. The selection primer will mutate a pre-existing single restriction site to a restriction site of specificity not found in the wild-type plasmid, thus allowing the selection of the mutated DNA strand. The most critical part of the mutagenesis procedure is primer design. It is possible to incorporate any number of mismatches in the primer, although primer length must be increased with the number of mismatches. Stability of annealing should evaluated carefully, and is quantifiable using any one of the large number of primer design software packages available. A good rule of thumb is to use at least 10 bp either side of the mismatch, with further elongation required if the regions adjacent to the mutation site are AT-rich. It is possible to design primers that incorporate insertions and deletions into the target DNA. The same principles apply to the design of such primers, although it is imperative that no frame-shift occurs as a result of the mutation, if one wishes to continue studying an intact Fv. The authors also note that this type of mutation is more facile when using PCR mutagenesis (*see* **Subheading 3.2.**).
2. The Qiaprep spin miniprep kit regularly yields pET plasmid DNA (high copy number) at a concentration of 100–150 ng/μL from 1.5 mL of culture. This can be checked by examining the A_{260} of the DNA sample.
3. Use an enzyme that is known to work efficiently. The use of excess amounts of enzyme (compared to that detailed in the kit protocol) is necessary to increase mutant yield, although it is important to avoid star activity if the total concentration of glycerol is greater than 5–10% (v/v) in the mixture (New England Biolabs supply their enzymes in 50% glycerol). Usually, the addition of 20–40 U yields complete digestion on unmutated DNA.
4. The chosen restriction enzyme should be active in the mixture, which contains 37.5 m*M* NaCl. If your chosen enzyme requires less or no salt, it is advisable to ethanol-precipitate the mixture and set up a restriction digest in the buffer supplied with the enzyme. If the enzyme requires much higher levels of NaCl—e.g., BamHI, 10X annealing buffer (which contains 500 m*M* NaCl) can be added at this stage.
5. Electroporation should not be used in the post-restriction step, as this appears to increase background wild-type DNA containing colonies. The use of the lower-efficiency chemical competent cells will exert added selection pressure on the linear wild-type DNA.
6. To prepare chemically competent BMH71-18 mutS cells, use either your own standard method, or as follows. Streak the cell stock onto an LB/tetracycline (50 μg/mL) plate. Incubate overnight at 37°C. Inoculate a single colony into 5 mL of LB/tetracycline and incubate at 37°C overnight with shaking. Use 0.1 mL of this culture to inoculate 10 mL of LB in a sterile 50-mL centrifuge tube. Grow at 37°C with shaking until the OD_{600} reaches 0.5. Chill tube on ice for 20 min, and collect cells by centrifugation at 2000*g* for 5 min at 4°C. Resuspend the cell pellet in 1 mL of ice-cold TSS solution.

7. Perform test restrictions using the original restriction enzyme and the enzyme that cuts the newly introduced site. Run products on a 1% agarose gel in 1X TBE buffer to evaluate selection-site mutagenesis. If this still demonstrates the presence of wild-type DNA, it may be necessary to incorporate another round of restriction digest and transformation. The method should preserve the integrity of the remainder of the plasmid-gene construct—e.g., no extra mutations should occur during the synthesis step. The difficulty arises from the need to screen a number of colonies for the desired mutation. Careful primer design is paramount to ensure that the first step incorporates enough mutagenic DNA into the first heteroduplex. Failure to produce mutants often results from the inability to anneal the mutagenic primers to the template.
8. Primers A and D are standard primers used for amplifying the gene that can contain restriction sites for further subcloning. Primers B and C are complementary, and contain the mutation required. They only need to be designed according to the same protocol outlined in **Note 1** for mutagenic primers.
9. If the mutation to be introduced is located at the terminal regions of the antibody gene, only one set of PCR is necessary (small PCR products <100 bp may be difficult to purify). Design primers A and/or D so that they incorporate the desired mutation followed by 10–12 complementary bases at the 3′-end. Perform PCR with the template and primers A and D as described in **Subheading 3.2., steps 1** and **3.**
10. Concentration of template may be reduced to optimize PCR.
11. Concentration of primers may be reduced or increased to optimize PCR.
12. Adjust annealing temperature (55°C) if required to optimize PCR.
13. The error rate of Taq polymerase is normally low (8×10^{-6}; **ref.** ***4***). However, if unintentional mutations are incorporated during PCR into your particular antibody gene, they can be minimized by using a proofreading DNA polymerase instead—e.g., Platinum *Pfx* DNA polymerase, with the following protocol:

    1 µL of Template DNA(100 ng/µL; *see* **Note 11**)
    1 µL of each primer (40 pmol/µL; *see* **Note 12**)
    6 µL of 2.5 m*M* dNTPs
    10 µL of 10X *Pfx* PCR buffer
    2 µL of 50 m*M* $MgSO_4$
    2.5 U (1 µL) of Platinum *Pfx* DNA polymerase
    H_2O to 100 µL.

    Overlay mixture with 40 µL of mineral oil (omit if using thermal cyclers with heated lid). Heat the reactions to 94°C for 5 min. Proceed with 20 cycles of 94°C (1 min), 55°C (1 min; *see* **Note 12**) and 68°C (1 min). Incubate at 68°C for 10 min for final extension.

References

1. Carter, P. (1991) Mutagenesis facilitated by the removal or introduction of unique restriction sites, in *Directed Mutagenesis—A Practical Approach* (McPherson, M. J., ed.), IRL Press, Oxford, UK, pp. 1–25.
2. Deng, W. P. and Nickoloff, J. A. (1992) Site-directed mutagenesis of virtually any plasmid by eliminating a unique site. *Anal. Biochem.* **200,** 81–88.

3. Horton, R. M., Cai, Z. L., Ho, S. N., and Pease, L. R. (1990) Gene splicing by overlap extension: tailor-made genes using the polymerase chain reaction. *Biotechniques* **8(5)**, 528–535.
4. Cline, J., Braman, J. C., and Hogrefe, H. H. (1996) PCR fidelity of *Pfu* DNA polymerase and other thermostable DNA polymerases. *Nucleic Acids Res.* **24(18),** 3546–3551.

19

Antibody Affinity Maturation by Chain Shuffling

James D. Marks

1. Introduction

Phage display can be used to increase the affinity of antibodies more than 1,000-fold *(1,2)*. The starting point is typically a specific antibody isolated from a phage antibody library (*see* **Chapter 8**). To accomplish affinity maturation (increased affinity), the sequence of the antibody is diversified, the mutant gene repertoire is displayed on filamentous phage, and higher-affinity binders are selected on antigen. Although the process is straightforward, the investigator should be reasonably certain that increasing the affinity of their particular antibody will lead to the desired biologic effect prior to undertaking in vitro affinity maturation. Higher affinity is especially important when attempting to neutralize a circulating toxin or growth factor in solution. In this case, the antibody distributes in the same compartment as the antigen, allowing the antigen-antibody interaction to proceed to equilibrium. The higher the affinity, the greater the amount of antigen that is bound. In contrast, affinity is only one determining factor in the amount of antibody fragment that will accumulate in a tumor in vivo *(3)*. Factors such as antibody fragment size, pharmacokinetics, and valency may have a more important impact *(4)*.

Mutations can be introduced into the antibody gene either randomly or at specific sites. Random introduction of mutations is the simplest approach, and makes no *a priori* assumptions as to which sites are the best to mutate in order to increase affinity. Random mutagenesis more closely mimics the in vivo process of somatic hypermutation. One common method for introducing random mutations has been "chain shuffling" *(5)*, in which one of the two chains (V_H and V_L) is fixed and combined with a repertoire of partner chains to yield a secondary library that can be searched for superior pairings. This approach takes advantage of "random" mutations that have been introduced into V_H and V_L germline genes in vivo. In the first described example of chain shuffling, V_H

From: *Methods in Molecular Biology, Vol. 248: Antibody Engineering: Methods and Protocols*
Edited by: B. K. C. Lo © Humana Press Inc., Totowa, NJ

or V_L repertoires were combined with the wild-type complementary chain using splicing by overlap extension to splice either the wild-type V_H or V_L to a gene repertoire of the complementary chain *(5,6)*. However, we found that this can artifactually generate a shortened linker sequence (between V_H and V_L) leading to scFv dimers (diabodies) *(7)*, which may be preferentially selected on the basis of avidity *(8)*. This occurs because of the repetitive nature of the linker. Thus, we now prefer to clone the wild-type V_H or V_L gene into a phage display vector that already contains a repertoire of the complementary chain. Such repertoires have been described *(8)*, and may be obtained from the author under a materials transfer agreement, saving considerable time. Alternatively, as described in the protocols that follow, the V_H and V_L gene repertoires can be obtained by amplifying from a pre-existing naïve single-chain variable fragment (scFv) library, or from human peripheral-blood lymphocyte (PBL) RNA using the procedures described in Chapter 6, **Subheadings 3.1.–3.3.** We typically shuffle the V_L gene of a scFv first, because more of the binding energy is usually contained within the V_H. If the affinity of the affinity-matured light-chain shuffled scFv is inadequate, we proceed with heavy-chain shuffling.

The protocols outlined here are organized into those required for light-chain shuffling (**Subheadings 3.1.** and **3.2.**), those required for heavy-chain shuffling (**Subheadings 3.3.** and **3.4.**), and a protocol for selection, identification, and characterization of higher-affinity chain-shuffled scFv (**Subheading 3.5.**), which is applicable to both heavy- and light-chain shuffling. An overview of the approach is shown in **Figs. 1** and **2**. A detailed view of the light-chain shuffling strategy is shown in **Fig. 3,** and a detailed view of heavy-chain shuffling is shown in **Fig. 4**.

2. Materials

2.1. Construction of Human V_L Gene-Repertoire Libraries

1. Vent DNA polymerase and 10X buffer (New England Biolabs).
2. 20X dNTPs: 5 m*M* each of deoxyadenosine 5′ triphosphate (dATP), deoxyguanosine 5′ triphosphate (dGTP), deoxycytidine 5′ triphosphate (dCTP), and deoxythymidine 5′ triphosphate (dTTP).
3. PCR thermocycler with heated lid.
4. Agarose, DNA-grade and high melting (Fisher Biotech; Cat. #BP164-500).
5. 10X TBE buffer: dissolve 108 g of Tris base and 55 g of boric acid in 900 mL of water, add 40 mL of 0.5 *M* ethylenediaminetetraacetic acid (EDTA), pH 8.0, and bring volume to 1 L.
6. 100-bp ladder DNA mass markers (New England Biolabs; Cat. #N3231S).
7. Double-stranded DNA template prepared from a naïve scFv gene repertoire in pHEN1 (*see* Chapter 6 for preparation) (10 ng/µL).
8. Geneclean kit (Qbiogene).
9. Wizard PCR purification kit (Promega).

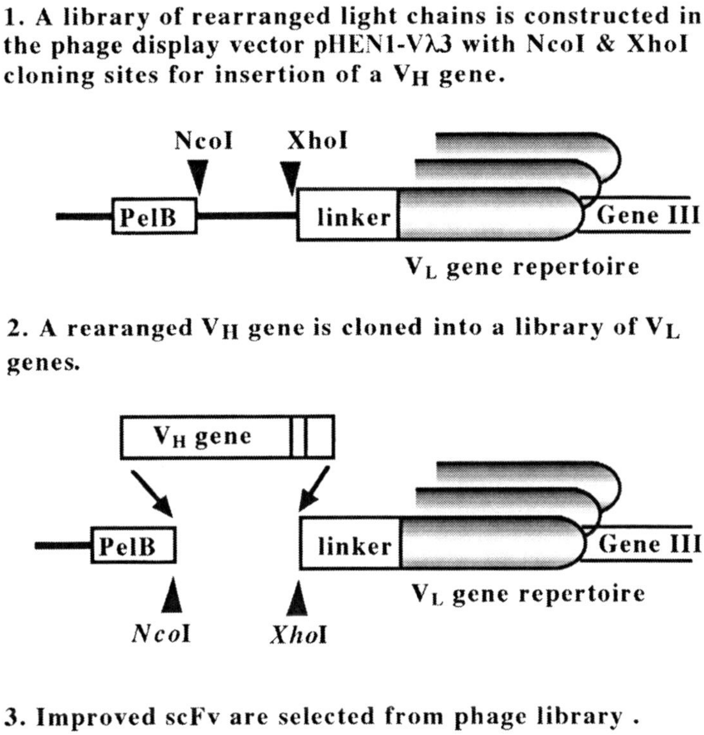
1. A library of rearranged light chains is constructed in the phage display vector pHEN1-Vλ3 with NcoI & XhoI cloning sites for insertion of a VH gene.
NcoI
XhoI
PelB
linker
Gene III
VL gene repertoire
2. A rearanged VH gene is cloned into a library of VL genes.
VH gene
PelB
linker
Gene III
VL gene repertoire
NcoI
XhoI
3. Improved scFv are selected from phage library .

Fig. 1. Overview of light-chain shuffling. Step 1: A library of rearranged light chains is created in the phage display vector pHEN1-Vλ3. Step 2: A wild-type V_H gene is cloned into the light-chain library to create a light chain-shuffled library. Higher-affinity scFv are isolated from the light chain-shuffled library by panning.

10. Back primers for amplification of linker and V_L repertoire (10 pmol/μL):
    a. RJH1/2Xho :5′-GGC ACC CTG GTC ACC GTC TCG AGT GGT GGA-3′.
    b. RJH3Xho :5′-GGG ACA ATG GTC ACC GTC TCG AGT GGT GGA-3′.
    c. RJH4/5Xho :5′-GGA ACC CTG GTC ACC GTC TCG AGT GGT GGA-3′.
    d. RJH6Xho :5′-GGG ACC ACG GTC ACC GTC TCG AGT GGT GGA-3′.
11. FdSEQ primer (10 pmol/μL): 5′- GAA TTT TCT GTA TGA GG-3′.
12. pHEN1-V_{λ}3 vector DNA (obtainable from the author under a materials transfer agreement).
13. *Xho*I and *Not*I restriction enzymes, NEB2 reaction buffer, and 100X BSA solution (New England Biolabs).
14. T4 DNA ligase and ligation buffer (New England Biolabs).
15. Electrocompetent *E. coli* TG1 cells (*see* Chapter 6, **Subheading 3.10.** for preparation).
16. 2X TY media: dissolve 16 g of bacto-tryptone, 10 g of yeast extract, and 5 g of NaCl in 1 L of distilled water; autoclave the solution.

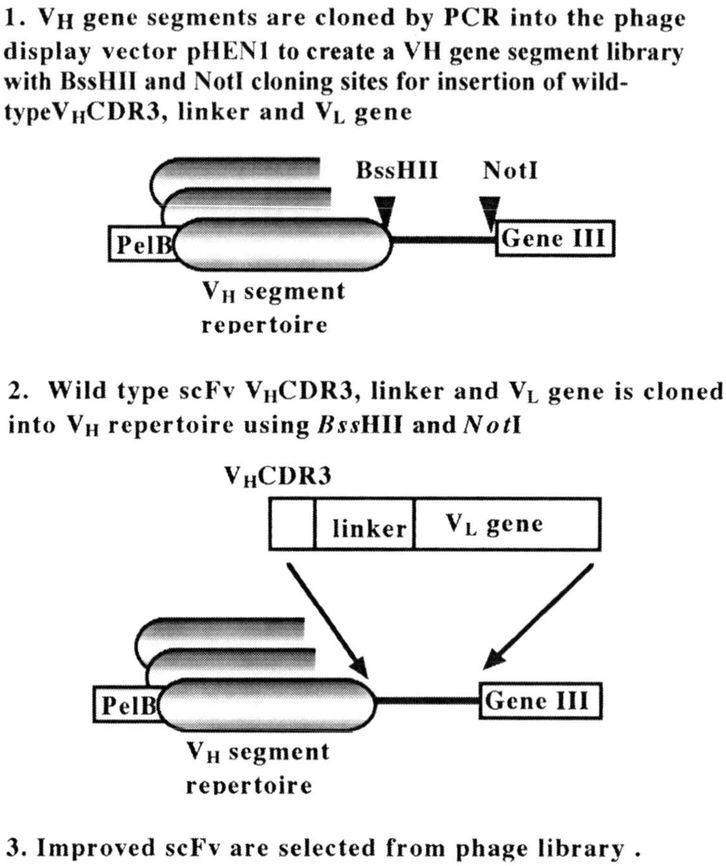

1. VH gene segments are cloned by PCR into the phage display vector pHEN1 to create a VH gene segment library with BssHII and NotI cloning sites for insertion of wild-typeVHCDR3, linker and VL gene
BssHII
NotI
PelB
Gene III
VH segment repertoire
2. Wild type scFv VHCDR3, linker and VL gene is cloned into VH repertoire using BssHII and NotI
VHCDR3
linker
VL gene
PelB
Gene III
VH segment repertoire
3. Improved scFv are selected from phage library .

Fig. 2. Overview of heavy-chain shuffling. Step 1: A library of rearranged V_H gene segments is created in the phage display vector pHEN1. Step 2: A wild-type VH CDR3, framework 4, scFv linker, and V_L are cloned into the V_H gene-segment library to create a heavy chain-shuffled library. Higher-affinity scFv are isolated from the heavy chain-shuffled library by panning.

17. 2X TY/amp/2%glu media: 2X TY media containing 100 μg/mL ampicillin and 2% glucose.
18. Plasmid midi-prep kit (Qiagen).

2.2. Construction of a Light Chain-Shuffled Phage Library

1. PCR reagents and equipment (*see* **Subheading 2.1., items 1–6**).
2. Double-stranded template DNA of vector pHEN1 containing your starting scFv gene (50 ng/μL).
3. scFvJH1-2XhoFOR primer (10 pmol/μL): 5′-GAG TCA TTC TCG TCT CGA GAC GGT GAC CAG GGT GCC-3′.

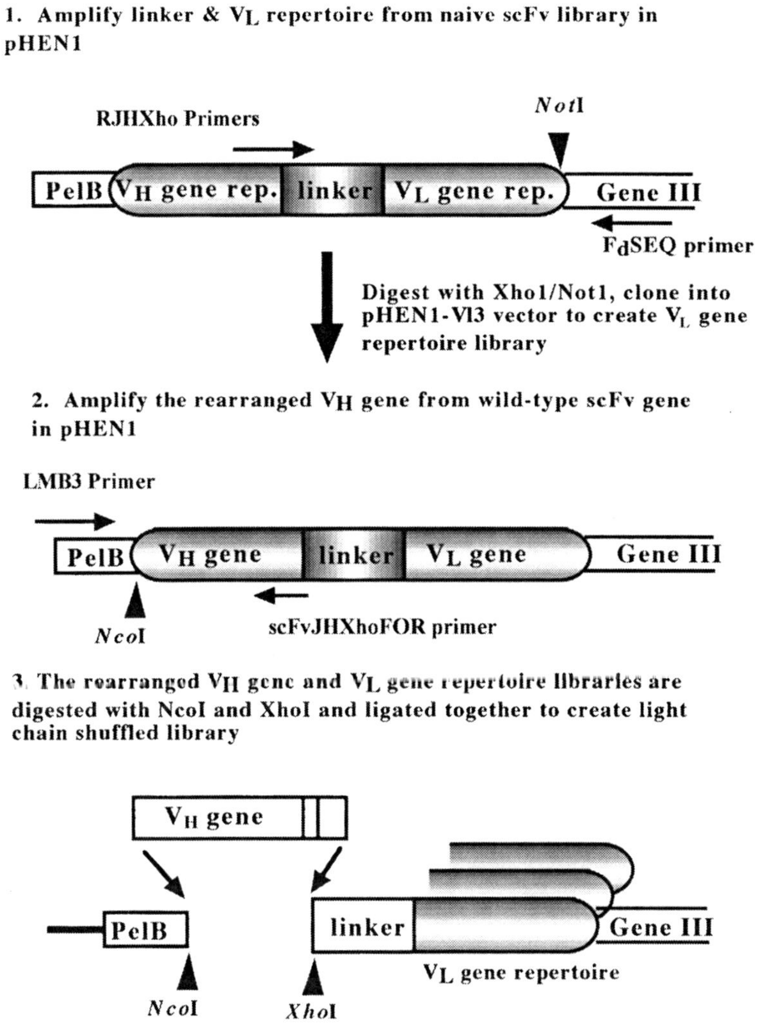
1. Amplify linker & VL repertoire from naive scFv library in pHEN1
RJHXho Primers
NotI
PelB
VH gene rep.
linker
VL gene rep.
Gene III
FdSEQ primer
Digest with XhoI/NotI, clone into pHEN1-VI3 vector to create VL gene repertoire library
2. Amplify the rearranged VH gene from wild-type scFv gene in pHEN1
LMB3 Primer
PelB
VH gene
linker
VL gene
Gene III
NcoI
scFvJHXhoFOR primer
3. The rearranged VH gene and VL gene repertoire libraries are digested with NcoI and XhoI and ligated together to create light chain shuffled library
VH gene
PelB
linker
Gene III
VL gene repertoire
NcoI
XhoI

Fig. 3. Overview of light-chain shuffling strategy. Step 1: the scFv linker and V_L repertoire are PCR-amplified from an existing naïve scFv library in pHEN1 using the primers RJHXHO and FdSEQ and cloned into pHEN1-V13 as *Xho*I-*Not*I fragments to create a library of rearranged light chains. Step 2: The wild-type rearranged V_H gene is PCR-amplified using the primers LMB3 and scFvJHXhoFor. Step 3: The PCR fragment is digested with *Nco*I and *Xho*I and ligated into the library of rearranged light chains (Step 1) to create a light chain-shuffled library.

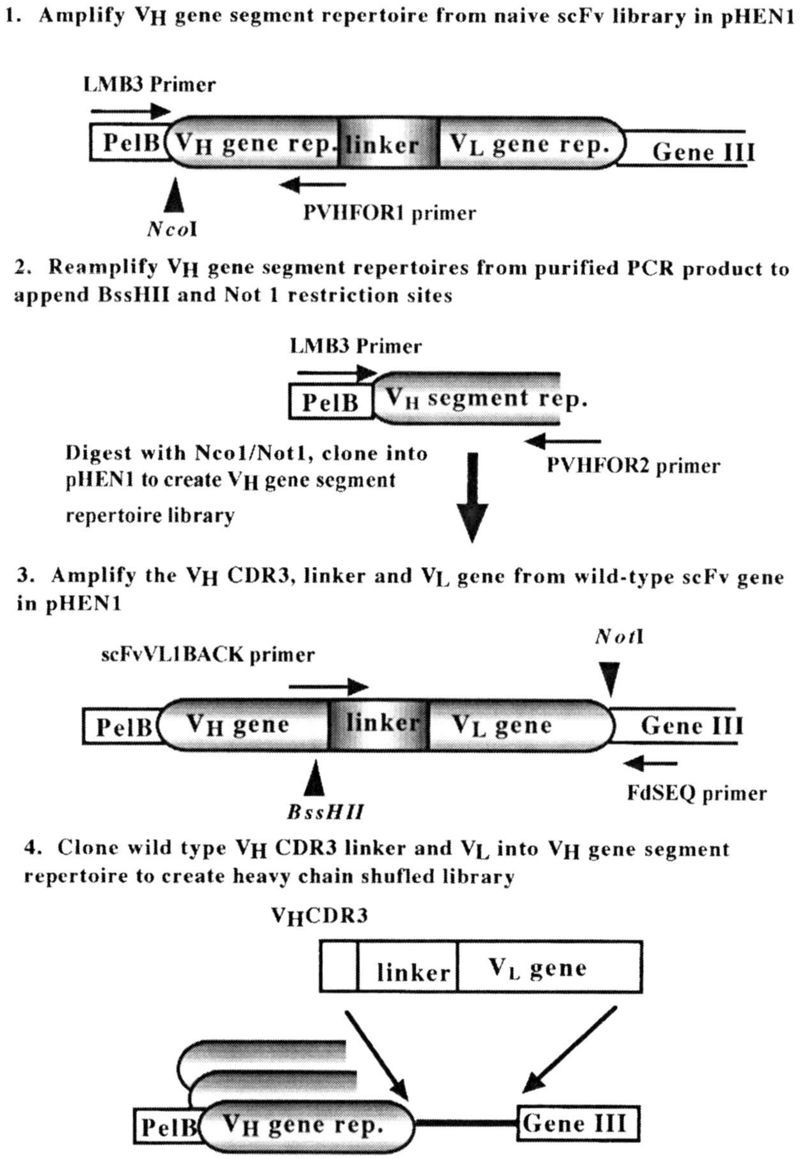
1. Amplify VH gene segment repertoire from naive scFv library in pHEN1
LMB3 Primer
PelB
VH gene rep.
linker
VL gene rep.
Gene III
PVHFOR1 primer
NcoI
2. Reamplify VH gene segment repertoires from purified PCR product to append BssHII and Not 1 restriction sites
LMB3 Primer
PelB
VH segment rep.
PVHFOR2 primer
Digest with Nco1/Not1, clone into pHEN1 to create VH gene segment repertoire library
3. Amplify the VH CDR3, linker and VL gene from wild-type scFv gene in pHEN1
NotI
scFvVL1BACK primer
PelB
VH gene
linker
VL gene
Gene III
FdSEQ primer
BssHII
4. Clone wild type VH CDR3 linker and VL into VH gene segment repertoire to create heavy chain shufled library
VHCDR3
linker
VL gene
PelB
VH gene rep.
Gene III

Fig. 4. Overview of heavy-chain shuffling strategy. Step 1: V_H gene-segment repertoires are PCR-amplified from an existing naïve scFv library in pHEN1 using the primers LMB3 and PVHFOR1. Step 2: The PCR product is re-amplified using the primers LMB3 and PVHFOR2 to append *Bss*HII and *Not*I restriction sites to the 3′ end of the gene. The PCR product is digested with *Nco*I and *Not*I and ligated into pHEN1 to create a library of V_H gene segments. Step 3: Wild-type VHCDR3, framework 4, scFv linker, and V_L are PCR-amplified using the primers scFvVL1BACK and FdSEQ. Step 4: The PCR product is digested with *Bss*HII and *Not*I and cloned into the V_H gene-segment library to create a heavy chain-shuffled library.

4. scFvJH3XhoFOR primer (10 pmol/μL): 5′-GAG TCA TTC TCG TCT CGA GAC GGT GAC CAT TGT CCC-3′.
5. scFvJH4–5XhoFOR primer (10 pmol/μL): 5′-GAG TCA TTC TCG TCT CGA GAC GGT GAC CAG GGT TCC-3′.
6. scFvJH6XhoFOR primer (10 pmol/μL): 5′-GAG TCA TTC TCG TCT CGA GAC GGT GAC CGT GGT CCC-3′.
7. LMB3 primer (10 pmol/μL): 5′-CAG GAA ACA GCT ATG AC-3′.
8. Geneclean kit.
9. *Xho*I and *Nco*I restriction enzymes, NEB2 digestion buffer, and 100X BSA solution (New England Biolabs).
10. V_L gene-repertoire vectors (from **Subheading 3.1.**).

2.3. Construction of Human V_H Gene Segment Repertoires

1. PCR reagents and equipment (**Subheading 2.1., items 1–6**).
2. ScFv gene-repertoire template DNA (double-stranded DNA) prepared from a naïve scFv library in pHEN1 (*see* Chapter 6 for preparation) (10 ng/μL).
3. Geneclean kit.
4. Wizard PCR purification kit.
5. FORWARD1 (FOR1) primers (10 pmol/μL) (*see* **Note 1**):
   a. PVH1FOR1 primer: 5′-TCG CGC GCA GTA ATA CAC GGC CGT GTC-3′;
   b. PVH3FOR1 primer: 5′-TCG CGC GCA GTA ATA CAC AGC CGT GTC CTC-3′;
   c. PVH5FOR1 primer: 5′-TCG CGC GCA GTA ATA CAT GGC GGT GTC CGA-3′.
6. FORWARD2 (FOR2) primers (10 pmol/μL) (*see* **Note 2**):
   a. PVH1FOR2 primer: 5′-GAG TCA TTC TCG ACT TGC GGC CGC TCG CGC GCA GTA ATA CAC GGC CGT GTC-3′;
   b. PVH3FOR2 primer: 5′-GAG TCA TTC TCG ACT TGC GGC CGC TCG CGC GCA GTA ATA CAC AGC CGT GTC CTC-3′;
   c. PVH5FOR2 primer: 5′-GAG TCA TTC TCG ACT TGC GGC CGC TCG CGC GCA GTA ATA CAT GGC GGT GTC CGA-3′.
7. LMB3 primer (*see* **Subheading 2.2.**).
8. *Nco*I and *Not*I restriction enzymes, NEB 3 buffer, and 100X BSA solution (New England Biolabs).
9. T4 DNA ligase and ligation buffer (New England Biolabs).
10. 2X TY/amp/2%glu media (*see* **Subheading 2.1.**, **item 17**).

2.4. Construction of a Heavy Chain-Shuffled Phage Library

1. PCR reagents and equipment (*see* **Subheading 2.1., items 1–6**).
2. Plasmid or single-stranded DNA containing starting scFv gene in the vector pHEN1.
3. Wizard PCR purification kit.
4. Geneclean kit.
5. FDSEQ primer (*see* **Subheading 2.1.**).
6. Custom BACK primer scFvVL1BACK (*see* **Note 3**).

7. *Bss*HII and *Not*I restriction enzyme, NEB3 buffer, and 100X BSA solution (New England Biolabs).
8. Plasmid DNA from the three V_H gene-segment phage libraries (from **Subheading 3.3.**).
9. T4 DNA ligase and ligation buffer.

2.5. Selection, Identification, and Characterization of Higher-Affinity Chain-Shuffled ScFv

1. Reagents for the preparation of phage antibodies (*see* **Chapter 8, Subheading 2.3.**).
2. Reagents for the selection of phage antibodies using soluble biotinylated antigen (*see* **Chapter 8, Subheading 2.2.**).
3. Reagents for phage enzyme-linked immunosorbent assay (ELISA) (*see* **Chapter 8, Subheadings 2.4.** and **2.5.**).
4. Reagents for soluble scFv ELISA (*see* **Chapter 8, Subheadings 2.6.** and **2.7.**).
5. BIAcore surface plasmon resonance (SPR) intrument (BIAcore).
6. CM5 sensor chip (BIAcore).
7. BIAcore EDC/NHS immobilization kit (BIAcore).
8. 10 m*M* sodium acetate buffer, pH 4.5.
9. HEPES-buffered saline (HBS): 10 m*M* HEPES, 150 m*M* NaCl, 3.4 m*M* EDTA, pH 7.4.
10. 2X TY/amp/2% glu media (*see* **Subheading 2.1., item 17**).
11. 2X TY/amp/0.1% glu media: 2X TY media (**Subheading 2.1, item 16**) containing 100 µg/mL ampicillin and 0.1% glucose.
12. 1 *M* isopropyl-β-D-thiogalactopyranoside (IPTG).
13. Periplasmic lysis buffer (PPB): 20% sucrose, 1 m*M* EDTA, 20 m*M* Tris-HCl, pH 8.0.
14. 5 m*M* $MgSO_4$.
15. CentriSep columns (Princeton Separations).
16. 4 *M* $MgCl_2$.

3. Methods

3.1. Construction of Human V_L Gene Repertoire Libraries (see Note 4)

1. Make up four separate 50-µL PCR mixes containing:

Water	35.5 µL
20X dNTPs	2.5 µL
10X Vent DNA polymerase buffer	5.0 µL
FdSEQ primer	25 pmol (2.5 µL)
BACK primers (*see* **Note 5**)	25 pmol (2.5 µL)
Double-stranded scFv template DNA	10 ng (1 µL)
Vent DNA polymerase	2 U (1 µL)

2. Heat the reaction mixes to 94°C for 5 min in a PCR thermocycler with a heated lid.
3. Cycle 25× to amplify the V_L genes at 94°C for 30 s, 42° C for 30 s, and 72° C for 1 min.

4. Visualize the PCR product by electrophoresis on a 1.5% agarose gel in 1X TBE buffer and purify the PCR product using the Wizard PCR purification kit.
5. To prepare the amplified V_L repertoire for cloning, digest the PCR product with the *Xho*I and *Not*I restriction enzymes in the supplied NEB2 buffer and BSA solution, under conditions recommended by the manufacturer (*see* **Chapter 6, Subheading 3.7.** for more detail).
6. Gel-purify the digested gene repertoire on a 1.5% agarose gel and extract the DNA using the Geneclean kit. Resuspend the product in 30 µL of water. Determine the DNA concentration by analysis on a 1.5% agarose gel with markers of known size and concentration.
7. Digest the pHEN1-$V_\lambda3$ vector DNA with the *Xho*I and *Not*I restriction enzymes in the supplied NEB2 buffer and 1X BSA solution, under conditions recommended by the manufacturer (*see* **Chapter 6, Subheading 3.7.** for more detail).
8. Purify the digested vector DNA on a 0.8% agarose gel. Extract the cut vector from the gel using the Geneclean kit.
9. Ligate the digested PCR products (from **step 6**) and the digested vector (from **step 8**) as described in **Chapter 6, Subheading 3.8.**
10. Electroporate the ligated DNA into electrocompetent *E. coli* TG1 cells to create four V_L gene repertoire phage libraries; determine the library size, and store the bacterial library stock at –70°C. (*see* **Chapter 6, Subheading 3.9.** for protocols).
11. Prepare DNA from each of the V_L gene repertoire libraries by inoculating 100 mL of 2X TY/amp/2% glu media with bacteria from the library glycerol stock (*see* **Note 6**). After overnight growth, perform a standard DNA midi plasmid preparation from the culture. For subsequent digestion of the libraries, DNA from the different repertoires can be combined.

3.2. Construction of a Light Chain-Shuffled Phage Library (see Note 7)

1. Make up 50-µL PCR mixes containing:

Water	34.5 µL
20X dNTPs	2.5 µL
10X Vent polymerase buffer	5.0 µL
scFv template DNA	100 ng (2.0 µL)
scFvJHnXhoFOR	25 pmol (2.5 µL)
LMB3 primer	25 pmol (2.5 µL)
Vent DNA polymerase	2 U (1.0 µL)

2. Heat the reaction mixes to 94°C for 5 min in a PCR thermocycler with a heated lid.
3. Cycle 25× to amplify the V_H gene at 94°C for 30 s, 42°C for 30 s, 72°C for 1 min.
4. Visualize the PCR product by electrophoresis on a 1.5% agarose gel in 1X TBE buffer and purify the PCR product using the Wizard purification kit.
5. Digest the PCR product with *Xho*I and *Nco*I restriction enzymes in the supplied NEB2 buffer and BSA solution, under conditions recommended by the manufacturer (*see* **Chapter 6, Subheading 3.7.** for more details).

6. Gel-purify the PCR product on a 1.5% agarose gel and extract the DNA using the Geneclean kit. Resuspend the product in 30 μL of water. Determine the DNA concentration by analysis on a 1.5% agarose gel with markers of known size and concentration.
7. Prepare the V_L gene-repertoire vector (generated from **Subheading 3.1.**) by digesting the vector with *Xho*I and *Nco*I restriction enzymes in the supplied NEB2 buffer and BSA solution, under conditions recommended by the manufacturer (*see* Chapter 6, **Subheading 3.7.** for more detail).
8. Purify the digested vector DNA on a 0.8% agarose gel. Extract the cut vector from the gel using the Geneclean kit.
9. Ligate the digested PCR products (from **step 6**) and the digested vector (from **step 8**), and electroporate the DNA into electrocompetent *E. coli* TG1 cells to create light chain-shuffled libraries; determine the library size, and store the bacterial library stock at –70°C (*see* **Chapter 6, Subheading 3.9.** for protocols).

3.3. Construction of Human V_H Gene-Segment Repertoires (see Note 8)

1. Make up three separate 50-μL PCR mixes containing:

Water	35.5 μL
20X dNTPs	2.5 μL
10X Vent polymerase buffer	5.0 μL
FOR1 primers	25 pmol (2.5 μL)
LMB3 primer	25 pmol (2.5 μL)
scFv gene template DNA	10 ng (1.0 μL)
Vent DNA polymerase	2 U (1.0 μL)

2. Heat to 94°C for 5 min in a PCR thermocycler with a heated lid.
3. Cycle 25× to amplify the V_H genes at 94°C for 30 s, 42°C for 30 s, and 72°C for 1 min.
4. Gel-purify the V_H gene repertoires by electrophoresis on a 1.5% agarose gel in 1X TBE buffer and extract the DNA using the Geneclean kit. Resuspend each product in 20 μL of water. Determine the DNA concentration by analysis on a 1.5% agarose gel with markers of known size and concentration.
5. Make up three 50-μL PCR mixes, one for each V_H gene-segment repertoire, containing:

Water	34.5 μL
20X dNTPs	2.5 μL
10X Vent polymerase buffer	5.0 μL
FOR2 primers	25 pmol (2.5 μL)
LMB3 primer	25 pmol (2.5 μL)
V_H gene-segment repertoires from **step 4**	50 ng (2.0 μL)
Vent DNA polymerase	2 U (1.0 μL)

6. Heat the reactions to 94°C for 5 min in a PCR thermocycler with a heated lid.
7. Cycle 25× to re-amplify the V_H gene repertoires at 94°C for 30 s, 42°C for 30 s, and 72°C for 1 min.

8. Visualize the PCR product by electrophoresis on a 1% agarose gel in 1X TBE buffer and purify the PCR product using the Wizard PCR purification kit.
9. To prepare the V_H gene repertoire for cloning, digest the PCR products with *Nco*I and *Not*I restriction enzymes in the supplied NEB3 buffer and BSA solution, under conditions recommended by the manufacturer (*see* **Chapter 6, Subheading 3.7.** for more detail).
10. Gel-purify the V_H gene repertoire by electrophoresis on a 1.5% agarose gel and extract the DNA using the Geneclean kit. Resuspend the product in 30 µL of water. Determine the DNA concentration by analysis on a 1.5% agarose gel with markers of known size and concentration.
11. Digest the vector pHEN1 with *Nco*I and *Not*I restriction enzymes in the supplied NEB3 buffer and BSA solution, under conditions recommended by the manufacturer (*see* **Chapter 6, Subheading 3.7.** for more details).
12. Purify the digested vector DNA by electrophoresis on a 0.8% agarose gel. Extract the cut vector from the gel using the Geneclean kit.
13. Ligate the digested PCR products (from **step 10**) and the digested vector (from **step 12**), and electroporate the DNA into *E. coli* TG1 cells to create three V_H gene-segment phage libraries. Determine the library size, and store the bacterial library stock at –70°C (*see* **Chapter 6, Subheading 3.9.** for protocols).
14. Prepare DNA from each of the V_H gene-segment repertoire libraries by inoculating 100 mL of 2X TY/amp/2%glu media with bacteria from the library glycerol stock (*see* **Note 6**). After overnight growth, perform a standard DNA midi plasmid preparation from the culture. For subsequent digestion of the libraries, the DNA from the different repertoires can be combined.

3.4. Construction of a Heavy Chain-Shuffled Phage Library

1. Make up 50-µL PCR mixes containing:

Water	34.5 µL
20X dNTPs	2.5 µL
10X Vent polymerase buffer	5.0 µL
FdSEQ1 primer	25 pmol (2.5 µL)
scFvVL1BACK primer	25 pmol (2.5 µL)
scFv template DNA	100 ng (2.0 µL)
Vent DNA polymerase	2 U (1.0 µL)

2. Heat the reactions to 94°C for 5 min in a PCR thermocycler with a heated lid.
3. Cycle 25× to amplify the V_H genes at 94°C for 30 s, 42°C for 30 s, 72°C for 1 min.
4. Visualize the PCR product by electrophoresis on a 1% agarose gel in 1X TBE buffer and purify the PCR product using the Wizard PCR purification kit.
5. Digest the PCR product with *Bss*HII and *Not*I restriction enzymes in the supplied NEB3 buffer and BSA solution, under conditions recommended by the manufacturer (*see* **Chapter 6, Subheading 3.7.** for more detail). First digest with *Not*I at 37°C, then add the *Bss*HII enzyme and transfer to 50°C.
6. Gel-purify the PCR product by electrophoresis on a 1.5% agarose gel and extract the DNA using the Geneclean kit. Resuspend the product in 30 µL of water. Deter-

mine the DNA concentration by analysis on a 1.5% agarose gel with markers of known size and concentration.

7. Prepare the recipient V_H gene segment repertoire vector (generated from **Subheading 3.3.**) by digesting the vector with *Bss*HII and *Not*I restriction enzymes in the supplied NEB3 buffer and BSA solution, under conditions recommended by the manufacturer (*see* **Chapter 6, Subheading 3.7.** for more detail). First digest with *Not*I at 37°C, then add the *Bss*HII enzyme and transfer to 50°C.
8. Purify the digested vector DNA by electrophoresis on a 0.8% agarose gel. Extract the cut vector from the gel using the Geneclean kit.
9. Ligate the digested the V_L gene PCR product (from **step 6**) with the digested V_H gene-segment phage library vectors (from **step 8**) and electroporate the DNA into electrocompetent *E. coli* TG1 cells to create V_H gene-shuffled phage libraries; determine the library size, and store the bacterial library stock at –70°C (*see* Chapter 6, **Subheading 3.9.** for protocols).

3.5. Selection, Identification, and Characterization of Higher-Affinity Chain-shuffled ScFv

1. Prepare chain-shuffled phage antibody library (light chain-shuffled: **Subheading 3.2.**; heavy-chain-shuffled: **Subheading 3.4.**) as described in **Chapter 8, Subheading 3.3.**, using a starting culture volume of 250 mL.
2. Select chain-shuffled phage antibody library on soluble biotinylated antigen as described in **Chapter 8, Subheading 3.2.** (*see* **Notes 9** and **10**) but using three different antigen concentrations for selection equal to: $K_d/1$, $K_d/10$, and $K_d/100$, where K_d = the affinity of the wild-type antibody for the antigen. Elute the phage from the beads using 100 µL of 100 m*M* HCl at room temperature for 10 min (*see* **Note 11**), then add 1 mL of 1 *M* Tris-HCl, pH 7.4.
3. Prepare phage for the next round of selection as described in **Chapter 8, Subheading 3.3.**, using a starting culture volume of 50 mL.
4. Repeat **steps 1–3** for a total of 4–5 rounds of selection. Select the phage output from a single antigen concentration (of the three utilized) for the next round of selection (*see* **Note 12**).
5. Prepare soluble scFv from selection rounds 3–5 as described in **Chapter 8, Subheading 3.6.**
6. Perform soluble scFv ELISA as described in **Chapter 8, Subheading 3.7.** to identify antigen-binding scFv.
7. Identify higher-affinity scFv by off-rate screening using SPR in a BIAcore as described in the following steps (*see* **Note 13**).
8. Select 30–50 scFv ELISA-positive clones for further study. For each clone, grow a 1-mL overnight culture in 2X TY/amp/2%glu media at 30°C, shaking at 250 rpm.
9. Add 50 µL of the overnight culture to 5 mL of 2X TY/amp/0.1%glu in a 50-mL tube and grow at 37°C shaking at 250 rpm for approx 2 h, to A_{600} of about 0.9.
10. Induce scFv expression by adding 5 µL of 1 *M* IPTG to each tube (final concentration of 500 µ*M*) and grow shaking (250 rpm) at 30°C for 4 h. Collect the cells by centrifuging in a 15-mL Falcon tube at 4,000*g* for 15 min.

11. To prepare a periplasmic extract, resuspend the bacterial pellet in 200 µL of PPB buffer, transfer the resuspended pellet to 1.5-mL tubes, and place them on ice for 20 min. This shrinks the cells, extracts some scFv protein, and makes the periplasm hyperosmotic, facilitating the swelling that occurs with addition of hyposomotic $MgCl_2$. Pellet the cells in a microcentrifuge at full speed (14,000 rpm) for 5 min. Discard the supernatant. Use the bacterial pellet for the osmotic shock preparation.
12. To prepare an osmotic shock fraction, resuspend the pellet in 200 µL of 5 m*M* $MgSO_4$ and incubate the resuspended pellet on ice for 20 min. Pellet the cells in a microcentrifuge at full speed (14,000 rpm) for 5 min. Save the supernatant, which is the osmotic shock fraction, in a new tube.
13. Change the buffer containing the scFv from the osmotic shock buffer to the BIAcore running buffer (HBS) using the CentriSep columns, as per the manufacturer's instructions (*see* **Note 14**)
14. Set up the BIAcore instrument according to the manufacturer's instructions. In the flow cell of the BIAcore instrument, immobilize the target antigen to a CM5 sensor chip using EDC/NHS chemistry as described by the manufacturer (*see* **Note 15**).
15. Inject the scFv preparation from **step 13** over the flow cell for 1 min, followed by 2–10 min observation of the dissociation phase at a constant flow rate of 15 µL/min, with HBS as the running buffer.
16. Regenerate the chip surface and analyze the next scFv (*see* **Note 16**). A program can be written to automate this process, allowing unattended analysis of 40–50 scFv overnight. An apparent k_{off} can be determined from the dissociation part of the sensorgram for each scFv analyzed, using the software provided by the BIAcore manufacturer (*see* **Note 17**).

4. Notes

1. We use primers that generate V_H gene-segment repertoires derived primarily from the V_H 1, 3, and 5 gene families, since we have not observed any scFv phage antibodies derived from the V_H 2, 4, or 6 families. Primers PVH1FOR1, PVH3FOR1, and PVH5FOR1 are designed to anneal to the consensus 3′-FR3 sequence of V_H1, V_H3, or, V_H5, respectively ***(9)***.
2. Primers PVH1FOR2, PVH3FOR3, and PVH5FOR5 contain the sequences of the PVH1FOR1, PVH3FOR1, and PVH5FOR1 primers and append a *Bss*HII site at the 3′-end of FR3 followed by a *Not*I restriction sites. The *Bss*HII site corresponds to V_H amino acid residues 93 and 94 (Kabat numbering [**ref.** ***14***] and does not change the amino acid sequence [alanine-arginine]).
3. The light-chain gene, linker DNA, and V_H CDR3 and FR4 from the starting scFv are amplified using PCR and cloned into the V_H gene-segment phage libraries generated from **Subheading 3.3.** For a starting scFv cloned into pHEN1, this is achieved using a specific BACK primer (scFvVL1BACK) designed for the particular scFv, and primer FdSEQ1, which anneals within the phage gene 3 in pHEN1. The BACK primer used here, in the example, scFvVL1BACK, anneals to the last 24 nucleotides of V_H framework 3 and the first 18 nucleotides of the CDR3. The primer also contains a *Bss*HII restriction site incorporated into the six nucleotides

encoding the last two amino acid residues of V_H framework 3 (amino acid residues 93 and 94, Kabat numbering [*14*]). For any given scFv, it is likely that a unique scFvVL1BACK primer will need to be designed, incorporating the features described here. In addition, a different 3′ primer may be required, depending on the vector backbone in which the starting scFv is cloned.

4. For light-chain shuffling, a library of light-chain genes is constructed in the vector pHEN1-V_λ3. This vector is based on pHEN1 and contains the pelB leader, an *Nco*I cloning site, a polylinker, a *Xho*I cloning site comprising the two C-terminal amino acids of the V_H framework 4, linker DNA and a single V_λ light chain ***(10)***. The *Xho*I site can be encoded at the end of the V_H FR4 without changing the amino acid sequence of residues 102 and 103 (serine-serine) (**ref. *14***). To create the libraries, V_κ and V_λ gene repertoires and linker DNA are amplified by PCR using as a template double-stranded DNA prepared from a naïve scFv phage antibody library prepared as described in Chapter 6. Four BACK primers are used that anneal to the first six nucleotides of the $(Gly_4Ser)_3$ linker and either the J_H1, 2, J_H3, J_H 4,5 or J_H6 segments, and contain the *Xho*I cloning site. The pHEN1-V_λ3 vector is digested with *Xho*I and *Not*I, removing the single V_λ gene, and the amplified repertoires are cloned into the digested vector. The resulting library contains rearranged human V_κ and V_λ gene repertoires, linker DNA, and cloning sites for inserting a rearranged V_H gene as an *Nco*I-*Xho*I fragment (**Fig. 3**).
5. Either RJH1/2Xho, RJH3Xho, RJH4/5Xho, or RJH6Xho (*see* **Subheading 2.1.**).
6. The innoculum should be large enough so that the number of bacteria inoculated is at least 5× larger than the library size.
7. Your specific rearranged V_H gene is amplified from the vector pHEN1 using primers LMB3 (anneals upstream of the beginning of the V_H gene) and a specific primer that anneals to the J_H gene of your scFv (either scFvJH1-2XhoFOR, scFvJH3XhoFOR, scFvJH4-5XhoFOR, or scFvJH6XhoFOR (*see* **Subheading 2.2., items 3–6**).
8. To facilitate heavy-chain shuffling, libraries are constructed in pHEN1 ***(5)*** containing human V_H gene-segment repertoires (FR1 to FR3) and a cloning site at the end of V_H FR3 for inserting the V_H CDR3, V_H FR4, linker DNA and light chain from a binding scFv as a *Bss*HII-*Not*I fragment (**Fig. 4**). We conserve the wild-type V_H CDR3 as it contains much of the antigen-binding energy. To create the libraries, three V_H gene-segment repertoires enriched for human V_H 1, V_H 3, and V_H 5 gene segments are amplified by PCR using as a template double-stranded DNA prepared from a naïve scFv phage antibody library created as described in Chapter 6.
9. Selections must be carefully designed in order to select antibodies that are of higher affinity rather than those clones that display better on phage, are less toxic to *E. coli,* are more stable, or multimerize, leading to a higher functional affinity from avidity. Two approaches have been used to select rare, higher-affinity scFv from a background of lower-affinity scFv or nonbinding scFv: selections based on binding kinetics and selections based on the equilibrium constant (K_d). In either case, it is important to use labeled antigen in solution rather than antigen adsorbed to a solid matrix. Use of soluble antigen biases toward selections based on binding affinity or

binding kinetics, rather than avidity. This is especially important when selecting scFv libraries, since it is known that some scFv can spontaneously dimerize in a sequence-dependent manner ***(8)***. Failure to use soluble antigen is likely to result in the selection of dimeric scFv with a monovalent binding constant no higher than wild-type. Even with the Fab format, soluble antigen should also be used to avoid selecting for phage that displays multiple copies of Fab that will have a higher functional affinity (avidity).

10. We typically utilize equilibrium selections in which phage are incubated with an antigen concentration below the equilibrium binding constant. This approach effectively selects for antibodies with improved equilibrium constants. Reduction of the antigen concentration also helps to ensure that selection for higher-affinity scFv occurs, rather than selection for scFv that express well on phage or are less toxic to *E. coli.* The ability to use soluble antigen for selections affords the control over antigen concentration required for equilibrium selections. Phage are allowed to bind to biotinylated antigen and then recovered with streptavidin magnetic beads ***(2,11)***. The optimal antigen concentration used for the selection can be estimated *a priori* ***(12)***, but we prefer to use several different concentrations of antigen for the first round of selection. As an alternative to equilibrium selections, one can use selections based on binding kinetics, also termed off-rate (k_{off}) selections. In this case, the phage population is allowed to saturate the labeled antigen before a large molar excess of unlabeled antigen is added to the mix for a given amount of time. The duration of the competition with unlabeled antigen is chosen to allow the majority of the bound wild-type antibodies to dissociate while the improved mutants remain bound ***(13)***. This approach effectively selects for slower off-rates. Since a reduction in k_{off} is typically the major kinetic mechanism resulting in higher affinity when V-genes are mutated, both in vivo and in vitro ***(5)***, this approach should generally result in the selection of scFv with improved K_ds.
11. Do not leave the phage in HCl longer than 10 min, or infectivity will be significantly reduced. The use of less stringent eluants may not remove the highest-affinity phage from the antigen ***(11)***. Alternatively, NHS-SS-biotin (Pierce) can be used for biotinylation and the phage eluted with 100 m*M* DTT (for details, *see* Chapter 8, **Subheading 3.2.**).
12. Analyze the results of the output titration for the selection to decide which of the selections (from the antigen concentrations = $K_d/1$, $K_d/10$, and $K_d/100$) to rescue for the next round of selection. As a rule of thumb, the decision can be based on the output titers: When the output titer drops significantly more than the antigen concentration was lowered (when comparing the outputs from the concentrations $K_d/1$, $K_d/10$, and $K_d/100$), it is generally a result of loss of binding, and phage from this selection concentration should not be used for the subsequent round. Instead, use one of the higher antigen concentrations. For large, randomized libraries (complexity $>10^7$), the output number of phage from the first round of selection should be 10^4–10^6 cfu. If the titers for all antigen concentrations are larger than 10^7, it is likely that the antigen concentration was too high or that the washing steps were inadequate. If the titer is below 5×10^4, either too many washes were performed, or

too little antigen was used. On average, we lower the antigen concentration about 10-fold per round and we often use fM concentrations for the last round of the affinity maturation. For additional data, phage can be rescued from all three antigen concentrations and antigen binding determined by ELISA, to confirm that the selection conditions were not too stringent.

13. Even after using the stringent selections described here, only a fraction of the ELISA positive clones will have a higher affinity than wild-type. The strength of the ELISA signal is a poor indicator of which clones have a lower K_d, and more typically correlates with expression level of the scFv. Thus, a technique is required to screen ELISA positive clones to identify those with a lower K_d. One such technique is a competition or inhibition ELISA. Alternatively, one can take advantage of the fact that a reduction in k_{off} is typically the major kinetic mechanism resulting in higher affinity, and thus antibodies with improved affinity can be identified by measuring the off-rate. Since k_{off} is concentration-independent (unlike k_{on}), it is possible to measure k_{off}—for example, using surface plasmon resonance (SPR) in instruments such as the BIAcore, without purifying the antibody fragment. We have found this a very useful technique for identifying higher affinity scFv, and we use it to rank affinity-maturated clones.
14. The buffer must be changed from the osmotic shock fraction to the BIAcore running buffer in order to avoid excessive refractive index change during the SPR analysis, which would obscure measurement of k_{off}.
15. The appropriate antigen concentration and buffer for immobilization varies considerably among antigens. We typically start with 10 µg/mL antigen in 10 m*M* sodium acetate buffer, pH 4.5, and adjust antigen concentration, buffer ionic strength and pH as appropriate for the antigen in question. The amount of antigen immobilized should result in approx 100–300 RU of scFv binding.
16. The appropriate reagent for regeneration of the sensor chip between samples must also be determined for each antigen. However, we find that 4 *M* $MgCl_2$ will regenerate most antigen surfaces without a significant change in the sensorgram baseline after analysis of more than 100 samples.
17. Those clones with the slowest k_{off} are next subcloned and purified, and the K_d is measured using SPR in a BIAcore. Note that in the case of scFv, the shape of the dissociation curve can indicate whether single or multiple k_{off}s are present. ScFv that have multiple k_{off}s (a curve vs a straight line when plotting ln R1/R0 vs T) are a mixture of monomer and dimer, and are best avoided for subsequent characterization.

References

1. Yang, W.-P., Green, K., Pinz-Sweeney, S., Briones, A. T., Burton, D. R., and Barbas, C. F. (1995) CDR walking mutagenesis for the affinity maturation of a potent human anti-HIV-1 antibody into the picomolar range. *J. Mol. Biol.* **254,** 392–403.
2. Schier, R., McCall, A., Adams, G. P., Marshall, K., Yim, M., Merritt, H., et al. (1996) Isolation of picomolar affinity anti-c-erbB2 single-chain Fv by molecular

evolution of the complementarity determining regions in the centre of the antibody combining site. *J. Mol. Biol.* **263,** 551–567.

3. Adams, G. P., Schier, R., Marshall, K., Wolf, E. J., McCall, A. M., Marks, J. D., et al. (1998) Influence of affinity on the in vitro and in vivo binding properties of human single chain Fv molecules directed against c-erbB-2. *Cancer Res.* **58,** 485–490.
4. Adams, G. P., Schier, R., McCall, A. M., Crawford, R. S., Wolf, E. J., Weiner, L. M., et al. (1998) Prolonged in vivo tumor retention of a human diabody targeting the extracellular domain of human HER2/neu. *Br. J. Cancer* **77,** 1405–1412.
5. Marks, J. D., Griffiths, A. D., Malmqvist, M., Clackson, T., Bye, J. M., and Winter, G. (1992) Bypassing immunisation: high affinity human antibodies by chain shuffling. *Bio/Technology* **10,** 779–783.
6. Clackson, T., Hoogenboom, H. R., Griffiths, A. D., and Winter, G. (1991) Making antibody fragments using phage display libraries. *Nature* **352,** 624–628.
7. Holliger, P., Prospero, T., and Winter, G. (1993) Diabodies': small bivalent and bispecific antibody fragments. *Proc. Natl. Acad. Sci. USA* **90,** 6444–6448.
8. Schier, R., Bye, J. M., Apell, G., McCall, A., Adams, G. P., Malmqvist, M., et al. (1996) Isolation of high affinity monomeric human anti-c-erbB-2 single chain Fv using affinity driven selection. *J. Mol. Biol.* **255,** 28–43.
9. Tomlinson, I. M., Walter, G., Marks, J. D., Llewelyn, M. B., and Winter, G. (1992) The repertoire of human germline VH sequences reveals about fifty groups of VH segments with different hypervariable loops. *J. Mol. Biol.* **227,** 776–798.
10. Hoogenboom, H. R. and Winter, G. (1992) Bypassing immunisation: human antibodies from synthetic repertoires of germ line VH-gene segments rearranged *in-vitro*. *J. Mol. Biol.* **227,** 381–388.
11. Schier, R. S. and Marks, J. D. (1996) Efficient in vitro selection of phage antibodies using BIAcore guided selections. Human antibodies and hybridomas. **7(3),** 97–105.
12. Boder, E. T. and Wittrup, K. D. (1998) Optimal screening of surface-displayed polypeptide libraries. *Biotechnol. Prog.* **14,** 55–62.
13. Hawkins, R. E., Russell, S. J., and Winter, G. (1992) Selection of phage antibodies by binding afinity: mimicking affinity maturation. *J. Mol. Biol.* **226,** 889–896.
14. Kabat, E. A., Wu, T. T., Perry, H. M., Gottesman, K. S., and Foeller, C. (1991) *Sequences of proteins of immunological interest.* US Department of Health and Human Services, US Government Printing Office.

20

Antibody Affinity Maturation by Random Mutagenesis

Ikuo Fujii

1. Introduction

The immune response to an antigen can be considered to occur in two stages. In the initial stage, low-affinity antibodies are generated from an existing pool of the B-cell repertoires available at the time of immunization. In the second stage, which is driven by antigen stimulation, high-affinity antibodies are produced as a result of affinity maturation, starting with the light- and heavy-chain variable region (V_L and V_H) genes selected in the primary response. The mechanism for affinity maturation consists of somatic hypermutation of V-region genes followed by clonal selection of B-cells that produce antibodies of the highest affinity. Recent advances in phage-displayed antibody technology have enabled the affinity maturation process to be mimicked in vitro to improve the antigen-binding affinity and even alter the binding specificity of recombinant antibodies ***(1–6)***.

Filamentous bacteriophage—viruses that infect bacteria—have been used to display antibody fragments such as the single-chain Fv (scFv) or Fab. These are fused to a minor coat protein (pIII) or to the major coat protein (pVIII) by inserting DNA encoding the antibody fragment into either gene III or VIII *(7)*. The process of somatic hypermutation can be mimicked in vitro by scattering the entire V-gene with random mutations. One approach involves the use of an error-prone DNA polymerase in the polymerase chain reaction (PCR) to introduce random point mutations into the V-gene of choice. In a different approach, synthetic oligonucleotides are used to focus mutations on residues likely to be involved in antigen binding and away from key structural residues—for example, in the antibody complementarity-determining regions (CDRs). The library of V-gene mutants is then displayed on phage, and mutant antibodies with improved affinities are selected by binding to antigen under stringent selection

From: *Methods in Molecular Biology, Vol. 248: Antibody Engineering: Methods and Protocols*
Edited by: B. K. C. Lo © Humana Press Inc., Totowa, NJ

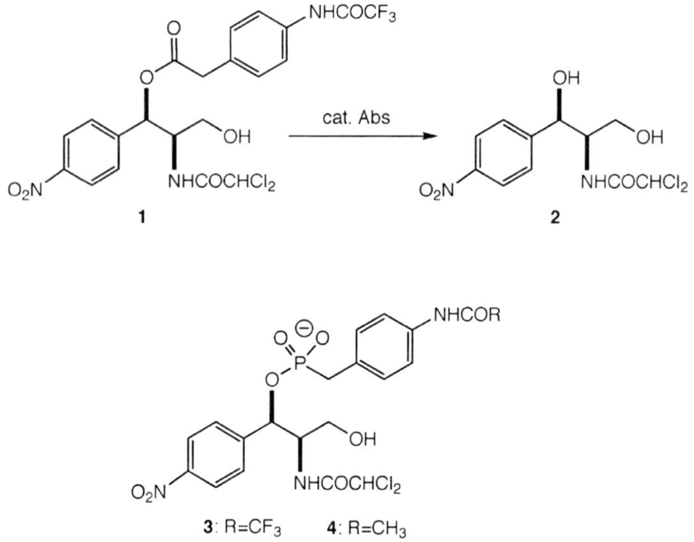

Fig. 1. Antibody-catalyzed hydrolysis. The chloramphenicol monoester derivative (**1**) was hydrolyzed by catalytic antibodies to produce chloramphenicol (**2**). Hapten (**3**), a transition-state analog of the hydrolysis, was used for immunization to produce the catalytic antibodies. Hapten (**4**) was used for panning selections of the phage-displayed catalytic antibodies.

conditions. Here we describe these two methods for in vitro affinity maturation using the catalytic antibody 6D9 as a model system *(8)*. This antibody was generated by the immunization of a transition-state analog **3** and it catalyzes the hydrolysis of ester **1** to provide chloramphenicol **2** *(9,10)* (**Fig. 1**).

A phage-displayed library of 6D9 was constructed by a modification of the pComb3 phagemid system *(6,11)* in which Fab fragments are fused to a minor coat protein (pIII) (*see* **Note 1**). The light-chain CDR1 (LCDR1) including the antigen-contacting residues was randomized by PCR with randomized primers (**Fig. 2**) (*see* **Note 2**) and the entire light chain was mutated by error-prone PCR *(12)*. By transforming *E. coli* with the randomized pComb3/6D9, a phage library of 8×10^7 clones was obtained. The phage particles were then screened for binding to antigen-coated plastic plates. After several rounds of panning with increasingly stringent washing conditions, antibody V-gene(s) from the binding phage(s) was subcloned into the pARA7 expression vector to produce soluble Fab fragments *(13)*.

2. Materials

2.1. Preparation of the M13 Helper Phage

1. VCSM13 interference-resistant helper phage (Stratagene).

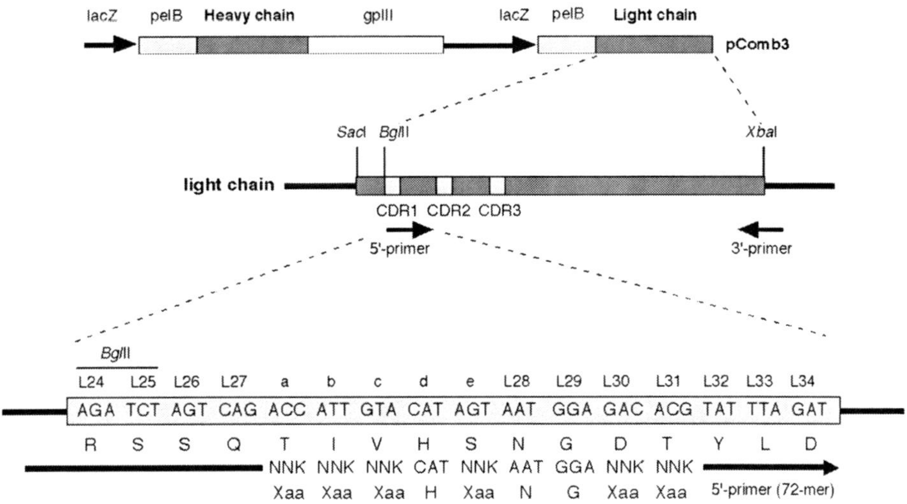

Fig. 2. Construction of the phage-displayed 6D9 light-chain CDR1 library in the pComb3 vector.

2. *E. coli* XL1-Blue strain (Stratagene).
3. Luria Bertani (LB) media: dissolve 10 g of Bacto tryptone, 5 g of Bacto yeast extract, and 10 g of NaCl in distilled water, and make up to 1 L; sterilize by autoclaving.
4. LB/tet media: LB media containing 10 µg/mL tetracycline (stock at 5 mg/mL in ethanol).
5. LB top agar: LB media containing 0.8% (w/v) agar.
6. LB plates: Prepare LB media containing 1.5% (w/v) agar; autoclave and pour into Petri dishes.
7. SB media: Dissolve 30 g of Bacto tryptone, 20 g of Bacto yeast extract, and 10 g of 3-(N-morpholino)propanesulfonic acid (MOPS) in distilled water, adjust the pH to 7.0 with NaOH and make up to 1 L; sterilize by autoclaving.
8. SB/tet media: SB media containing 10 µg/mL tetracycline.
9. SB/tet/kan media: SB media containing 10 µg/mL tetracycline and 70 µg/mL kanamycin (stock at 70 mg/mL in sterile, distilled water).
10. Millex-GP 0.22-µm filter (Millipore).
11. Polyethylene glycol, MW 8,000.
12. NaCl.
13. Phosphate-buffered saline (PBS): 137 m*M* NaCl, 2.7 m*M* KCl, 8.1 m*M* Na_2HPO_4, 1.5 m*M* KH_2PO_4, pH 7.4.

2.2. Preparation of pComb3 Phagemid Vector

1. QIAprep spin miniprep kit (Qiagen).
2. *E. coli* MC1061 strain (Promega).

3. LB media (*see* **Subheading 2.1., item 3**).
4. LB/amp media: LB media containing 100 μg/mL ampicillin (stock at 100 mg/mL in distilled water.
5. LB/amp plates: LB plates (**Subheading 2.1., item 6**) containing 100 μg/mL ampicillin. The antibiotic is added only when the autoclaved LB-agar solution cools to below 50°C.

2.3. Library Construction

2.3.1. LCDR1 Randomization (Focus Mutations)

1. TaKaRa Ex Taq DNA polymerase and 10X Taq buffer (TaKaRa, Japan).
2. dNTPs (2.5 m*M* each of deoxyadenosine 5′ triphosphate (dATP), deoxythymidine 5′ triphosphate (dTTP), deoxycytidine 5′ triphosphate (dCTP) and deoxyguanosine 5′ triphosphate (dGTP).
3. pComb3-6D9 vector ***(8)***.
4. 5′-Randomization primer: 5′– ATCTCTTGC*AGATCT*AGTCAGNNKNNKNNKC ATNNKAATGGANNKNNKTATTTAGATTGGTTCCTGCAGAAA-3′ (the *Bgl*II site is underlined), where N = G, A, T, or C, and K = T or G.
5. 3′-Randomization primer: 5′– GCTTAAC*TCTAGA*ATTAACACTCATTCCTGTT GAA-3′ (the *Xba*I site is underlined).
6. Sterile distilled water.
7. 3 *M* sodium acetate.
8. Ethachinmate (Wako Pure Chemical Industries, Japan; Cat. #31-0179).
9. Ethanol.
10. 50X TAE buffer: dissolve 242 g of Tris(hydroxymethyl)aminomethane, 57.1 mL of acetic acid, and 7.43 g of EDTA2Na in distilled water; make up to 1 L.
11. 2% agarose gel: 2% (w/v) agarose in 1X TAE buffer.
12. QIAquick gel extraction kit (Qiagen).
13. 10X T buffer (TaKaRa): 330 m*M* Tris-acetate pH 7.9, 100 m*M* magnesium acetate, 5 m*M* dithiothreitol (DTT) and 660 m*M* potassium acetate.
14. *Bgl*II (TaKaRa).
15. *Xba*I (TaKaRa).
16. QIAquick PCR purification kit (Qiagen).

2.3.2. Light-Chain Randomization (Error-Prone PCR)

1. Error-prone PCR mixture: 10 m*M* Tris-HCl, pH 9.0, 50 m*M* KCl, 7 m*M* $MgCl_2$, 0.5 m*M* $MnCl_2$, 0.2 m*M* dATP, 0.2 m*M* dGTP, 1.0 m*M* dCTP, 1.0 m*M* dTTP, and 0.1% (v/v) Triton.
2. pComb3/6D9 vector ***(8)***.
3. 5′-primer: 5′-CCAGATGTGAGCTCGTGATGACCCAGACTCCA-3′.
4. 3′-primer: 5′-GCTTAAC*TCTAGA*ATTAACACTCATTCCTGTTGAA-3′ (the *Xba*I site is underlined).
5. Taq DNA polymerase (Promega).
6. 50X TAE buffer (*see* **Subheading 2.3.1., item 10**).

7. 2% agarose gel (*see* **Subheading 2.3.1., item 11**).
8. QIAquick gel extraction kit.
9. 10X T buffer (*see* **Subheading 2.3.1., item 13**).
10. QIAquick PCR purification kit.
11. *Bgl*II.
12. *Xba*I.

2.3.3. Preparation of Randomized pComb3/6D9 Vector

1. pComb3/6D9 vector *(8)*.
2. 10x T buffer (*see* **Subheading 2.3.1., item 13**).
3. Sterile distilled water.
4. *Bgl*II.
5. *Xba*I.
6. MicroSpin s-400HR® (Amersham Biosciences).
7. Ligation High T4 DNA ligase (Toyobo, Japan).
8. QIAquick PCR purification kit.

2.3.4. Preparation of Phage Library

1. XL1-Blue electroporation-competent cells, Epicurian Coli® (Stratagene).
2. Electroporator and 0.1 cm-gap electroporation cuvets.
3. Sterile distilled water.
4. SOB media: dissolve 20.0 g of tryptone; 5.0 g of yeast extract, and 0.5 g of NaCl in distilled water to a total volume of 1 L; sterilize by autoclaving. Add 10 mL of filter-sterilized 1 *M* $MgCl_2$ and 10 mL of 1 *M* $MgSO_4$ per L of SOB media prior to use.
5. SOC media: add 1 mL of filter-sterilized 2 *M* glucose solution to SOB media prior to use and filter-sterilize.
6. SB/tet/amp media: SB media (*see* **Subheading 2.1., item 7**) containing 10 μg/mL tetracycline and 100 μg/mL ampicillin.
7. SB/tet/amp/kan media: SB media containing 10 μg/mL tetracycline, 100 μg/mL ampicillin, and 70 μg/mL kanamycin.
8. VCSM13 helper phage (from **Subheading 3.1.**).
9. Glycol solution: 20% (w/v) polyethylene glycol, mW 8,000 and 15% (w/v) NaCl in sterile distilled water.
10. PBS.

2.4. Selection of the Phage-Displayed Library

2.4.1. Coating Antigens on Microtiter Plates

1. F8 Nunc-Immuno Maxisorp modules (Nunc).
2. 10 μg/mL of antigen in PBS (or in other appropriate buffers).
3. PBS-Tween: PBS containing 0.05% (v/v) Tween 20.
4. Blocking buffer: 1% (w/v) skim milk, 0.03% (w/v) sodium azide in PBS.

2.4.2. Panning of the Phage-Displayed Library

1. Microtiter plate coated with the antigen (from **Subheading 3.4.1.**).
2. Phage libraries (from **Subheading 3.3.4.**).
3. Elution buffer: 0.1 *M* glycine-HCl, pH 2.2.
4. Neutralization buffer: 1.0 *M* Tris-HCl, pH 9.0.
5. *E. coli* XL1-Blue strain.
6. SB/tet media (*see* **Subheading 2.1., item 8**).
7. 100 mg/mL ampicillin in sterile distilled water.
8. Helper phage VCSM13 (from **Subheading 3.1.**).
9. SB/tet/amp/kan media (*see* **Subheading 2.3.4., item 7**).
10. Millex-GP 0.22-μm filter.
11. Polyethylene glycol, mW 8,000.
12. NaCl.
13. PBS-Tween (*see* **Subheading 2.4.1., item 3**).

2.4.3. Titration of Output Phage Particles

1. *E. coli* XL1-Blue culture infected with output phage (from **Subheading 3.4.2., step 5**).
2. LB/tet/amp plates: LB plate (**Subheading 2.1., item 6**) containing 10 μg/mL tetracycline and 100 μg/mL ampicillin.

2.4.4. Phage ELISA

1. *E. coli* XL1-Blue colonies harboring output phagemids (from **Subheading 3.4.3., step 2**).
2. SB/tet/amp media (*see* **Subheading 2.3.4., item 6**).
3. VCSM13 helper phage (from **Subheading 3.1.**).
4. 70 mg/mL kanamycin in sterile distilled water.
5. Glycol solution (*see* **Subheading 2.3.4., item 9**).
6. PBS.
7. Antigen-coated, pre-blocked microtiter plates (from **Subheading 3.4.1.**).
8. PBS-Tween (*see* **Subheading 2.4.1., item 3**).
9. HRP/Anti-M13 monoclonal conjugate (Amersham Biosciences).
10. TMB-H_2O_2 peroxidase kit (Bio-Rad).

2.4.5. pComb3 Plasmid Purification

1. *E. coli* XL1-Blue colonies harboring output phagemids (from **Subheading 3.4.3., step 2**).
2. SB/tet/amp media (*see* **Subheading 2.3.4., item 6**).
3. QIAprep spin miniprep kit.

2.5. Soluble Fab Expression

2.5.1. Subcloning into pARA7 Expression Vector

1. pARA7/6D9 vector ***(13)***.

2. 10X *M* buffer (TaKaRa): 500 m*M* Tris-HCl, pH 7.5, 100 m*M* $MgCl_2$, 10 m*M* DTT, 1 *M* NaCl.
3. *Sac*I.
4. *Xba*I.
5. Sterile distilled water.
6. QIAquick gel extraction kit.
7. MinElute PCR purification kit (Qiagen).
8. Ligation High T4 DNA ligase.

2.5.2. Production of Soluble Fab

1. *E. coli* MC1061 competent cells (prepared using standard methods).
2. SOC media (*see* **Subheading 2.3.4., item 5**).
3. SB media (*see* **Subheading 2.1., item 7**).
4. LB/amp plates (*see* **Subheading 2.2., item 5**).
5. Circlegrow media (Q-BIO gene) containing 100 μg/mL ampicillin.
6. 20% Arabinose in sterile distilled water.
7. STERICUP 0.22-μm filter (Millipore).
8. Ammonium sulfate.
9. Dialysis buffer: 20 m*M* Tris-HCl, 0.05% (v/v) Tween 20, pH 8.0.
10. Q Sepaharose ff® (Amersham Biosciences).
11. Ion-exchange elution buffer: 20 m*M* Tris-HCl, 0.2 *M* NaCl, 0.05% (v/v) Tween 20, pH 8.0.
12. Anti-mouse IgG $F(ab')_2$ (Rockland).
13. HiTrap NHS-activated HP media or prepacked columns (Amersham Biosciences).
14. 0.1 *M* glycine-HCl buffer, pH 2.6.
15. 1.0 *M* Tris-HCl buffer, pH 8.0.

3. Methods

3.1. Preparation of M13 Helper Phage

1. Inoculate a single colony of *E. coli* XL-1 Blue into 3 mL of LB/tet media and incubate shaking at 37°C overnight.
2. The next day, inoculate 40 μL of the overnight culture into 2 mL of LB/tet media and incubate shaking at 37°C until $OD_{600} = 0.8$ (~3 h).
3. Aliquot 100-μL volumes into tubes.
4. Add 10 μL of the VCSM13 helper phage stock and serial dilutions ($1/10^7$, $1/10^8$, $1/10^9$ and $1/10^{10}$) into the tubes containing *E. coli* and incubate shaking at 37°C for 15 min.
5. Transfer the culture into 5 mL of LB top agar (50°C) and then pour onto LB plates.
6. Incubate the plates at 37°C overnight for plaques to form.
7. Infect a single plaque into 2 mL of precultured *E. coli* XL-1 Blue ($OD_{600} = 0.4$), and then incubate shaking at 37°C for 2 h.
8. Transfer the culture to 50 mL of SB/tet/kan media and incubate shaking at 37°C overnight.

9. Centrifuge at 2,000*g* for 15 min at room temperature.
10. Filter the supernatant through a Millex-GP 0.22-µm filter and incubate at 70°C for 20 min.
11. Centrifuge at 2,000*g* at room temperature for 15 min.
12. Add 2 g of polyethylene glycol mW 8,000 and 1.5 g of NaCl to the supernatant, and then incubate on ice for 30 min.
13. Centrifuge at 8,500*g* for 30 min at 4°C.
14. Discard the supernatant and resuspend pellet in 12 mL of PBS.
15. Filter the helper phage solution through a Millex-GP 0.22-µm filter and store at 4°C (*see* **Note 3**).

3.2. Preparation of pComb3 Phagemid Vector

1. Inoculate a single colony of *E. coli* MC1061 harboring the pComb3 phagemid vector from a LB/amp plate into 3 mL of LB/amp media, and incubate shaking (200 rpm) at 37°C for 6 h (*see* **Note 4**).
2. Transfer the culture to 50 mL of LB/amp media, and incubate shaking (200 rpm) at 37°C overnight.
3. Centrifuge the culture at 2,500*g* for 10 min at room temperature.
4. Prepare plasmid DNA using the QIAprep spin miniprep kit as per manufacturer's instructions.
5. Store the purified pComb3 vector at –30°C.

3.3. Library Construction

A combinatorial library of the light chain in 6D9 was prepared by the pComb3 phagemid system *(8)*. The light chain CDR1 (LCDR1) was randomized by PCR with randomized primers (**Subheading 3.3.1.**) (**Fig. 2**), and in a separate experiment, the entire V_L was mutated by error-prone PCR (**Subheading 3.3.2.**). In order to prevent wild-type contamination in the LCDR1 library, the pComb3/6D9 vector was modified to replace L31Thr (ACG) with a stop codon (TAA) and used as the template for PCR. The randomized PCR products were cloned via *Bgl*II-*Xba*I sites back into the pComb3-6D9 vector. The ligation mixture was transformed into *E. coli* XL1-blue by electroporation. These methods are equally applicable to construct libraries for heavy chains, and indeed, for any CDR of interest.

3.3.1. LCDR1 Randomization (Focus Mutations)

1. Prepare the following PCR mixture in a PCR tube:

10X Taq buffer	10 µL
dNTPs	8 µL
pComb3/6D9 vector	50 ng
5′-Randomization primer	100 pmol
3′-Randomization primer	100 pmol
TaKaRa Ex Taq DNA polymerase	2.5 U (1 µL)
Sterile distilled water to 100 µL.	

2. Carry out 25 rounds of PCR: 95°C for 30 s, 60°C for 30 s, and 72°C for 30 s.
3. Add 3.3 μL of 3 *M* sodium acetate, 1 μL of Ethachinmate, and 250 μL of ethanol to the PCR tube.
4. Centrifuge at 20,000*g* at 4°C for 5 min and discard the supernatant.
5. Resuspend the precipitate in 40 μL of sterile distilled water.
6. Purify the randomized DNA by electrophoresis on a 2% agarose gel in 1X TAE buffer and extract using the QIAquick gel extraction kit.
7. Add 10 μL of 10X T buffer to 250 ng of the purified DNA, and then add sterile distilled water to make a solution of 90 μL.
8. Add 50 U (5 μL) of *Bgl*II and 50 U (5 μL) of *Xba*I to the tube.
9. Incubate at 37°C for 4 h.
10. Purify the digested DNAs using the QIAquick PCR purification kit. Proceed to **Subheadings 3.3.3.** and **3.3.4.** for library preparation.

3.3.2. Light-Chain Randomization (Error-prone PCR)

1. Prepare the following PCR mixture in a PCR tube:

Error-prone PCR mixture	100 μL
pComb3/6D9 vector	50 ng
5′-primer	50 pmol (0.5 μL)
3′-primer	50 pmol (0.5 μL)
Taq DNA polymerase	2.5 U (1 μL)

2. Carry out 30 rounds of PCR: 95°C for 1 min, 60°C for 1 min, and 72°C for 4 min.
3. Purify the randomized DNA on a 2% agarose gel in 1X TAE buffer and extract using the QIAquick gel extraction kit.
4. Subject 1 ng of the purified DNA to one more round of error-prone PCR (these PCR conditions have been shown to introduce an average of 3–4 amino acid substitutions in the entire light-chain product).
5. Dissolve 250 ng of the purified DNA in 10 μL of 10X T buffer, and then add distilled water to make a solution of 90 μL.
6. Add 50 U (5 μL) of *Bgl*II and 50 U (5 μL) of *Xba*I to the tube.
7. Incubate at 37°C for 4 h.
8. Purify the digested DNAs using the QIAquick PCR purification kit.

3.3.3. Preparation of Randomized pComb3/6D9 Vector

After the randomization of LCDR1 or the entire V_L, the library of mutants is subcloned into the pComb3/6D9 vector for the preparation of phage particles for selection.

1. Dissolve 3 μg of pComb3/6D9 vector in 5 μL of 10X T buffer.
2. Add 25 U (2.5 μL) of *Bgl*II and 25 U (2.5 μL) of *Xba*I to the tube, then add sterile distilled water to 50 μL.
3. Incubate at 37°C for 4 h.
4. Transfer the reaction mixture to MicroSpin s-400HR and centrifuge at 800*g* at 4°C for 1 min.

5. Add 1 μg of the digested pComb3/6D9 vector and 250 ng of the randomized DNA insert (from **Subheadings 3.3.1.** or **3.3.2.**) to 80 μL of Ligation High T4 DNA ligase.
6. Incubate at 16°C overnight.
7. Purify the ligated vector using the QIAquick PCR purification kit.

3.3.4. Preparation of Phage Library

1. Add the ligation product in 10 μL of sterile distilled water to 240 μL of XL1-Blue electroporation-competent cells. Transfer to a 0.1-cm-gap electroporation cuvet.
2. Electroporate cells at 1.8 kV, 25 μF, and 200 W.
3. Add 6 mL of SOC media to the cells and incubate shaking at 37°C for 1 h.
4. Centrifuge at 2,500*g* for 15 min at 4°C and discard the supernatant.
5. Resuspend the pellet in 8 mL of SB/tet/amp media and incubate shaking at 37°C for 1 h.
6. Add 10^{12} pfu (300 μL) of VCSM13 helper phage into the culture and incubate shaking at 37°C for 1 h.
7. Centrifuge at 2,000*g* for 15 min at room temperature and discard the supernatant.
8. Transfer the cells to 100 mL of SB/tet/amp/kan media and incubate shaking at 25°C overnight.
9. Centrifuge at 2,500*g* for 15 min at 4°C (4 × 25 mL).
10. Add 20 mL of glycol solution to 80 mL of the supernatant, mix gently, and then incubate on ice for 45 min.
11. Centrifuge at 9,000*g* for 45 min at 4°C, and discard the supernatant.
12. Resuspend phage pellet in 1 mL of PBS and then use for selection (**Subheading 3.4.**).

3.4. Selection of the Phage-displayed Library

Affinity selection of the light-chain randomized libraries for the BSA-conjugated transition-state analog **4** (**Fig. 1**) is performed on polystyrene plates.

3.4.1. Coating Antigens on Microtiter Plates

1. Add 50 μL of an antigen solution (10 μg/mL) to a well on a microtiter plate and incubate at 4°C overnight.
2. Wash the well twice with 300 μL of PBS-Tween.
3. Add 150 μL of blocking buffer to the well and incubate at room temperature for 1 h.
4. Wash well twice with 300 μL of PBS-Tween.
5. Store the plate at 4°C.

3.4.2. Panning of the Phage-Displayed Library

1. Add 50 μL of the freshly prepared phage particles (approx 1×10^{10} cfu) in PBS into a single well of microtiter plates precoated with the antigen and incubate at 37°C for 2 h.
2. Wash the wells with 300 μL of PBS-Tween. The number of washings is increased gradually at each round of panning (first round: 3×, second round: 5×; third round: 10×; and fourth round: 20×).

3. Add 50 µL of elution buffer to the well and let stand at room temperature for 10 min, and then elute the attached phage particles into a tube.
4. Neutralize the eluted phage with 50 µL of neutralization buffer.
5. Infect the phage particles into 3.5 mL of *E. coli* XL1-blue cells in SB/tet media and incubate shaking at 37°C for 30 min. Obtain the titer of the output phage by plating serial dilutions (*see* **Subheading 3.4.3.**).
6. Add ampicillin to a concentration of 100 µg/mL into the culture and incubate shaking at 37°C until OD_{600} = 0.4 (~ 1 h).
7. Add 20 µL of VCSM13 helper phage (1×10^{12} pfu/mL) in PBS into the culture and incubate shaking at 37°C for 1 h.
8. Centrifuge the culture at 800*g* for 15 min at room temperature and discard the supernatant.
9. Suspend the cells in 20 mL of SB/tet/amp/kan media and incubate shaking at 37°C for 1 h.
10. Incubate the culture shaking at 30°C overnight.
11. The next day, centrifuge the culture at 800*g* for 15 min at room temperature.
12. Filter the supernatant through a Millex-GP 0.22-µm filter.
13. Add 0.8 g of polyethylene glycol, mW 8,000 and 0.6 g of NaCl to the supernatant, then incubate on ice for 45 min.
14. Centrifuge at 9,000*g* for 45 min at 4°C.
15. Discard the supernatant and resuspend in 1 mL of PBS.
16. Filter the supernatant through a Millex-GP 0.22-µm filter.
17. Repeat this panning process 4× (*see* **Note 5**).

3.4.3. Titration of Output Phage Particles

The ratio of the number of colonies between the input phage and the output phage (output/input) indicates the efficiency of the panning process.

1. Inoculate serial dilutions of *E. coli* XL1-blue cells infected with the selected phage particles (**Subheading 3.4.2., step 5**) on LB/tet/amp plates.
2. Incubate the plates at 37°C overnight and count the number of colonies.
3. After the final round of the panning, the colonies are screened by phage ELISA (**Subheading 3.4.4.**) and the binding clones are subjected to pComb3 plasmid purification.

3.4.4. Phage ELISA

1. Inoculate a single colony of the phage-infected *E. coli* XL1-blue cells (from **Subheading 3.4.3., step 2**) into 3 mL of SB/tet/amp media and incubate shaking at 37°C for 6 h.
2. Add 2 µL of VCSM13 helper phage (1×10^{12} pfu/mL) in PBS into the culture and incubate shaking at 37°C for 1 h.
3. Add kanamycin to a final concentration of 70 µg/mL and incubate shaking at 30°C overnight.
4. Centrifuge the culture at 800*g* for 15 min at room temperature.

5. Add 400 μL of glycol solution to 1.6 mL of the supernatant, and incubate on ice for 45 min.
6. Centrifuge at 9,000*g* for 45 min at 4°C, discard the supernatant, and resuspend the phage pellet in 1 mL of 1.0% (w/v) skim milk in PBS.
7. Apply 50 μL of the phage solution to an antigen-coated microtiter plate and incubate at room temperature for 2 h.
8. Wash wells 3× with 300 μL of PBS-Tween.
9. Add 50 μL of HRP/Anti-M13 monoclonal Conjugate in PBS (at 1/1,000 dilution) and incubate at room temperature for 1 h.
10. Wash wells 5× with 300 μL of PBS-Tween.
11. Detect bound phage using the TMB-H_2O_2 peroxidase kit and measure optical density (O.D.) at 415 nm.

3.4.5. pComb3 Plasmid Purification

After phage ELISA screening, phagemid DNA is purified from colonies producing antigen-binding Fabs for DNA sequencing.

1. Inoculate a single colony of *E. coli* XL1-blue cell infected with the binding phage (*see* **Subheading 3.4.3., step 2**) into 3 mL of SB/tet/amp media and incubate shaking at 37°C overnight.
2. Purify pComb3 plasmid using the QIAprep spin miniprep kit.

3.5. Soluble Fab Expression

The gene that encodes the light chain of the selected pComb3 phagemid is subcloned into the *Sac*I-*Xba*I sites of the expression plasmid pARA7-6D9. The soluble Fab fragment of the selected clone is expressed in *E. coli* MC1061 cultures with L-arabinose induction ***(12)***.

3.5.1. Subcloning into pARA7 Expression Vector

1. Prepare the following restriction digestion mixture for the selected pComb3 phagemid:

10X M buffer	2 μL
Selected pComb3/6D9 phagemid DNA	100 ng
*Sac*I	5 U (0.5 μL)
*Xba*I	5 U (0.5 μL)
Sterile distilled water	to 20 μL.

2. Incubate the reaction mixture at 37°C for 2 h.
3. Purify the digested DNA by electrophoresis on a 2% agarose gel in 1X TAE buffer and extract using the QIAquick gel extraction kit.
4. Prepare the following restriction digestion mixture for the expression vector:

10X M buffer	2 μL
pARA7/6D9 vector	100 ng
*Sac*I	5 U (0.5 μL)
*Xba*I	5 U (0.5 μL)
Sterile distilled water	to 20 μL.

5. Incubate the reaction mixture at 37°C for 2 h.
6. Purify the digested DNA using the MinElute PCR purification kit.
7. Transfer 5 µL of the purified DNAs to 5 µL of a solution of Ligation High T4 DNA ligase.
8. Incubate at 16°C overnight.

3.5.2. Production of Soluble Fab

1. Add 10 µL of the ligation product (**Subheading 3.5.1.**) to 50 µL of competent *E. coli* MC1061 cells in 50 m*M* $CaCl_2$ (15% glycerol) on ice.
2. Stir the solution gently and incubate on ice for 1 h.
3. Warm the solution to 42°C for 45 s.
4. Cool the solution on ice for 3 min.
5. Add 450 µL of SOC media to the transformed cells and incubate shaking at 37°C for 1 h.
6. Centrifuge at 800*g* for 2 min at 4°C.
7. Discard the supernatant and resuspend in 100 µL of SB media.
8. Plate cells onto a LB/amp plate and incubate at 37°C overnight.
9. Inoculate a single colony to 3 mL of Circlegrow media with 100 µg/mL ampicillin and incubate shaking at 37°C until OD_{600} = 0.6 (~2 h).
10. Transfer the culture to 200 mL of Circlegrow media with 100 µg/mL ampicillin and incubate shaking at 37°C until OD_{600} = 0.2 (~1 h).
11. Add 2 mL of 20% arabinose in sterile distilled water to the culture and incubate shaking (100–200 rpm) at 23°C for 3 d.
12. Centrifuge at 8,500*g* for 30 min.
13. Filter the supernatant through a Stericup 0.22-µm filter.
14. Add 96.4 g of ammonium sulfate and stir for 1 h on ice.
15. Centrifuge at 30,000*g* for 30 min and discard the supernatant.
16. Resuspend the precipitate in 20 mL of dialysis buffer and and dialyze (mW cut-off 12,000–14,000 Daltons) at 4°C overnight.
17. Apply the Fab protein to a Q-Sepharose ff® ion-exchange column and elute with ion-exchange elution buffer.
18. Further purify the Fab fragment by applying the eluent to an affinity column, which is prepared by crosslinking anti-mouse $F(ab')_2$ antibodies to a Hi-Trap NHS-activated HP media, as per manufacturer's instructions. Elute with 0.1 *M* glycine-HCl buffer, pH 2.6, and neutralize the eluent with 1.0 *M* Tris-HCl buffer, pH 8.0 (*see* **Note 6**).

4. Notes

1. The Fd and κ-chain gene fragments of 6D9 were amplified by PCR as previously described ***(14)***. Amplified PCR products were digested with *Spe*I-*Xho*I for the Fd gene fragment and *Xba*I-*Sac*I for the κ-chain gene fragment, respectively, and were ligated into the pComb3 vector (pComb3/6D9) ***(8)*** and the pARA7 vector (pARA7/6D9) ***(13)***.
2. The X-ray structural analysis and site-directed mutagenesis of antibody 6D9 defined a catalytic residue, histidine at 27d in the light chain CDR1—His^{L27d}, num-

bering according to Kabat, et al. ***(15)***—which was placed in a key position to form a hydrogen bond to the transition-state structure ***(16)***. Based on the kinetic and structural information, we have designed phage-displayed libraries of 6D9 to direct antibodies toward those with higher activity.

3. The VCSM13 helper phage is titrated as follows:
   a. Incubate 20 μL of freshly prepared *E. coli* XL-1 Blue overnight culture in 2 mL of SB/tet media at 37°C for 2 h.
   b. Aliquot 100-μL volumes into tubes.
   c. Make serial dilutions of VCSM13 helper phage in LB/tet media.
   d. Add 10 μL of the VCSM13 dilutions into the tubes and incubate shaking at 37°C for 15 min.
   e. Transfer the culture into 5 mL of LB top agar and then pour onto LB plates.
   f. Incubate the plates at 37°C overnight.
   g. Count the number of plaques to determine titer (typically 10^{12-13} pfu/mL).
4. The pComb3 phagemid vector was a gift from Drs. Carlos Barbas III and Richard Lerner at the Scripps Research Institute, La Jolla, CA.
5. The panning is usually repeated several times until the ratio of the number of colonies between the input phage and the output phage (output/input) becomes constant.
6. The Fab concentration is measured at OD_{280} using ε = 1.4 (0.1%, 1 cm) and Mr = 50,000. The purity and the concentration of the Fabs are confirmed by sodium dodecyl sulfate-polyacrylamide gel electrophoresis (SDS-PAGE).

References

1. Hawkins, R. E., Russell, S. J., and Winter, G. (1992) Selection of phage antibodies by binding affinity mimicking affinity maturation. *J. Med. Biol.* **226,** 889–896.
2. Barbas III, C. F., Bain, J. D., Hoekstra, D. M., and Lerner, R. A. (1992) Semisynthetic combinatorial antibody libraries: a chemical solution to the diversity problem. *Proc. Natl. Acad. Sci. USA* **89,** 4457–4461.
3. Marks, J. D., Hoogenboom, H. H., Griffiths, A. D., and Winter, G. (1992) Molecular evolution of proteins on filamentous phage. *J. Biol. Chem.* **267,** 16,007–16,010.
4. Jackson, J. R., Sathe, G., Rosenberg, A., and Sweet, R. (1995) In vitro antibody maturation. *J. Immunol.* **154,** 3310–3319.
5. Barbas, C. F. and Burton, D. R. (1996) Selection and evolution of high-affinity human anti-viral antibodies. *TIBTECH* **14,** 230–234.
6. Fujii, I. Fukuyama, S., Iwabuchi, Y., and Tanimura, R. (1998) R. Evolving catalytic antibodies in a phage-displayed combinatorial library. *Nat. Biotechnol.* **16,** 463–467.
7. Kay, B. K., Winetr, J., and McCafferty, J. (1996) *Phage Display of Peptides and Protein,* Academic Press.
8. Takahahi, N., Kakinuma, H., Liu, L., Nishi, Y., and Fujii, I. (2001) In vitro abzyme evolution to optimize antibody recognition for catalysis. *Nat. Biotechnol.* **19,** 563–567.
9. Miyashita, H., Karaki, Y., Kikuchi, M., and Fujii, I. (1993) Prodrug activation via catalytic antibodies. *Proc. Natl. Acad. Sci. USA* **90,** 5337–5340.

10. Fujii, I., Tanaka, F., Miyashita, H., Tanimura, R., and Kinoshita, K. (1995) Correlation between antigen-combining-site structures and functions within a panel of catalytic antibodies generated against a single transition state analog. *J. Am. Chem. Soc.* **117,** 6199–6209.
11. Barbas III, C. F., Kang, A. S., Lerner, R. A., and Benkovic, S. J. (1991) Assembly of combinatorial antibody libraries on phage surface: the gene III site. *Proc. Natl. Acad. Sci. USA* **88,** 7978–7982.
12. Gram, H., Marconi, L. A., Barbas III, C. F., Collet, T. A., Lerner, R. A., and Kang, A. S. (1992) *In vitro* selection and affinity maturation of antibodies from a naïve combinatorial immunoglobulin library. *Proc. Natl. Acad. Sci. USA* **89,** 3576–3580.
13. Miyashita, H., Hara, T., Tanimura, R., and Fujii, I. (1997) Site-directed mutagenesis of active site contact residues in a hydrolytic abzyme: evidence for an essential histidine involved in transition state stabilization. *J. Mol. Biol.* **267,** 1247–1257.
14. Miyashita, H., Hara, T., Tanimura, R., Tanaka, F., Kikuchi, M., and Fujii, I. (1994) A common ancestry for multiple catalytic antibodies generated against a single transition-state analog. *Proc. Natl. Acad. Sci. USA* **91,** 6045–6049.
15. Kabat, E. A., Wu, T. T., Perry, H. M., Gottesman, K. S., and Foeller, C. in *Sequences of Proteins of Immunological Interest,* 5th ed. (United States Department of Health and Human Services, National Institutes of Health, Bethesda, MD; 1991).
16. Kristensen, O., Vassylyev, D. G., Tanaka, F., Morikawa, K., and Fujii, I. (1998) A structural basis for transition-state stabilization in antibody-catalyzed hydrolysis: crystal structures of an abzyme at 1.8 Å resolution. *J. Mol. Biol.* **281,** 501–511.

21

Developing a Minimally Immunogenic Humanized Antibody by SDR Grafting

Syed V. S. Kashmiri, Roberto De Pascalis, and Noreen R. Gonzales

1. Introduction

Since the advent of hybridoma technology a quarter century ago, a large number of murine monoclonal antibodies (MAbs) have been developed that are potentially useful clinical reagents against human infectious diseases and cancers. However, the clinical value of murine antibodies is limited because of the human anti-murine antibody (HAMA) response they evoke in patients *(1–4)*. Early attempts to reduce the HAMA response led to the development of mouse-human chimeric MAbs that are generated by replacing the constant regions of the heavy and light chains of the murine antibodies with those of the human antibodies *(5)*. Another approach to reducing the immunogenicity of a murine antibody is to resurface or veneer its variable domains. This is accomplished by replacing the exposed residues in the framework region of the murine antibody with the residues that are present in the corresponding positions of human antibodies *(6)*. A more commonly used procedure for the reduction of HAMA response involves grafting of the complementarity-determining regions (CDRs) of the xenogeneic antibody onto the human antibody frameworks, while retaining those residues of the xenogeneic framework regions that are considered essential for antibody reactivity to its antigen *(7)* (*see* Chapter 7). Following this approach, several xenogeneic antibodies have been successfully humanized *(8)*.

The humanized antibodies are likely to be less immunogenic in patients, except that the recipients of such antibodies may still elicit an anti-idiotypic (anti-Id) response against the potentially immunogenic murine CDRs. Several investigators have found that some humanized antibodies remain immunogenic in both subhuman primates *(9–11)* and humans *(12,13)*, and that the humoral response of the host is directed against the variable regions of these MAbs.

From: *Methods in Molecular Biology, Vol. 248: Antibody Engineering: Methods and Protocols*
Edited by: B. K. C. Lo © Humana Press Inc., Totowa, NJ

Attempts have been made to identify the amino acid residues of the antibody variable regions that constitute the idiotopes that are targets of patients' immune response. Contributions of different CDRs and the individual residues to idiotope structures have been evaluated by the mutational approach. Schneider et al. *(11)* found that the predominant antibody response to the humanized anti-Tac MAb was directed against idiotopes composed wholly or in part of the heavy-chain (H) CDR2, HCDR3, and light-chain (L) CDR3. The immunoreactivity of the CDR-substitution mutants of humanized CC49 (HuCC49) to serum from a patient who received murine CC49 showed that the patient's anti-Id responses were directed mostly against LCDR3 and moderately against LCDR1 and HCDR2 of the HuCC49 *(14)*.

The current procedure for the humanization of a murine antibody is based on grafting all the CDRs of a murine antibody onto the human antibody. However, experimental evidence from various laboratories suggests that not all the CDRs of an antibody are equally important, or even essential, for antigen binding. Glaser et al. *(15)* showed that HCDR1, HCDR3, and LCDR3 of the human anti-Tac antibody, HAT, are essential for antigen binding, but the involvement of the other three CDRs in the antigen binding, if it occurs, is of a lesser degree. For an antibody specific for p-azophenylarsonate, it was shown that its antigen-binding affinity does not suffer a loss when its LCDR2 is replaced with polyglycine *(16)*. An evaluation of the antigen-binding affinity of the CDR-replacement variants of the HuCC49 antibody showed that LCDR1 and LCDR2 of the antibody are dispensable, and the other CDRs are essential for the antigen-binding reactivity *(14)*. The reason for this disparity between different CDRs of an antibody for ligand binding is that only 20–33% of CDR residues are involved in the antigen contact *(17)*. Only those CDRs of an antibody that contain the residues that are critical to the complementarity of antigen/antibody surfaces are essential for antigen binding. A comprehensive analysis of the available data of the sequences and the three-dimensional (3D) structures of antibody-combining sites has helped to identify the CDR residues that may be most critical in the antigen-antibody interaction *(18)*. These residues have been designated as specificity-determining residues (SDRs). SDRs, which are usually unique to each antibody, are most commonly located at positions that display high variability. They may be identified either by the determination of the 3D structure of the antigen-antibody complex or by the genetic manipulation of the antibody-combining site.

A new approach to reducing the immunogenicity of the xenogeneic antibodies described here, is based on the rationale that humanization of an antibody can be accomplished by transplanting only the SDRs of a xenogeneic antibody onto the human antibody frameworks. This approach, described earlier for the prototypic humanization of MAb CC49 *(19)*, involves prior identification of the residues of the hypervariable regions that are most critical to the antigen-antibody interaction. Although only clinical trials based on administering the humanized antibody in patients could evaluate the humanized antibody's immunogenicity, its reactivity to sera from patients who had been immunized with the parental antibody, if available, could be evaluated in vitro. Sera reactivity is a reasonable indication of the immunogenic potential of the newly designed antibody.

2. Materials

2.1. Pre-Adsorption of Patients' Sera

1. An antibody that reacts with an epitope of the antigen that is different from the one recognized by the target antibody and of the same isotype.
2. Reacti-gel HW65F (Pierce).
3. Sera from patients who were administered the target MAb.

2.2. Surface Plasmon Resonance (SPR)-Based Competition Assay to Measure Sera Reactivity of Antibodies

1. SPR biosensor (BIAcore X system).
2. Carboxymethylated dextran chips CM5 (BIAcore).
3. Pre-adsorbed patients' sera (from **Subheading 3.2.2.**).
4. Parental (target) and variant MAbs.
5. *N*-ethyl-N′-(3-dimethylaminopropyl)carbodiimide hydrochloride (EDC), from the BIAcore amine coupling kit.
6. *N*-hydroxysuccinimide (NHS), from the BIAcore amine coupling kit.
7. 10 m*M* sodium acetate buffer, pH 5.0.
8. 1 *M* ethanolamine, pH 8.5.
9. Rabbit gamma globulin or bovine serum albumin (BSA).
10. Running buffer: 10 m*M* [2 hydroxyethyl] piperazine N′ [2-ethanesulfonic acid] (HEPES), pH 7.4, 150 m*M* NaCl, 3 m*M* EDTA, and 0.005% Tween 20.
11. 10 m*M* glycine, pH 2.0.

3. Methods

3.1. Antibody Humanization by SDR Grafting

3.1.1. Choosing the Human Template and Identification of Framework Residues Crucial for Antigen Binding

Both the SDR- and CDR-grafting procedures require selection of the most appropriate human frameworks to be used as scaffolds. A more detailed discussion of this topic can be found in Chapter 7; only the relevant features are presented here.

1. Search human immunoglobulin sequences with closest sequence homology with the target antibody (*see* **Note 1**). The nucleotide or protein databases can be

accessed through the web addresses of GenBank *(24)*, EMBL *(25)*, SwissProt *(26)*, and PIR *(27)* (*see* **Note 2**). *See* Chapter 1 for more information on available resources for homology searches.

2. Choose the human framework sequence that shows the highest overall homology or highest framework homology with the target antibody sequence; use the highest homology sequence as the template (*see* **Notes 3** and **4**).
3. Identify crucial framework residues of the target antibody to be retained (*see* **Note 5**). The crucial framework residues are those that may be involved in V_L-V_H contact or interaction with the CDRs. The residues may also be crucial if they are buried and affect the overall structure of the combining site.
4. Using the information obtained from **steps 2** and **3,** and the procedure described in Chapter 7, develop a CDR-grafted humanized antibody. This antibody could be used as a control in characterizing the SDR-grafted humanized antibody. The construct that encodes the CDR-grafted antibody may be needed to carry out mutational analysis of the CDR residues to check whether they are ligand contact residues.

3.1.2. Identification of Potential SDRs Using the PDB Database

1. Using information from the Protein Data Bank (PDB) database *(18)* of known crystal structures of antibody-ligand complexes (*see* **Notes 6** and **7**), determine which residues of the target antibody are involved in ligand contact. These are the potential SDRs (*see* **Note 8**).
2. Identify which CDR residues of the target antibody are different from those of the human template (*see* **Note 9**).
3. Mark those potential SDRs of the target antibody that are different from those present in the template antibody at the corresponding positions.

3.1.3. Designing a Panel of Variants to Validate SDRs

Since the combining site of an antibody is unique, it is advisable to experimentally validate whether the CDR residues of the target antibody that are different from their counterparts in the template antibody and are designated as potential SDRs based on the database search are indeed SDRs. This validation is all the more important for those potential SDRs that are found to have a low frequency of ligand contact in the database. The most reliable experimental validation is based on changing, by mutation, CDR residues of the target antibody to their human counterparts and testing the mutant antibodies for their antigen-binding reactivity (*see* **Notes 10** and **11**). To this end, use the constructs that encode the CDR-grafted humanized V_L and V_H regions and carry out mutational analysis as follows:

1. Using primer-induced mutagenesis, generate a panel of genes that encode variants of each of the CDR-grafted V_L and V_H genes. In each of the variants, one or more of the murine CDR residues to be validated as an SDR is replaced with its human counterpart (*see* **Note 12**).

2. Develop appropriate expression constructs of the genes that encode the variant antibodies, and introduce them into mammalian or insect cells.
3. Test the antigen-binding activity of the panel of variant antibodies by enzyme-linked immunosorbent assay (ELISA) and/or competition radioimmunoassay.

For further details about generating the residue substitution variants, *see* Chapters 6 and 18.

3.1.4. Generation and Characterization of SDR-Grafted Humanized Antibody

Once the appropriate human antibody frameworks to be used as scaffolds are chosen, the crucial framework residues of the target antibody are identified, and the CDR residues that are involved in ligand contact using database and experimental validation are identified, proceed to design and generate the final variant of the SDR-grafted humanized antibody as follows:

1. Design the amino acid sequence of the V_L and V_H regions of the SDR-grafted humanized antibody, incorporating:
   a. framework sequence of the human template antibody, while retaining the framework residues of the target antibody that are deemed crucial for antigen binding,
   b. all CDR residues that are identical in the target and the template antibodies,
   c. those CDR residues of the template antibody that are located at the positions that are designated as non-SDRs, and
   d. any CDR residues of the target antibody that are identified as SDRs and are different from their counterparts in the template antibody.
2. Determine nucleotide sequence from the designed amino acid sequence of each of the V_L and V_H regions.
3. Refine the nucleotide sequence to provide high-frequency usage of codons for the efficient expression of the genes in the cell system of choice.
4. Eliminate any self-annealing regions and any internal restriction endonuclease sites that may become troublesome for cloning the gene in the desired vector, using software programs such as mFOLD and MAPSORT (GCG Wisconsin Package; Accelrys, Inc.)
5. Design alternating long oligonucleotides (Watson and Crick) that encompass the entire sequence of the V_L or V_H gene (*see* **Note 13**). The alternating oligonucleotides should have overlapping flanks (~ 17–21 bp long) to facilitate their annealing (*see* Chapter 7 for gene-synthesis protocol).
6. Anneal the oligonucleotides together, fill in the gaps using *Taq* DNA polymerase, and PCR amplify the entire sequence using appropriate end primers. The end primers for PCR amplification should carry, at their flanks, the appropriate restriction endonuclease recognition sites to facilitate cloning of the construct into a vector.
7. Clone the PCR product into a cloning vector, and sequence the insert to check the fidelity of the PCR product to the sequence of the designed gene.

8. Insert the PCR-amplified DNA encoding the V_L or V_H region in the expression cassette carrying the sequences encoding the respective constant region and the appropriate leader peptides.
9. Introduce the expression constructs into either mammalian or insect-cell lines. Screen the transfectants for the secretion of the antigen-specific antibody by ELISA, and purify the antibody from the highest-producing clone.
10. Evaluate the antigen-binding affinity of the expressed antibodies by competition radioimmunoassay or using commercially available biosensors.

Protocols for the expression of recombinant antibodies (Chapters 14–17), their purification (Chapter 22) and affinity measurements (Chapters 23–25) are detailed elsewhere in this volume.

3.2. In Vitro Sera Reactivity

The most important characterization of the SDR-grafted humanized antibody is to evaluate its potential immunogenicity. In vitro reactivity of the humanized antibody to the sera from patients injected with the parental antibody is as reasonable a measure of its immunogenicity as one could get without administering it in patients. Sera reactivity of a variant antibody is measured by its ability to compete with the parental antibody for binding to the anti-variable region antibodies that are present in a patient's serum in response to the administered antibody. To that end, use the SPR-based competition assay ***(35)*** (*see* **Notes 14** and **15**) described here, using a commercially available Biacore X system.

3.2.1. Immobilization of Proteins on Sensor Chips

Activate the dextran layer of the CM5 sensor chips by injecting 35 µL of a mixture of EDC and NHS, at a flow rate of 5 µL/min.

1. Inject the proteins, diluted in 10 m*M* sodium acetate buffer, pH 5.0 at a concentration of 100 µg/mL, into the flow cells until surfaces of 5,000 resonance units (RUs) are obtained. Either the murine antibody or the CDR-grafted humanized antibody is injected in flow cell 1, while a reference protein (e.g., rabbit gamma globulin or BSA) is injected in flow cell 2.
2. Block the remaining reactive groups on the surfaces by injecting 35 µL of 1 *M* ethanolamine, pH 8.5.

3.2.2. Competition Assay

1. Inject 85 µL of serial dilutions of a patient's serum in running buffer (10 m*M* HEPES, pH 7.4, 150 m*M* NaCl, 3 m*M* EDTA, and 0.005% Tween 20) (*see* **Note 16**) to determine the appropriate dilutions of the serum to be used for the competition experiments.
2. For each dilution, let the mobile sample flow across both cells at 5 µL/min and measure the binding at 25°C until the total volume is applied (*see* **Note 17**). The

difference between the binding signals generated by the two surfaces corresponds to the specific binding of the serum anti-variable region antibodies to the parental antibody immobilized on the sensor chip.

3. Regenerate the surfaces using a 1 min injection of 10 m*M* glycine, pH 2.0, before injecting the next serum dilution. Repeat **steps 2** and **3** until the binding of all samples has been measured.
4. Plot the response differences between flow cells 1 and 2 as a function of time for each dilution and use the serum dilution that generates a response difference of optimum resonance (250–400 RU) for the competition experiment.
5. By repeating **steps 1–3,** described previously, measure the binding of serum anti-variable region antibodies to the immobilized parental antibody (e.g., protein **A**) using the appropriate dilution of serum prior to, and after equilibration with different concentrations of the competitor antibodies (e.g., proteins **A** and **B**) as the mobile reactant. The binding of a competitor to the serum anti-variable region antibodies results in a decreased binding signal compared to the signal generated by the binding of serum alone (without competitor) to the immobilized parental antibody (*see* **Note 18**).

3.2.3. Analysis of the Sera Reactivity Data

1. Determine the slopes of the response difference curves of the competition of protein **A** and protein **B** for binding of serum anti-variable region antibodies to immobilized protein **A;** the slope is measured for a defined and narrow range of time—e.g., between 800 and 1,000 s.
2. Calculate the percentage of binding at a certain concentration of the competitor as % binding = [(slope of the signal obtained at a certain concentration of competitor) / (slope of the signal obtained without competitor)] × 100. **Fig. 1** shows hypothetical binding sensorgrams of the anti-variable region antibodies to the immobilized parental antibody using serum alone or after its equilibration with increasing concentrations of the competitor proteins **A** and **B.**
3. Plot the percent binding vs the competitor concentration to yield the competition profile of the antibody. **Fig. 2** shows the competition profiles of **A** and **B** from the sensorgrams shown in **Fig. 1.**
4. Calculate the IC_{50} for each antibody, the concentration required for 50% inhibition of the binding of the serum to the immoblized parental antibody. In the example given here, the IC_{50} values calculated from the curves for competitors **A** and **B** are 8 nm and 23 nm, respectively. This corresponds to a threefold lower serum reactivity for **B** relative to **A.**

4. Notes

1. Another approach that has been used for the selection of human frameworks is based on the strategy that relies on the known X-ray structures of templates *(20,21)*. This involves searching the Protein Data Bank (PDB) database to identify the human antibody structure that shows the closest overall identity to the antibody to be humanized *(22,23)*.

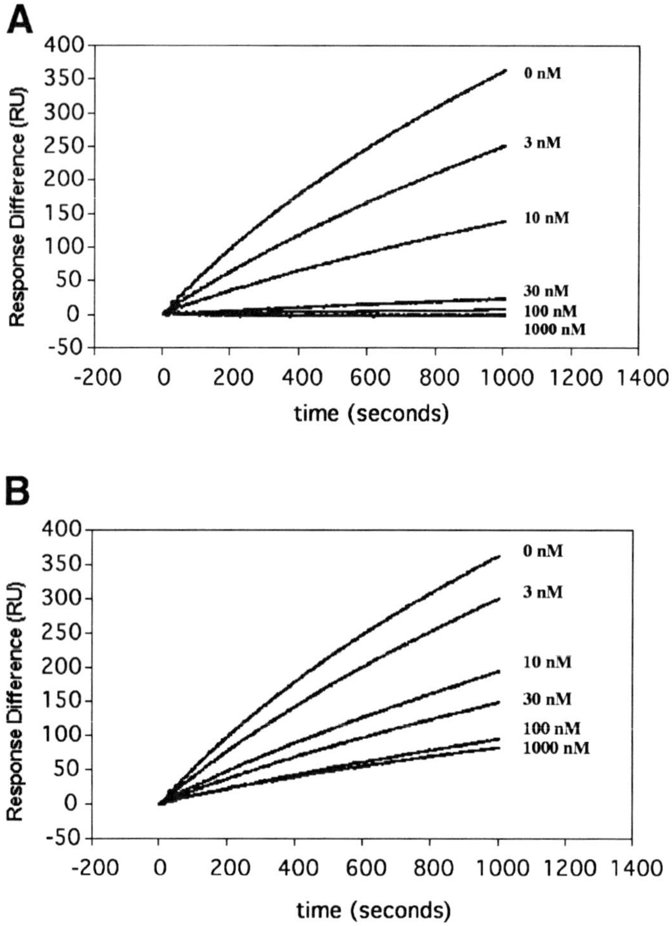

Fig. 1. Difference sensorgrams showing the competition of protein **A** (Panel A) and protein **B** (Panel B) for binding of the serum to immobilized parental antibody. Different concentrations of the competitors (as indicated) are equilibrated with the serum before they are injected into the sensor chip to generate the binding signals.

2. The nucleotide and protein databases can be accessed through their web addresses: GenBank via the NCBI homepage, http://www.ncbi.nlm.nih.gov ***(24)***; EMBL via http://www.ebi.ac.uk/embl/ ***(25)***; SWISS-PROT via http://www.expasy.ch/sprot/; and http://www.ebi.ac.uk/swissprot/ ***(26)***; and PIR via http://pir.georgetown.edu/ ***(27)***.
3. It has been suggested that restricting the choice of the human templates to the germline sequences precludes the potential immunogenicity of the residues that are generated as a result of somatic mutations ***(28,29)***.
4. Frameworks of two different human antibodies could be used as templates for the humanization of the V_L and V_H domains of the target antibody ***(9)***.

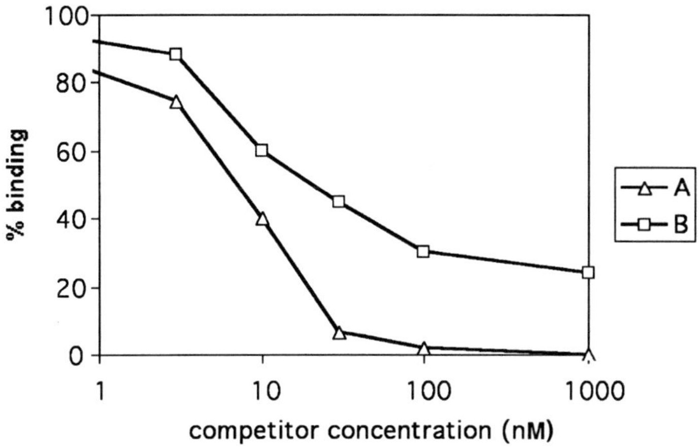

Fig. 2. Competition profiles of proteins **A** *(open triangle)* and **B** *(open square)* for binding of patient's serum to immobilized parental antibody. Increasing concentrations of the competitors A and B are used to compete with the patient's serum for binding to the parental antibody immobilized on the sensor chip. The percent binding at a certain concentration of the competitor was calculated relative to that of serum alone from the sensorgrams shown in *Fig. 1,* and plotted as a function of competitor concentration.

5. Crucial framework residues may be identified if a 3D structure of the antigen-binding region, based on X-ray crystallography, is available. Alternatively, molecular modeling or experimental strategies, such as studying the effect of site-specific mutations ***(30,31)*** on ligand-binding activity, may be employed. An examination of the known structures of the antigen-antibody complexes of the closely related molecules may also help identify the crucial framework residues of the target antibody ***(17)*** (*see* Chapter 7).
6. When the 3D structure of the antigen-antibody complex, based on X-ray crystallographic studies, is known, the residues of the combining site that are directly involved in ligand contact can be identified directly without database search.
7. **Tables 1** and **2** are based on the updated search of the database showing the V_L and V_H residues that are in contact with the ligand in murine antibody-ligand complexes of known 3D structures.
8. Unfortunately, the database does not always help in reaching a definite conclusion to designate some CDR residues of the target antibody as ligand-contact residues. There is no cut-off number (in terms of the frequency of ligand contact) to use in making this judgment. Even when the high frequency of ligand contact in the database is compelling evidence to designate a certain residue as an SDR, the choice is not always infallible.
9. **Table 3** illustrates, using MAb CC49 as an example, the CDR residues that are denoted as potential SDRs, and the CDR residues of the target antibody that are different from those of the templates at the corresponding positions. Residue positions

Table 1
V_L Residues That Are in Contact with the Ligand in Murine Antibody-Ligand Complexes of Known 3D Structures

		CDR1		CDR2		CDR3		
	10 20	27abcdef 30	40	50	60 70 80	90 95ab	100	PDB Code
D1.3	....	....------....Y..	..............Y	Y..T...	....	..fWs..--..	....	1FDL
HyHEL-5	....	....------...NY.Y	....	D......	....	..WgR.---P.	....	2HFL
HyHEL-10	....	....------..gNN..	....	Y..Q...	....	..SN.W.--Y.	....	3HFM
D44.1	....	....------....N..	....	Y......	....	..sNsW.--R.	....	1MLC_1
D11.15	....	....------..S....	....	....	....	...NEY.--W.	....	1JHL
F9.13.7	....	....------....Y..	....	....	....	..gYTL.--..	....	1FBI_1
NC41	....	....------.......	..............Y	W.ST.HI	....	..hYSP.--W.	....	1NCA
NC10	....	...Q------..S.Y..	....	....	....	..dFTL.--..	....	1NMB
N10	....	.......T--SSF....	..............K	Y......	....	...WEI.--Y.	....	1NSN
Jel42	....	.......Hg-N...Y..	....	....	....	.......--Y.	....	1JEL
B13I2	....	.........-D.D.Y..	....	....	....	..g..V.--P.	....	2IGF
17/9	....	..............Y..	....	....	....	..DysN.--L.	....	1IFH
26/9	....	..............F..	....	....	....	..DYSH.--L.	....	1FRG
50.1	....	.......---......H	.Y.........L..Y	.....I.	....	Q.S..D.--L.	....	1GGI_1
59.1	....	......dS..Y...F..	....	....	....	..NNED.--..	....	1ACY
TE33	....	.......H.-sgN.Y..	....	K......	....	..gs...--F.	....	1TET
Se155-4	........-..............	.......---....H..	....	....	....	..W..N.--W.	....	1MFA
BV04-01	....	.......HS-....Y.H	....	....	....	..StHV.--..	....	1CBV
DB3	....	.......H.-.......	....	....	....	..s.hVP--..	....	1DBB
26-10	....	.........-.......	....	....	....	..T....--P.	....	1IGJ_1
4-4-20	....	.......H.-....Y.R	....	....	....	..S....--W.	....	4FAB
McPC603	....	....	....	....	....	..DH.Y.--L.	....	2MCP
AN02	....	....-------...Y.Y	....	D......	....	Q.W.....-I.	....	1BAF
CHA255	........-..............	.......---....Y..	....	....	....	..W....--W.	....	1IND
1F7	....	.....-----.......	....	....	....	.....Y.--..	....	1FIG
NC6.8	....	.......H.-N...Y.H	....	....	....	..g....--Y.	....	2CGR
40-50	....	.......T--S......	....	....	....	..Sr.Y.--L.	....	1IBG
C3	....	.......H.-N.K.Y..	....	....	....	..s....--Y.	....	1FPT
R45-45-11	....	....------....Y..	....	Y......	....	..gsRI.--..	....	1IKF
Jel103	....	.........-N.N.Y..	....	K......	....	..S....--..	....	1MRD
CNJ206	....	....------.......	....	....	....	L.Y....--Y.	....	1KNO_1

The amino acid residues which contact the ligand are identified by the one-letter code. Capital letters are used when the contact involves the side chains whereas small letters are used when only main chain atoms are involved in the interaction. Dots represent residues that are not in contact with the ligand and dashes indicate missing residues at the indicated positions.

Table 2
V_H Residues That Are in Contact with the Ligand in Murine Antibody-Ligand Complexes of Known 3D Structures

		CDR1		CDR2		CDR3		
	10 20 30	35ab	40	52abc 60	70 82abc 90	100abcdefghi	110	PDB Code
D1.3	....	gY...--	....	..W---gD...........	... .	..DYR.---------..	....	1FDL
HyHEL-5	....	..W.E--	...........W..	E.L.--.SgStN.......	... .	gnY..----------..	....	2HFL
HyHEL-10	.............................T	SDY..--	....	Y.S---YS.S.Y.......	... .	W..------------..	....	3HFM
D44.1	.............................s	TYW.E--	....	E.L.--.SgS.Y.......	... .	gdg..----------..	....	1MLC_1
D11.15	....	S.W..--	....	..Y.--.D.Y.........	... .	D.NY...--------..	....	1JHL
F9.13.7	.............................t	S.W..--	....	E.D.--SD.Y.N.......	... .	....gTS...-----..	....	1FBI_1
NC41	.............................t	NY...--	....	..N.--N............	... .	.EDNF.SL.------..	....	1NCA
NC10	....	.....--	....	..Y.--gN.DtS.......	... .	....Y.YD...----..	....	1NMB
N10	..........................Y...	S.....-	....	Y.T---YS.T.........	... .	.N-------------..	....	1NSN
Jel42	.............................T	TyA..--	....	L.SP--SS.Y.........	... .	.MgE.Y.--------..	....	1JEL
B13I2	....	r.A..--	....	.iSS--g.SY.F.......	... .	Y...PF..-------..	....	2IGF
17/9	....	.....--	....	T.SN--g.gY.Y.......	... .	ReR..E.g.------..	....	1IFH
26/9	....	.....--	....	T.SN--gggY.Y.......	... .	R.R..E.g.------..	....	1FRG
50.1	....	....	...........W..	H.F---WD.D.R.......	... .	EgY------------I.	....	1GGI_1
59.1	....	..n.C..	....	R.C---YE.S.........	... .	..HM..T..------..	....	1ACY
TE33	.............................t	tYg..--	....	W.NT--Y............	... .	RSW..----------..	....	1TET
Se155-4	....	..W.H--	....	A...--....F........	... .	.gHg...--------..	....	1MFA
BV04-01	....	tnA..--	....	R.RS..N............	... .	DqTgtaW.-------..	....	1CBV
DB3	....	..g.N--	....	W...--.............	... .	gdY..W.F-------..	....	1DBB
26-10	....	..Y.N--	...........Y..	Y...--.............	... .	S....WaM-------..	....	1IGJ_1
4-4-20	....	..W..--	....	......Y............	... .	SYY..----------..	....	4FAB
McPC603	....	..Y..--	....	..R................	... .	N.....W..------..	....	2MCP
AN02	....	......-	....	...--..............	... .	.WP.-----------..	....	1BAF
CHA255	....	..T..--	....	T.L.--...F.F.......	... .	HR.------------..	....	1IND
1F7	....	.hN.N--	....	N...--.............	... .r	R.d..Y.F-------..	....	1FIG
NC6.8	....	..W.E--	....	E...--...R.N.......	... .	.YSsM----------..	....	2CGR
40-50	....	....H--	....	L.W---.............	... .	F.F..YY.Y------V.	....	1IBG
C3	....	..L..--	....	V.N.--.S.g.D.......	... .	DfYDYD...------..	....	1FPT
R45-45-11	....	d.Y.Y--	....	F..N--.............	... .	H.L..T.ygNYP.W...	....	1IKF
Jel103	....	.....--	....	...--..............	... .	.R...----------..	....	1MRD
CNJ206	....	....H--	.V.........W..	...--..............	... .	gdYY...--------..	W..........	1KNO_1

The amino acid residues which contact the ligand are identified by the one-letter code. Capital letters are used when the contact involves the side chains whereas small letters are used when only main chain atoms are involved in the interaction. Dots represent residues that are not in contact with the ligand and dashes indicate missing residues at the indicated positions.

Table 3
CDR Sequences of Murine MAb CC49 and Human Template MAbs LEN and 21/28′CL[a]

Light chain																	
CDR1	24	25	26	27	a	b	c	**d**[b]	**e**	**f**	**28**	**29**	**30**	**31**	**32**	**33**	**34**
CC49	Lys	Ser	Ser	Gln	Ser	Leu	Leu	Tyr	Ser	Gly	Asn	Gln	Lys	Asn	Tyr	Leu	Ala
LEN	Lys	Ser	Ser	Gln	Ser	**Val**[c]	Leu	Tyr	Ser	**Ser**	Asn	**Ser**	Lys	Asn	Tyr	Leu	Ala
CDR2	**50**	**51**	**52**	**53**	**54**	**55**	56										
CC49	Trp	Ala	Ser	Ala	Arg	Glu	Ser										
LEN	Trp	Ala	Ser	**Thr**	Arg	Glu	Ser										
CDR3	**89**	**90**	**91**	**92**	**93**	**94**	**95**	**96**	97								
CC49	Gln	Gln	Tyr	Tyr	Ser	Tyr	Pro	Leu	Thr								
LEN	Gln	Gln	Tyr	Tyr	Ser	**Thr**	Pro	**Tyr**	**Ser**								
Heavy chain																	
CDR1	**31**	**32**	**33**	**34**	**35**												
CC49	Asp	His	Ala	Ile	His												
21/28'CL	**Ser**	**Tyr**	Ala	**Met**	His												
CDR2	**50**	**51**	**52**	**a**	**53**	**54**	**55**	**56**	**57**	**58**	59	60	61	62	63	64	65
CC49	Tyr	Phe	Ser	Pro	Gly	Asn	Asp	Asp	Phe	Lys	Tyr	Asn	Glu	Arg	Phe	Lys	Gly
21/28'CL	**Trp**	**Ile**	**Asn**	**Ala**	Gly	Asn	**Gly**	**Asn**	**Thr**	Lys	Tyr	**Ser**	**Gln**	**Lys**	Phe	**Gln**	Gly
CDR3	**95**	**96**	**97**	**98**	**99**	**100**	**a**	**b**	**101**	102							
CC49	Ser	Leu	Asn	Met	Ala	-	-	-	-	Tyr							
21/28'CL	**Gly**	**Gly**	**Tyr**	**Tyr**	**Gly**	**Ser**	**Gly**	**Ser**	**Asn**	Tyr							

[a] Amino acid residues are numbered according to Kabat et al. *(39)*.
[b] Residue positions shown in bold denote the specificity-determining residues (SDRs) ***(18)***.
[c] Residues of human MAbs LEN and 21/28′CL shown in bold indicate differences among MAb CC49 residues at the corresponding positions ***(19)***.

shown in bold denote the SDRs identified by the PDB database search ***(18)*** of known crystal structures of antibody ligand complexes. Residues of the human antibodies MAb LEN and 21/28′CL shown in bold indicate those template antibody residues that are different from the residues of the target antibody at the corresponding positions.

10. A conservative approach to developing an SDR-grafted humanized antibody would entail replacing, with the corresponding human residues, only those CDR residues that are found not to be interacting with the ligand in any of the complexes in the database. This approach does not require experimental validation of the SDRs.
11. Another conservative approach to humanize an antibody, which precludes experimental validation of SDRs, is to graft "abbreviated CDRs," the cluster of CDR residues that contain all the SDRs, of the target antibody onto the frameworks of the human template ***(18,32)***.
12. Primer-induced mutagenesis may be carried out by a two-step PCR method ***(33)***. For the first step of the PCR, a mutagenic primer is used as a 3′ primer, and a 20-nucleotide long primer is used as a 5′ primer. The product of the first PCR is gel-purified and used as a 5′ primer for the second PCR in which an appropriate primer was used as a 3′ primer (*see* Chapter 18 for protocol).

13. We have routinely used 121- to 126-bp-long oligonucleotides.
14. An HPLC-based competition assay ***(19)*** may also be used to compare the sera reactivity of a variant to its parental antibody. Essentially, the assay measures the ability of the variant to inhibit complex formation between the radiolabeled parental antibody and anti-variable region antibodies in the patient's serum, preadsorbed to remove any circulating antigen and anti-Fc antibodies. Complex formation is determined by an alteration of the retention time of the radiolabeled parental antibody after it has been incubated with the patient's serum before subjecting it to size-exclusion HPLC. When a cold competitor (parental or variant antibody) is also included with the serum and radiolabeled parental antibody in the incubation mixture, the complex formation is not inhibited, partially inhibited, or completely inhibited depending on the degree of reactivity of the competitor to the anti-variable region antibodies in the patient's serum.
15. Sera reactivity of a variant to its parental antibody may be further compared by a competition assay dubbed as double-antigen (DAB) assay ***(34)***. To carry out the assay, the plates are coated with the parental antibody to be used as the solid-phase catcher antibody. Serial dilutions of the patient's serum containing anti-variable region antibodies are added to the wells of the plate. After a 24-h incubation, radiolabeled variant or radiolabeled parental antibody is added. After proper incubation and washing, the bound radioactivity is determined. A comparison of the bound radiolabeled variant to that of the bound radiolabeled parental antibody provides a comparison of their sera reactivity.
16. Sera from patients who received the parental antibody in a clinical trial contain anti-Fc antibodies and often circulating antigen, which might interfere with the binding of the parental antibody and its variants to the anti-variable region antibodies present in sera. To overcome this difficulty, the circulating antigen and the anti-Fc antibodies should be removed from the sera by immunoadsorption prior to checking the sera reactivity of the parental antibody and its variants. The procedure, which has been previously described in detail ***(14,19)***, involves pre-adsorption of serum using an antibody that reacts with an epitope of the antigen that is distinct from the one recognized by the parental antibody and matches its isotype. This antibody is coupled to Reacti-gel HW65F ***(36)*** and the serum is added to an equivalent volume of the antibody-coupled gel (wet-packed volume). After overnight incubation at 4°C with end-over-end rotation, the samples are centrifuged at 1,000*g* for 5 min, and the supernatant is saved and stored at –20°C.
17. When using small sample volumes, one could take advantage of a recently developed sample application technique ***(37)*** in which the microfluidics control of the instrument is replaced by an externally installed computer-controlled syringe pump with stepping motor (model 402 from Gilson, Inc.). A tubing is inserted into the open port of the connector block in order to serve as an inlet port, through which the sample can be aspirated; the port previously designated as running buffer inlet is now connected to the syringe pump ***(37,38)***. Using this new injection technique, only 5–20 µL of sample is required ***(35)***. The microfluidics system is rinsed and filled with running buffer. This is followed by sequential aspiration of 2 µL of air,

0.3 μL of sample (serum ± antibody as competitors), 2 μL of air, 5–6 μL of sample, 2 μL of air, 0.3 μL of sample, and 2 μL of air into the inlet tubing at a rate of 20 μL/min. The sample is centered across the sensor surfaces, and an oscillatory flow is applied at a rate of 20 μL/min. This flow ensures efficient mass transfer of the sample to the surface and allows for a very long contact time without net displacement of the sample. After measuring the binding for 1,000 s, the unbound samples are removed from the surfaces by washing with running buffer using a flow rate of 100 μL/min, and the surfaces are regenerated.

18. As a control, human IgG may be used as a competitor, which should have no effect on the binding of sera anti-variable region antibodies to the immobilized parental antibody.

Acknowledgments

We thank Dr. Eduardo A. Padlan of NIDDK, NIH for providing **Tables 1** and **2.** We also thank Debra Weingarten for her editorial assistance.

References

1. Seccamani, E., Tattanelli, M., Mariani, M., Spranzi, E., Scassellati, G. A., and Siccardi, A. G. (1989) A simple quantitative determination of human antibodies to murine immunoglobulins (HAMA) in serum samples. *Nucl. Med. Biol.* **16,** 167–170.
2. Reynolds, J. C., DelVecchio, S., Sakahara, H., and Lora, M. (1989) Anti-murine response to mouse monoclonal antibodies. *Nucl. Med. Biol.* **16,** 121–125.
3. Colcher, D., Milenic, D. E., Ferroni, P., Carrasquillo, J. A., Reynolds, J. C., Roselli, M., et al. (1990) *In vivo* fate of monoclonal antibody B72.3 in patients with colorectal cancer. *J. Nucl. Med.* **31,** 1133–1142.
4. Blanco, I., Kawatsu, R., Harrison., K., Leichner, P., Augustine, S., Baranowska-Kortylewicz, J., Tempero, M., and Colcher, D. (1997) Antiidiotypic response against murine monoclonal antibodies reactive with tumor-associated antigen TAG-72. *J. Clin. Immunol.* **17,** 96–106.
5. Morrison, S. L. and Schlom, J. (1990) Recombinant chimeric monoclonal antibodies in *Important Advances in Oncology* (Rosenberg, S. A., ed.), J. B. Lippincott, Philadelphia, PA, pp. 3–18.
6. Padlan, E. A. (1991) A possible procedure for reducing the immunogenicity of antibody variable domains while preserving their ligand-binding properties. *Mol. Immunol.* **28,** 489–498.
7. Winter, G. and Harris, W. J. (1993) Humanized antibodies. *Immunol. Today.* **14,** 243–246.
8. Carter, P. (2001) Improving the efficacy of antibody-based cancer therapies. *Nature Rev./Cancer* **1,** 118–129.
9. Singer, I. I., Kawka, D. W., DeMartino, J. A., Daugherty, B. L., Elliston, K. O., Alves, K., et al. (1993) Optimal humanization of 1B4, an anti-CD18 murine monoclonal antibody, is achieved by correct choice of human V-region framework sequences. *J. Immunol.* **150,** 2844–2857.

10. Hakimi, K., Chizzonite, R., Luke, D. R., Familletti, P. C., Bailon, P., Kondas, J. A., et al. (1991) Reduced immunogenicity and improved phamacokinetics of humanized anti-Tac in cynomolgus monkeys. *J. Immunol.* **147,** 1352–1359.
11. Schneider, W. P., Glaser, S. M., Kondas, J. A., and Hakimi, J. (1993) The anti-idiotypic response by cynomolgus monkeys to humanized anti-Tac is primarily directed to complementarity-determining regions H1, H2, and L3. *J. Immunol.* **150,** 3086–3090.
12. Stephens, S., Emitage, S., Vetterlein, O., Chaplin, L., Bebbinton, C., Nesbitt, A., et al. (1995) Comprehensive pharmacokinetics of a humanized antibody and analysis of residual anti-idiotypic responses. *Immunology* **85,** 668–674.
13. Sharkey, R. M., Juweid, M., Shevitz, J., Behr, T., Dunn, R., Swayne, L. C., et al. (1995) Evaluation of a complementarity-determining region-grafted (humanized) anti-carcinoembryonic antigen monoclonal antibody in preclicinal and clinical studies. *Cancer Res.* **55,** 5935s–5945s.
14. Iwahashi, M., Milenic, D. E., Padlan, E. A., Bei, R., Schlom, J., and Kashmiri, S.V.S. (1999) CDR substitutions of a humanized monoclonal antibody (CC49): Contributions of individual CDRs to antigen binding and immunogenicity. *Mol. Immunol.* **36,** 1079–1091.
15. Glaser, S. M., Vasquez, M., Payne, P. W., and Schneider, P. W. (1992) Dissection of the combining site in a humanized anti-Tac antibody. *J. Immunol.* **149,** 2607–2614.
16. Sompuram, S. R., Den, W., and Sharon, J. (1996) Analysis of antigen binding and idiotypic expression by antibodies with polyglycine-replaced complementarity-determining regions. *J. Immunol.* **156,** 1071–1081.
17. Padlan, E. A. (1994) Anatomy of an antibody molecule. *Mol. Immunol.* **31,** 169–217.
18. Padlan, E. A., Abergel, C., and Tipper, J. P. (1995) Identification of specificity-determining residues in antibodies. *FASEB J.* **9,** 133–139.
19. Tamura, M., Milenic, D. E., Iwahashi, M., Padlan, E. A., Schlom, J., and Kashmiri, S.V.S. (2000) Structural correlates of an anticarcinoma antibody: Identification of specificity-determining residues (SDRs) and development of a minimally immunogenic antibody variant by retention of SDRs only. *J. Immunol.* **164,** 1432–1441.
20. Jones, P. T., Dear, P. H., Foote, J., Neuberger, M. S., and Winter, G. (1986) Replacing the complementarity-determining regions in a human antibody with those from a mouse. *Nature* **321,** 522–525.
21. Riechman, L., Clark, M., Waldmann, H., and Winter, G. (1988) Reshaping human antibodies for therapy. *Nature* **332,** 323–327.
22. Bernstein, F. C., Koetzle, T. F., Williams, G. J., Meyer, E. F. Jr., Brice, M. D., Rodgers, J. R., et al. (1977) The Protein Data Bank: a computer-based archival file for macromolecular structures. *J. Mol. Biol.* **112,** 535–542.
23. Abola, E. E., Sussman, J. L., Prilusky, J., and Manning, N. O. (1997) Protein Data Bank archives of three-dimensional macromolecular structures. *Methods Enzymol.* **277,** 556–571.
24. Benson, D. A., Karsch-Mizrachi, I., Lipman, D. J., Ostell, J., Rapp, B. A., and Wheeler, D. L. (2002) GenBank. *Nucleic Acids Res.* **30,** 17–20.

25. Stoesser, G., Baker, W., van den Broek, A., Camon, E., Garcia-Pastor, M., Kanz, C., et al. (2002) The EMBL Nucleotide Sequence Database. *Nucleic Acids Res.* **30,** 21–26.
26. Bairoch, A. and Apweiler, R. (2000) The SWISS-PROT protein sequence database and its supplement TrEMBL in 2000. *Nucleic Acids Res.* **28,** 45–48.
27. Wu, C. H., Huang, H., Arminski, L., Castro-Alvear, J., Chen, Y., Hu, Z. Z., et al. (2002) The Protein Information Resource: an integrated public resource of functional annotation of proteins. *Nucleic Acids Res.* **30,** 35–37.
28. Rosok, M. J., Yelton, D. E., Harris, L. J., Bajorath, J., Hellstrom, K. E., Hellstrom, I., et al. (1996) A combinatorial library strategy for the rapid humanization of anticarcinoma BR96 Fab. *J. Biol. Chem.* **271,** 22,611–22,618.
29. Caldas, C., Coelho, V. P., Rigden, D. J., Neschich, G., Moro, A. M., and Brigido, M. M. (2000) Design and synthesis of germline-based hemi-humanized single-chain Fv against the CD18 surface antigen. *Protein Eng.* **13,** 353–360.
30. Queen, C., Schneider, W. P., Selick, H. E., Payne, P. W., Landolfi, N. F., Duncan, J. F., et al. (1989) A humanized antibody that binds to the interleukin 2 receptor. *Proc. Natl. Acad. USA* **86,** 10,029–10,033.
31. Tempest, P. R., White, P., Buttle, M., Carr, F. J., and Harris, W. J. (1995) Identification of framework residues required to restore antigen binding during reshaping of a monoclonal antibody against the glycoprotein gB of human cytomegalovirus. *Int. J. Biol. Macromol.* **17,** 37–42.
32. De Pascalis, R., Makoto, I., Tamura, M., Padlan, E. A., Gonzales, N. R., Santos, A. D., et al. (2002) Grafting of "abbreviated" CDRs containing specificity determining residues (SDRs) essential for ligand contact to engineer a less immunogenic humanized mAb. *J. Immunol.* **169,** 3076–3084.
33. Landt, O., Grunert, H.-P., and Hahn, U. (1990) A general method for rapid site-directed mutagenesis using the polymerase chain reaction. *Gene* **96,** 125–128.
34. Pavlinkova, G., Colcher, D., Booth, B.J.M., Goel, A., and Batra, S. K. (2001) Effects of humanization and gene shuffling on immunogenicity and antigen binding of anti-TAG-72 single-chain Fvs. *Int. J. Cancer* **94,** 717–726.
35. Gonzales, N. R., Schuck, P., Schlom, J., and Kashmiri, S.V.S. (2002) Surface plasmon resonance-based competition assay to assess the sera reactivity of variants of humanized antibodies. *J. Immunol. Methods* **268,** 197–210.
36. Hearn, M. T., Bethell, G. S., Ayers, J. S., and Hancock, W. S. (1979) Application of 1,1′-carbonyldiimidazole-activated agarose for the purification of proteins. II. The use of an activated matrix devoid of additional charged groups for the purification of thyroid proteins. *J. Chromatogr.* **185,** 463–470.
37. Abrantes, M., Magone, M. T., Boyd, L. F., and Schuck, P. (2001) Adaptation of a surface plasmon resonance biosensor with microfluidics for use with small sample volumes and long contact times. *Anal. Chem.* **73,** 2828–2835.
38. Gilligan, J. J., Schuck, P., and Yergey, A. L. (2002) Mass spectrometry after capture and small volume elution of analyte from a surface plasmon resonance biosensor surface. *Anal. Chem.* **74,** 2041–2047.
39. Kabat, E. A., Wu, T. T., Perry, H. M., Gottesman, K. S., and Foeller, C. (eds.) (1991) *Sequences of Proteins of Immunological Interests,* 5th ed., US Department of Health and Human Service, National Institute of Health, Bethesda, MD (NIH Publication No. 91-3242).

IV

ANTIBODY CHARACTERIZATION AND NOVEL APPLICATIONS

22

Antibody Purification by Column Chromatography

**Andrea Murray, C. Rosamund L. Graves,
Kevin Brady, and Benny K. C. Lo**

1. Introduction

The provision of an antibody or antibody fragment for use as an immunological reagent necessitates—at the very least—partial, but preferably total, purification of that antibody or antibody fragment from its source fluid. Numerous methods exist for this process, all designed to separate contaminating proteins from the antibody of interest. The majority of these methods utilize column chromatography of one form or another (as expanded upon here) and therefore require some form of preliminary step in order to eliminate particulate matter (such as cell debris) from the feed-stock. This is often achieved by centrifugation, ultrafiltration, or ammonium sulfate precipitation.

1.1. Gel Filtration

With regard to the purification of antibodies, gel filtration is most effective as a simple but effective desalting step, and as such is often used after ion exchange, hydrophobic interaction, or affinity chromatography to reconstitute the antibody in the storage buffer of choice. This relatively crude purification modality relies on size-dependent separation of molecules, and is achieved by passing the solution to be resolved through a column packed with an essentially inert chromatographic matrix. In essence, the smaller the molecule, the further into the matrix it is able to diffuse, and the more convoluted its passage through the column. Therefore, large molecules elute from the column first, followed in ever-decreasing size, by smaller molecules. Gels manufactured for this purpose come in a range of pore sizes, allowing separation of molecules over several ranges of molecular size.

1.2. Hydrophobic Interaction

Hydrophobic-interaction chromatography can be used directly after ammonium sulfate precipitation. Separation is determined by the relative hydrophobic-

From: *Methods in Molecular Biology, Vol. 248: Antibody Engineering: Methods and Protocols*
Edited by: B. K. C. Lo © Humana Press Inc., Totowa, NJ

ities of the proteins that are present in the solution. The chromatographic matrix used has a hydrophobic surface that is capable of interacting reversibly with proteins, depending on the particular environment induced by the chromatographic buffer. Samples are applied in a buffer of high ionic strength, and eluted using a stepwise or continuous gradient of any one—or a combination of—decreasing salt concentration, pH, chaotropic agents, or decreasing polarity.

1.3. Ion Exchange

It is possible to use this method for either negative purification—in which contaminants bind to the matrix while the antibody passes through—or the opposing positive format, in which the antibody interacts with the matrix and contaminants are allowed to flow through. Separation by ion exchange is dependent on charge and is aided by the ability to alter the overall charge of a protein by adjusting the pH of its environment. If the pH of the antibody-containing solution is above the pI of the antibody, an overall negative charge will be imposed. To induce a positive charge, the pH of the solution is lowered to a level below the pI of the antibody. As the net charge increases, so does the strength of adsorption to the matrix. Two types of matrix are available for ion exchange—one that enables anion exchange, and the other, cation. For antibody purification, cation exchange enables positive chromatographic purification when the pH is less than the pI. Anion exchange at or about neutral pH is suitable for negative purification of antibodies. Elution is normally achieved using a continuous gradient of salt concentration or pH. The shallower the gradient, the longer the column required to achieve separation. Purity of the final product is not guaranteed to be absolute, as proteins with similar pI may be co-purified.

1.4. Affinity Purification Techniques

Purification by affinity interactions is used to positively purify molecules against a specific ligand or binding agent, and can be conveniently applied to the purification of antibodies and antibody fragments. Affinity columns can be purchased for common purifications—for instance, when the binding agent is an anti-mouse antibody that is capable of purifying any mouse antibody from solution. Alternatively, activated chromatography matrices can be purchased to which more specific binding agents can be permanently bound using simple chemistries.

Common binding agents on "off the shelf" affinity columns for generic antibody purification include Protein A, Protein G, Protein L, anti-species antibody, avidin, and anti-Tag antibody (for the purification of tagged recombinant fragments). More specialized binding agents, with affinity for the purification of a specific antibody, require immobilization to activated matrices by the researcher and include antigen, epitope, mimotope, and other moieties that are specific to the paratope of an antibody.

The basic principle of affinity chromatography is common to all binding agents. The sample is applied to the column in an environment buffered appropriately for the binding of antibody to occur. Non-binding contaminants are rigorously washed from the column prior to the application of an elution solution, leading to the disruption of the antibody complex and allowing antibody to be eluted from the column in pure form. In general, the only further purification step required is desalting. Application of the eluting solution to the column can be conducted by stepwise increments (for instance, when antibodies of differing affinities for the binding agent are to be purified from the same starting material), by a single step (when a monoclonal antibody [MAb] is being purified) or by a continuous gradient (when the least harsh disruption conditions are required). A final desalting step, such as gel filtration, is then performed in order to return the antibody to a suitably buffered state for downstream applications.

Protein A and G columns are suitable for whole IgG antibodies, and protein L columns may be used for the purification of antibodies and fragments containing kappa light chains. Furthermore, protein L does not bind goat, sheep, or bovine antibodies, and can therefore be used to purify MAbs from media supplemented with bovine or fetal calf serum (FCS). Elution is achieved by altering the pH or chaotropic agents.

The purification of genetically engineered antibody fragments can be facilitated by the addition of specifically designed "tags" created by gene fusions. Many of these tags have been developed, and are often marketed with the appropriate affinity matrix or column. Examples of tag/binding agent pairings are:

1. E-Tag/anti-E-Tag system (Amersham Biosciences), in which the tag is an epitope and the binding agent is an antibody to that epitope;
2. Hexahistidine tag (Invitrogen, Qiagen and Novagen), in which the binding system is immobilized metal affinity chromatography (IMAC) (such as HiTrap Chelating matrix from Amersham Biosciences, Probond resin from Invitrogen and Ni-NTA resin from Qiagen);
3. Biotinylation tag (PinPoint; Promega) in which the tag is an in vivo biotinylated peptide and the binding agent is avidin or an avidin derivative.

Antibody purification by immobilized anti-species antibody can be performed—for instance, enabling murine antibodies to be purified from spent hybridoma-culture supernatant. In this example, anti-mouse antibodies would act as the immobilized binding agent, extracting murine antibodies from solution. Elution would be achieved by a step or continuous gradient of either chaotropic agents or pH.

Finally, antigen, epitope, or mimotope affinity purification can be achieved by immobilizing the specific antigen, epitope, or mimotope onto a column matrix and using this to capture antibody from the feedstock (spent hybridoma-culture supernatant; diluted sera; culture broth). When epitopes are known and are linear in nature, preference should be given, if possible, to the use of epitope or mimotope rather than whole molecule antigen, as these smaller molecules appear to demonstrate a lowered matrix degeneration rate on repeated usage. Elution is achieved by a step or continuous gradient of either chaotropic agents or pH. These particular purification mechanisms effectively purify and concentrate the desired antibody from other contaminating proteins. Furthermore, by using a purification modality that requires antibody/antigen interaction, the resulting antibody preparation will be of high specific activity, because the process has removed, as a contaminant, any inactive antibody.

In the authors' opinion, affinity and ion-exchange purification techniques are the most effective methods for purification of antibody from biological feedstocks because they offer the best properties in terms of ease, robustness, scale, and quality of the final product. General methodologies based on these two techniques are described in the following section.

2. Materials

All buffers for column chromatography should be filtered (0.45 μm) and preferably de-gassed (by sonication or vacuum suction), especially if purification is to be carried out at room temperature.

2.1. Purification of Monoclonal Antibody (MAb) by Positive Ion-Exchange Chromatography

1. Express-Ion Mini Columns (Whatman), or other method scouting columns (*see* **Note 1**).
2. Equilibration/binding buffer: e.g., 0.1 *M* sodium phosphate buffer or 0.1 *M* sodium acetate buffer (*see* **Note 2**).
3. Elution buffers: 0.1–0.5 *M* NaCl, 0.1 to 0.5 *M* in 0.1 *M* steps made up in equilibration buffer.
4. Express-Ion S (Whatman) or other cation-exchange matrix (*see* **Note 3**).
5. 0.1 *M* phosphate-buffered saline (PBS), pH 7.4: 1.34 g $Na_2HPO_4.2H_2O$, 0.39 g NaH_2PO_4 and 8.5 g NaCl made up to 1 L with distilled water.
6. Desalting column containing Sephadex G25—e.g., PD10 column (Amersham Biosciences) or dialysis device.
7. Syringes.
8. Peristaltic pump, associated tubings, and valves.
9. Automated liquid chromatography system—e.g., fast protein liquid chromatograph (FPLC) or ÄKTA (Amersham Biosciences).

2.2. Purification of MAb by Peptide Epitope-Affinity Chromatography

1. Guard column containing Sepharose 4B (Amersham Biosciences).
2. Peptide epitope affinity column (*see* **Note 4**).

3. Equilibration buffer: PBS containing 0.02% (w/v) sodium azide.
4. Wash buffer: 0.5 *M* NaCl, pH 7.4 in distilled water.
5. Elution buffer: 3 *M* sodium thiocyanate, pH 7.4 in distilled water.
6. Desalting column containing Sephadex G25—e.g., PD10 column (Amersham Biosciences) or dialysis device.
7. Automated liquid chromatography system—e.g., FPLC or ÄKTA (Amersham Biosciences).

2.3. Purification of MAbs by Protein A Affinity Chromatography

1. HiTrap Protein A HP columns, 1 or 5 mL (Binding capacity ≈ 20 mg human IgG/mL of drained gel matrix; Amersham Biosciences).
2. Binding buffer: 20 m*M* sodium phosphate buffer, pH 7.0 (12.2 m*M* Na_2HPO_4, 7.8 m*M* NaH_2PO_4 in distilled water).
3. Elution buffer: 100 m*M* citric acid buffer, pH 3.0 (82 m*M* citric acid, 18 m*M* trisodium citrate in distilled water).
4. 1 *M* Tris-HCl, pH 8.0.
5. Desalting column containing Sephadex G25—e.g., PD10 column (Amersham Biosciences) or dialysis device.
6. Peristaltic pump, associated tubings, and valves.
7. Automated liquid chromatography system—e.g., FPLC or ÄKTA (Amersham Biosciences).

2.4. Purification of Recombinant Antibody Fragments by IMAC

1. HiTrap Chelating HP column, 5 mL (binding capacity ≈ 12 mg His-tagged protein/mL of gel; Amersham Biosciences).
2. Metal buffer: 0.2 *M* $NiSO_4$/$ZnSO_4$/$CuSO_4$ in distilled water (*see* **Note 5**).
3. Binding buffer: 20 m*M* sodium phosphate buffer, 500 m*M* NaCl, 0.5% (v/v) Tween 20, pH 7.5.
4. Elution buffer: 20 m*M* sodium phosphate buffer, 500 m*M* NaCl, pH 7.5, 0.5% (v/v) Tween 20, 500 m*M* imidazole, pH 7.5.
5. Dialysis equipment.
6. Automated liquid chromatography system, e.g., FPLC or ÄKTA (Amersham Biosciences).

3. Methods

3.1. Purification of MAbs by Positive Ion-Exchange Chromatography

3.1.1. Method Scouting

1. Equilibrate an Express-Ion S mini column with 5 mL of 0.1 *M* sodium acetate buffer pH 5.0 (*see* **Note 2**) at approx 0.5 mL/min using a syringe.
2. Apply a 1-mL aliquot of preferably (*see* **Note 6**) purified antibody (200 μg/mL in sodium phosphate buffer, pH 7.0) at approx 0.5 mL/min. Collect eluate for analysis **(step 5)**.

3. Wash the column with 10 mL of 0.1*M* sodium phosphate buffer, pH 7.0 at approx 0.5 mL/min. Collect eluate for analysis **(step 5)**.
4. Elute bound material with a range of elution buffers (1 mL). Begin with 0.1 *M* NaCl and increase the concentration sequentially through to 0.5 *M* NaCl. Collect 1-mL fractions throughout for analysis **(step 5)**.
5. Analyze loading, washing, and elution samples for antibody content (*see* **Note 6**) and calculate the percentage recovery of antibody.
6. Repeat **steps 1–5** with the other equilibration buffers to determine the optimal binding conditions.

3.1.2. Laboratory-Scale Purification

1. Equilibrate approx 25 mL of Express-Ion S matrix (or other cation-exchange matrix) with the most appropriate buffer as determined from method scouting studies **(Subheading 3.1.1.)** for 30 min at room temperature and pack matrix into a column with 1.6-cm internal diameter.
2. Adjust the pH of the pretreated antibody feed-stock (*see* **Note 7**) to match that of the equilibration buffer and apply to the column at a rate of 5 mL/min.
3. Wash the column with 250 mL of equilibration buffer.
4. Elute bound antibody with a step or linear gradient from 0 to 0.5 *M* NaCl in equilibration buffer (*see* **Note 8**). Pool fractions containing the purified antibody according to the peak of absorbance at 280 nm on the UV trace (*see* **Note 9**).
5. Buffer-exchange the purified antibody into a suitable storage buffer such as PBS, pH 7.4 using a desalting column or by dialysis.

3.2. Purification of MAbs by Epitope-Affinity Chromatography

1. Purge the chromatography system **(Fig. 1)** of air and prime with equilibration buffer.
2. Connect the guard column in series before the peptide epitope affinity column and equilibrate the system with at least ten column volumes (*see* **Note 10**) of equilibration buffer at 0.5 mL/min (*see* **Note 8**).
3. Load the pretreated antibody feedstock (*see* **Note 7**) onto the system at 0.5 mL/min (*see* **Note 11**).
4. Wash the chromatography system with equilibration buffer until the UV trace at 280 nm returns to baseline, then wash with an additional three column volumes.
5. **Optional:** In order to remove material that is bound non-specifically to the affinity matrix, the column may be washed with 0.5 *M* NaCl until the UV trace returns to baseline.
6. Connect the G25 desalting column in series after the affinity column and desorb the antibody from the matrix by the application of 5–10 column volumes of elution buffer. The chaotrope will be immediately separated from the purified antibody by gel filtration as the eluate passes down the desalting column.
7. Pool fractions containing the purified antibody according to the peak of absorbance at 280 nm on the UV trace (*see* **Note 9**).

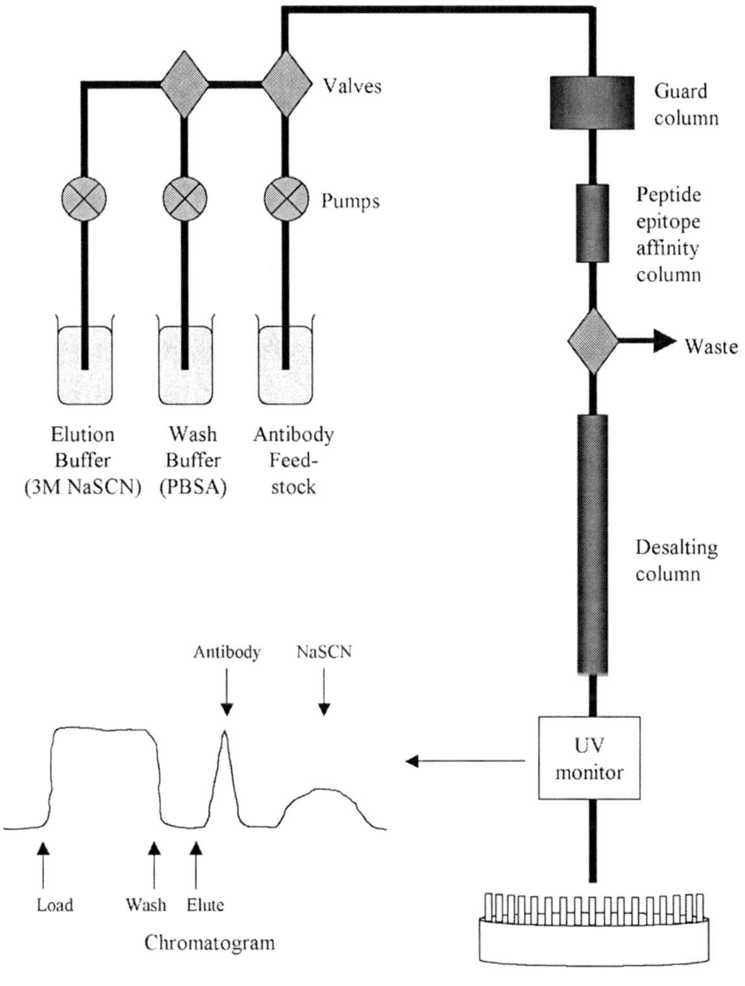

Fig. 1. Diagrammatic representation of a liquid chromatography system suitable for the purification of antibodies from biological feed-stocks such as hybridoma supernatant using peptide epitope-affinity chromatography.

3.3. Purification of IgG Antibodies by Protein A Affinity Chromatography

1. Equilibrate a HiTrap Protein A HP column with ten column volumes of binding buffer at a flow rate of 1 mL/min using a peristaltic pump.
2. Load pretreated (*see* **Note 7**) antibody feed-stock onto the column at 1–2 mL/min.

3. Transfer the column assembly to an automated liquid chromatography system and wash column by passing 20 column volumes of binding buffer at 1 mL/min, until the UV trace at 280 nm returns to the baseline.
4. Elute bound IgG antibodies by passing five column volumes of elution buffer at 1 mL/min. Collect 0.5-mL eluted fractions in tubes containing 0.5 mL Tris-HCl, pH 8.0, to neutralize the eluted antibody.
5. Pool fractions containing the purified antibody according to the peak of absorbance at 280 nm on the UV trace.
6. Buffer-exchange the purified antibody into a suitable storage buffer such as PBS, pH 7.4 using a desalting column or by dialysis.

3.4. Purification of Recombinant Antibody Fragments by IMAC

1. Resuspend *E. coli* cell pellet in 1/10 culture volume of binding buffer (*see* **Note 12** and **13**), and lyse cells using sonication or French press. Clarify soluble cytoplasmic extract by centrifugation at 30,000*g* for 30 min (*see* **Note 14**).
2. Connect a HiTrap Chelating column to the automated liquid chromatography system. Wash column with ten column volumes of distilled water at 2.5 mL/min.
3. Load 5 mL of the chosen metal solution onto column at 1 mL/min, and wash with ten column volumes of distilled water at 2.5 mL/min.
4. Equilibrate the column in ten column volumes of binding buffer at 2.5 mL/min.
5. Load clarified cell extract onto the column using the injection loop, or superloop for larger volumes, at 1 mL/min.
6. Wash with binding buffer at 2.5 mL/min until the UV trace at 280 nm returns to the baseline.
7. Elute protein using either a gradient (e.g., 0–100% elution buffer over 30 min) or step elution (e.g., 5% elution buffer, followed by 50%, then 100%) at 1 mL/min (*see* **Note 15**).
8. Pool fractions and dialyze in the chosen buffer—e.g. 20 m*M* sodium phosphate, pH 7.5 (*see* **Note 16**).

4. Notes

1. Method scouting columns are small, prepacked columns that allow the optimal buffer conditions for binding and elution to be defined before moving on to laboratory-scale purification ***(1)***.
2. Buffers with a range of pHs such as phosphate buffer (pH 5.8–7.0) and sodium acetate buffer (pH 5.0–5.8) can be chosen. The authors have found that sodium acetate buffer pH 5.0 is effective for the purification of a murine IgG3 from hybridoma supernatant.
3. Antibodies have been reported to have isoelectric points within the range of 6.0–8.0 ***(2)***. If a cation-exchange matrix is used and the pH of the separation system is lower than the pI, antibodies would be adsorbed on the column (a positive chromatographic step). Alternatively, the contaminating proteins with acidic isoelectric points such as albumin can be adsorbed to an anion-exchange matrix and at pH 7.0, the antibody is allowed to pass unretained through the column (a negative chromatographic step).

4. Peptide epitope-affinity matrices are prepared and columns are packed according to the manufacturer's instructions. The authors have found that cyanogen bromide-activated Sepharose (Amersham Biosciences) is a robust matrix for immobilization of peptide ligands. However, matrices with other activation chemistries are available, and may be more appropriate for individual applications.
5. It is advisable to perform small-scale experiments using different metal ions on the column. Metal leaching and specificity of the metal for the hexahistidine tail usually follows the pattern of $Ni^{2+} > Cu^{2+} > Zn^{2+}$ for leaching and $Ni^{2+} < Cu^{2+} < Zn^{2+}$ for affinity.
6. Analysis of fractions for antibody content is straightforward if purified antibody is used for method scouting studies, since it can be achieved by measuring the absorbance of the sample at 280 nm against the appropriate buffer blank. If impure sources of antibody such as tissue-culture supernatant are used, an alternative method for comparing antibody content in fractions, such as ELISA, must be used.
7. Pretreatment of antibody feedstocks such as hybridoma supernatant or serum usually takes the form of removal of particulate matter—e.g., cellular debris, and is achieved by centrifugation (4,800*g* for 1 h) and ultrafiltration (0.45 μm).
8. Peristaltic pumps and suitable valves are adequate for low-pressure liquid chromatography; however, the use of an automated system such as FPLC or the ÄKTA system allows stricter control over the run conditions including the generation of gradients for elution, as well as facilitating data analysis.
9. Protein-containing fractions are detected using a flow cell with 280-nm filters attached to a UV monitor. However, NaSCN also absorbs strongly at 280 nm, but this will elute from the desalting column much later because it is retarded, while antibody elutes in the void volume.
10. The guard column is used to prevent particulates from blocking the flow of the affinity column. Column volume refers to the packed volume of the peptide epitope-affinity column.
11. The amount of antibody that binds to an affinity ligand (i.e., recovery) is a function of its concentration in the feed-stock, affinity for the ligand, and time in contact with the ligand. Reducing the flow rate may increase the latter.
12. Reducing agents such as dithiothreitol (DTT) are incompatible with IMAC. If reducing agents are necessary, use 1–5 m*M* β-mercaptoethanol.
13. These protocols are designed for the purification of soluble antibody fragments, either exported out of the cell, prepared from periplasmic extracts, or expressed in the cytoplasm. It is common to incorporate EDTA into buffers when preparing periplasmic extracts (and cytoplasmic extracts). Obviously, this must be removed either by dialysis or buffer exchange on a PD-10 column (Amersham Biosciences) before application to the metal affinity column.
14. In some cases, antibody expression will lead to large, insoluble aggregates forming within the cells. These inclusion bodies often contain large amounts of pure protein. These can be dissolved in 8 *M* urea or 6 *M* guanidine hydrochloride and purified in the denatured state using IMAC. In this case, the chosen denaturant (usually urea because of cost) must be included in the binding buffer. The protein can be

eluted from the column in the denatured state (and thus also add denaturant to elution buffer) for refolding, or a gradual reduction of denaturant can be performed while the protein is still bound to the column, and eluted afterwards (called matrix-assisted or on-column refolding) *(3)*.

15. When performing initial experiments for determining chromatographic behavior of the His-tagged protein, it is advisable to use gradient elutions. Once the concentration of imidazole required to successfully separate contaminants from the His-tagged protein is known, a step elution can be used. Most expression vectors allow N-terminal or C-terminal tagging of the protein of interest. In some cases, a particular orientation is more favorable because of potential problems arising either in purification or activity evaluations. Also note that imidazole absorbs at 280 nm; a 0.5 *M* solution has an absorbance of approx 0.5 AU.
16. At this stage, the protein is usually of 95% or greater purity, which can be sufficient for biochemical or immunological studies. However, higher purity is often required—e.g., for crystallography, and thus, a further gel-filtration chromatography step is required.

References

1. Denton, G., Murray, A., Price, M. R., and Levison, P. R. (2001) Direct isolation of monoclonal antibodies from tissue culture supernatant using cation-exchange cellulose Express-Ion S. *J. Chromatogr. A* **908,** 223–234.
2. Danielsson, A., Ljunglof, H., and Lindblom, J. J. (1988) One-step purification of monoclonal IgG antibodies from mouse ascites—an evaluation of different adsorption techniques using high- performance liquid-chromatography. *J. Immunol. Methods* **115,** 79–88.
3. Kobayashi, H., Morioka, H., Tobisawa, K., Torizawa, T., Kato, K., Shimada, I., et al. (1999) Probing the interaction between a high-affinity single-chain Fv and a pyrimidine (6–4) pyrimidone photodimer by site-directed mutagenesis. *Biochemistry* **38(2),** 532–539.

23

Affinity Measurement Using Surface Plasmon Resonance

Robert Karlsson and Anita Larsson

1. Introduction

The change in free energy for a reaction, ΔG, can be determined from the dissociation affinity constant, K_D, since ΔG is equal to RTlnK_D. The affinity constant provides no information on the rate of the reaction or the energy required to reach the transition state of the reaction (**Fig. 1**).

The affinity constant is usually determined from titration experiments in which the concentration of reactants or products at equilibrium is measured. Numerous techniques *(1,2)*, including radioimmunoassay (RIA), enzyme-linked immunosorbent assay (ELISA), chromatography, and isothermal titration calorimetry, can be used to estimate or determine the affinity constant of a reaction. With a direct technique, the molecular interaction can be monitored without the use of labels, whereas for an indirect technique a label such as a radioactive tag, an enzyme, or a fluorophore is necessary. Surface plasmon resonance (SPR) *(3)* is a direct technique. It allows the analysis of interactions between analytes in solution and a ligand attached to a sensor chip surface, providing a continuous readout of complex formation and dissociation. In principle, the signal is directly proportional to the increase in molecular mass on the sensor surface *(4)*. SPR technology can thus be generally applied to affinity and kinetic analysis of protein-protein, protein-peptide, protein-DNA, and protein-small molecule interactions. The information gathered from a few binding curves is sufficient for the determination of both the association- and dissociation-rate constants, k_a and k_d, respectively. The affinity, K_D, can then be calculated from the ratio between k_d and k_a. By measuring the interaction over a range of temperatures, the kinetic data can be used to determine activation (transition state) energies *(5,6)*.

From: *Methods in Molecular Biology, Vol. 248: Antibody Engineering: Methods and Protocols*
Edited by: B. K. C. Lo © Humana Press Inc., Totowa, NJ

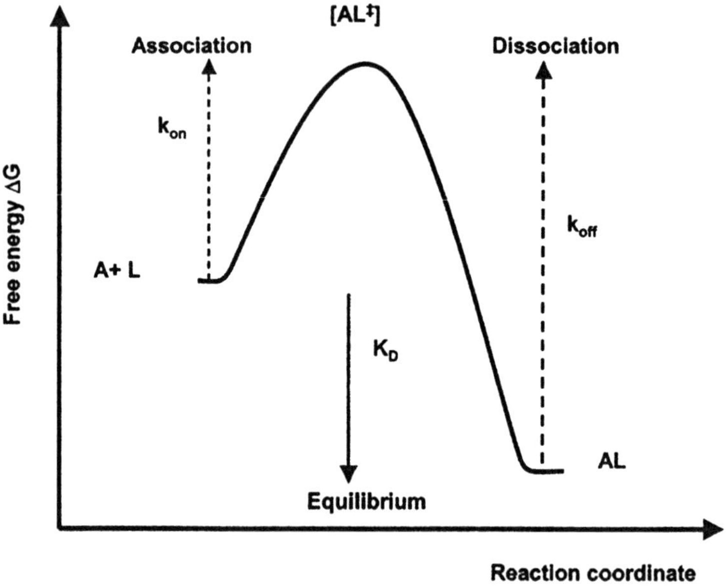

Fig. 1. Energy vs reaction coordinates. A = analyte, L = ligand.

The increasing popularity of SPR biosensors is probably related to the ease with which comparisons can be made between the interaction properties of different analyte molecules and immobilized ligands *(7)*. The search terms "Antibody AND (SPR OR Biacore OR Optical Biosensor)" in PubMed return 541 references for the period from January 1989 to April 2002. SPR biosensors enable the characterization of genetically engineered antibody and antigen constructs *(8–19)*. This is also useful for antibody screening and selection procedures *(20–27)* and detailed structure:activity relationship studies *(28–31)*. A contributing factor to the utility of SPR biosensors is the wide range of affinity and kinetic constants that can be determined. Affinity constants can be determined from the millimolar to picomolar range; association-rate constants, from 10^3 to 10^8 M^{-1} s^{-1}; and dissociation-range constants, from 10^{-5} to 1 s^{-1} *(32)*. This chapter provides an introduction to the mechanisms of SPR biosensor technology and to interaction kinetics. We describe the basic steps in experimental design and evaluation of a typical kinetic experiment. Rate and affinity constants derived from such experiments can be used later in thermodynamic analysis, but this is not described here.

1.1. SPR Technology

Several instrument systems that involve SPR or related evanescent-wave techniques are commercially available *(33)*. Of these, Biacore systems are by far the

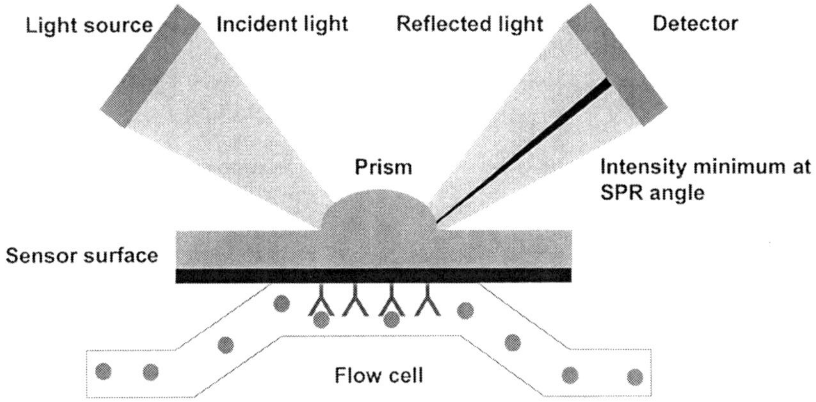

Fig. 2. SPR setup. A light source illuminates the back side of the sensor surface. At a certain angle of incident light, the electrons in the gold film are excited. At this SPR angle, an intensity minimum in the reflected light is observed. The SPR angle is shifted when analyte injected in the flowcell bind to immobilized ligand. The shift in SPR angle is proportional to the increase in mass on the sensor surface.

most widely used. They are the most sensitive instruments available, and can easily detect the binding of small molecules with a mol wt of as little as 100 Daltons. They are also adapted to kinetic analysis by using a flow system rather than a cuvet for sample delivery. The basic outline of a Biacore system is illustrated in **Fig. 2.**

Samples are delivered to the sensor surface in Biacore using a micro-fluidic system. This ensures reproducible sample delivery and low sample consumption. Accurate and reproducible sample delivery with immediate switching between buffer and sample is essential for reliable kinetic measurements. At a high flow rate (>30 μL/min), the transition from 0–100% sample concentration is achieved in less than 1 s in the flow system, allowing rapid kinetic processes to be detected and measured. The sensor surface is a glass slide coated with a thin 50-nm gold film mounted in a plastic carrier. The gold surface is derivatized to allow simple covalent attachment of molecules using well-defined chemistry *(34–37)*. Changes in the refractive index at the sensor surface are measured. The sensor surface can be divided into several sensing areas or spots, and a signal can be obtained from each spot. In the same injection, a sample can therefore pass over both active and reference spots. Before sample injection, running buffer is pumped over the surface and a baseline is recorded. During sample injection, any change in refractive index is detected. A change can either be caused by a binding event or by a difference in refractive index between sample solution and running buffer. The signal related to ligand binding alone is obtained by subtracting the signal obtained on the reference spot from that obtained on the ligand spot **(Fig. 3)**. When the injection is terminated,

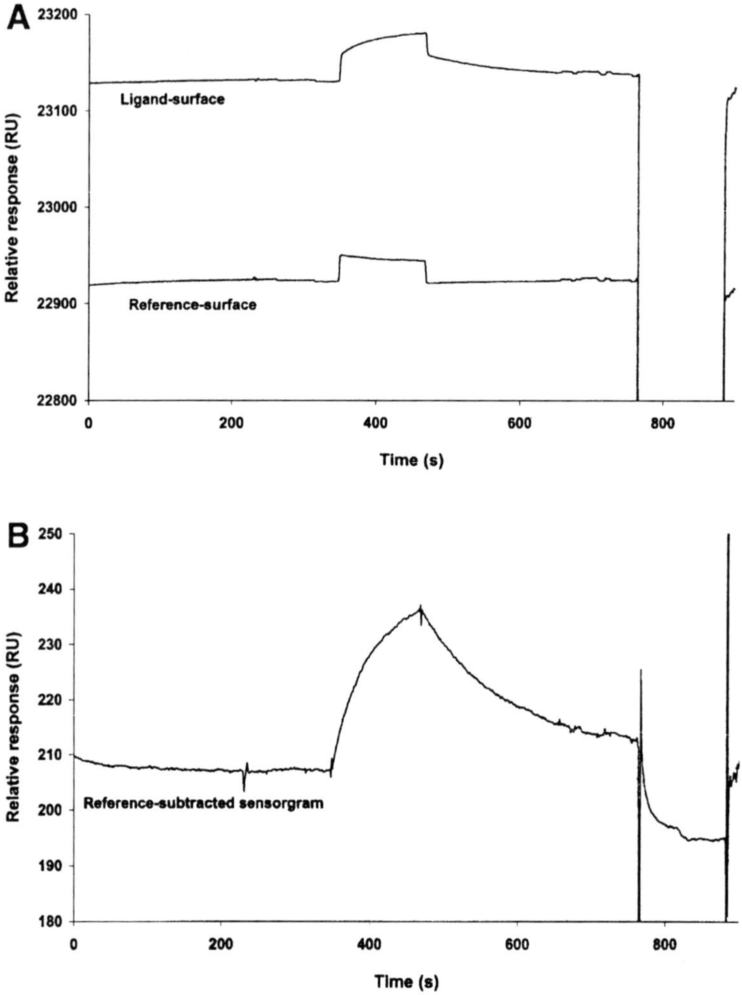

Fig. 3. (**A**) Injection of a recombinant camel-antibody fab-fragment over a surface immobilized with lysozyme (ligand-surface) and a non-modified surface (reference-surface). (**B**) shows the sensorgram after reference-subtraction.

running buffer again flows over the surface, and dissociation of the analyte:ligand complex can be observed. The output from the SPR detector is called a sensorgram, and is a plot of the SPR response vs time. It is generated in real-time, and is presented on a computer screen. The SPR response is expressed in resonance units (RU), where one RU corresponds to 10^{-6} refractive index units.

SPR technology has many practical advantages: i) detection is instantaneous, allowing continuous real-time monitoring of molecular binding; ii) detection can be localized to individual spots on a sensor surface; iii) no labels are required on any of the molecules involved; and iv) the light used in detection is reflected from the back of the sensor surface and does not penetrate the sample. Analysis can thus be performed on colored, turbid, or opaque samples with no interference from absorption or scattering.

1.2. Interaction Kinetics

Reaction rates in solution:

When the two proteins A and B react, the rate of complex, AB formation is described by the equation:

$$dAB/dt = k_a*A*B\text{-}k_d*AB. \tag{1}$$

The concentrations of A, B, and AB are initially A_{total}, B_{total}, and zero. A, B, and AB vary over time and eventually reach their equilibrium values.

For protein:protein interactions, k_a values are normally between 10^3 and 10^9 M^{-1} s^{-1}, and k_d values are often greater than 10^{-4} s^{-1}.

1.3. Reaction Rates and Kinetics in Biacore

The SPR response in Biacore is directly related to the concentration of formed complex. The following equation system (2a–2c) describes the interaction between injected analyte, A and immobilized ligand, B:

$$dA_s/dt = f^{1/3}*\ k_t*\ (Conc\text{-}\ A_s) - (k_a*A_s*B - k_d*AB);\ \text{initial value } A_s[0] = 0 \tag{2a}$$

$$dB/dt = -\ (k_a*A_s*B - k_d*AB);\ \text{initial value } B[0] = R_{max} \tag{2b}$$

$$dAB/dt = (k_a*A_s*B - k_d*AB);\ \text{initial value } AB[0] = 0 \tag{2c}$$

Equation 2a describes how the concentration of analyte, A_s, close to the sensor surface varies over time. The analyte is injected at a certain concentration, Conc, and is continuously replenished during injection. The concentration in the middle of the flow cell is constant over time and equal to the injected concentration, Conc. In a laminar flow system such as Biacore, the concentration of analyte in close proximity to the sensor surface, A_s, can vary over time. Here, the concentration is determined by the balance between the reaction rate and the diffusion of analyte through the unstirred layer of the flow system. By introducing a flow-independent transport coefficient, k_t and considering the difference between the injected concentration and the concentration close to the sensor surface, $f^{1/3}*k_t*(Conc\text{-}A_s)$ as well as the rate of complex formation, $k_a*A_s*B - k_d*AB$, the change in A_s as a function of time can be calculated. When A_s is significantly lower than Conc, the reaction is said to be "transport limited," and binding rates are influenced by transport effects. Note that equilibrium values are unaffected by transport effects. The transport coefficient depends on the diffusion properties of the analyte and on the geometry of the flow cell. When the ratio between $k_a{*}R_{max}$, and $f^{1/3}{*}k_t$ becomes greater than 5, the reaction rate, for practical purposes, can be considered fully transport limited. The intrinsic binding rate is much faster than the transport of analyte to the surface through the diffusion layer. The binding curves that arise from these analyses can be used for affinity analysis, but the data is not sufficiently reliable for kinetic analysis *(32)*. **Equation 2b** describes how fast the ligand B is consumed by the reaction. The saturation response R_{max} reflects the total concentration of immobilized ligand. **Equation 2c** describes the rate of complex formation, and it has the same form as the equation describing the interaction in solution (**Equation 1**). Further descriptions of this interaction model can be found in the literature *(32, 38–43)*. Note that **Equations 2a–2c** are valid only when a single interaction site is present on the analyte molecule. When an intact antibody is used as analyte, the reaction cannot be described by these equations.

At this point, antibody valency will be important, since avidity effects have a profound effect on the apparent stability of the complex. Avidity effects are smaller when the ligand density is reduced, but are difficult to completely eliminate. Therefore, strict evaluation of interaction kinetics, is not always possible, although binding curves can be used for ranking purposes. When surfaces with low binding capacity (antibody saturation response at 10–50 RU) are used, a good approximation of the affinity may be obtained by fitting the data to the bivalent analyte model included in Biaevaluation software.

1.4. Steady-State Values and Affinity

For interactions in which binding curves reach equilibrium during injection, the affinity can be obtained from a plot of the equilibrium response, R_{eq}, vs analyte concentration, C.

$$\mathrm{Req} = C{*}R_{max}/(C + K_D) \qquad (3)$$

To obtain reliable K_D values it is important to determine R_{eq} values for concentrations in the range from $0.1{*}K_D$ to $10{*}K_D$.

2. Materials

2.1 Ligand and Analyte

1. The ligand should be present at 10–100 μg/mL in immobilization buffer (the composition of the immobilization buffer varies between ligands—see Biacore manual for more detail).
2. The analyte should typically be present in concentrations from 1 n*M* to 100 μ*M* in running buffer (running buffer varies between experiments—see Biacore manual for more details).

2.2. Amine Coupling

1. N-hydroxy-succinimide (NHS) (Biacore AB).

2. 1-ethyl-3-(3-dimethylaminopropyl) carbodiimide (EDC) (Biacore AB).
3. 1 *M* ethanolamine-HCl, pH 8.5 (Biacore AB).

2.3. Surface Thiol Coupling

2.3.1. Ligand Modification using PDEA

1. 2-(2-pyridinyldithio)ethaneamine (PDEA) (Biacore AB).
2. 0.1 *M* MES buffer, pH 5.0: dissolve 21.3 g of morpholinoethanesulfonic acid, monohydrate in 900 mL of water; adjust the pH to 5.0 with 1 *M* NaOH, and make up to 1 L with water.
3. EDC.

2.3.2. Ligand Immobilization on the Chip Surface

1. NHS.
2. EDC.
3. 40 m*M* cystamine in sodium borate buffer, pH 8.5: dissolve 0.45 g of cystamine dihydrochloride and 0.31 g of boric acid in 25 mL of water. Adjust the pH to 8.5 with 1 *M* NaOH, and make up to 50 mL with water. Store frozen.
4. 100 m*M* dithiothreitol (DTT) or dithioerythritol (DTE) in sodium borate buffer, pH 8.5: dissolve 0.77 g of DTE or DTT and 0.31 g of boric acid in 25 mL of water. Adjust the pH to 8.5 with 1 *M* NaOH and make up to 50 mL with water. Store frozen.
5. 0.1 *M* sodium acetate buffer, pH 4.3: add 5.72 mL of acetic acid (98–100% purity) to 900 mL of water. Adjust the pH to 4.3 with 1 *M* NaOH and make up to 1 L with water.
6. PDEA in 0.1 *M* sodium acetate buffer, pH 4.3 containing 1 *M* NaCl: dissolve 1.2 mg of PDEA and 14 mg of NaCl in 250 µL of 0.1 *M* sodium acetate buffer, pH 4.3.

2.4. Ligand Thiol Coupling

1. NHS.
2. EDC.
3. 0.1 *M* sodium borate buffer: dissolve 6.18 mg of boric acid in 900 mL of water. Adjust the pH to 8.5 with 1 *M* NaOH and make up to 1 L with water.
4. 80 m*M* PDEA in 0.1 *M* sodium borate buffer pH 8.5: dissolve 4.5 mg of PDEA in 250 µL of 0.1 *M* sodium borate buffer. **Note:** Use within 1 h.
5. 0.1 *M* sodium formate buffer: add 3.77 mL of formic acid (100% purity) to 900 mL of water. Adjust the pH to 4.3 with 1 *M* NaOH and make up to 1 L with water.
6. 50 m*M* cysteine + 1 *M* NaCl in 0.1 *M* sodium formate buffer pH 4.3: dissolve 1.5 mg of cysteine and 14 mg of NaCl in 250 µL of sodium formate buffer. **Note:** Use within 1 h.

2.5. Aldehyde Coupling

1. NHS.
2. EDC.
3. 5 m*M* hydrazine hydroxide or carbohydrazide in water (**Note:** hydrazine is extremely toxic. Carbohydrazide is a less toxic alternative that is equally effective for aldehyde coupling).

4. 1 *M* ethanolamine-HCl, pH 8.5.
5. 0.1 *M* sodium acetate buffer, pH 4.0: add 5.72 mL of acetic acid (98–100% purity) to 900 mL of water. Adjust the pH to 4.0 with 1 *M* NaOH and make up to 1 L with water.
6. 100 m*M* sodium cyanoborohydride in 100 m*M* sodium acetate buffer, pH 4.0.

3. Methods

The design of a kinetic experiment starts with the immobilization of the ligand. For a single interaction pair, this involves: i) selection of which molecule in the pair to immobilize; ii) selection of immobilization chemistry; iii) selection of buffer conditions; iv) assessment of the activity of the immobilized ligand; v) identification of a suitable reference surface; and vi) identification of suitable regeneration conditions (if necessary). Sufficient experimental data for reliable calculation of k_a, k_d, $R_{\max}$, and k_t must be gathered. To achieve this, the following experimental parameters should be considered vii) what temperature to use; viii) what level(s) of immobilization to use in the kinetic experiment; ix) the injection time; x) the dissociation time; xi) what concentrations of analyte to use and; xii) what flow rate to use. In addition the experimental design should include: xiii) blank injections for elimination of systematic instrument variations; xiv) repeated injections of at least one analyte concentration for evaluation of any variation in ligand activity over time, and; xv) start up cycles to condition the flow system and the sensor surface. All these steps can usually be performed within 1 d.

3.1. Choice of Ligand vs Analyte

The choice of which interactant to use as an analyte can be crucial for the success of kinetic measurements. If the interactions of several analytes with a common ligand are to be compared, it is usually best to immobilize the common ligand. This will save time and sensor surfaces, and will enable comparisons (**Fig. 4**).

The *mol wt* of the interactants is an important factor because the response obtained from analyte binding is proportional to its mass. When the mol wts of the interactants are similar (within a factor 10), the selection of analyte and ligand is less important. If the mol wts of the ligand and analyte differ greatly, using the smaller molecule as analyte will reduce the expected response level because the changes in mass concentration will be lower. However, it is easier to immobilize a macromolecule such as a protein than to derivatize a small molecule for immobilization. *Purity* is another important factor. If the ligand preparation contains impurities, these may be immobilized together with the ligand when general coupling methods such as amine or thiol coupling are used. Impurities in the analyte preparation can affect the accuracy with which

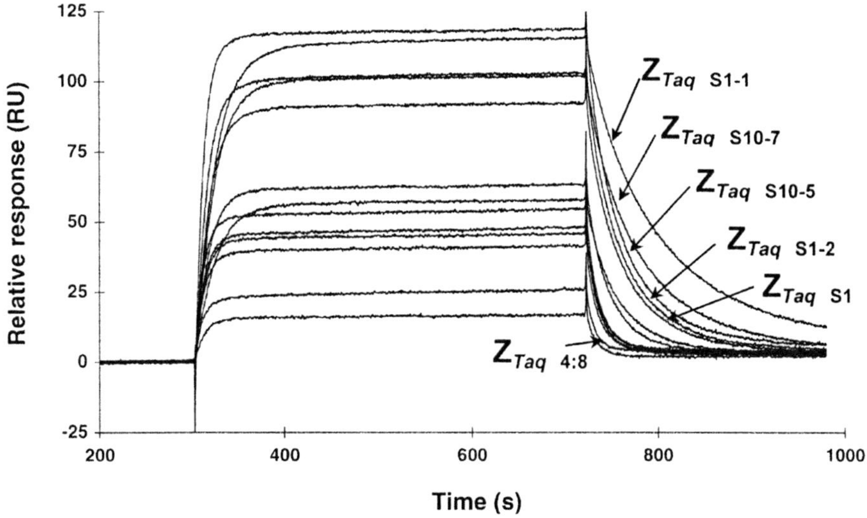

Fig. 4. Affinity ranking of twelve affibody variants. Overlay plot of sensorgrams obtained for *Taq* DNA polymerase specific affibodies when injected over a sensor chip surface immobilized with *Taq* DNA polymerase. Reprinted with the kind permission of E. Gunneriusson, K. Nord, M. Uhlén, and P-A Nygren, Department of Biotechnology, Royal Institute of Technology (KTH) *(47)*.

the analyte concentration can be determined. This in turn will influence the validity of calculated association-rate constants. Partially purified material can be used as either ligand or analyte, but impurities can complicate results by introducing nonspecific binding. The purity of the interactants is often a compromise between the effort involved in rigorous purification and the level of ambition in the kinetic measurements. For careful kinetic measurements, it is important that both the ligand and the analyte are as *homogeneous* as possible. Many preparations are by nature heterogeneous—for example, polyclonal antibodies. Heterogeneity in an otherwise pure preparation can arise through aggregation, partial denaturation, or degradation as a result of purification or storage conditions. Always try to establish the homogeneity of the preparation using other techniques such as chromatography or electrophoresis.

A prerequisite for studying 1:1 kinetics described in **Equation 2** is that the analyte has *a single binding site* for the immobilized ligand. Molecules with multiple binding sites can be used as ligands, provided that the binding sites are independent. Molecules with multiple binding sites are less suitable as analytes, since this leads to avidity effects in which one analyte molecule binds to multiple binding sites on the surface. Avidity is seen as a significant reduction in the rate of dissociation as compared to single-site binding. Antibodies are a

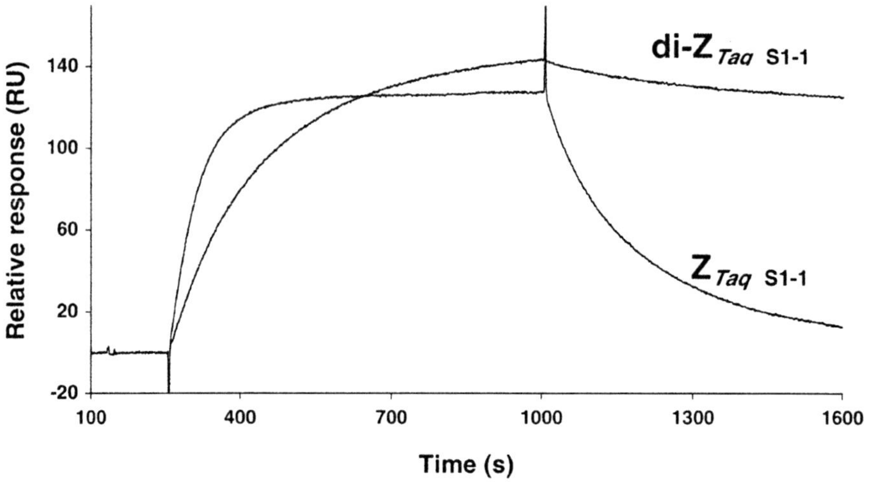

Fig. 5. Effect of dimerization of an affibody. Overlay plot of sensorgrams obtained after injection of $Z_{TaqS1-1}$ and di-$Z_{\mathrm{Taq\ S1-1}}$. Reprinted with the kind permission of E. Gunneriusson, K. Nord, M. Uhlén, and P-A Nygren, Department of Biotechnology, Royal Institute of Technology (KTH) *(47)*.

common example of an interactant with multiple binding sites. When monovalent antibody fragments are used as analyte, note that some fragments may have a tendency to dimerize in solution. Contaminating dimers in the analyte solution can have a detrimental effect on the results (**Fig. 5**).

The *concentration* of interactant is important in two respects for the choice of ligand vs analyte: i) the concentration of analyte must be accurately known in order to calculate correct association-rate constants. It is not necessary to accurately determine the concentration of ligand; ii) analyte preparations must be available over a suitable concentration range for kinetic measurements. The range of available ligand concentrations is less important. Analyte stock solutions must be at a relatively high concentration (usually in the mg/mL range) in order to allow dilution to a concentration series for analysis. Ligands are generally immobilized at low concentrations (typically 10 µg/mL for amine and thiol coupling). If the concentration of stock solution is low, it may be necessary to perform a buffer exchange on the ligand prior to immobilization.

3.2. Designing the Active Surface

A range of immobilization chemistries can be used by exploiting amine groups, carboxyl groups, thiol groups, sugar moieties, or tags that are present on the ligand. All types of immobilizations can be performed directly in the Biacore system, and the process is usually less than 30 min. Immobilization can be performed using a low flow rate (5–10 µL/min) to minimize consumption of ligand.

The concentration of ligand used for immobilization is typically 1–100 μg/mL. Biacore AB offers a wide choice of sensor surfaces and coupling chemistries for ligand immobilization. Sensor chip surfaces with a carboxymethyl (CM) dextran matrix allow a flexible choice of immobilization chemistries and provide a hydrophilic environment for biomolecular interactions.

3.2.1. Immobilization Approaches

Immobilization is initiated with activation of the carboxyl groups on the sensor surface. These are converted to reactive esters through a reaction with a mixture of carbodiimide/hydroxysuccinimide (EDC/NHS). The esters can then be reacted directly with proteins or with molecules that introduce new surface functionality—for instance, a thiol group, a reactive disulfide, or a hydrazine.

3.2.1.1. Amine Coupling

This method uses carbodiimide/hydroxysuccinimide chemistry to immobilize ligand via free amine groups such as the N-terminus or lysine residues. This is the most generally applicable method for protein ligands.

1. Mix equal volumes of 0.4 *M* EDC and 0.1 *M* NHS and inject the mixture over the sensor surface (suggested contact time 1–10 min).
2. Inject ligand solution.
3. Inject ethanolamine-HCl (suggested contact time 7 min) to deactivate excess reactive groups.

3.2.1.2. Surface Thiol Coupling

This method uses thiol-disulfide exchange between thiol groups on the matrix surface and reactive disulfide introduced into carboxy-groups on the ligand. This process requires prior modification of the ligand, and may not always work when the analyte requires reducing conditions. The following steps are also illustrated in **Fig. 6.**

1. To modify the ligand, prepare a solution of ligand at 1 mg/mL in 0.1 *M* MES buffer, pH 5.0. A volume of 0.5 mL is sufficient.
2. Prepare a solution of PDEA at 15 mg/mL in 0.1 *M* MES buffer, pH 5.0. A volume of 0.3 mL is sufficient.
3. Add 0.25 mL of PDEA solution to 0.5 mL of ligand solution. Incubate at 25°C for 5 min.
4. Add 25 μL of 0.4 *M* EDC in water. Incubate at 25°C for 10 min.
5. Remove the reagents by gel filtration, dialysis, or an equivalent technique.
6. To immobilze the ligand, mix equal volumes of 0.4 *M* EDC and 0.1 *M* NHS. Inject over the sensor surface (suggested contact time 2 min).
7. Inject cystamine (suggested contact time 3 min).
8. Inject DTE or DTT (suggested contact time 3 min) to ensure that thiol groups are reduced.

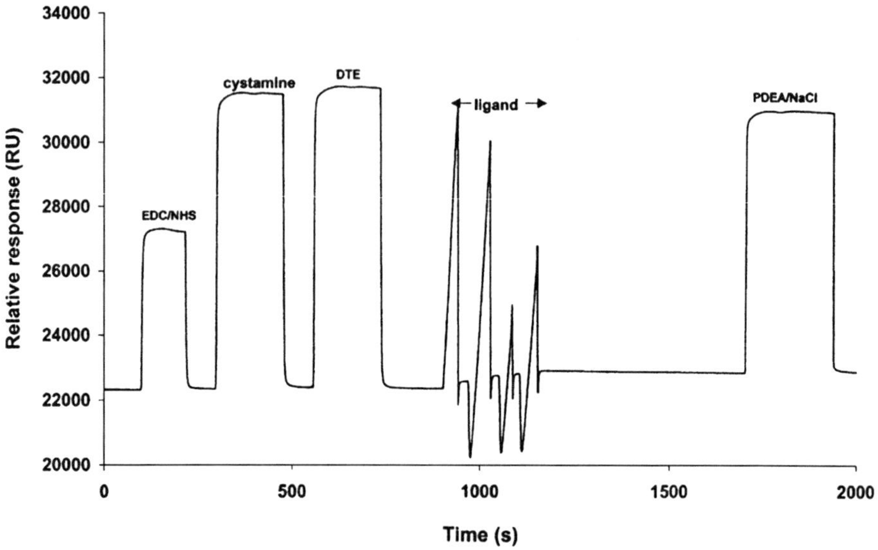

Fig. 6. Surface thiol coupling of lysozyme. The ligand is injected repeatedly until the desired immobilization level is reached (500 RU).

9. Inject ligand solution.
10. Inject PDEA (suggested contact time 4 min) to deactivate excess thiol groups.

3.2.1.3. Ligand Thiol Coupling

This method uses thiol-disulfide exchange between reactive disulfide groups and thiol groups (typically cysteine residues) in the ligand. This approach may not always work when the analyte requires reducing conditions.

1. Mix equal volumes of 0.4 *M* EDC and 0.1 *M* NHS. Inject the mixture over the sensor surface (suggested contact time 2 min).
2. Inject PDEA solution (suggested contact time 4 min).
3. Inject ligand solution.
4. Inject cysteine-HCl (suggested contact time 4 min) to deactivate excess PDEA groups.

3.2.1.4. Aldehyde Coupling

This method uses hydrazine chemistry to immobilize ligand via aldehyde groups on the ligand molecule. Although aldehyde groups are not common in native ligand molecules, they may be introduced by oxidation of ligands containing *cis*-diol groups. This approach can be useful for immobilization of carbohydrates and glycoproteins.

1. Oxidize the ligand, if necessary, to introduce aldehyde groups. Reagents and conditions for this step will depend on the particular ligand.

2. To immobilize the ligand, mix equal volumes of 0.4 *M* EDC and 0.1 *M* NHS. Inject the mixture over the sensor surface (suggested contact time 1–5 min).
3. Inject hydrazine or carbohydrazide (suggested contact time 7 min).
4. Inject ethanolamine-HCl (suggested contact time 7 min) to inactivate remaining reactive groups.
5. Inject ligand solution.
6. Inject sodium cyanoborohydride (suggested contact time 20 min). This will reduce the hydrazone bond and stabilize the immobilized ligand.

3.2.1.5. High-Affinity Capture

This method relies on high-affinity binding of the ligand to a capturing molecule immobilized on the sensor surface. Capture works as a nanoscale-affinity purification of ligand, which means that the ligand preparation does not have to be purified. Regeneration usually removes both ligand and bound analyte, and fresh ligand is captured for each analysis cycle, increasing the consumption of ligand. For kinetic analysis, it is important that the ligand dissociates as little as possible from the surface during measurement and that the ligand-capture step is highly reproducible so that conditions for each cycle are comparable.

1. Immobilize the capturing molecule by amine coupling (*see* **Subheading 3.2.1.1.**).
2. Inject ligand solution (suggested contact time 1–3 min) and allow the surface to stabilize (typically 5–10 min).

3.3. Buffer Conditions

The analyte buffer should have the same composition as the running buffer. This will minimize differences in the refractive index and ensure that association and dissociation phases are monitored under the same conditions. If no special requirements must be considered, 10–50 m*M* HEPES or phosphate buffers are recommended. Typically, these buffers should include at least 100 m*M* NaCl to avoid electrostatic interactions between carboxyl groups on the surface and positively charged groups on the analyte. A detergent such as Tween 20 at a concentration of between 0.005% and 0.1% is recommended, and can prevent adsorption of analyte to vials and tubing.

3.4. Test of Surface Activity

1. Prepare analyte solutions at concentrations of 0.001, 0.01, 0.1, 1, 10, and 100 μ*M*.
2. Inject these solutions over the immobilized surface. Use a flow rate of 30 μL/min and 1-min injections. Start with the lowest concentration and inject in concentration order.
3. When the highest concentration has been injected, observe dissociation for at least 20 min.

This experiment will provide a first indication of whether the analyte binds to the immobilized ligand, and also whether the kinetics are in the measurable range.

For measurable kinetics, the progress of binding and dissociation should be readily apparent. It should also be easy to determine whether the surface is saturated or if higher concentrations are needed to reach saturation. If the signal rapidly returns to baseline, no regeneration will be required. If the baseline has not been reached after 20 min of dissociation, this indicates that regeneration will be needed to speed up the analysis.

The saturation response gives an indication of the apparent activity of the immobilized ligand.

Ligand level = R_{max} * ligand mW/analyte mW * 1/binding stoichiometry

For example, for a ligand with a mol wt of 150 kDa, analyte mol wt of 50 kDa, and 1,500 RU immobilized ligand, a saturation response of 300 RU represents an apparent activity of 300/(1500*50/150)*100 = 60%. If the apparent activity is significantly higher than 100%, this may indicate that the analyte is binding to multiple sites on the ligand or that the analyte is binding as multimers or aggregates. If the maximum binding capacity indicates that ligand activity is low, it may still be possible to perform kinetic experiments. The level of immobilized ligand should then be increased to compensate for the low activity.

3.5. Reference Surface

Kinetic analysis requires a reference surface to correct for bulk response. The reference surface may be prepared in one of three ways: i) unmodified surface; ii) activated-deactivated; or iii) immobilized with a dummy ligand. In general, test experiments are necessary to determine which type of reference surface is most appropriate. As a prerequisite, the analyte must show no or low levels of nonspecific binding to the reference surface.

1. Inject the analyte over the unmodified sensor chip to check for nonspecific binding. Use a high concentration of analyte—for example, 10 μ*M* for typical protein analytes. A flow rate of 30 μL/min or higher and a contact time of 5 min are recommended.
2. Remove bound protein and refresh the surface with a 1-min injection of 50 m*M* NaOH.
3. Nonspecific binding appears as the response obtained from injection of analyte over the unmodified surface. Evaluate the level of nonspecific binding in relation to the response expected from analyte binding to the ligand. Nonspecific binding levels up to 5% of the expected R_{max} can sometimes be tolerated. If nonspecific binding is high at the analyte concentration used, test lower concentrations in order to judge the binding in relation to the expected measurements. If nonspecific binding is unacceptably high, the following measures may be taken:
   a. Test nonspecific binding to a surface that has been activated and deactivated. For example, for amine coupling, activate the surface with EDC/NHS and then deactivate directly with ethanolamine. If this reduces the amount of nonspecific

binding to acceptable levels, use an activated-deactivated surface as a reference in your kinetic experiment;

b. Try a different buffer—e.g., Tris-HCl or phosphate instead of HEPES;
c. Reverse the roles of ligand and analyte if other considerations allow this. Nonspecific binding may be lower for one interactant than for the other; or
d. Try a different sensor surface. Several surfaces with varying charge and hydrophilicity properties are available.

3.6. Establish Regeneration Conditions

If surface activity experiments demonstrate that the complex is stable and that complete dissociation will not be obtained within a reasonable time, a regeneration procedure will have to be determined. Successful regeneration must cause rapid dissociation of the complex and maintain ligand activity.

To find regeneration conditions, test regeneration with repeated cycles of injection of analyte followed by injection of the regeneration solution. Note the level of analyte binding and the baseline after regeneration for each cycle. Use a high analyte concentration to achieve a high level of analyte binding. This will make it easier to detect incomplete regeneration or loss of binding capacity. Always start with the mildest regeneration conditions and proceed successively to harsher conditions. This will reduce the risk of the surface being destroyed during the establishment of an optimum procedure. Evaluate the results on the basis of trends in baseline and analyte response values from repeated analyses, not on single sensorgrams **(Fig. 7)**.

1. Inject analyte for 1 min, then inject regeneration solution for 1 min.
2. Repeat the cycle of analyte injection and regeneration 5×.
3. Inspect trend plots (response against cycle number) for baseline just before analyte injection and analyte response.
4. Optimal conditions for regeneration are specific for each ligand-analyte interaction. However, a narrow range of conditions is effective for a wide range of interactants. The following solutions are suggested as a starting point in searching for effective regeneration conditions. Test the solutions in the order given until you find acceptable conditions: low pH (10 m*M* glycine-HCl buffers recommended), ethylene glycol (add 50, 75 and 100%), high pH (1–100 m*M* NaOH) and $MgCl_2$ (1–4 *M*). Test each solution with a contact time of 1 min. The contact time can be varied if necessary to obtain optimal regeneration.
5. For result interpretation, follow the trend in analyte-binding response (relative response measured from the baseline just before injection) for replicate tests of each regeneration solution. Follow the absolute baseline response after regeneration as well, although trends in the baseline level are less informative, they can help to clarify and confirm interpretation of trends in response levels. In an ideal regeneration, the analyte response is constant within ± 10% of the level reached in the first injection. When conditions are too mild, the analyte response decreases and the baseline level increases. Test slightly harsher conditions or change the type of

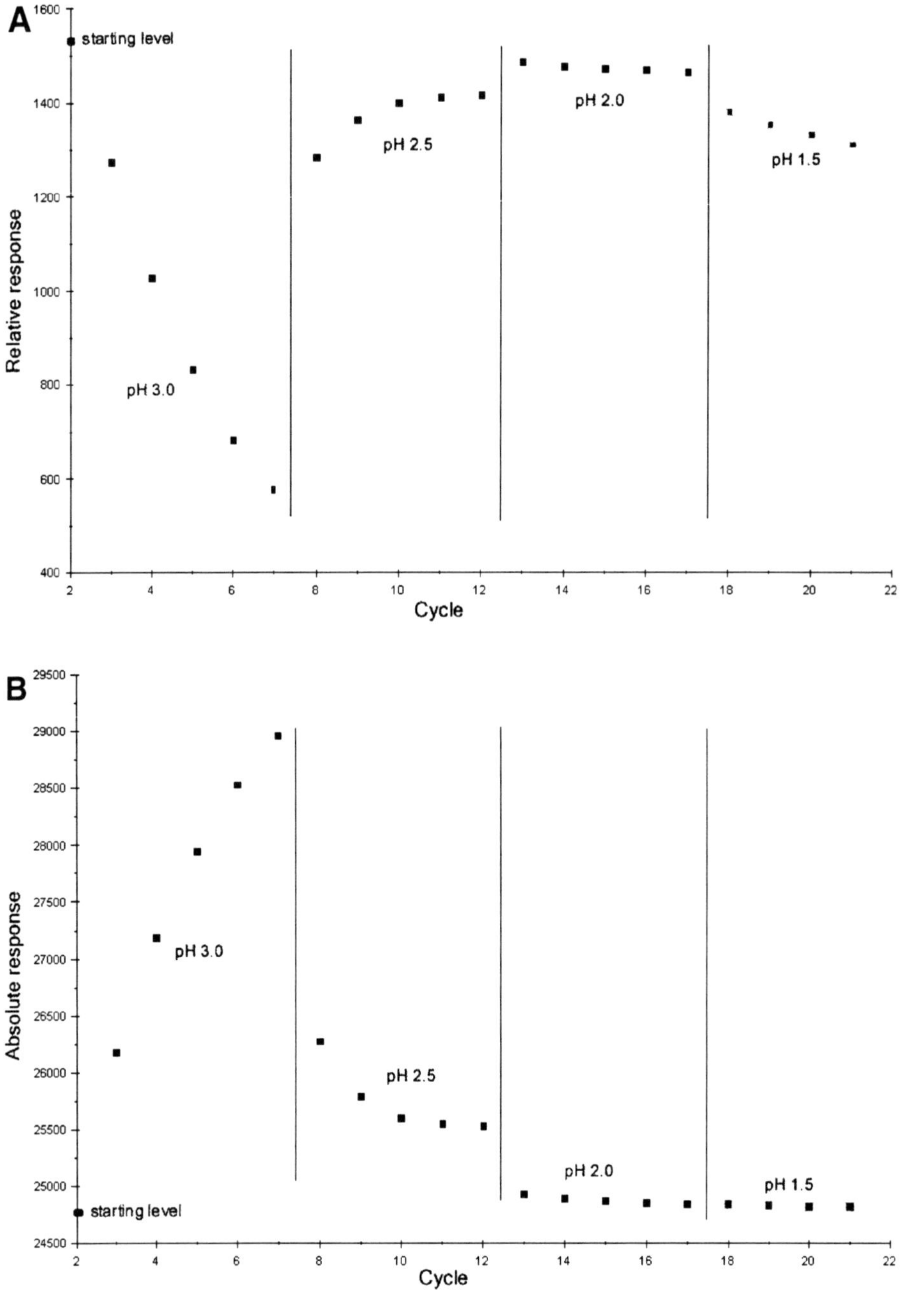

Fig. 7. Analyte response (**A**) and baseline (**B**) after each analyte injection for replicate tests of 10 m*M* glycine-HCl with pH 3.0, 2.5, 2.0, and 1.5. In this example, regeneration at pH 2.0 works well.

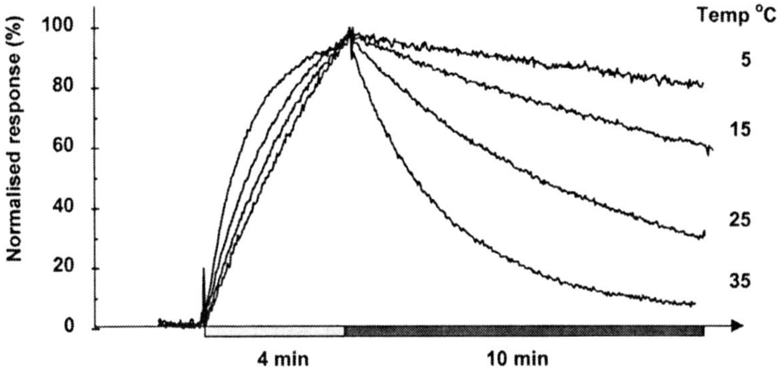

Fig. 8. A change in temperature can have drastic effects on interaction properties. Here, dissociation half time, $t_{1/2}$ was reduced from 38 min to 80 s when the interaction temperature was altered from 5–35°C.

regeneration solution. When conditions are too harsh, the analyte response decreases at the same time that the baseline level is either constant or decreases. A decrease in analyte response without a corresponding increase in baseline indicates that the surface is losing analyte-binding capacity. In this case, change the type of regeneration solution. You may need to prepare a new surface if the ligand has been too heavily damaged.

6. When you have found conditions for regeneration, run repeated cycles to ensure that ligand activity is fully retained throughout multiple analysis cycles.

3.7. Temperature

Most kinetic experiments are performed at room temperature. Both complex formation and dissociation are more rapid at elevated temperatures. Dissociation is often more influenced by a change in temperature, which leads to lower affinities at higher temperatures (**Fig. 8**). An interaction that is difficult to study at room temperature because of high complex stability and difficult regeneration can be more suitable for analysis at higher temperatures. From kinetic data obtained at different temperatures, reaction enthalpy and entropy can be determined. For thermodynamic analysis, a total of five temperatures in the range from 5–40°C are recommended.

3.8. Experimental Parameters for Kinetic Analysis

1. Immobilize ligand to a suitable level. To obtain a favorable balance between the intrinsic reaction rate and the transport of analyte to the surface, an *immobilization level* that yields a low saturation level is recommended. Use the surface activity studies to calculate the immobilization level that would be required to yield a saturation response of 20–50 RU. Prepare surfaces with this immobilization level. A low level of immobilization is needed when interactions with high k_a values are investigated. In cases in which many analytes are to be compared, it is likely that

some will have a low affinity for the immobilized ligand. In these cases, it can be necessary to use a higher level of immobilization.

2. Run three to five repeated cycles consisting of injection of analyte followed by regeneration to confirm that the surface performance is reproducible through multiple analysis cycles and in order to condition the flow system and the surface. The analyte should be injected at a high concentration so that binding approaches saturation, if possible. This experiment will also show whether the response returns to baseline immediately after regeneration in each cycle. In some cases, it may be necessary to wait for the baseline to stabilize after regeneration.
3. Perform analyses over a wide range of *analyte concentrations,* ideally 100-fold or more. Results from the surface activity experiments should indicate a suitable concentration range. At least one and preferably more of the highest concentrations should reach steady state during the sample injection. The lowest concentrations should show measurable binding rates, but will probably not reach steady state. An *analyte injection time* of 1–3 min is normally sufficient. Longer injection times may be required in cases in which the reaction is slow and high analyte concentrations are not readily available. For accurate determination of the dissociation-rate constant, significant *dissociation* must be observed. The surface activity experiment in which dissociation is followed for 20 min can serve as a guideline. It is usually not practical to extend dissociation time to more than 1 h. *Repeat* at least one analyte concentration in duplicate or triplicate during a concentration series. In this way, the repeatability of the system as well as any change in surface activity can be monitored over time. Ideally, the sample should reach the active and reference surfaces at the same time. This situation is approached in Biacore systems by using flow rates of 30 μL/min or more. A high *flow rate* is also beneficial for two other reasons—it reduces sample dispersion and improves mass transport.
4. Inject running buffer (blank sensorgram). When the signal from the reference surface has been subtracted, there may still be artefacts in the remaining data. Artefacts can arise from small temperature changes (0.01°C) or from pressure jumps. These artefacts will also usually be present when buffer is used instead of a sample. If they occur systematically, subtracting a blank sensorgram can improve binding data by reducing or even eliminating these disturbances.

3.9. Data Analysis

There are three steps in data analysis: i) preparation of data; ii) calculation of rate constants; and iii) interpretation of results.

3.9.1. Preparation of Data

1. Baseline adjustment: set the baseline recorded immediately before the injection of analyte to zero (*see* **Note 1**). Align sensograms with respect to time to allow substraction of data from the reference surface and subtraction of a blank sensogram. This cleanup procedure, illustrated in **Fig. 9,** compensates for differences in refractive index between sample and running buffer and reduces systematic instrument errors *(44)*.

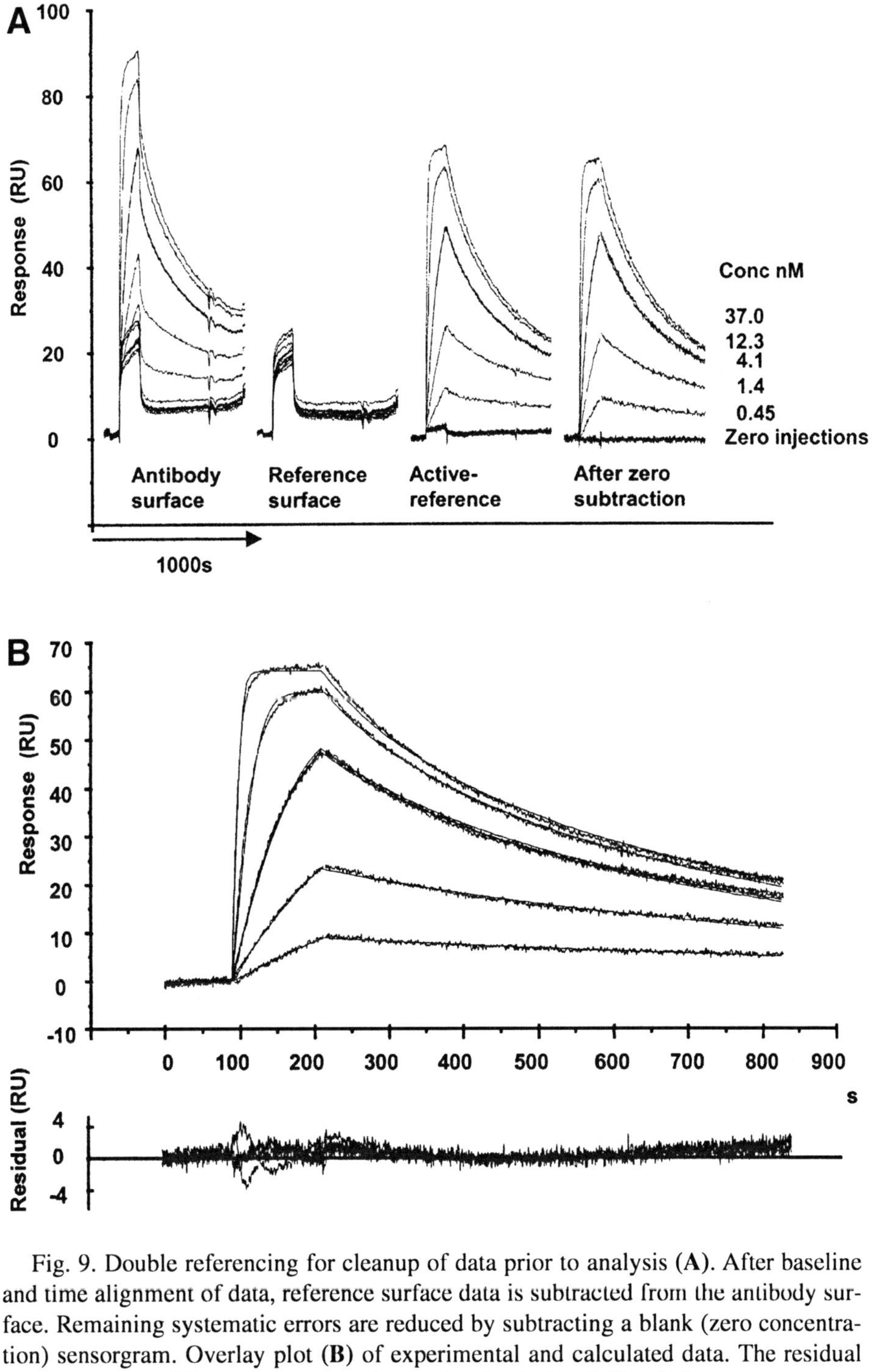

Fig. 9. Double referencing for cleanup of data prior to analysis (**A**). After baseline and time alignment of data, reference surface data is subtracted from the antibody surface. Remaining systematic errors are reduced by subtracting a blank (zero concentration) sensorgram. Overlay plot (**B**) of experimental and calculated data. The residual plot records the difference between observed and calculated (fitted) data.

2. When analyte injections cover a wide range of concentrations, it is normal to exclude some sensorgrams from analysis (*see* **Notes 2** and **3**). A very high concentration can be excluded when the injection leads to almost instantaneous saturation. In this case, the injection phase will not provide any kinetic information. A low concentration may not give significant binding data, and should then be excluded from the data set.

3.9.2. Calculation of Rate Constants

1. Include binding curves obtained at four to six analyte concentrations in final data analysis.
2. Select interaction model and start the fitting procedure (*see* **Note 4**). By using a set of default starting values, for parameters to be determined, the evaluation software calculates theoretical binding curves and compares them with experimental data. The difference between observed and calculated data sets is minimized by the software in an iterative manner. The program then returns the parameter values, k_a, k_d, R_{max}, k_t, that best describe the experimental data (*see* **Note 5**) *(45,46)*.
3. The analysis results in an overlay plot of experimental and calculated data, a residual plot (e.g., a plot of the difference between observed and calculated data) and a list of parameter values.

3.9.3. Interpretation of Results

1. Inspect the overlay plot. It compares observed and calculated data on the same scale and provides an immediate visual evaluation of the fit. Large differences between calculated and observed data indicate a poor fit, but when binding curves overlap, data is usually reliable.
2. Inspect the residual plot. It provides finer details. In an ideal case, the residual plot represents instrument noise and is without systematic trends. In reality, small residuals are often observed, particularly for data collected during the injection of analyte. Here, residuals can reflect errors in concentration (dilution errors). The size of the residual should therefore be viewed in relation to the absolute response. The setting of the injection start and stop can be critical. These are identified from the sensorgram by the user. Time errors are usually less than 1 s, but can result in significant residuals when binding rates are high. In the dissociation phase, residuals are normally less significant, since dissociation data is independent of the concentration of injected analyte. Absolute guidelines for interpretation of the size and shape of residuals are difficult to provide, but large deviations between observed and calculated data indicate that parameter values are uncertain (*see* **Note 6**).
3. Inspect parameter values.
   a. R_{max}. The R_{max} value, which is expressed in RU, is the calculated saturation level. The reliability of this value is easily checked by comparing it to experimental data. For the data in **Fig. 9,** the calculated R_{max} value was 64 RU, in close agreement with the observed signal.
   b. k_t. The k_t value, RU/(M*s), for proteins is normally between 10^8 and 5×10^9, but depends on the dimensions of the flow system, the flow rate and the mol wt of

the analyte. For reactions in which the intrinsic binding rate is much lower than the transport rate, the k_t value cannot be determined. In this case, computer analysis of the data returns a much higher k_t value that is often $>10^{11}$. For the data in **Fig. 9,** k_t was determined as 2.4×10^8, indicating a transport-limited interaction.

c. K_d. The validity of the k_d value can be checked by inspecting signal levels. For a reaction that is not transport-limited, the time it takes for 50% of the complex to dissociate is $\ln 2/k_d$. For a transport-limited reaction, this time is extended. In **Fig. 9,** k_d was 3.2×10^{-3} S^{-1}, and the corresponding half-time is 217 s. From the top curves, 50% dissociation is reached just after 300 s, which is in agreement with a transport-limited reaction. A time much shorter than 217 s would have been treated with suspicion. In general, k_d values higher than 0.1–1 and lower than 10^{-5} s^{-1} cannot be determined using SPR, and values outside this range must be treated with skepticism.

d. K_D. The affinity constant K_D is obtained from the ratio between rate constants, k_d/k_a. The K_D concentration gives an equilibrium response, R_{eq}, which is equal to half the saturation response. In **Fig. 9,** the K_D was 0.46 n*M*. The lowest concentration injected was 0.45 n*M*. Inspection and extrapolation of the injection phase by eye supports the finding of a K_D lower than 1 n*M*.

In cases in which equilibrium is reached during injection, K_D can be determined directly from equilibrium binding levels using **Equation 3.** This allows a direct comparison of K_D values obtained from kinetic analysis and from steady-state analysis. These two K_D values are based on the same interaction model, and should therefore be identical **(Fig. 9)**.

4. Notes

The use of more complex interaction models in data analysis is possible in principle. Ligand or analyte heterogeneity, avidity, or conformation change models can be used. In these cases, however, the results from data analysis require far more consideration and the interpretation of multiple rate constants is not straightforward. A more complex model can be used in cases in which there is a biochemical reason for it, or when fitting of data to the 1:1 interaction model is clearly inappropriate. However, there can be many causes for poor curve fits, which should be considered when the results are interpreted. This section identifies some of the common causes of poor curve-fitting in kinetic evaluation.

1. Baseline not adjusted: Curve-fitting algorithms in the evaluation software assume that the baseline for all curves is set to zero. If you have forgotten to adjust the sensorgram baseline to zero, you will not be able to fit the curves **(Fig. 10)**.
2. Incorrect analyte concentrations: If the wrong analyte concentrations are entered for the sensorgrams, a poor fit will be obtained. Just two deviant curves with an acceptable fit for the remaining curves may indicate that the two concentrations have been reversed **(Fig. 11)**.

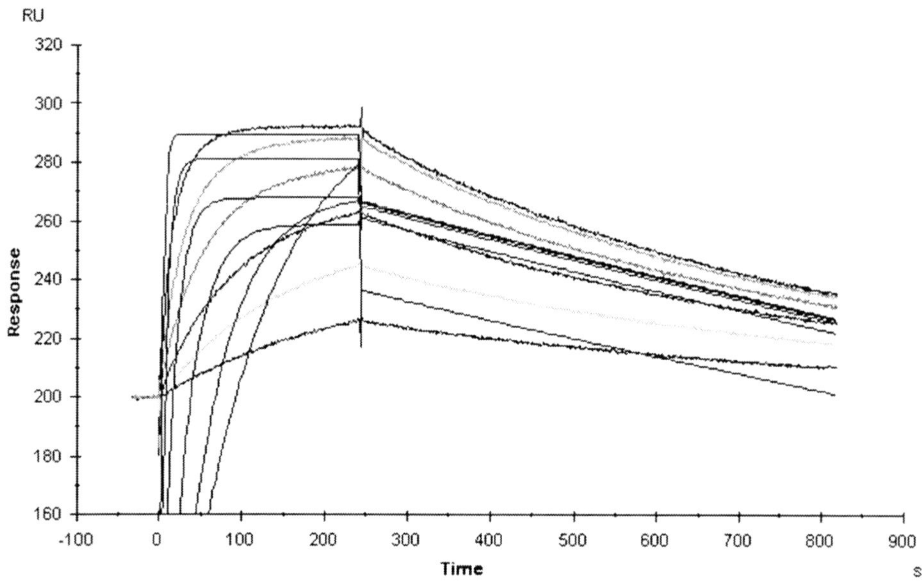

Fig. 10. Fitting to sensorgrams with the baseline at 200 RU.

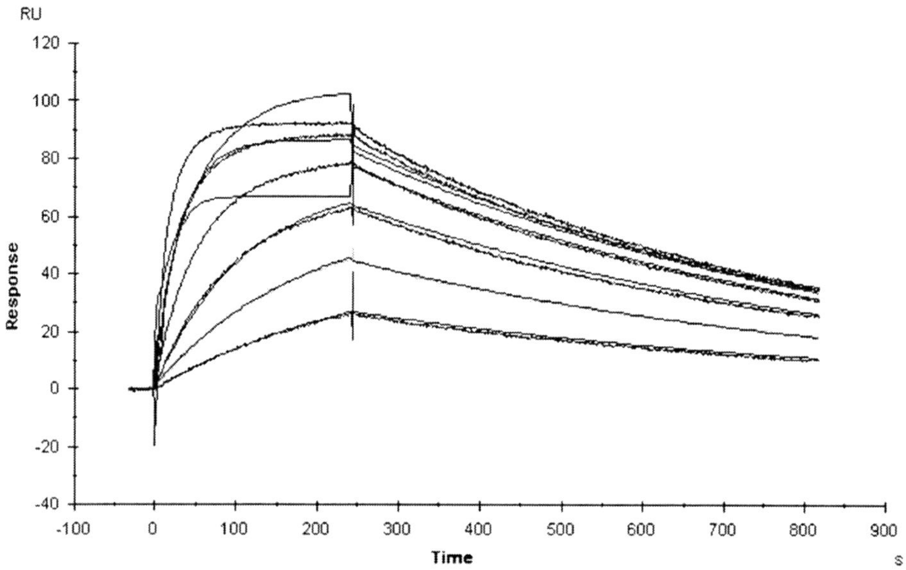

Fig. 11. Fitting to sensorgrams with two analyte concentrations interchanged.

3. Wrong order of magnitude: If concentrations are entered in the evaluation software with an incorrect or omitted order of magnitude (e.g., entering 10 instead of 10n for 10 nanomolar), the software may be unable to find a fit for the curves because the starting values for the fitting procedure are out of range. If the software is able to

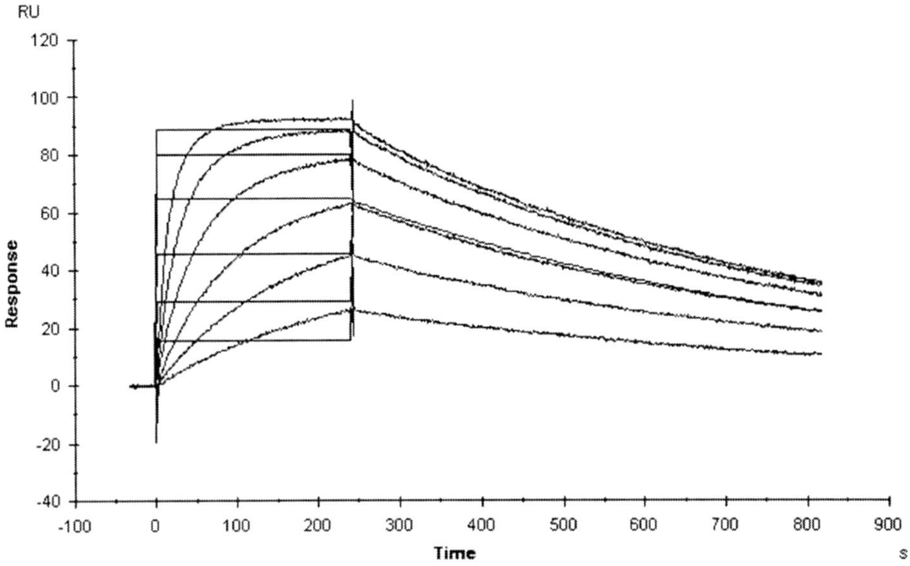

Fig. 12. Fitting to sensorgrams with analyte concentrations entered as molar instead of nanomolar.

find a fit with the wrong order of magnitude for the analyte concentration, the fit will be valid, but an incorrect value will be obtained for the association-rate constant k_a. The dissociation-rate constant k_d is independent of analyte concentration, and will not be affected (**Fig. 12**).

4. Unsuitable model: Fitting data from a complex interaction to a simple model will give a more or less poor fit, depending on the nature and the extent of interaction complexity. Note: fitting data from a simple interaction to a complex model will generally not give a poor fit, but some of the values obtained will lack significance.
5. Inappropriate starting values: Fitting algorithms in the evaluation software require starting values for all variables in the model. If the starting values provided differ too widely (usually by several orders of magnitude) from the best-fit value, the software may not be able to find a fit. This type of error typically results in fitted curves that consist of a series of straight-line segments that deviate greatly from the sensorgrams (similar to those in **Fig. 12**). If you observe this kind of behavior, check first that the analyte concentration values are correct. If this does not help, try changing the starting values for k_a and/or k_d either up or down by a factor 10 or more.
6. Inappropriate bulk response values: Most models in the evaluation software assign a locally variable parameter RI to the bulk response. The algorithms used sometimes assign unnecessarily high values to this parameter in the fitted curves. If you are evaluating data that has been properly reference-subtracted and find that the curves show response shifts at the beginning and end of injection, try setting the bulk response parameter to a constant value of 0.

References

1. Phizicky, E. M. and Fields, S. (1995) Protein-protein interactions: methods for detection and analysis. *Microbiol Rev.* **59(1),** 94–123. (Review.)
2. van Regenmortel, M. H. and Azimzadeh, A. (2000) Determination of antibody affinity. *J. Immunoass.* **21(2–3),** 211–234. (Review.)
3. Liedberg, B., Nylander, C., and Lundstrom, I. (1995) Biosensing with surface plasmon resonance—how it all started. *Biosens. Bioelectron.* **10(8),** i–ix.
4. Stenberg, E., Persson, P., Roos, H., and Urbaniczky, C. (1991) Quantitative determination of surface concentration of protein with surface plasmon resonance using radiolabeled proteins. *J. Colloid Interface Sci.* **143,** 513–526.
5. Roos, H., Karlsson, R., Nilshans, H., and Persson, A. (1998) Thermodynamic analysis of protein interactions with biosensor technology. *J. Mol. Recognit.* **11(1–6),** 204–210.
6. Lipschultz, C. A., Yee, A., Mohan, S., Li, Y., and Smith-Gill, S. J. (2002) Temperature differentially affects encounter and docking thermodynamics of antibody—antigen association. *J. Mol. Recognit.* **15(1),** 44–52.
7. McDonnell, J. M. (2001) Surface plasmon resonance: towards an understanding of the mechanisms of biological molecular recognition. *Curr. Opin. Chem. Biol.* **5(5),** 572–577. (Review.)
8. Goel, A., Colcher, D., Koo, J. S., Booth, B. J., Pavlinkova, G., and Batra, S. K. (2000) Relative position of the hexahistidine tag effects binding properties of a tumor-associated single-chain Fv construct. *Biochim. Biophys. Acta.* **1523(1),** 13–20.
9. Nakayashiki, N., Yoshikawa, K., Nakamura, K., Hanai, N., Okamoto, K., Okamoto, S., et al. (2000) Production of a single-chain variable fragment antibody recognizing type III mutant epidermal growth factor receptor. *Jpn. J. Cancer Res.* **91(10),** 1035–1043.
10. Nielsen, U. B., Adams, G. P., Weiner, L. M., and Marks, J. D. (2000) Targeting of bivalent anti-ErbB2 diabody antibody fragments to tumor cells is independent of the intrinsic antibody affinity. *Cancer Res.* **60(22),** 6434–6440.
11. Coeffier, E., Clement, J. M., Cussac, V., Khodaei-Boorane, N., Jehanno, M., Rojas, M., et al. (2000) Antigenicity and immunogenicity of the HIV-1 gp41 epitope ELDKWA inserted into permissive sites of the MalE protein. *Vaccine* **19(7–8),** 684–693.
12. Goel, A., Colcher, D., Baranowska-Kortylewicz, J., Augustine, S., Booth, B. J., Pavlinkova, G., et al. (2000) Genetically engineered tetravalent single-chain Fv of the pancarcinoma monoclonal antibody CC49: improved biodistribution and potential for therapeutic application. *Cancer Res* **60(24),** 696,469–696,471
13. Shaw, D. M., Embleton, M. J., Westwater, C., Ryan, M. G., Myers, K. A., Kingsman, S. M., et al. (2000) Isolation of a high affinity scFv from a monoclonal antibody recognising the oncofoetal antigen 5T4. *Biochim. Biophys. Acta.* **1524(2–3),** 238–246.
14. Willuda, J., Kubetzko, S., Waibel, R., Schubiger, P. A., Zangemeister-Wittke, U., and Pluckthun, A. (2001) Tumor targeting of mono-, di-, and tetravalent anti-p185(HER-2) miniantibodies multimerized by self-associating peptides. *J. Biol. Chem.* **276(17),** 14,385–14,392.

15. Roovers, R. C., van der Linden, E., de Bruine, A. P., Arends, J. W., and Hoogenboom, H. R. (2001) In vitro characterisation of a monovalent and bivalent form of a fully human anti Ep-CAM phage antibody. *Cancer Immunol. Immunother.* **50(1),** 51–59.
16. Bijnens, A. P., Ngo, T. H., Gils, A., Dewaele, J., Knockaert, I., Stassen, J. M., et al. (2001) Elucidation of the binding regions of PAI-1 neutralizing antibodies using chimeric variants of human and rat PAI-1. *Thromb. Haemostasis* **85(5),** 866–874.
17. Houimel, M., Corthesy-Theulaz, I., Fisch, I., Wong, C., Corthesy, B., Mach, J., et al. (2001) Selection of human single chain Fv antibody fragments binding and inhibiting Helicobacter pylori urease. *Tumour Biol.* **22(1),** 36–44.
18. Power, B. E., Caine, J. M., Burns, J. E., Shapira, D. R, Hattarki, M. K., Tahtis, K., et al. (2001) Construction, expression and characterisation of a single-chain diabody derived from a humanised anti-Lewis Y cancer targeting antibody using a heat-inducible bacterial secretion vector. *Cancer Immunol. Immunother.* **50(5),** 241–250.
19. Cupit, P. M., Lorenzen, N., Strachan, G., Kemp, G. J, Secombes, C. J., and Cunningham, C. (2001) Neutralisation and binding of VHS virus by monovalent antibody fragments. *Virus Res.* **81(1–2),** 47–56.
20. Horn, I. R., Wittinghofer, A., de Bruine, A. P., and Hoogenboom, H. R. (1999) Selection of phage-displayed fab antibodies on the active conformation of ras yields a high affinity conformation-specific antibody preventing the binding of c-Raf kinase to Ras. *FEBS Lett.* **463(1–2),** 115–120.
21. Hanes, J., Schaffitzel, C., Knappik, A., and Pluckthun, A. (2000) Picomolar affinity antibodies from a fully synthetic naive library selected and evolved by ribosome display. *Nat. Biotechnol.* **18(12),** 1287–1292.
22. Su, J. L., McKee, D. D., Ellis, B., Kadwell, S. H., Wisely, G. B., Moore, L. B., et al. (2000) Production and characterization of an estrogen receptor beta subtype-specific mouse monoclonal antibody. *Hybridoma* **19(6),** 481–487
23. Houimel, M., Schneider, P., Terskikh, A., and Mach, J. P. (2001) Selection of peptides and synthesis of pentameric peptabody molecules reacting specifically with ErbB-2 receptor. *Int. J. Cancer* **92(5),** 748–755.
24. van Remoortere, A., van Dam, G. J., Hokke, C. H., van den Eijnden, D. H., van Die, I., and Deelder, A. M. (2001) Profiles of immunoglobulin M (IgM) and IgG antibodies against defined carbohydrate epitopes in sera of Schistosoma-infected individuals determined by surface plasmon resonance. *Infect. Immun.* **69(4),** 2396–2401.
25. Raum, T., Gruber, R., Riethmuller, G., and Kufer, P. (2001) Anti-self antibodies selected from a human IgD heavy chain repertoire: a novel approach to generate therapeutic human antibodies against tumor-associated differentiation antigens. *Cancer Immunol. Immunother.* **50(3),** 141–150.
26. Rozemuller, H., Chowdhury, P. S., Pastan, I., and Kreitman, R. J. (2001) Isolation of new anti-CD30 scFvs from DNA-immunized mice by phage display and biologic activity of recombinant immunotoxins produced by fusion with truncated pseudomonas exotoxin. *Int. J. Cancer* **92(6),** 861–870.

27. Tanha, J., Xu, P., Chen, Z., Ni, F., Kaplan, H., Narang, S. A., et al. (2001) Optimal design features of camelized human single-domain antibody libraries. *J. Biol. Chem.* **276(27),** 24,774–24,780.
28. Novick, D., Nabioullin, R. R., Ragsdale, W., McKenna, S., Weiser, W., Garone, L., et al. (2000) The neutralization of type I IFN biologic actions by anti-IFNAR-2 monoclonal antibodies is not entirely due to inhibition of Jak-Stat tyrosine phosphorylation. *J. Interferon Cytokine Res.* **20(11),** 971–982.
29. Uthaipibull, C., Aufiero, B., Syed, S. E., Hansen, B., Guevara Patino, J. A., Angov, E., et al. (2001) Inhibitory and blocking monoclonal antibody epitopes on merozoite surface protein 1 of the malaria parasite Plasmodium falciparum. *J. Mol. Biol.* **307(5),** 1381–1394.
30. Nagumo, Y., Oguri, H., Shindo, Y., Sasaki, S., Oishi, T., Hirama, M., et al. (2001) Concise synthesis of ciguatoxin ABC-ring fragments and surface plasmon resonance study of the interaction of their BSA conjugates with monoclonal antibodies. *Bioorg. Med. Chem. Lett.* **11(15),** 2037–2040.
31. Thomas, R., Patenaude, S. I., MacKenzie, C. R., To, R., Hirama, T., Young, N. M., et al. (2002) Structure of an anti-blood group A Fv and improvement of its binding affinity without loss of specificity. *J. Biol. Chem.* **277(3),** 2059–2064.
32. Karlsson, R. (1999) Affinity analysis of non-steady-state data obtained under mass transport limited conditions using BIAcore technology. *J. Mol. Recognit.* **12(5),** 285–292.
33. Baird, C. L. and Myszka, D. G. (2001) Current and emerging commercial optical biosensors. *J. Mol. Recognit.* **14(5),** 261–268. Review.
34. Johnsson, B., Lofas, S., and Lindquist, G. (1991) Immobilization of proteins to a carboxymethyldextran-modified gold surface for biospecific interaction analysis in surface plasmon resonance sensors. *Anal. Biochem.* **198(2),** 268–277.
35. Löfås, S., Johnsson, B., Tegendal, K., and Rönnberg, I. (1993) Dextran modified gold surfaces for surface plasmon resonance sensors: immunoreactivity of immobilized antibodies and antibody-surface interaction studies. *Colloids and Surfaces B: Biointerfaces* **1,** 83–89.
36. Johnsson, B., Lofas, S., Lindquist, G., Edstrom, A., Muller Hillgren, R.-M., and Hansson, A. (1995) Comparison of methods for immobilization to carboxymethyl dextran sensor surfaces by analysis of the specific activity of monoclonal antibodies. *J. Mol. Recognit.* **8(1–2),** 125–131.
37. Löfås, S., Johnsson, B., Edström, ÅHansson, A., Lindquist, G., Müller Hillgren, R.-M., et al. (1995), Methods for site controlled coupling to carboxymethyldextran surfaces in surface plasmon resonance sensors. *Biosens. Bioelectron.* **10,** 813–822.
38. Glaser, R. W. (1993) Antigen-antibody binding and mass transport by convection and diffusion to a surface: a two-dimensional computer model of binding and dissociation kinetics. *Anal. Biochem.* **213(1),** 152–161.
39. Karlsson, R., Roos, H., Fägerstam, L., and Persson, B. (1994) Kinetic and concentration analysis using BIA technology. Methods: *A companion to methods in enzymology* **6,** 99–110.

40. Schuck, P. (1996) Kinetics of ligand binding to receptor immobilized in a polymer matrix, as detected with an evanescent wave biosensor. I. A computer simulation of the influence of mass transport. *Biophys. J.* **70(3),** 1230–1249.
41. Myszka, D. G., Morton, T. A., Doyle, M. L., and Chaiken, I. M. (1997) Kinetic analysis of a protein antigen-antibody interaction limited by mass transport on an optical biosensor. *Biophys. Chem.* **64(1–3),** 127–137.
42. Karlsson, R. and Falt, A. (1997) Experimental design for kinetic analysis of protein- protein interactions with surface plasmon resonance biosensors. *J. Immunol. Methods* **200(1–2),** 121–133.
43. Wofsy, C., and Goldstein, B. (2002) Effective rate models for receptors distributed in a layer above a surface: application to cells and biacore. *Biophys. J.* **82(4),** 1743–1755.
44. Morton, T. A. and Myszka, D. G. (1998) Kinetic analysis of macromolecular interactions using surface plasmon resonance biosensors. *Methods Enzymol.* **295,** 268–294.
45. Morton, T. A., Myszka, D. G., and Chaiken, I. M. (1995) Interpreting complex binding kinetics from optical biosensors: a comparison of analysis by linearization, the integrated rate equation, and numerical integration. *Anal. Biochem.* **227(1),** 176–185.
46. Roden, L. D. and Myszka, D. G. (1996) Global analysis of a macromolecular interaction measured on BIAcore. *Biochem. Biophys. Res. Commun.* **225(3),** 1073–1077.
47. Gunneriusson, E., Nord, K., Uhlén, M., and Nygren, P.-A. (1999) Affinity maturation of a *Taq* DNA polymerase specific affibody by helix shuffling. *Protein Eng.* **12 no. 10,** 873–878.

24

Kinetic Exclusion Assays to Study High-Affinity Binding Interactions in Homogeneous Solutions

Robert C. Blake II and Diane A. Blake

1. Introduction

High-affinity equilibrium binding constants for antibody-antigen interactions are inherently difficult to measure. Depending on the method used, difficulties arise from either the loss in signal-to-noise at the low concentrations that are necessary for measuring tight binding affinities, or from very slow off-rate kinetics. Methods for the quantification of antibody-binding interactions may be arbitrarily divided into two categories, homogeneous and heterogeneous. Homogeneous methods are those in which all of the binding reagents are evenly dispersed in solution. These methods require that some quantifiable property of the antibody or the antigen be changed as a result of the interaction. Examples include absorbance, fluorescence, fluorescence polarization, and sedimentation coefficient. In general, homogeneous methods cannot be readily applied to the majority of antibody-antigen interactions. In those few instances in which such methods do apply, they rarely exhibit great enough sensitivity to study the high-affinity interactions that are common for antibody-antigen interactions, in which the value of the equilibrium dissociation constant, K_d, is often orders of magnitude less than micromolar. In contrast, a number of heterogeneous methods that do exhibit high sensitivity have been described for the quantification of high-affinity antibody-binding interactions (**refs.** ***1–4***; *see* Chapter 23). A heterogeneous method is defined here as one in which one or more of the binding reagents is immobilized on a surface. Investigators have expressed concerns about the problems of limiting mass transport to the immobile binding phase ***(5)***, immobilization in multiple conformations ***(6)***, and steric hindrance on the surface *(7)*.

This chapter describes a new heterogeneous method to study antibody-antigen interactions in which the immobilized binding partner is merely a tool used

From: *Methods in Molecular Biology, Vol. 248: Antibody Engineering: Methods and Protocols*
Edited by: B. K. C. Lo © Humana Press Inc., Totowa, NJ

to capture and quantify a portion of the free, uncomplexed antibodies that are present in homogeneous reaction mixtures ***(8–10)***. This method is conducted on the KinExA™ 3000 immunoassay instrument, a computer-controlled flow fluorimeter designed to achieve the rapid separation and quantification of free, unbound antibody-binding sites present in reaction mixtures of free antibody, free antigen, and antibody-antigen complexes. In the best of circumstances, an assay to quantify antibody-antigen binding interactions should be accurate, sensitive, and convenient. Although only the first of these properties is essential, the latter two are certainly desirable. The kinetic exclusion assays described here meet all three criteria. When measuring binding interactions in the KinExA, the principal conditions that must be met are as follows: i) binding of the immobilized antigen and the corresponding soluble antigen must be mutually exclusive; and ii) binding to the immobilized antigen must be sufficiently tight to permit efficient protein capture, leading to an instrument response with acceptable signal-to-noise characteristics. Once these conditions are met, it is anticipated that this method may be applied to a wide variety of macromolecular interactions.

The operating principles of the KinExA instrument are illustrated schematically in **Fig. 1.** Briefly, the KinExA is comprised of an arrangement of tubing, connectors, valves, syringes, and pumps, and the purpose of these is to deliver accurate quantities of soluble or suspended reagents to the observation cell of the fluorimeter. Liquid samples and suspensions are drawn into the fluid-handling system of the instrument under negative pressure via a series of flexible plastic tubes (inner diameter = 0.5 mm) that are assigned to individual ports in a rotary valve with a position that is controlled by the instrument software. The capillary flow/observation cell (inner diameter = 1.6 mm) of the instrument is fitted with a microporous screen with an average pore size of 53 μm. Uniform particles larger than the pore size of the screen are precoated with antigen and deposited above the screen in a packed bed. A small volume of the reaction mixture—typically 0.5–5.0 mL—is then percolated through the packed bed of antigen-coated microbeads under negative pressure. Antibodies with unoccupied antigen-binding sites are available to bind to the immobilized antigen coated on the surface of the microbeads; antibodies with both binding sites occupied by soluble antigen are not. Exposure of the soluble binding mixture to the immobilized antigen is sufficiently brief (~480 ms for a flow rate of 0.5 mL/min) to ensure that negligible dissociation of the soluble antibody-antigen complex occurs during the time of exposure to the beads. The antibodies with binding sites that are occupied by antigens with slow unimolecular dissociation-rate constants are thus kinetically excluded from interactions with the immobilized antigen. The soluble reagents are removed from the beads by an immediate buffer wash. Quantifi-

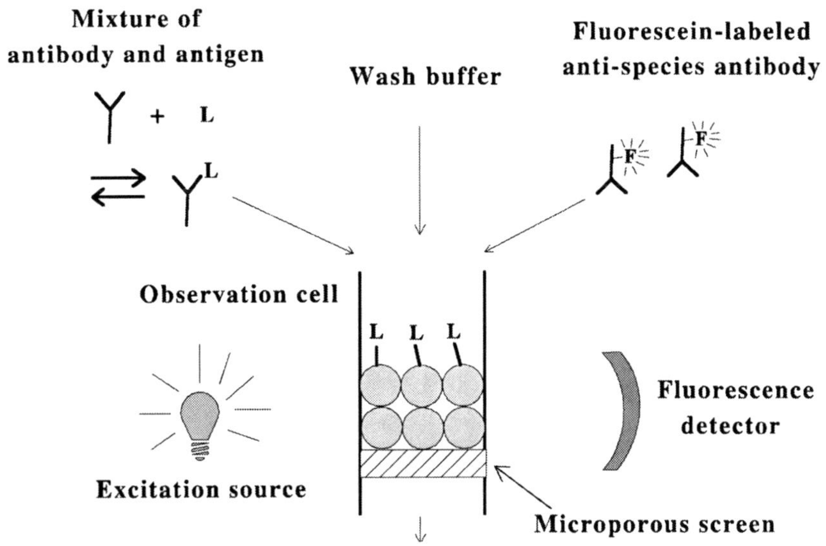

Fig. 1. Schematic of KinExA immunoassay instrument.

cation of the primary antibody thus captured on the immobilized antigen can subsequently be achieved by the brief exposure of the particles to a fluorescently labeled anti-species secondary antibody directed against the primary antibody, followed by measurement of the resulting fluorescence from the particles after removal of excess unbound reagents.

The features of this new method are illustrated herein using a model system comprised of digoxin and a mouse MAb directed against digoxin. The antibody exhibits an equilibrium dissociation constant for soluble digoxin of 8.6 pM and a bimolecular association-rate constant for binding of 4.8×10^7 M^{-1} s^{-1}. Unlike other heterogeneous methods, the values of the equilibrium and rate constants obtained by kinetic exclusion assays are those for the binding reaction in homogeneous solution. We have found the KinExA to be a remarkably sensitive tool for the characterization of antibody-antigen interactions. As part of a functional characterization of a MAb directed against chelated ionic cadmium, our laboratories determined 25 individual equilibrium and rate constants while consuming less than 500 µg of purified antibody *(8)*. Typical assay conditions that employed 0.5 n*M* antibody generated fluorescence signals of approx 1.0 V with 1–2 mV noise. Furthermore, the cost of disposable reagents was minimal compared with other heterogeneous methods of measuring binding. It is hoped that these KinExA assays will enjoy widespread use in the efforts of life scientists to understand complex biochemical processes at the molecular level.

2. Materials

1. Mouse monoclonal anti-digoxin antibody, clone number 91286 (Fitzgerald Industries International).
2. Digoxin-bovine serum albumin (BSA) conjugate (digoxin-BSA) (Fitzgerald Industries International).
3. Goat anti-mouse polyclonal antibodies covalently labeled with R-phycoerythrin (RPE) (Jackson ImmunoResearch).
4. 10% (w/v) normal goat serum (Jackson ImmunoResearch).
5. Digoxin (12β-hydroxydigitoxin) (Sigma).
6. Phosphate-buffered saline (PBS): 137 m*M* NaCl, 3.0 m*M* KCl, 10 m*M* sodium phosphate, pH 7.4, 0.03% (w/v) sodium azide.
7. Polymethylmethacrylate beads (140–170 mesh, 98 μm in diameter) (Sapidyne Instruments).
8. The KinExA™ 3000 immunoassay instrument (Sapidyne Instruments).

3. Methods

3.1. Determination of Equilibrium Dissociation Constants

3.1.1. Preparation of Reagents

1. Prepare antigen immobilized on polymethylmethacrylate beads by suspending 200 mg (dry weight) of beads in 1.0 mL of PBS that contains 100 μg of the digoxin-BSA conjugate. Agitate the suspension gently at 37°C for 1 h, then centrifuge the beads and decant the supernatant solution.
2. Block any nonspecific protein-binding sites that remain on the beads by incubating the beads with 1.0 mL of 10% (w/v) goat serum in PBS that contains 0.03% (w/v) sodium azide at 37°C for an additional hour. The blocked beads may be stored at 4°C in the blocking solution for up to 1 mo before use.
3. Prepare five reaction mixtures that contain at least 1.0 mL of 10 pM anti-digoxin antibody (20 pM in binding sites) and different concentrations of soluble digoxin (0–10 n*M*) in PBS.

3.1.2. Preparation of the KinExA Instrument to Assay Equilibrium Binding Mixtures

1. Transfer the suspension of 200 mg of beads coated with digoxin-BSA and blocked with goat serum (~1.0 mL of goat serum with ~0.5 mL of bead volume), along with 30 mL of PBS, to the bead reservoir on the instrument.
2. Assign, via the plastic tubing, each of the five reaction mixtures to ports 2 through 6, inclusive.
3. Assign a solution of RPE-conjugated goat anti-mouse antibody (1.0 μg/mL) to port 14 on the instrument.
4. Follow the KinExA instrument manual to instruct the KinExA software to perform the sample-handling timing routine summarized by the lines of code contained in the "bead-handling" and "equilibrium analysis" sections of **Table 1.** Each row in **Table 1** represents a command in which the instrument is instructed to conduct a specific

Table 1
Sample-Handling Timing Routines for the Determination of Free Antibody in Either Equilibrium or Kinetic Reaction Mixtures

				Draw		Inject			
Line no.	Material	Port	Time (s)	Vol. (µL)	Rate (µL/min)	Vol. (µL)	Rate (µL/min)	Loop	Stir
Bead-handling:									
1	Backflush	–	30	–	–	–	–	No	Off
2	Buffer	1	20	500	1.5	–	–	No	On
3	Polybeads	–	27	675	1.5	–	–	No	On
4	Buffer	1	40	1000	1.5	–	–	No	Off
5	Waste	–	10	50	0.3	–	–	No	Off
6	Buffer	1	20	0	0	–	–	No	Off
7	Buffer	1	6	150	1.5	–	–	No	Off
Equilibrium analysis (5 samples):									
8a	Sample 1	2	600	5000	0.5	–	–	Yes	Off
8b	Sample 2	3	600	5000	0.5	–	–	Yes	Off
8c	Sample 3	4	600	5000	0.5	–	–	Yes	Off
8d	Sample 4	5	600	5000	0.5	–	–	Yes	Off
8e	Sample 5	6	600	5000	0.5	–	–	Yes	Off
9	Buffer	1	20	166	0.5	–	–	No	Off
10	2° antibody	14	120	1000	0.5	–	–	No	Off
11	Buffer	1	120	3000	1.5	–	–	No	Off
Kinetic analysis (5 samples):									
8a	Sample 1	2	60	250	0.25	250	0.25	Yes	Off
8b	Sample 2	2	60	250	0.25	250	0.25	Yes	Off
8c	Sample 3	2	60	250	0.25	250	0.25	Yes	Off
8d	Sample 4	2	60	250	0.25	250	0.25	Yes	Off
8e	Sample 5	2	60	250	0.25	250	0.25	Yes	Off
9	Buffer	1	20	166	0.5	–	–	No	Off
10	2° antibody	14	120	1000	0.5	–	–	No	Off
11	Buffer	1	120	3000	1.5	–	–	No	Off

This table was adapted from the KinExA™ software display.

fluid-handling operation. The lines of command are executed in numerical order for each assay. Lines 1–7, inclusive, comprise the bead-handling routine. At the conclusion of line 7, a packed bed of antigen-coated beads is ready for the introduction of sample. A new set of disposable beads is deposited in the observation cell for each individual KinExA assay. Lines 8–11, inclusive, comprise the sample-handling routine. In the "equilibrium analysis" example given in **Table 1,** five individual assays are specified, corresponding to the analyses of samples 1–5, inclusive. A different line 8 (*a* through *e*) is executed per individual assay. Lines 9–11 are executed for each of the five assays. The operation specified by each line in **Table 1** is as follows:

a. Line 1, backflush: A peristaltic pump is activated to flush the current contents of the observation cell to a waste receptacle.
b. Line 2, buffer: 500 µL of PBS are drawn through the observation cell to equilibrate the relevant tubing and chambers to the solution conditions of the assay. The mechanical stirrer is activated to create a homogeneous suspension of the beads in the reservoir immediately prior to their withdrawal.
c. Line 3, polybeads: 675 µL of the bead suspension are drawn into the observation cell, where the beads are retained upstream from the microporous screen to create a bed of packed beads approx 4 mm high.
d. Line 4, buffer: 1.0 mL of PBS is drawn past the packed beads to remove excess goat serum and any remaining soluble digoxin-BSA.
e. Line 5, waste: 50 µL are forced upstream through the observation cell to "lift" the beads from their initial packed bed into a transient suspension.
f. Line 6, waste: 20 s of inactivity permit the beads, which are denser than the PBS, to settle via gravity back into a loosely packed bed on top of the microporous screen. Experience has shown that beads deposited in this fashion behave much more reproducibly in the subsequent assay than those that are initially deposited at the conclusion of line 3.
g. Line 7, buffer: 150 µL of PBS are drawn through the observation cell to create a tighter beadpack in preparation for the introduction of sample.
h. Line 8a, sample 1: 5.0 mL of sample 1 are drawn through the observation cell via port 2. Any free, uncomplexed anti-digoxin antibody present in the sample has the opportunity to bind to the immobilized digoxin on the beads (*see* **Note 1**). Line 8a is only executed for the first assay, corresponding to the analysis of sample number 1. Any line that contains a "Yes" in the Loop column of the software display is executed for only one assay. Thus, Line 8b is executed for a subsequent assay of sample 2 and Line 8c is executed for a subsequent assay of sample 3.
i. Line 9, buffer: 166 µL of PBS are drawn through the observation cell to remove excess unbound reagents that are present in the equilibrium reaction mixture.
j. Line 10, secondary antibody: 1.0 mL of the fluorescently labeled anti-species antibody is drawn through the observation cell to bind to primary antibody captured on the beads during the execution of line 8 (*see* **Note 2**).
k. Line 11, buffer: 3.0 mL of PBS are drawn through the observation cell to remove excess unbound secondary antibody. Any line that contains a "No" in the Loop column of the display is executed for each assay. Thus, lines 9, 10, and 11 specify operations that are performed for all the five assays.

3.1.3. Assay Samples and Analyze Equilibrium Binding Data

1. Initiate data acquisition in the KinExA software. The instrument conducts fluorescence readings at the rate of one datum per s at the beginning of line 8. **Fig. 2** shows examples of the time-courses for the fluorescence signal when the quantities of free antibody present in different equilibrium mixtures of anti-digoxin and soluble

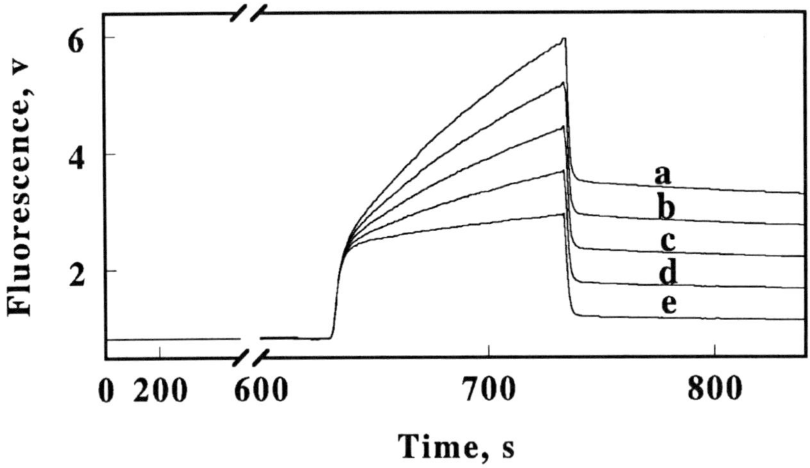

Fig. 2. Time-courses of individual fluorescence responses observed when different equilibrium mixtures of digoxin and anti-digoxin MAb were exposed to beads coated with an excess of digoxin-BSA, followed by goat anti-mouse antibodies covalently conjugated with RPE. Initial concentrations of soluble digoxin were zero, 8.7 pM, 29.2 pM, 41.7 pM, and 10.0 n*M* in experiments *a* through *e*, respectively. Anti-digoxin primary antibody and RPE-labeled secondary antibody concentrations were 10 pM (20 pM in total binding sites) and 1.0 μg/mL, respectively.

digoxin were determined. The instrumental response from 0–630 s corresponded to the background signal generated during the time that the unlabeled equilibrium mixture was exposed to and subsequently washed out of the packed microbead column. The beads were then exposed to a solution of RPE-labeled goat anti-mouse antibodies (630–730 s). Excess unbound labeled secondary antibody was then removed from the beads with a buffer wash (730–840 s).

2. When the equilibrium mixture contained a saturating concentration of the free digoxin *(curve e)*, the instrument response approximated a square wave corresponding to the fluorescence of the labeled secondary antibody during its transient passage past the beads in the observation cell. The signal failed to return to that of the background, indicating a nonspecific binding of the RPE-labeled antibody of 3.5%. When soluble digoxin was omitted from the equilibrium reaction mixture *(curve a)*, the instrument response from 630–730 s reflected the sum of two contributions: the fluorescence of the unbound labeled antibody in the interstitial regions among the beads and that of the labeled secondary antibody that had bound to anti-digoxin antibody captured by the immobilized digoxin on the beads. Binding of the secondary antibody was an ongoing process that produced a positive slope in this portion of the curve. When the excess unbound label was removed from the beads, the signal that remained was the sum of that from the nonspecifically bound antibody plus that of the labeled anti-mouse goat antibody that specifically bound to

the anti-digoxin antibody captured on the beads. Equilibrium mixtures comprised of soluble ligand present at concentrations intermediate between those of zero and saturation thus provided intermediate instrument responses (*curves b–d*) from which the concentration of free anti-digoxin in each reaction mixture could be determined.

3. Analyze the fluorescence data. In principle, the concentration of free antibody can be calculated from the slopes of the curves in the 630–730-s interval, from the average value of a portion of the plateau in the 750–840-s interval, or from the corresponding integrals of the areas under selected portions of the curves. The latter alternative is illustrated herein. The fraction of occupied binding sites on the soluble anti-digoxin MAb may be taken as:

$$\text{Fraction of occupied binding sites} = \frac{I_0 - I_{exp}}{I_0 - I_\infty} \quad (1)$$

where I represents the integral of each time-course such as those in **Fig. 2** over the interval of 750–840 s, and the subscripts 0, exp, and ∞ refer to time-courses corresponding to a soluble digoxin concentration of zero, an intermediate digoxin concentration, and a saturating concentration of digoxin, respectively. A plot of the fraction of occupied binding sites on the anti-digoxin antibody as a function of the concentration of soluble digoxin is given in **Fig. 3.** The value of the apparent equilibrium dissociation constant may be obtained from a nonlinear regression fit of the following quadratic equation to the data:

$$\text{Fraction of occupied binding sites} = \frac{([Ab]+[Ag]+K_d) - \{([Ab]+[Ag]+K_d)^2 - 4[Ab][Ag]\}^{1/2}}{2} \quad (2)$$

where [Ab] and [Ag] are the total concentrations of anti-digoxin and digoxin, respectively, and K_d is the equilibrium dissociation constant. **Equation 2** corresponds to a one-site homogeneous binding model where the binding event causes a significant depletion in the concentration of both of the free, unbound reagents (*see* **Note 3**). The equilibrium dissociation constant for the binding of digoxin to the anti-digoxin antibody in homogeneous solution was determined on the KinExA 3000 to be 8.6 ± 1.7 pM (*see* **Note 4**).

3.2. Determination of Individual Kinetic Constants

Values for the association and dissociation-rate constants for the binding reaction between digoxin and the anti-digoxin antibody may also be determined by kinetic exclusion assays on the KinExA. Instead of exposing an equilibrium mixture of the binding partners to the immobilized digoxin on the beads, a solution of the anti-digoxin MAb is injected into and mixed with a stream of soluble digoxin that reacts for 7 s before the mixture encounters the packed beads. This is physically accomplished by filling the injection syringe on the instrument with

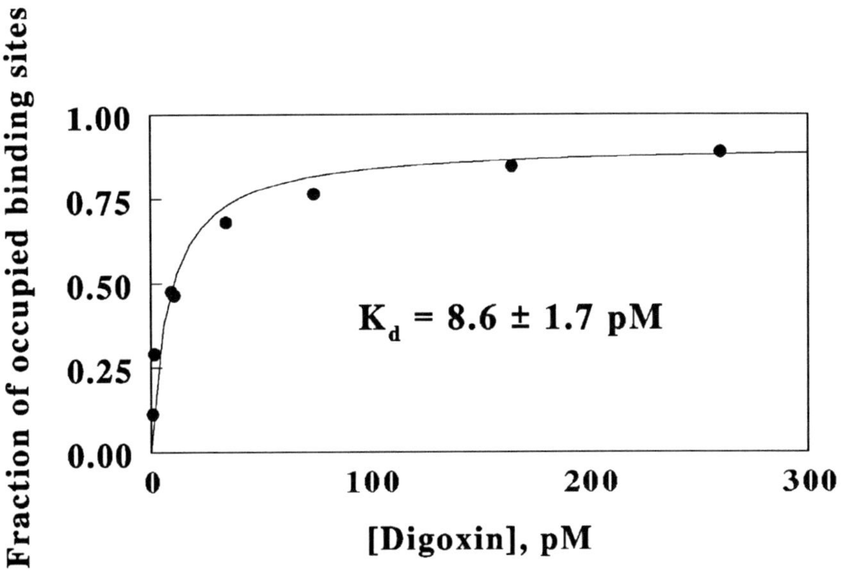

Fig. 3. Determination of the equilibrium dissociation constant for the binding of the anti-digoxin antibody to soluble digoxin. The concentration of occupied antibody-binding sites, determined from measurements of the unbound antibody such as those illustrated in **Fig. 2**, was expressed as a fraction of the concentration of total antibody-binding sites and plotted versus the concentration of the free antigen in solution. The parameters for the *curve* drawn through the data points were determined by nonlinear regression analysis using **Equation 2** in the text.

an appropriate solution of anti-digoxin, and then injecting the antibody solution by positive pressure into a moving stream of soluble digoxin (which is simultaneously being drawn by negative pressure from port 2).

3.2.1. Preparation of Reagents

1. Repeat **Subheading 3.1.1., steps 1** and **2** to prepare immobilized antigen on blocked beads.
2. Prepare 5 mL of 120 pM anti-digoxin antibody (240 pM in binding sites) in PBS.
3. Prepare five solutions that contain different concentrations of soluble digoxin (0–10 n*M*) in PBS.

3.2.2. Preparation of the KinExA Instrument for Kinetic Analysis

1. Repeat **Subheading 3.1.2., step 1** to transfer beads coated with immobilized antigen to the bead reservoir on the instrument.
2. Assign the antibody solution to the injection port on the instrument.
3. Assign the five antigen solutions to ports 2–6, inclusive.

4. Follow the KinExA instrument manual to instruct the KinExA software to perform the sample-handling timing routine summarized by the lines of code contained in the "kinetic analysis" section in **Table 1.** Line 8 in the equilibrium routine is replaced with one in which the time is 60 s and the Draw and Inject volumes are 250 μL each (which necessarily dictates that the corresponding rates must be 0.25 mL/min in both categories). The effect of this command is to mix the two solutions at the point of injection, and to subsequently draw the mixture past the packed bed of beads in the observation cell. In the example presented here, the final concentration of anti-digoxin (after mixing 1:1 with the digoxin solution) was 120 pM in the binding sites. The final concentrations of soluble digoxin ranged from 5.0–60 n*M*.

3.2.3. Assay Samples and Analyze Kinetic Data

1. Initiate data acquisition in the KinExA software. The primary data derived from these kinetic experiments resembled those obtained from the equilibrium experiments (for example, see the traces in **Fig. 2**), because in either case the KinExA assay was measuring the number of unoccupied binding sites that remained in the population of antibody molecules. However, when performing kinetic experiments, these data represented the concentration of unoccupied binding sites plucked from a mixture of binding partners that was only 7 s old and far from equilibrium.
2. Analyze the fluorescence data. The extent of the binding reaction during the 7 s that the antibody and the antigen are mixed prior to encountering the immobilized antigen on the beads is controlled by varying the concentration of the reagent in excess, the soluble digoxin. The fraction of unoccupied binding sites that remains after 7 s of reaction is taken as:

$$\text{Fraction of binding sites remaining} = \frac{I_{exp} - I_{\infty}}{I_0 - I_{\infty}} \qquad (3)$$

Fig. 4 shows a secondary plot of the remaining unoccupied binding sites as a function of the digoxin in the final kinetic mixture. A single exponential function of the concentration of digoxin was fit to the data in **Fig. 4:**

$$I_{exp} = (I_0 - I_{\infty})\, e^{-k1\,[\text{digoxin}]\,t} + I_{\infty} \qquad (4)$$

where t is 7 s and k_1 is the second-order rate constant for the bimolecular binding of digoxin to the anti-digoxin antibody. A value of $4.8 \pm 0.3 \times 10^7\ M^{-1}\ s^{-1}$ was obtained for the bimolecular association of anti-digoxin with digoxin from the nonlinear regression fit of **Equation 4** to the data in **Fig. 4.** The value of the corresponding unimolecular dissociation-rate constant, k_2, for the antibody-antigen complex was obtained from the KinExA data using the identity $k_2 = K_d \times k_1$, or $k_2 = 4.1 \pm 0.9 \times 10^{-4}\ s^{-1}$ (*see* **Note 5**).

4. Notes

1. Unlike other automated instrumentation devoted to the study of protein-binding interactions in which the interaction to be quantified is that between a soluble and an

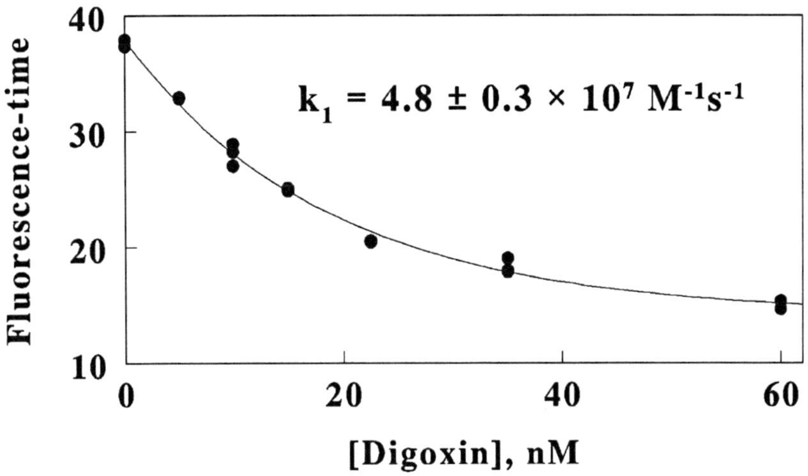

Fig. 4. Determination of the association-rate constant for the bimolecular combination of the anti-digoxin antibody with soluble digoxin. Anti-digoxin antibody and different concentrations of digoxin were incubated for 7 s before separation of the primary antibodies in the mixture into bound and free fractions. Each determination was performed in duplicate; in some cases the error in the data was less than the diameter of the plotted points.

immobilized binding partner, the equilibrium and kinetic-rate constants determined on the KinExA are those obtained for the binding reaction in homogeneous solution. This is an important distinction. The very act of covalently immobilizing one of the binding partners can physically interfere with the normal mode of binding in which small ligands are involved or introduce site-specific heterogeneity in the presentation of large proteinaceous ligands. The immobilized reactant in the KinExA is merely a tool used to separate and quantify the free proteins with unoccupied binding sites that are present in the solution reaction mixture. The actual kinetic properties of binding to the immobilized reagent are immaterial as long as: i) capture from the solution phase is sufficiently rapid to generate a quantifiable signal and ii) a constant percentage of the free protein is separated from each reaction mixture, regardless of the concentration of the competing soluble ligand. The latter condition is readily achieved by ensuring that the effective concentration of the immobilized ligand is in a 10-fold or greater molar excess to that of the soluble protein ***(8,10)***. Binding of the free protein to the immobilized ligand under these conditions is a pseudo-first-order process, in which the rate and thus the extent of binding during the limited contact time between the free protein and the immobilized ligand are dictated by the concentration of the reagent in excess (the immobilized ligand) and are independent of the concentration of the limiting component, the free protein.

2. Although it may appear that the requirement to create a fluorescence signal that is proportional to the quantity of free protein captured on the beads might limit the

usefulness of this procedure, that is not usually the case. When the primary binding reaction in solution is that between an antibody and an antigen, these KinExA assays are very convenient because anti-species secondary antibodies labeled with a variety of fluorescent molecules are available from many commercial sources. In those cases in which the primary binding reaction involves proteins other than antibodies, poly- or MAbs to many physiologically significant proteins of interest are widely available, such antibodies may be readily covalently coupled to fluorescent molecules using commercially available kits. It is anticipated that kinetic exclusion assays conducted on the KinExA instrument can be regarded as an attractive alternative to more labor-intensive, time-consuming methods for studying protein-binding interactions.

3. Regardless of the experimental approach used to quantify binding, the ability to quantify very high-affinity binding interactions (K_d much less than 1.0 n*M*) is limited by two constraints: i) the time required to achieve practical equilibrium, which depends primarily upon the value of the unimolecular dissociation constant for the bimolecular binding reaction. This is a serious practical limitation in the study of high-affinity interactions, since a binding reaction in which dissociation of the complex occurs with a rate constant of 10^{-5} s^{-1} will only be 87.5%, or three half-lives, of the way to completion after 48 h of incubation; and ii) the necessity of generating a fluorescence signal sufficient for reliable quantification. This is a less serious problem because free protein is accumulated and concentrated on the beads. In order to achieve reliable equilibrium-binding data, it is necessary to ensure that the total concentration of antibody is equivalent to or less than the equilibrium dissociation constant for the binding reaction. For high-affinity interactions in which one must use a low concentration of the binding protein, an acceptable fluorescence signal can be obtained in the KinExA with very dilute antibodies by simply increasing the volume of the dilute equilibrium reaction mixture drawn past the beads. Our laboratories have quantified equilibrium dissociation constants in the low pM range using individual KinExA assays of less than 30 min each (**ref. *11***, and data not shown). Studies on interactions of even higher affinity become problematic because of the time required to achieve practical equilibrium in solution before analysis.
4. The ability to quantify low-affinity binding interactions in solution is equally broad. On the one hand, the rapid dissociation of low-affinity complexes means that proteins complexed with soluble ligands will dissociate and be able to bind to the high-affinity ligand immobilized on the solid phase during the short time that each portion of the sample is exposed to the beadpack. However, the high concentration of soluble ligand required to produce a particular equilibrium concentration of low-affinity protein-ligand complexes means that the soluble ligand can effectively compete with the immobilized ligand by rapidly rebinding the newly dissociated protein. It is anticipated that individual components in a low-affinity solution reaction mixture would remain in rapid equilibrium as small amounts of free protein were removed by the high-affinity binding to immobilized ligand on the beads. The strategy that would yield the most reliable value for the bimolecular association-

rate constant for the solution binding reaction under these conditions is thus one in which the amount of free protein captured on the beadpack was as small a portion of the total protein as possible (thus, the solution reaction would be perturbed as little as possible). This could be accomplished experimentally by increasing the total protein concentration, decreasing the concentration of immobilized ligand on the beads, and/or decreasing the time of exposure to the beads (by increasing the flow rate). Our laboratories have quantified equilibrium dissociation constants in the mM range, although the error in the data appears to increase as the affinity decreases (data not shown).

5. One limitation of this method is its inability to quantify bimolecular association-rate constants for low-affinity binding interactions, because of the finite time required to transport the newly mixed binding reagents from the point of mixing to the beadpack. The difficulty in studying the kinetics of low-affinity interactions is that the combination of high unimolecular dissociation-rate constants and the high concentrations of soluble ligand required to form protein-ligand complexes dictate that equilibrium is achieved in solution reaction mixtures well within the 7 s needed for the newly formed mixture to encounter the beadpack. Kinetic studies on binding reactions with unimolecular dissociation-rate constants of 1.0 s^{-1} or higher (corresponding roughly to equilibrium dissociation constants on the order of 1.0 μM or higher for reactions with association-rate constants on or around 10^6 $M^{-1}s^{-1}$) are usually not possible with this method. In principle, one could extend the range of binding reactions amenable to kinetic studies by moving the observation cell closer to the point of mixing and by increasing the flow rate over the beadpack, although the latter change is of limited value because it would also serve to decrease the magnitude of the fluorescence signal. It is unlikely that these proposed changes in the current configuration of the KinExA instrument would extend the range of dissociation constants of value to kinetic studies by more than a single order of magnitude.

Acknowledgments

This research was supported by the Office of Science (BER), U.S. Department of Energy, Grants No. DE-FG02-98ER62704 and DE-FG-02-02ER63459 and by the U.S. National Institutes of Health Grant No. GM08008-26S1.

References

1. Chaiken, I., Rose, S., and Karlsson, R. (1992) Analysis of macromolecular interactions using immobilized ligands. *Anal. Biochem.* **201,** 197–210.
2. Dill, K., Lin, M., Poteras, C., Fraser, C., Hafeman, D. G., Owicki, J. C., et al. (1994) Antibody-antigen binding constants determined in solution phase with the Threshold membrane capture system: binding constants for anti-fluorescein, anti-saxitoxin, and anti-ricin antibodies. *Anal. Biochem.* **217,** 128–138.
3. Friguet, B., Chaffotte, A. F., Djavaki-Ohaniance, L., and Goldberg, M. (1985) Measurement of the true affinity constant in solution of antigen-antibody complexes by enzyme-linked immunosorbent assay. *J. Immunol. Methods* **77,** 305–319.

4. Carter, R. M., Mekalanos, J. J., Jacobs, M. B., Lubrano, G. J., and Guilbault, G. G. (1995) Quartz crystal microbalance detection of *Vibrio cholerae* O139 serotype. *J. Immunol. Methods* **187,** 121–129.
5. Glaser, R. W. (1993) Antigen-antibody binding and mass transport by convection and diffusion to a surface: a two-dimensional computer model of binding and dissociation kinetics. *Anal. Biochem.* **213,** 152–161.
6. O'Shannessy, D. J. and Winzor, D. J. (1996) Interpretation of deviations from pseudo-first-order kinetic behavior in the characterization of ligand binding by biosensor technology. *Anal. Biochem.* **236,** 275–283.
7. Edwards, P. R., Gill, A., Pollard-Knight, D. V., Hoare, M., and Buckle, P. E. (1995) Kinetics of protein-protein interactions at the surface of an optical biosensor. *Anal. Biochem.* **231,** 210–217.
8. Blake, D. A., Chakrabarti, P., Khosraviani, M., Hatcher, F. M., Westhoff, C. M., Goebel, P., et al. (1996) Metal binding properties of a monoclonal antibody directed toward metal-chelate complexes. *J. Biol. Chem.* **271,** 27,677–27,685.
9. Blake, D. A., Khosraviani, M., Pavlov, A. R., and Blake, II, R. C. (1997) Characterization of a metal-specific monoclonal antibody, in *Immunochemical Technology for Environmental Applications* (Aga, D. S. and Thurman, E. M., eds.), American Chemical Society, pp. 49–60.
10. Blake, II, R. C., Pavlov, A. R., and Blake, D. A. (1999) Automated kinetic exclusion assays to quantify protein binding interactions in homogeneous solution. *Anal. Biochem.* **272,** 123–134.
11. Kaplan, B. E., Hefta, L. J., Blake, II, R. C., Swiderek, K. M., and Shively, J. E. (1998) Solid phase synthesis and characterization of carcinoembryonic antigen (CEA) domains. *J. Peptide Res.* **52,** 249–260.

25

Characterization of Antibody–Antigen Interactions by Fluorescence Spectroscopy

Sotiris Missailidis and Kevin Brady

1. Introduction

Fluorescence spectroscopy is a widely used technique for the characterization of molecular interactions in biological systems. It is highly sensitive, and allows measurements at low sample concentrations (down to 10^{-7} *M*). It is also available at a reasonably low equipment cost. These features make fluorescence spectroscopy an attractive technique as compared to other forms of optical spectroscopy.

Fluorescence is a three-stage process that occurs in certain molecules known as fluorophores or fluorescent dyes (generally polyaromatic hydrocarbons or heterocycles). The process leading to fluorescence emission from such molecules is illustrated by the Jablonski diagram **(Fig. 1)**. The first stage is the excitation step, in which the fluorophore absorbs a photon of energy $h\nu_{EX}$ and becomes an excited electronic singlet state (S_1'). The excitation wavelength is usually the same as the absorption wavelength of the fluorophore. The second stage is the excited-state lifetime, when the excited state exists for a finite period of time (typically 1–10 × 10^{-9} s). During this period, the fluorophore undergoes conformational changes and is also subject to possible interactions with its molecular environment, with two important consequences. First, the energy of S_1' is partially dissipated, yielding a relaxed singlet excited state (S_1) from which fluorescence emission originates. Second, not all the molecules initially excited by absorption (stage 1) return to the ground state (S_0) by fluorescence emission, as other processes such as collisional quenching, fluorescence energy transfer, and intersystem crossing may also depopulate S_1. The third stage is fluorescence emission, when a photon of energy $h\nu_{EM}$ is emitted, returning the fluorophore to its ground state S_0. Because of energy dissipation during the excited-state lifetime, the energy of this photon is lower, and thus of

From: *Methods in Molecular Biology, Vol. 248: Antibody Engineering: Methods and Protocols*
Edited by: B. K. C. Lo © Humana Press Inc., Totowa, NJ

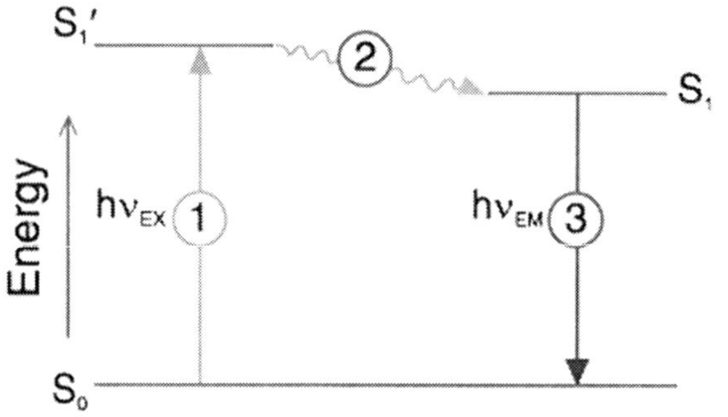

Fig. 1. The Jablonski diagram illustrates the three stages involved in the creation of an excited electronic singlet state by optical absorption and subsequent emission of fluorescence.

a longer wavelength, than the excitation photon $h\nu_{EX}$. Consequently, the ratio of the number of fluorescence photons emitted (stage 3) to the number of photons absorbed (stage 1) represents the *fluorescence quantum yield.*

The fluorescence process recurs, and the same fluorophore can be repeatedly excited and detected. For polyatomic molecules in solution, the discrete electronic transitions represented by $h\nu_{EX}$ and $h\nu_{EM}$ in **Fig. 1** are replaced by broad energy spectra known as the fluorescence excitation and fluorescence emission spectrum, respectively. With few exceptions, the excitation spectrum of a single species in dilute solution is identical to its absorption spectrum. Under the same conditions, the fluorescence emission spectrum is independent of the excitation wavelength, because of the partial dissipation of excitation energy during the excited-state lifetime. The emission intensity is proportional to the amplitude of the fluorescence excitation spectrum at the excitation wavelength. Fluorescence emission intensity depends on the same parameters as absorbance—defined by the Beer-Lambert law as the product of the molar extinction coefficient, optical pathlength, and concentration—as well as on the fluorescence quantum yield, the intensity of the excitation source, and the efficiency of the instrument and, in dilute solutions, is linearly proportional to these parameters.

Fluorescence is an absorption/emission phenomenon that has been associated with a number of biological molecules and extensively reviewed in the literature *(1)*. The induction of fluorescence is determined by the presence of certain chromophores within a molecule. When examining proteins, these chromophores are found in the aromatic amino acids tryptophan, tyrosine, and phenylalanine, which have high densities of delocalized electrons. Of the three aromatic amino acids, tryptophan has by far the greatest contribution to the fluorescence of proteins. Its excitation wavelength is approx 295 nm. The emission wavelength varies according to the local environment of the tryptophans—e.g., 320 nm for buried residues, and 360 nm for solvent-exposed residues.

Alterations in the intensity of the tryptophan emission spectrum can be studied when an antibody is involved in a binding event. A reduction in the intensity of the fluorescence emission is observed upon binding of the antibody to its antigen. However, such a reduction is dependent upon the presence of tryptophan residues in the vicinity of the antigen-binding pocket that would make them susceptible to the minor alterations in the antibody structure upon binding, thus resulting in the reduction of their fluorescence emission. This is known as fluorescence quenching, and is especially useful when the binding partner, the antigen, contains no fluorescent chromophores and is thus invisible to the fluorimeter. When studying the intensity of the antibody fluorescent signal, the changes observed are solely caused by the change in the environment of the tryptophans within the antibody.

Two methods are described here for the study of antibody-antigen interactions using fluorescence quenching. The first method is based on keeping the antibody concentration constant and varying the antigen concentration accordingly to achieve various antibody-antigen ratios. This is an inverse titration, in which the first point measured is that of the antibody-antigen complex, when the antigen is in excess and maximum binding has been achieved. The antigen concentration is reduced at subsequent steps as aliquots of the antibody-antigen complex are replaced by aliquots of free antibody at the same concentration. A final point of free antibody in the absence of antigen is measured.

The second method is based on continuously increasing antigen concentrations throughout the titration. This method is a straightforward titration, in which the first point is that of free antibody, to which aliquots from a stock antigen solution are added until antigen is in excess in solution and saturation of all antigen-binding pockets of the antibody has been achieved. The data must then be normalized as the volume, and thus the concentration of the antibody changes every time an antigen aliquot is added to the solution. This method presents the significant advantage of using much smaller amounts of antibody material. However, normalizing data for volume addition has the potential to incorporate errors in a system that is so sensitive to volume changes. It is therefore imperative to know how sensitive your system is to changes in volume/concentration before you use this method of analysis. Each of the two methods therefore presents certain advantages and disadvantages (*see* **Notes 1** and **2**), which should be considered before an appropriate choice of method is made. Furthermore, both methods have a general applicability and can be used with other spectroscopic techniques for the analysis of molecular interactions, including those of antibody-antigen interactions (*see* **Note 3**).

The analysis of the data is sometimes as important as taking the experimental measurements. In the past, many workers have employed the Scatchard analysis plots for determining equilibrium association constants K_A. Scatchard analysis was first introduced *(2)* to provide data from straight-line plots produced for systems that were completely homogeneous and monomeric. An examination of most of the published uses of Scatchard plots makes it clear that many workers fit linear plots through nonlinear data. These situations oversimplify the binding event, and can incorporate large errors in the determination of binding constants. We therefore recommend (*see* **Note 4**) that users analyze their data using **Equation 1** *(3,4)* or variations of it using nonlinear least squares fitting, which would alleviate such problems of data fitting:

$$\mathrm{Acalc(P0)} = \left[\left[\frac{(\mathrm{E1} - \mathrm{E2})\left[(1 + \mathrm{K} \cdot \mathrm{D} + \mathrm{K} \cdot \mathrm{P0}) - \sqrt{(1 + \mathrm{K} \cdot \mathrm{D} + \mathrm{K} \cdot \mathrm{P0})^2 - 4 \cdot \mathrm{K} \cdot \mathrm{K} \cdot \mathrm{D} \cdot \mathrm{P0}}\right]}{2 \cdot \mathrm{K}}\right] + \mathrm{E2} \cdot \mathrm{D}\right] \cdot \frac{1}{\mathrm{D}}$$

Equation 1: Quadratic equation for fitting binding curves.

where K is the equilibrium association constant, D is the concentration of antibody, E1 is the maximum value of Acalc(P0), and E2 is the ratio F_0/F at F equal to F_0 (no antigen present) and therefore equals 1. Acalc(P0) is the calculated parameter representing F_0/F, the maximum fluorescence of the free antibody divided by the fluorescent signals obtained in the presence of quenching antigen. P0 is the total antigen concentration. Values of F_0/F for each antigen concentration vs antigen concentration at each titration point will have to be imported. Values of K and E1 are iteratively manipulated until the best fit of line to data is achieved or computer-calculated for best fit.

2. Materials

1. Phosphate-buffered saline (PBS): dissolve one PBS tablet (Sigma; Cat. #P4417) in 200 mL of distilled water to give yield 0.01 *M* phosphate buffer, 0.0027 *M* KCl, and 0.137 *M* NaCl, pH 7.4; sterile-filter through a 0.22-μm filter.
2. Antibody of interest: store the antibody at –20°C in small aliquots to avoid denaturation or inactivation from repeated freeze-thaw cycles.
3. Antigen of interest: Depending on the nature of your antigen—if it is a small peptide, keep under dry conditions as solid and dissolve in buffer as needed; whereas if it is a protein in solution, keep frozen in small aliquots to avoid denaturation or inactivation through repeated freeze-thaw cycles.
4. Measurements can be taken in any fluorimeter available in your laboratory, using a fluorescence quartz cuvet. We have used a Perkin-Elmer LS-5 luminescence spectrometer.

3. Methods

3.1. Using Fixed Antibody Concentration

When using a fixed concentration of the antibody and varying concentrations of the antigen, one can produce a binding curve from which equilibrium constants K_A and K_D can be derived. This method was previously employed by Reedstrom et al. *(5)* for the study of protein-DNA interactions, but it can equally be applied to the study of antibody-antigen interactions as well as most intermolecular interactions. The method of using fixed antibody concentration is in fact a reverse titration, in which the initial measurement is of antibody fully saturated by antigen and the last is of free antibody.

1. Prepare in a fluorescence quartz cuvet a 2-mL solution of the antibody/antigen mixture. The initial concentration of the antibody should be 0.26 µ*M*, expressed in antigen-binding pockets (or 0.13 µ*M* of whole IgG molecules); the initial concentration of antigen epitopes should be 1.5 µ*M*. If a multimeric antigen is used, be sure to calculate the concentration per epitope and not for the whole antigen. It is generally believed that antibody-antigen interactions are straightforward, one-to-one interactions. However, this is not always the case, and care should be taken in calculating the concentration of antigen per epitope and of the antibody per antigen-binding pocket. Cooperativity effects can be observed with multimeric antigens with many epitopes, and this should be taken into consideration when calculating the association constants (*see* **Note 5**). The excess concentration of antigen epitopes (~5.8:1) ensures total saturation of all antigen-binding pockets on the antibody.
2. Prepare a separate free antibody solution, of 10 mL in volume and at the same concentration as that in the cuvet (0.26 µ*M* expressed in antigen-binding pockets).
3. Adjust the fluorimeter settings to: excitation slit width at 5 nm; emission slit width at 10 nm; excitation wavelength at 290 nm; and emission wavelength at 345 nm.
4. Measure the fluorescence emission of the solution (F; observed intensity of fluorescence in the presence of varying amounts of antigen).
5. Remove from the cuvet 750 µL of antibody/antigen complex solution and replace it with 750 µL of the free antibody solution. Allow the mixture to equilibrate for 1 min and measure the fluorescence emission.
6. Repeat **step 5** for ten consecutive times and record the data. During **steps 5** and **6** the antigen concentration is reduced exponentially over the ten replacements of the antibody-antigen mixture with free antibody, while keeping the antibody concentration unchanged throughout the titration (**Fig. 2**).
7. Clean the cuvet and fill it with the remainder of the pure antibody solution. Take the measurement of the emission. This value will give you the fluorescence emission of free antibody in solution (F_0). Fit the data using **Equation 1** *(3,4)* or variations of it using nonlinear least squares fitting. You can use this equation in any curve-fitting program available in your laboratory. We have used this equation in two different programs, MathCad and Origin. If you are using MathCad for fitting your data, then you should perform iterative, manual manipulations to achieve the best fit. If

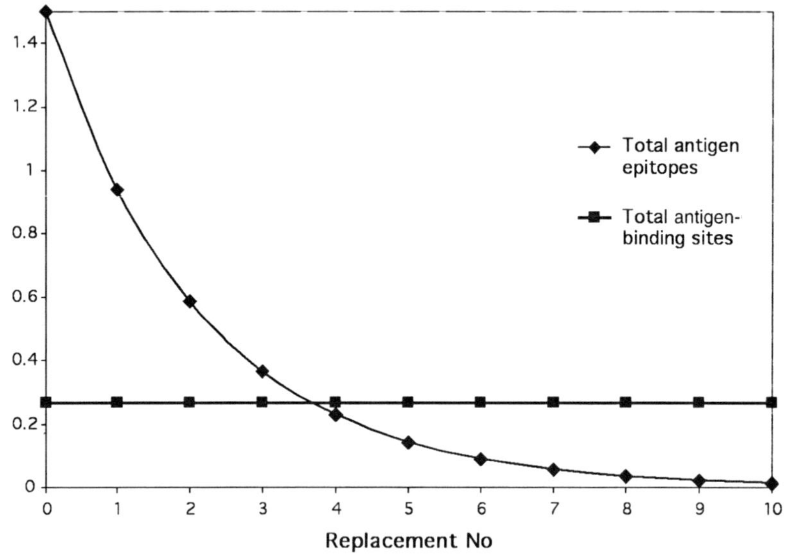

Fig. 2. Relationship between the number of total antigen epitopes and total antigen-binding sites over the course of a flourescence quenching experiment (method one). The antibody concentration remains constant throughout, and the antigen concentration decreases exponentially. Initially there is an excess of antigen epitopes (ratio of 5.8:1, antigen:antibody). At about replacement number 4, there is almost a 1:1 ratio of antigen to antibody. Finally, the antigen concentration decreases to an extent in which there is a large excess of antibody binding sites (ratio of 1:19.6, antigen:antibody). In this way, the occupancy of binding pockets by antigen changes from fully saturated at the start to about 5% occupancy at the end of the experiment.

you are using Origin, the program will automatically calculate the values for E1 and K to achieve the best fit.

3.2. Using Increasing Amounts of Antigen

A direct titration, in which increasing amounts of antigen are added to an antibody solution and the change in fluorescence is measured at each addition point, can also be employed for the identification of equilibrium binding constants. This method is based on the work of Parker *(6)*, in which small aliquots of antigen are added to a solution of antibody, and the signal is monitored. The signal is normalized for the dilution effect of adding the antigen to the antibody solution. K_A is then determined by plotting the data. This is a direct titration, in which the first point is that of free antibody and the last point is with the antibody fully saturated by antigen.

1. Prepare 2.5 mL of an antibody solution at a concentration of 0.12 µ*M* in PBS, pH 7.4 (0.018 mg/mL for an IgG antibody). This equals an antigen-binding pocket concentration of 0.24 µ*M*.
2. Place 2.0 mL of the antibody solution in a 3-mL fluorescence quartz cuvet of 1-cm cell path length.
3. Adjust the fluorimeter settings to: excitation slit width at 5 nm; emission slit width at 10 nm; excitation wavelength at 290 nm; and emission wavelength at 345 nm.
4. Measure the fluorescence emission of the free antibody solution (F_0; observed intensity of antibody fluorescence in the absence of its antigen).
5. Prepare an antigen solution in PBS at a concentration of 60 µ*M*. The large difference in concentration between the antigen and antibody lessens the need for adding large volumes to the antibody solution in **step 6,** thus minimizing the dilution effect on fluorescence emission.
6. Add 1-µL aliquots of the antigen solution to the cuvet. Mix well and measure the fluorescence emission (F: observed fluorescence in the presence of varying amounts of antigen). Each addition of an antigen aliquot will add a concentration of 0.03 µ*M* of antigen to the antibody solution.
7. It will take you eight additions (experimental points) before you almost reach the one-to-one ratio (antigen epitope to antigen-binding pocket). Make sure you continue your titration to an excess of antigen to antibody, until you achieve several points where the fluorescence does not change (or decreases very slightly because of dilution effects) from measurement to measurement. If your antigen is not as soluble and you have difficulties achieving such concentrations, prepare your antigen in as concentrated a form as possible, making sure to adjust the volumes of the antigen aliquots added to the antibody solution for each measurement.
8. Record the fluorescence emission (F; observed intensity of fluorescence in the presence of varying amounts of antigen) at each step of your titration.
9. Correct for dilution effects resulting from the addition of antigen to your antibody solution. Dilution effects of titrating antigen solution into the antibody solution should be determined by titrating PBS into 2 mL of antibody solution as a control.
10. Plot your data F_0/F vs the antigen concentration, using **Equation 1** (*see* **Subheading 1.**) or any of the available equations for the fitting of binding data, based on nonlinear squares fitting, that would result in the calculation of the association equilibrium constant that represents the binding affinity of your antibody for the given antigen (*see* **Subheading 3.1., step 7**). For fitting data resulting from this method, be sure to consider the dilution in antigen concentration caused by the change in votal volume after each addition of antibody solution.

4. Notes

1. Method one, in which fixed antibody colcentration is used throughout the titration, has an obvious advantage—the concentration of the antibody does not change, thus resulting in more accurate data and equilibrium-constant determination. Therefore, method one should be used, especially if the system is susceptible to changes depending on concentration, or is concentration/volume sensitive in any way.

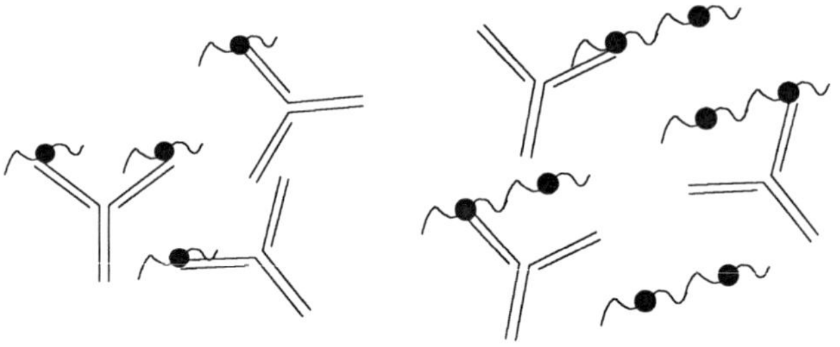

Fig. 3. Monovalent binding of antibody to antigen.

2. Method two presents an advantage over method one because much smaller amounts of antibody are utilized. If sample volume is of the essence, then method two is obviously preferred. However, if at all possible, you would be advised to compare at least once the two methods in your system to determine if and how much concentration/volume changes are affecting your measurements.
3. Both methods described here could be used for any other titrations using optical spectroscopy, such as ultraviolet (UV) or Circular Dichroism (CD) spectroscopy. The only difference is that these techniques require higher concentrations, and you should adjust your concentrations accordingly to get the required signal. A 10-fold higher antibody concentration (approx 2.3×10^{-6} *M* or 0.35 mg/mL for an IgG antibody) is usually required to obtain a sufficient signal in the UV and CD spectrophotometer in order to perform a titration and obtain equilibrium binding constants. Furthermore, it should be noted here that if you have access to a CD spectrophotometer, you can also obtain structural information about your antibody, or changes that occur upon antigen binding, by analyzing the CD spectra in any of the software programs available for CD data analysis.
4. Avoid using Scatchard analysis for fitting your data. It is an outdated and oversimplified representation of the binding event, and can incorporate significant errors in the calculations of the equilibrium binding constants.
5. It is generally accepted that antibody-antigen interactions are classic cases of one-to-one interactions without any cooperativity effects. The antigen-binding site is believed to accommodate a single hapten molecule, and studies also indicate that binding sites in immunoglobulins do not interact in binding of haptens in solution. However, this is not always the case, and avidity can play a part in some experiments in which polymeric antigens with more than one epitope bind to the same antibody's binding sites. This should be taken into account when analyzing the data obtained from such experiments. When one epitope is available per antigen, the association constant reflects that of the single interaction between one epitope and one binding site, with no avidity effects occurring and comparable to association constants described in the literature for antibodies interacting with single epitope molecules. Such an interaction is illustrated in **Fig. 3.** Polymeric antigens with many epitopes may allow two epitopes on the same molecule to interact simultaneously with the same antibody. The possibility of avidity effects on the binding inter-

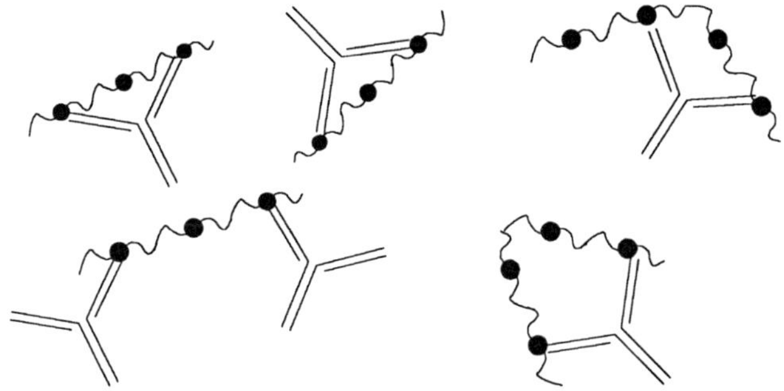

Fig. 4. Monovalent and multivalent binding of polymeric antigens with multiple epitopes.

actions thus exists. It is important to utilize an accurate binding model that would account for such an interaction (**Fig. 4**). The binding models for mono- and multivalent binding interactions are shown in **Equation 2.** For monovalent interactions, K_A, the equilibrium association constant can be described as:

$$Ab + Ag \overset{K_A}{\Leftrightarrow} AbAg \qquad (2)$$

$$K_A = \frac{[AgAb]}{[Ag][Ab]}$$

Multivalent interactions require a more complex model. The following example shows the interaction between an antibody $Ab_{(2)}$ with two binding pockets and an antigen $Ag_{(3)}$ with three epitopes. (AbAg) represents a binding pocket/epitope complex. Numbers in subscript indicate unbound binding pockets/epitopes.

$$Ab_{(2)} + 2Ag_{(3)} \overset{K_{A1}}{\Leftrightarrow} Ab_{(1)}(AbAg)Ag_{(2)} + Ag_{(3)} \overset{K_{A1}}{\Leftrightarrow} 2(AbAg)Ag_{(2)} \qquad Excess[Ag]$$

Or

$$Ab_{(2)} + Ag_{(3)} \overset{K_{A1}}{\Leftrightarrow} Ab_{(1)}(AbAg)Ag_{(2)} \overset{K_{A1}}{\Leftrightarrow} (2(AbAg))Ag_{(1)} \qquad Excess[Ab]$$

$$K_{A2} > K_{A1}$$

$$K_{A1} = \frac{[Ab_{(1)}(AbAg)Ag_{(2)}]}{[Ab_{(2)}][Ag_{(3)}]} \quad \text{and} \quad K_{A1} = \frac{[(AbAg)Ag_{(2)}]^2}{[Ab_{(1)}(AbAg)Ag_{(2)}][Ag_{(3)}]}$$

Rearrange to substitute out the $Ab_{(1)}$ (AbAg)$Ag_{(2)}$ term to give

$$K_{A1} = \frac{\left[(AbAg)Ag_{(2)}\right]}{\sqrt{\left[Ab_{(2)}\right]\left[Ag_{(3)}\right]}} \quad (3)$$

Then

$$K_{A2} = \frac{\left[(2(AbAg))Ag_{(1)}\right]}{\left[Ab_{(1)}(AbAg)Ag_{(2)}\right]} \quad (4)$$

Substitution into **Equation 4** also gives:

$$K_{A2} = \frac{\left[(2(AbAg))Ag_{(1)}\right]}{K_{A1} \times \sqrt{\left[Ab_{(2)}\right]\left[Ag_{(3)}\right]}} \quad (5)$$

There are two equilibrium association constants K_{A1} and K_{A2} depending on the concentration of antigen in the mixture. In antigen excess, there is monovalent binding with each antibody that associates with one epitope on two separate antigen molecules. This gives the association constant which represents the true affinity of the interaction, K_{A1}. In antibody excess, there is bivalent binding of one antibody molecule to two epitopes on the same antigen molecule. This yields the association constant, which incorporates in part the true affinity, as well as avidity, K_{A2}. So, when analyzing the data obtained, it is important to do two fittings. For points measured at antigen excess, the graph is produced using the monovalent binding model. For points at antibody excess, the graph is fitted using the bivalent model. K_{A2} is greater than K_{AI} and the association rate is higher for K_{A2}. This is because the second epitope has lost entropy as a result of the complex formed on the same antigen molecule by the first epitope. In effect, the binding of one epitope restricts the conformational space the rest of the antigen molecule can explore before finding a free antigen-binding pocket. The likelihood of associating with the remaining free antigen-binding pocket on the same antibody molecule is much higher because of closer proximity.

References

1. Eftink, M. R. (1997) in *Methods in Enzymology: Fluorescence Spectroscopy* (Brand, L. and Johnson, M. L., eds.), Academic Press, New York, NY, pp. 278, pp. 221–257.
2. Scatchard, J. (1949) The attractions of proteins for small molecules and ions. *Ann. NY Acad. Sci.* **51,** 660.
3. Missailidis, S., Cannon, W. V., Drake, A., Wang, X. Y., and Buck, M. (1997) Analysis of the architecture of the transcription factor sigma(N) (sigma(54)) and its domains by circular dichroism. *Mol. Microbiol.* **24(3),** 653–664.
4. Spencer, D.I.R., Missailidis, S., Denton, G., Murray, A., Brady, K., Matteis, C.I.D., et al. (1999) Structure activity studies of the anti-MUC1 monoclonal antibody C595

and synthetic MUC1 mucin-core-related peptides and glycopeptides. *Biospectroscopy* **5(2),** 79–91.
5. Parker, C. W. (1978) Spectrofluorometric methods, in *Experimental Immunology,* 3rd ed., Blackwell Scientific Publications, Oxford UK, pp. 1–25.
6. Reedstrom, R. J., Brown, M. P., Grillo, A., Roen, D., and Royer, C. A. (1997) Affinity and specificity of trp repressor-DNA interactions studied with fluorescent oligonucleotides. *J. Mol. Biol.* **273(3),** 572–585.

26

Antibody Epitope Mapping Using Arrays of Synthetic Peptides

Ulrich Reineke

1. Introduction

The identification of antibody epitopes and their characterization at the amino acid level is extremely important. Understanding antibody specificity at the molecular level provides the key to optimizing their use as research or diagnostic tools as well as their application as therapeutic agents. Among other techniques, the use of chemically prepared arrays of protein sequence-derived short peptides has emerged as a powerful tool to identify and characterize antibody epitopes. In particular, the SPOT synthesis technique *(1,2)*, which is described in detail in this chapter, is extremely well-suited for this purpose, and has been widely used for epitope mapping and a variety of other applications. Three different types of peptide libraries are described: i) Protein sequence-derived scans of overlapping peptides (peptide scans) *(3,4)* used to locate and identify antibody epitopes (**Subheading 3.4.1.**). ii) Truncation libraries used to identify the minimal peptide length required for antibody binding. In these libraries, amino acids from the termini of a peptide previously identified by a peptide scan are systematically omitted (**Subheading 3.4.2.**). iii) Finally, complete substitutional analyses to identify the key residues important for antibody binding (**Subheading 3.4.3.**)*(5)*.

1.1. The Principle of the SPOT Synthesis Technique

The SPOT synthesis concept, a highly parallel and technically simple approach, was developed by Ronald Frank *(1,2)*. This method is very flexible and economical in comparison to other multiple solid-phase procedures, particularly with regard to miniaturization and array geometries. The basic principle involves the positionally addressed delivery of small volumes of activated amino acid solutions directly onto a coherent membrane sheet. The areas wet by the resulting droplets can be considered as micro-reactors provided that a

From: *Methods in Molecular Biology, Vol. 248: Antibody Engineering: Methods and Protocols*
Edited by: B. K. C. Lo © Humana Press Inc., Totowa, NJ

nonvolatile solvent is used. One standard membrane support material is cellulose, but other membranes are also used (**Subheading 3.2.**) The SPOT synthesis of peptide arrays uses the Fmoc strategy following the general scheme outlined in **Fig. 1.** The peptide libraries can be synthesized manually without great effort in non-specialized laboratories simply by pipetting activated amino acids onto predefined spots on the functionalized membrane. Depending on the viscosity of the amino acid solution and the type of membrane used, a drop of 1 μL results in a spot of approx 0.7 cm in diameter. Washing, capping, and deprotection steps—which are the same for all the peptides on the membrane—are carried out in a stainless steel dish by rinsing and shaking with the appropriate solvents and reagents. In order to facilitate the numerous pipetting steps required for SPOT synthesis, an automated SPOT synthesizer was developed and is briefly described in **Subheading 3.8**.

Usually, peptides between 4 and 15 amino acids in length are synthesized. Peptides of this length are certainly sufficient to identify linear—although not usually discontinuous—antibody epitopes, and have purities similar to peptides synthesized by solid-phase methods in reactors. However, longer peptides can also be synthesized, and give reliable screening results *(6)*.

Since its introduction, the SPOT method has become a widely used approach, mainly because of the following advantages: i) The highly parallel synthesis format permits rapid simultaneous synthesis of many different peptides. Several hundred peptides can be prepared manually within 2–3 days. ii) The technique is very economical compared to other solid-phase synthesis methods because only very small amounts of reagents are used. Depending on the peptide length and the spot size, each spot represents between 5 and 100 nmol of the respective peptide, an amount sufficient to detect antibody binding. iii) The peptide arrays are synthesized on cellulose membranes that are compatible with many biological screening systems—for example, conventional enzyme-linked immunosorbent assay (ELISA) or Western blot procedures. All the array peptides are probed directly and simultaneously, e.g., for antibody binding, on the synthesis support. The sequence of any active peptide is already recorded on the sequence list for the array synthesis, so no coding or decoding procedures are necessary. iv) Usually, the peptide membrane can be reused several times. v) In addition to the proteinogenic amino acids, a variety of unnatural building blocks such as D-amino acids or β-amino acids can be used. Furthermore, arrays of cyclic as well as branched peptides and peptidomimetics have been described.

1.2. Linear vs Discontinuous Epitopes

Two different types of epitopes must be considered. In linear (continuous) binding sites, the key amino acids that mediate the contacts to the antibody are located within one part of the primary structure, usually not exceeding 15 amino

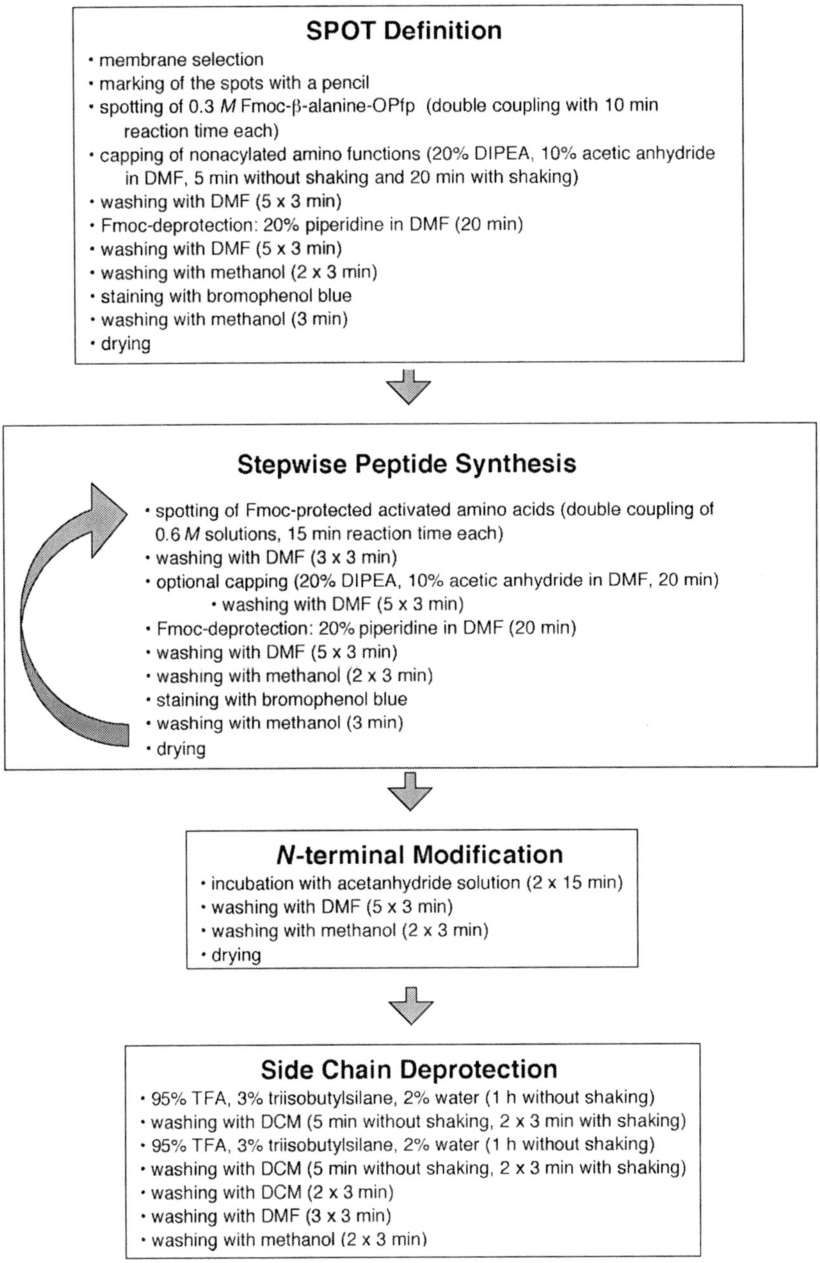
SPOT Definition
• membrane selection
• marking of the spots with a pencil
• spotting of 0.3 *M* Fmoc-β-alanine-OPfp (double coupling with 10 min reaction time each)
• capping of nonacylated amino functions (20% DIPEA, 10% acetic anhydride in DMF, 5 min without shaking and 20 min with shaking)
• washing with DMF (5 x 3 min)
• Fmoc-deprotection: 20% piperidine in DMF (20 min)
• washing with DMF (5 x 3 min)
• washing with methanol (2 x 3 min)
• staining with bromophenol blue
• washing with methanol (3 min)
• drying
Stepwise Peptide Synthesis
• spotting of Fmoc-protected activated amino acids (double coupling of 0.6 *M* solutions, 15 min reaction time each)
• washing with DMF (3 x 3 min)
• optional capping (20% DIPEA, 10% acetic anhydride in DMF, 20 min)
• washing with DMF (5 x 3 min)
• Fmoc-deprotection: 20% piperidine in DMF (20 min)
• washing with DMF (5 x 3 min)
• washing with methanol (2 x 3 min)
• staining with bromophenol blue
• washing with methanol (3 min)
• drying
N-terminal Modification
• incubation with acetanhydride solution (2 x 15 min)
• washing with DMF (5 x 3 min)
• washing with methanol (2 x 3 min)
• drying
Side Chain Deprotection
• 95% TFA, 3% triisobutylsilane, 2% water (1 h without shaking)
• washing with DCM (5 min without shaking, 2 x 3 min with shaking)
• 95% TFA, 3% triisobutylsilane, 2% water (1 h without shaking)
• washing with DCM (5 min without shaking, 2 x 3 min with shaking)
• washing with DCM (2 x 3 min)
• washing with DMF (3 x 3 min)
• washing with methanol (2 x 3 min)

Fig. 1. General scheme summarizing the SPOT synthesis process.

acids in length. Peptides that span these sequences have affinities to the antibody that are within the range shown by the entire protein antigen. In discontinuous (conformational) binding sites, the key residues are distributed over two or more binding regions that are separated in the primary structure. Upon folding, these binding regions are brought together on the protein surface to form a composite epitope. Even if the complete epitope forms a high-affinity interaction, peptides that cover only one binding region—as synthesized in a scan of overlapping peptides—have very low affinities, which often cannot be measured in normal ELISA or surface plasmon resonance (SPR) experiments. The mapping of linear epitopes using the protocols described here is an easy and straightforward approach. The incubation of a peptide scan results in the identification of one or a few consecutive overlapping peptides that are able to bind the antibody under investigation. On the other hand, the identification of discontinuous epitopes can fail because of the low affinities described here. Several publications have described the mapping of these epitopes *(7)*, in which two or more binding regions are identified in the peptide scan. In general, the mapping of discontinuous epitopes is facilitated by employing an enzyme-labeled primary antibody (**Subheadings 3.9., 3.10.,** and **3.11.**). An alternative procedure for mapping conformational epitopes is described in **Chapter 27.**

1.3. Polyclonal vs Monoclonal Antibodies (MAbs)

The techniques described here are used for epitope mapping of monoclonal as well as polyclonal antibodies. Two or more antibody binding regions may be observed for polyclonal antibodies, but in this case it is not possible to determine whether the active peptides represent linear epitopes or different binding regions of discontinuous epitopes.

1.4. Other Applications

This chapter addresses only the synthesis and application of peptide arrays prepared by SPOT synthesis for the mapping of antibody epitopes. During the last decade, many other applications have emerged, including paratope mapping; mapping of protein-protein interactions in general; identification of kinase and protease substrates and inhibitors; identification of metal ion, DNA, or co-factor binding peptides; and many others. These were accompanied by the development of novel types of peptide libraries and array-based assay techniques. These diverse applications are reviewed in detail elsewhere *(8–11)*.

2. Materials

2.1. General Equipment

1. Stainless-steel dish with a good closing lid that is slightly bigger than the peptide array membrane.

2. A rocker table, e.g. Rocky (Fröbel Labortechnik, Lindau, Germany).
3. A pipet adjustable from 0.5–10 µL with suitable plastic tips. A multistep version is highly recommended.
4. A fume cupboard for handling solvents and reagents. The rocker table must fit into the fume cupboard.
5. A pH-meter for adjusting the buffer solutions.

2.1.1. General Remarks for Organic Solvents

The purity of *N,N*-dimethylformamide (DMF) and 1-methyl-2-pyrrolidone (NMP) is critical for the peptide synthesis process, since their degradation can result in free amines that lead to premature deprotection of the Fmoc group or decomposition of amino acid active esters. This will reduce the yield of full-length peptide, and can cause byproduct formation. In order to ensure that DMF or NMP free of amines is used, add 10 µL of 1% bromophenol blue solution in DMF to 1 mL of NMP or DMF in a 1.5-mL tube and mix thoroughly. Let this stand for 5 min and then observe the color: yellow indicates that it is satisfactory to use, and yellow/green or blue/green means that it should not be used.

2.2. Membranes for Peptide Arrays

1. Amino-functionalized membranes are commercially available from AIMS Scientific Products (http://www.aims-scientific-products.de).

2.3. Array Formatting

1. Fmoc-β-alanine-OPfp solution: 0.3 *M* Fmoc-β-alanine-OPfp in dimethyl sulfoxide (DMSO).
2. DMF, peptide synthesis grade. **Caution:** DMF is toxic. It may cause harm to an unborn child, is harmful through inhalation and through contact with the skin, and is irritating to the eyes.
3. Acetanhydride solution: 10% acetanhydride, 20% *N,N*-diisopropylethylamine (DIPEA) in DMF. Freshly prepare this solution prior to use.
4. Piperidine solution: 20% piperidine in DMF.
5. Methanol.
6. Staining solution: 0.01% (w/v) bromophenol blue in methanol. This solution should be yellow or orange, and must be discarded if it becomes green or green/blue.

2.4. Library Design

1. Standard word processing software can be used.

2.5. Stepwise Peptide Synthesis

1. All amino acids are applied as 9-fluorenylmethoxycarbonyl (Fmoc)-protected active esters (*see* **Table 1**). Prepare 0.6 *M* stock solutions of all amino acids except

Table 1
L-amino Acid Derivatives for SPOT Synthesis

	Amino Acid	mW [g/mol]	0.6 *M* 200 µL	0.6 *M* 400 µL	0.6 *M* 1.0 mL
A	Fmoc-Ala-OPfp	477.4	57.3	114.6	286.4
C	Fmoc-Cys(Trt)-OPfp	751.8	90.2	180.4	451.1
D	Fmoc-Asp(OtBu)-OPfp	577.5	69.3	138.6	346.5
E	Fmoc-Glu(OtBu)-OPfp	591.5	71.0	142.0	354.9
F	Fmoc-Phe-OPfp	553.5	66.4	132.8	332.1
G	Fmoc-Gly-OPfp	463.2	55.6	111.2	277.9
H	Fmoc-His(Boc)-OPfp	643.6	77.2	154.5	386.2
I	Fmoc-Ile-OPfp	519.5	62.3	124.7	311.7
K	Fmoc-Lys-(tBoc)-OPfp	634.6	76.1	152.3	380.8
L	Fmoc-Leu-OPfp	519.5	62.3	124.7	311.7
M	Fmoc-Met-OPfp	537.5	64.5	129.0	322.5
N	Fmoc-Asn(Trt)-OPfp	762.8	91.5	183.1	457.7
P	Fmoc-Pro-OPfp	503.4	60.4	120.8	302.0
Q	Fmoc-Gln(Trt)-OPfp	776.8	93.2	186.4	466.1
R	Fmoc-Arg(Pbf)-OPfp	814.9	97.8	195.6	488.9
S	Fmoc-Ser(tBu)-OPfp	549.5	65.9	131.9	329.7
T	Fmoc-Thr(tBu)-OPfp	563.5	67.6	135.2	338.1
V	Fmoc-Val-OPfp	505.4	60.6	121.3	303.2
W	Fmoc-Trp(tBoc)-OPfp	692.4	83.1	166.2	415.4
Y	Fmoc-Tyr(tBu)-OPfp	625.6	75.1	150.1	375.4

The amounts for the preparation of 200 µL, 400 µL, and 1 mL of the 0.6 *M* stock solutions are given.

for arginine. Use NMP as solvent except for serine and threonine. These amino acids should be dissolved in DMF. Use 0.5-mL or 1.5-mL reaction tubes. Test NMP and DMF regularly for free amines as described in **Subheading 2.1.1.** The solutions are stable at –20°C for several days. The arginine solution must be prepared freshly each working day. Calculate the consumption for each peptide array carefully, because the reagents are quite expensive. If necessary, vortex or sonicate to dissolve the amino acid derivatives. Store at –20°C. Allow to warm up to room temperature prior to use. Discard the stock solutions if precipitates are observed.

2. DMF.
3. Piperidine solution: 20% piperidine in DMF.
4. Methanol.
5. Staining solution: 0.01% (w/v) bromophenol blue in methanol. This solution should be yellow or orange, and must be discarded if it becomes green or green/blue.

2.6. Acetylation of the N-terminus

1. Acetanhydride solution: 10% acetanhydride, 20% DIPEA in DMF. Prepare this solution freshly prior to use.
2. DMF.
3. Methanol.

2.7. Side-Chain Deprotection

1. Deprotection solution: 95% trifluoroacetic acid (TFA), 3% triisobutylsilane, and 2% water. TFA is toxic and very corrosive, and should be handled with the greatest caution. Do not mix TFA and DMF waste, as it can undergo an exothermic and explosive reaction. Consult your safety officer for approved handling and disposal procedures.
2. Dichloromethane (DCM).
3. DMF.
4. Methanol.

2.8. Automated Spot Synthesis

1. Automated SPOT synthesizer (INTAVIS Bioanalytical Instruments; http://www.intavis.com/html/autospot.html).

2.9. Screening of Peptide Arrays

1. Methanol.
2. Tris-buffered saline (TBS): 50 m*M* Tris-HCl, pH 8.0, 137 m*M* NaCl, 2.7 m*M* KCl.
3. T-TBS: TBS containing 0.05% Tween 20.
4. Blocking buffer: e.g., 1 equivalent of the blocking reagent delivered together with the BM chemiluminescence blotting substrate (POD) (Roche Diagnostics; Cat. #1 500 708) to 9 equivalents of TBS.
5. Primary antibody solution: Dilute the monoclonal antibody of interest to 0.1–3.0 μg/mL in blocking buffer. Polyclonal sera should be diluted between 1:100 and 1:10,000 in blocking buffer.

2.10. Detection of Antibody Binding by Chromogenic Substrates

1. T-TBS: (*see* **Subheading 2.9.**).
2. Blocking buffer (*see* **Subheading 2.9.**).
3. Secondary antibody solution (alkaline phosphatase conjugated): dilute the antibody to 1.0 μg/mL in blocking buffer or follow the instructions of the supplier for Western blotting protocols. Make sure that the secondary antibody corresponds to the species and antibody class of the primary antibody under investigation.
4. Nitroblue tetrazolium (NBT) stock solution: dissolve 500 mg NBT in 10 mL of 70% DMF in water. The stock solution is stable for at least 1 yr if stored at 4°C.
5. Bromochloroindolyl phosphate (BCIP) stock solution: dissolve 500 mg BCIP disodium salt in 10 mL of DMF. The stock solution is stable for at least 1 yr if stored at 4°C.

6. Alkaline phosphatase buffer: 100 m*M* NaCl, 5 m*M* $MgCl_2$, 100 m*M* Tris-HCl, pH 9.5.
7. Enzyme substrate solution: add 330 µL of NBT stock solution to 50 mL of alkaline phosphatase buffer. Mix well and add 165 µL of BCIP stock solution. The enzyme substrate solution must be used within 1 h.
8. Stop solution: 20 m*M* EDTA in phosphate-buffered saline (PBS: 9.2 m*M* Na_2HPO_4, 1.6 m*M* NaH_2PO_4, 150 m*M* NaCl, pH 7.4).

2.11. Detection of Antibody Binding by Chemiluminescence

1. T-TBS (see **Subheading 2.9.** and **Note 1**).
2. Blocking buffer (*see* **Subheading 2.9.**).
3. Secondary antibody solution (peroxidase-conjugated): dilute the antibody to 1.0 µg/mL in blocking buffer or follow the instructions of the supplier for Western blotting protocols. Make sure that the secondary antibody corresponds to the species and antibody class of the primary antibody under investigation (*see* **Note 1**).
4. Chemiluminescence substrate [e.g., BM Chemiluminescence Blotting Substrate (POD), Roche Diagnostics; Cat. #1 500 708]: Mix luminescence substrate solution A and starting solution B 100:1 just before developing the peptide array. Other chemiluminescence substrates can also be used.
5. X-ray films, film cassette, and developing equipment or a chemiluminescence imager, e.g., LumiImager (Roche Diagnostics).

2.12. Reutilization of Peptide Arrays

1. Regeneration buffer I: 62.5 m*M* Tris-HCl, pH 6.7, 2% SDS in water. Add 70 µL of 2-mercaptoethanol per 10 mL of SDS buffer prior to regeneration. Use this buffer only in an operating fume cupboard because of the pungent smell of 2-mercaptoethanol.
2. 10X T-TBS: 500 m*M* Tris-HCl, pH 8.0, 1.37 *M* NaCl, 27 m*M* KCl, 0.5% Tween 20.
3. Regeneration buffer IIA: 8 *M* urea, 1% SDS, 0.1% 2-mercaptoethanol in water. Use this buffer only in an operating fume cupboard because of the pungent smell of 2-mercaptoethanol.
4. Regeneration buffer IIB: for 1 L, mix 400 mL of water, 500 mL of ethanol, and 100 mL of acetic acid.

3. Methods

3.1. General Overview

The protocols described in this chapter fall into two categories. In **Subheadings 3.2–3.8.,** the synthesis of different protein sequence-derived peptide library arrays is described. In **Subheadings 3.9.–3.13.,** all processes for screening, regeneration, and storage of the peptide arrays are addressed. Solvents or solutions used in washing or incubation steps are always gently agitated on a rocker table at room temperature unless otherwise stated. During incubations and washings, the dishes are closed with a lid. Calculate the consumption of solvents and reagents carefully before starting the synthesis.

3.2. Membranes for Peptide Arrays

Membranes for the synthesis of peptide arrays are commercially available from AIMS Scientific Products GmbH, Mascheroder Weg 1B, D-38124 Braunschweig, Germany (http://www.aims-scientific-products.de). Amino-functionalized cellulose APEG-membranes with a peptide loading of 400 nmol/cm^2 are recommended and are supplied in different sizes, e.g., 8 × 12 cm or 10 × 15 cm. As an alternative, self-prepared synthesis membranes based on Whatman 50 filter paper can be used. Protocols for membrane preparation have already been described *(1,8,12)*. The membranes must be handled with gloves during the entire process described here. In addition, tweezers are recommended for handling wet membranes because they simplify the handling and avoid tearing.

3.3. Array Formatting

The SPOT method is particularly flexible with respect to spot numbers and layout of the array. Any desired array can be designed to fit the individual needs of an experiment. If necessary, the membrane can easily be cut with scissors to obtain a suitable membrane size. A standard format for the arrays is the 8 × 12 array that corresponds to the 96-well microtiter plates.

All washing steps must be carried out on a rocker table that must be set up in an operating fume cupboard for safety reasons.

1. Mark as many spots as needed for the peptide library in a suitable configuration on the membrane support using a soft pencil. Graphite is stable against all the chemicals and solvents used during the synthesis and screening steps. The spots should be spaced at a distance of 1.5 cm in order to avoid cross-contamination of neighboring spots during synthesis. To facilitate the stepwise synthesis process (**Subheading 3.5.**), the spots can also be numbered using the pencil.
2. Use a 0.3 *M* Fmoc-β-alanine-OPfp solution in DMSO (*see* **Note 2**) for the first coupling step. A volume of 1.0 μL must be spotted on each predefined position of the peptide array (*see* **Note 3**). A multistep pipet facilitates this procedure. Repeat this spotting step once after a 10-min reaction time to ensure a complete coupling. Go to **step 3** after a second 10-min reaction time.
3. Slowly place the membrane in a stainless-steel dish with acetanhydride solution (*see* **Note 4**). Avoid shaking and air bubbles. After 5 min, pour off the solution and incubate the membrane with a sufficient volume of fresh acetanhydride solution for 20 min. Shake the membrane on a rocker table. This step is carried out to acetylate the remaining amino functions that did not react with the Fmoc-β-alanine-OPfp solution in **step 2.** Thus, defined spots for the peptide library are achieved.
4. Wash the membrane 5 × with DMF for 3 min each.

5. Cleave the Fmoc protecting groups by treatment of the membrane with piperidine solution for 20 min.
6. Wash the membrane 5× with DMF for 3 min each.
7. Wash the membrane twice with methanol for 3 min.
8. Rinse the membrane with staining solution (bromophenol blue is an indicator for free amino functions). The bromophenol blue solution should remain yellow and the spots should become blue, leaving the surrounding membrane white. Treat the membrane until an equal blue staining of the spots is achieved.
9. Wash the membrane with methanol for 3 min to remove the remaining staining solution.
10. Air-dry the membrane. The process can be accelerated by carefully using a hair-dryer, operating with cold air.

3.4. Library Design

Three different types of protein sequence-derived peptide libraries for antibody epitope identification and characterization are described: scans of overlapping peptides *(3,4)*, truncation analyses, and complete substitutional analyses *(5)*.

3.4.1. Scans of Overlapping Peptides

A protein sequence-derived scan of overlapping peptide is used to identify the epitope region recognized by an antibody raised against a certain protein. Therefore, the entire protein sequence or a region of special interest is dissected into short linear peptides. Fifteen-mer peptides are recommended, since most linear binding sites do not exceed this range *(13)*. Consecutive peptides should overlap by 12 amino acids. Shorter overlaps can indicate that important peptides are overlooked. In peptide scans derived from proteins containing disulfide bonds or free cysteine residues, these residues should be exchanged by a similar amino acid such as serine to avoid dimerization and oligomerization of the peptides or covalent linkage to thiols in the screening antibody. As an example, a sequence list for a human interleukin-10 (huIL-10)-derived scan of overlapping peptides as well as the screening results for the anti-huIL-10 mab CB/RS/3 are shown in **Fig. 2.** The synthesis of a peptide scan requires a high degree of concentration because of the irregular order of pipetting steps in each synthesis cycle. It is recommended to prepare a pipetting scheme for the synthesis cycle and to check off those spotting steps that have been carried out.

3.4.2. Truncation Analyses

Peptide epitopes identified from scans of overlapping peptides often comprise dispensable positions resulting from the predefined peptide length (15-mers). Thus, truncation libraries are used to narrow down the epitope to the key interaction residues. An example of a sequence list for a truncation library

Human Interleukin-10 Sequence

SPGQGTQSENS**C**THFPGNLPNMLRDLRDAFSRVKTFFQMKDQLDNLLLKESLLEDFKGYLG**C**QALSEMIQFYLEE
VMPQAENQDPDIKAHVNSLGENLKTLRLRLRR**C**HRFLP**C**ENKSKAVEQVKNAFNKLQEKGIYKAMSEFDIFINYI
EAYMTMKIRN

Sequence list for a hIL-10-derived scan of overlapping peptides (15-mers, 12 amino acids overlap). Cysteine residues in the wild-type sequence (bold and underlined) are substituted by serine in the sequence list of overlapping peptides.

01. SPGQGTQSENSSTH
02. QGTQSENSSTHFPG
03. QSENSSTHFPGNLP
04. NSSTHFPGNLPNML
05. THFPGNLPNMLRDL
06. PGNLPNMLRDLRDA
07. LPNMLRDLRDAFSR
08. MLRDLRDAFSRVKT
09. DLRDAFSRVKTFFQ
10. DAFSRVKTFFQMKD
11. SRVKTFFQMKDQLD
12. KTFFQMKDQLDNLL
13. FQMKDQLDNLLLKE
14. KDQLDNLLLKESLL
15. LDNLLLKESLLEDF
16. LLLKESLLEDFKGY
17. KESLLEDFKGYLGS
18. LLEDFKGYLGSQAL
19. DFKGYLGSQALSEM
20. GYLGSQALSEMIQF
21. GSQALSEMIQFYLE
22. ALSEMIQFYLEEVM
23. EMIQFYLEEVMPQA
24. QFYLEEVMPQAENQ
25. LEEVMPQAENQDPD
26. VMPQAENQDPDIKA
27. QAENQDPDIKAHVN
28. NQDPDIKAHVNSLG
29. PDIKAHVNSLGENL
30. KAHVNSLGENLKTL
31. VNSLGENLKTLRLR
32. LGENLKTLRLRLRR
33. NLKTLRLRLRRSHR
34. TLRLRLRRSHRFLP
35. LRLRRSHRFLPSEN
36. RRSHRFLPSENKSK
37. HRFLPSENKSKAVE
38. LPSENKSKAVEQVK
39. ENKSKAVEQVKNAF
40. SKAVEQVKNAFNKL
41. VEQVKNAFNKLQEK
42. VKNAFNKLQEKGIY
43. AFNKLQEKGIYKAM
44. KLQEKGIYKAMSEF
45. EKGIYKAMSEFDIF
46. IYKAMSEFDIFINY
47. AMSEFDIFINYIEA
48. EFDIFINYIEAYMT
49. IFINYIEAYMTMKI
50. INYIEAYMTMKIRN

Incubation of an Interleukin-10-Derived Peptide Scan with MAb CB/RS/3

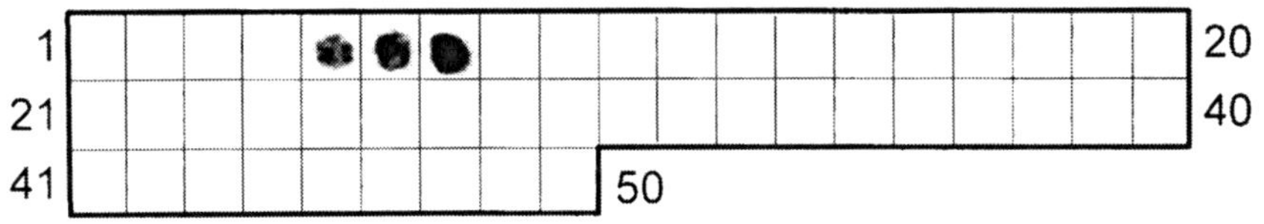

5.	THFPGN	**LPNMLRDL**	
6.	PGN	**LPNMLRDL**	RDA
7.		**LPNMLRDL**	RDAFSR

Fig. 2. Scan of overlapping peptides. The entire human interleukin-10 (huIL-10) sequence *(top)* is dissected into 15-mer overlapping peptides (12 amino acids overlap), resulting in a sequence list for the SPOT synthesis process of a total of 50 peptides *(middle)*. The resulting peptide library array was incubated with the anti-huIL-10 mab CB/RS/3 (mouse IgG). Subsequently, peptide-bound antibody was detected using a peroxidase-labeled secondary antibody in combination with a chemiluminescence substrate and an imaging system *(lower panel)*. Three consecutive peptides were identified that bind mab CB/RS/3. The sequence common to these peptides is framed.

```
N-terminal truncation
 1.     PGNLPNMLRDLRDA
 2.     -GNLPNMLRDLRDA
 3.     --NLPNMLRDLRDA
 4.     ---LPNMLRDLRDA
 5.     ----PNMLRDLRDA
 6.     -----NMLRDLRDA
 7.     ------MLRDLRDA

C-terminal truncation
 8.     PGNLPNMLRDLRD-
 9.     PGNLPNMLRDLR--
 10.    PGNLPNMLRDL---
 11.    PGNLPNMLRD----
 12.    PGNLPNMLR-----
 13.    PGNLPNML------

N-/C-terminal truncations
 14.    -GNLPNMLRDLRD-
 15.    --NLPNMLRDLR--
 16.    ---LPNMLRDL---
 17.    ----PNMLRD----
```

Fig. 3. Scheme for a truncation analysis peptide library. The active peptide number 6 from the huIL-10-derived scan of overlapping peptides (*see* **Fig. 2**) was subjected to *N*-, *C*- and bi-directional truncations.

designed for peptide 6 from the huIL-10-derived scan of overlapping peptides is shown in **Fig. 3.** The truncation library comprises peptides that omit *N*-, *C*-, or *N*- and *C*-terminal amino acids.

3.4.3. Complete Substitutional Analyses

Complete substitutional analysis is carried out in order to identify the key residues of a peptide epitope—i.e., those amino acids that mediate the contact with the antibody that cannot be substituted without loss of binding. This type of library consists of all possible single-site substitution analogs. A substitutional analysis library requires an array of 21 columns and as many rows as there are amino acids in the peptide of interest (**Fig. 4**). The starting peptide is synthesized in each row of the first column. In the other columns, each position is sequentially substituted by all 20 genetically encoded amino acids. In **Fig. 4,** the substitutions are arranged in alphabetical order following their one-letter code. Alternatively, the amino acids can be grouped according to their physicochemical properties (e.g., [DE], [KRH], [NQST], [FYW], [ILVM], [APG]). Carry out the pipetting steps as described under **Subheading 3.5.,** and follow these recommendations for the pipetting order:

wt	Ala (A)	Cys (C)	Asp (D)	G H I ...
NLPNMLRDLRDAF	**A**LPNMLRDLRDAF	**C**LPNMLRDLRDAF	**D**LPNMLRDLRDAF	..
NLPNMLRDLRDAF	N**A**PNMLRDLRDAF	N**C**PNMLRDLRDAF	N**D**PNMLRDLRDAF	..
NLPNMLRDLRDAF	NL**A**NMLRDLRDAF	NL**C**NMLRDLRDAF	NL**D**NMLRDLRDAF	..
NLPNMLRDLRDAF	NLP**A**MLRDLRDAF	NLP**C**MLRDLRDAF	NLP**D**MLRDLRDAF	..
NLPNMLRDLRDAF	NLPN**A**LRDLRDAF	NLPN**C**LRDLRDAF	NLPN**D**LRDLRDAF	..
NLPNMLRDLRDAF	NLPNM**A**RDLRDAF	NLPNM**C**RDLRDAF	NLPNM**D**RDLRDAF	..
NLPNMLRDLRDAF	NLPNML**A**DLRDAF	NLPNML**C**DLRDAF	NLPNML**D**DLRDAF	..
NLPNMLRDLRDAF	NLPNMLR**A**LRDAF	NLPNMLR**C**LRDAF	NLPNMLR**D**LRDAF	..
NLPNMLRDLRDAF	NLPNMLRD**A**RDAF	NLPNMLRD**C**RDAF	NLPNMLRD**D**RDAF	..
NLPNMLRDLRDAF	NLPNMLRDL**A**DAF	NLPNMLRDL**C**DAF	NLPNMLRDL**D**DAF	..
NLPNMLRDLRDAF	NLPNMLRDLR**A**AF	NLPNMLRDLR**C**AF	NLPNMLRDLR**D**AF	..
NLPNMLRDLRDAF	NLPNMLRDLRD**A**F	NLPNMLRDLRD**C**F	NLPNMLRDLRD**D**F	..
NLPNMLRDLRDAF	NLPNMLRDLRDA**A**	NLPNMLRDLRDA**C**	NLPNMLRDLRDA**D**	..

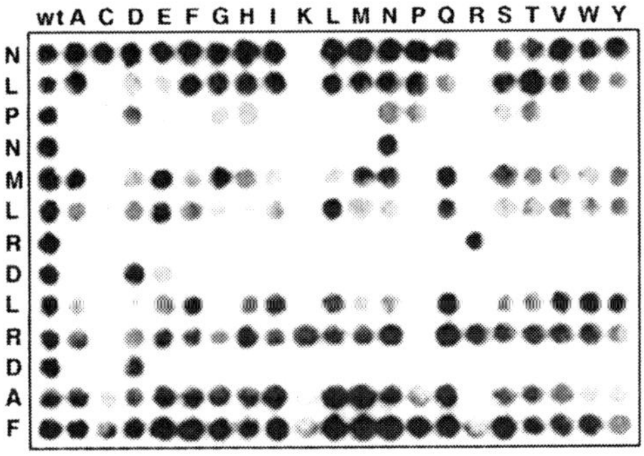

Fig. 4. Complete substitutional analysis library of the mab CB/RS/3 epitope. In this experiment, all possible single-site substitution analogs were synthesized and screened for mab CB/RS/3 binding as described in **Fig. 2.** As an example, the peptide sequences of the four left-hand columns are shown at the top. The results after screening for MAb CB/RS/3 binding are shown in the Figure. Four key residues were identified—namely, asparagine at position 4, arginine at position 7, and the two aspartic acid residues at positions 8 and 11.

1. In the first coupling step, pipet the *C*-terminal amino acid of the starting peptides onto all rows except the last one.
2. Pipet the same amino acid onto the first spot of the bottom row.
3. Spot all 20 amino acids successively onto the remaining 20 spots in alphabetical order.
4. Use an analogous procedure for the remaining synthesis steps.

3.5. Stepwise Peptide Synthesis

The synthesis of membrane-bound peptides is carried out in an iterative process **(Fig. 1)**. This cycle includes double-coupling of activated amino acid solutions, washing with DMF to remove excess amino acids, Fmoc deprotection with piperidine solution, washing with DMF and methanol, staining, washing with methanol, and drying. Subsequently, the next coupling step is carried out. All peptide sequences are written from the *N*-terminus (left) to the *C*-terminus (right). However, the peptides are chemically synthesized from *C*- to the *N*-terminus—e.g., for the peptide NLPNMLRDLRDAF **(Fig. 4)** the synthesis steps are:

F→A→D→R→L→D→R→L→M→N→P→L→N.

All washing steps must be carried out on a rocker table that must be set up in an operating fume cupboard for safety reasons.

1. Spot 1.5 µL of the activated amino acids onto the respective spots. Use a clean pipet tip for each amino acid. As soon as the droplets of activated amino acid solutions are added to the spots, coupling proceeds with a conversion of free *N*-terminal amino groups to amide bonds. During elongation of the peptides, a larger volume is spotted compared to array formatting (*see* **Subheading 3.3., step 2**). Thus, incomplete coupling at the edges is avoided. Repeat this step after a reaction time of 15 min. Go to the next step after an additional reaction time of 15 min. The coupling reaction can be followed by a color change from blue to green/blue or even yellow. Because the solvent within the spots is slowly evaporating over the reaction time, additional droplets may be added onto the same position without enlarging the spots and risking overlap with neighboring spots. Thus, difficult coupling reactions may be brought to completion by double or triple couplings.
2. Wash the membrane 3× with DMF for 3 min each (*see* **Note 4**).
3. Cleave the Fmoc protecting groups by treatment of the membrane with piperidine solution for 20 min.
4. Wash the membrane 5× with DMF for 3 min each.
5. Wash the membrane twice with methanol for 3 min.
6. Rinse the membrane with staining solution. The bromophenol blue solution should remain yellow, and the spots should become blue. Treat the membrane until an equal blue staining of the spots is achieved (*see* **Note 5**).
7. Wash the membrane with methanol for 3 min to remove the remaining staining solution.
8. Air-dry the membrane. The process can be accelerated by carefully using a hair-dryer, operating with cold air.
9. Go to **step 1** for coupling the next amino acid. The synthesis can also be stopped at this step to proceed for another day. In this case, the membrane should be placed in a plastic bag, sealed, and stored at –20°C. **Note:** In the coupling step for the last amino acid, do not stain with bromophenol blue.

3.6. Acetylation of the N-Terminus

This capping step is carried out because acetylation at the *N*-terminus stabilizes the peptide against proteolytic degradation. Furthermore, this counteracts the *N*-terminal positive charge, which is an artifact caused by dissecting the protein sequence into short peptides. All steps of this subheading must be carried out on a rocker table that must be set up in an operating fume cupboard for safety reasons.

1. Incubate the membrane twice with acetanhydride solution for 15 min each.
2. Wash the membrane 5× with DMF for 3 min.
3. Wash the membrane twice with methanol for 3 min.
4. Dry the membrane.

3.7. Side-Chain Deprotection

In the final synthesis step, all side-chain protecting groups are cleaved off. All steps of this subheading must be carried out in an operating fume cupboard for safety reasons.

1. Place the membrane carefully in a stainless steel dish with deprotection solution (*see* **Note 4**). Follow the safety instructions (**Subheading 2.7.**). Close the dish tightly with an appropriate lid, place the entire dish into a plastic bag, and seal it. Do not use a rocker table in **steps 1–3.** Incubate the membrane for 1 h and remove the deprotection solution. Now, the membrane is mechanically extremely unstable and must be handled carefully to avoid tearing.
2. Wash the membrane once with DCM without agitation for 5 min. Pour the solvent into the dish at an edge very slowly to avoid subjecting the membrane to mechanical stress. Wash the membrane twice with DCM under gentle agitation for 3 min.
3. Repeat **steps 1** and **2** once.
4. Wash the membrane two more times with DCM under gentle agitation for 3 min.
5. Wash the membrane 3× with DMF for 3 min.
6. Wash the membrane twice with methanol for 3 min.
7. Dry the membrane. Proceed with the screening (**Subheading 3.9.**), or store the membrane at –20°C.

3.8. Automated SPOT Synthesis

Although the SPOT synthesis technique is a robust and easy-to-use technique that can be carried out even in nonspecialized laboratories, the disadvantages of manual synthesis are obvious: the numerous pipetting and synthesis steps are rather time-consuming, and it is difficult to ensure the required precision during the manual pipetting steps. The laborious pipetting steps involved have led to the development of an automated SPOT synthesizer, providing a maximum amount of precision and reliability for the synthesis. This synthesizer (SPOT synthesizer ASP 222) was introduced by ABIMED Analysen-Technik GmbH (Langenfeld, Germany) and is now available from INTAVIS Bioanalytical Instruments AG (Bergisch Gladbach, Germany). Spot diameters are between 1.5 mm (50 nL dispensing volume) and 1.0 cm (1–2-µL dispensing volume). This unit facilitates consecutive delivery of peptide building blocks from a rack of 44 reservoirs onto the membrane surface. However, washing, capping, and deprotection steps still must be performed manually. In order to manage the increasing amount of data required to control the automated synthetic process, a software package that allows the development of appropriate sequence and chemistry files for a variety of library types has been devised together with the Auto-SPOT Robot ASP 222. For detailed protocols, please refer to the manufacturer's instructions.

3.9. Screening of Peptide Arrays

The screening strategy depends on whether the antibody under investigation is available in an enzyme-labeled form or if an enzyme-labeled secondary antibody must be used. The protocols described in **Subheadings 3.9., 3.10.,** and **3.11.** are applicable in both cases. It is helpful to know whether the primary antibody is active in Western blot experiments. If it is, this is a strong indication that it recognizes a linear epitope (*see* **Subheading 1.2.**), and implies a relatively high-affinity antibody-peptide interaction. Thus, a low concentration of the primary antibody can be used in **step 4.**

If a secondary antibody is used, it is essential to ensure that there is no detectable binding of this antibody to the peptides. Thus, for the detection by chromogenic substrates, the membrane must be incubated according to the protocols outlined in **Subheadings 3.9.** and **3.10.**—the only difference is that **step 4** of **Subheading 3.9.** is carried out with no primary antibody. The membrane should be incubated with the substrate solution (**Subheading 3.10., step 3**) for 30 min. Similarly, for the detection by chemiluminescence, the membrane must be incubated according to the protocols outlined in **Subheadings 3.9.** and **3.11.,** except that **step 4** of **Subheading 3.9.** is carried out with no primary antibody. Expose the X-ray film for at least 1 h (**Subheading 3.11., step 5**). A second exposure can be omitted. If a chemiluminescence imager is used, record an image for 30 min. Spots that give signals in these control experiments must be excluded as false-positives in the final assessment. Perform the control experiments before the main incubation.

All washing and incubation steps of the protocols described in **Subheadings 3.9., 3.10.,** and **3.11.** should be carried out on a rocker table at room temperature.

1. Rinse the peptide array membrane with a small volume of methanol for 1 min. This is done to facilitate the wetting of hydrophobic peptides for the following aqueous washing and incubation steps.
2. Wash the membrane 3× with an appropriate volume of TBS for 10 min each. The membrane should be sufficiently covered by the solution.
3. Block the membrane with a sufficient volume of blocking buffer for 3 h at room temperature (*see* **Note 4**).

4. Incubate the membrane with the same volume of 0.1–3.0 μg/mL primary antibody solution (*see* **Notes 7** and **8**) in blocking buffer for 3 h. If the primary antibody is directly labeled with alkaline phosphatase (*see* **Note 9**) go to **step 2** in **Subheading 3.10.** or **step 2** in **Subheading 3.11.** If not, proceed with the following steps:
5. Wash the membrane 3× with T-TBS for 10 min each.
6. Go to **Subheading 3.10.** or **3.11.**

3.10. Detection of Antibody Binding by Chromogenic Substrates

If a chemiluminescence imager or a darkroom is available the protocol for detecting antibody binding by chemiluminescence is recommended because of its higher sensitivity (**Subheading 3.11.**). Otherwise, the following alkaline phosphatase protocol should be used. The BCIP/NBT substrate generates an intense black-purple precipitate in which the enzyme-conjugated antibody has been bound.

1. Incubate the membrane with a sufficient volume of alkaline phosphatase-labeled secondary antibody solution (specific for the primary antibody that was used for the screening) at a concentration of 1 μg/mL in blocking buffer for 2 h, or follow the supplier's instructions for Western blotting (*see* **Notes 7** and **8**).
2. Wash the membrane 3× with T-TBS for 10 min each.
3. Incubate the membrane with the enzyme substrate solution. Go to **step 4** when the spots have turned suitably dark. This depends on the membrane-bound enzyme activity, and takes between 1 and 30 min.
4. Stop the reaction by rinsing the membrane with the stop solution 3× for 3 min each. The Mg^{2+} ions that are essential for alkaline phosphatase activity are complexed by ethylenediaminetetraacetic acid (EDTA) in the stop solution. For long-term documentation of the results, the membrane should be photographed or scanned. If the peptide array is not to be used further the membrane itself can stored as permanent record: wash the membrane 3× with water for 5 min and twice with methanol for 3 min, and dry. If the membrane is to be used again, wash the membrane 3× with T-TBS to remove excess stop solution and proceed with the regeneration procedures described in **Subheading 3.12.**

3.11. Detection of Antibody Binding by Chemiluminescence

Detection by chemiluminescence is highly sensitive, and has short imaging times ranging from a few seconds to 1 h. The reaction reaches its maximum after 1–2 min, and is relatively constant for 20–30 min. After 1 h, the signal intensity decreases to approx 60–70% of the maximum. (*see* **Note 1**)

1. Incubate the membrane with a sufficient volume of peroxidase-labeled secondary antibody solution (specific for the primary antibody that was used for the screening) at a concentration of 1 μg/mL in blocking buffer for 2 h or follow the supplier's instructions for Western blotting (*see* **Notes 7** and **8**).
2. Wash the membrane 3× with T-TBS for 10 min each.

3. The following steps must be carried out in a dark room. If a chemiluminescence imager is available, go to **step 9.** Rinse the membrane with the chemiluminescence substrate for about 1 min. The membrane has to be completely covered with the detection solution, which roughly corresponds to 50–100 µL per cm^2. Use a pipet to spread the solution repeatedly over the membrane.
4. Place the membrane in an X-ray film cassette that is lined with plastic film and cover it with film, which must be transparent.
5. Turn darkroom light to red and place an X-ray film onto the membrane. Close the cassette and expose for 60 s.
6. Upon opening the cassette, immediately replace the exposed film with a new one, then close and set aside the cassette. Develop the first exposed film at once.
7. Expose the second film for a suitable period (up to 45 min) estimated from the signal intensity on the first film (*see* **Notes 7, 8,** and **10**).
8. The films can be digitized and processed using a standard scanner.
9. Procedure if a chemiluminescence imager is available: rinse the membrane with the chemiluminescence substrate for about 1 min. The membrane must be completely covered with the detection solution, which roughly corresponds to 50–100 µL per cm^2. Use a pipet to spread the solution repeatedly over the membrane.
10. Place the membrane on the imager and record an image for 30 s. Make additional images with recording times adjusted according to the signal intensities and the signal-to-noise ratio of the first image.
11. Wash the membrane 3× with T-TBS to remove excess substrate solution. Proceed with the regeneration procedures described in **Subheading 3.12.**

3.12. Reutilization of Peptide Arrays

In order to reuse the peptide arrays, peptide-bound antibody must be completely removed. Usually, the peptide membranes can be used several times, but in a few cases regeneration fails as a result of strong binding of mature or denatured antibodies to the peptides or the cellulose. Two different protocols described in **Subheadings 3.12.1.** and **3.12.2.** are applicable. If the chromogenic substrate precipitate is not removed by these protocols, wash the membrane with DMF overnight, then 3× with methanol and dry.

3.12.1. Regeneration Protocol I

1. Wash the membrane 3× with water for 10 min.
2. Wash the membrane 3× for 10 min with regeneration buffer I at 50°C. Temperatures above 50°C can harm the membrane and/or the peptides.
3. Wash the membranes 3× for 10 min with 10X T-TBS at room temperature.
4. Wash the membranes 3× for 10 min with T-TBS at room temperature.
5. Proceed with the next incubation or store the membrane after washing 3× with water, twice with methanol, and drying.

3.12.2. Regeneration Protocol II

1. Wash the membrane twice with water for 10 min.
2. Incubate the membrane 3× with regeneration buffer IIA for 10 min.

3. Incubate the membrane 3× with regeneration buffer IIB for 10 min.
4. Wash the membrane with water for 10 min.
5. Wash the membrane 3× with T-TBS.
6. Proceed with the next incubation or store the membrane after washing 3× with water, twice with methanol, and drying.

3.12.3. Regeneration Controls

1. If the membrane was incubated with a directly labeled antibody, check the success of the regeneration by rinsing the membrane in substrate solution and then exposing it at least as long as the original exposure if the chemiluminescence method was used, or develop with the chromogenic substrates as in the main experiment. If spots are still detected, repeat regeneration protocol I (**step 2** can be prolonged), or go to regeneration protocol II.
2. If the membrane was incubated with a primary antibody in combination with an enzyme-labeled secondary antibody, re-incubate the membrane with the secondary antibody and substrate solution and make an exposure at least as long as the original exposure to show that the primary antibody is completely removed. Perform an analogous procedure if the binding was detected by chromogenic substrates. If spots are still detectable, repeat regeneration protocol I (**step 2** can be prolonged), or go to regeneration protocol II.

3.13. Storage of Peptide Membranes

1. New membranes should be stored at –20°C until use, where the membranes are placed in a plastic bag and sealed.
2. Incubated membranes that will be used again in only a few days should be washed 3× with T-TBS for 10 min and kept with a small volume of T-TBS in a Petri dish at 4°C. Drying out of the membrane sometimes leads to poor results in subsequent experiments.
3. Incubated membranes that will be stored for a longer period should be regenerated according to **Subheading 3.12.**, washed with methanol twice, air-dried, and kept at –20°C.

4. Notes

1. Do not use sodium azide as a preservative in buffers with peroxidase because it is an inhibitor of the enzyme. The presence of azide will greatly reduce or eliminate the signal.
2. If spots with a white center and a dark ring ("ring spots") are obtained after incubation and development of the array, a peptide membrane with a lower peptide loading is recommended ***(14)***. This can be achieved by mixing 0.3 *M* Fmoc-β-alanine-OPfp and 0.3 *M* Ac-β-alanine-OPfp 1:9 and using this solution for array formatting (**Subheading 3.3., step 2**).
3. The spotting process is most conveniently done by first dispensing the volume to be spotted and then gently touching the pipet tip to the center of each marked spot. This allows the most accurate liquid handling.
4. Do not overlay two or more peptide membranes in one incubation dish.

5. Bromophenol blue staining (**Subheading 3.5., step 6**): the intensity of staining varies depending on the last spotted amino acid. Some amino acids—such as cysteine, aspartic acid, glutamic acid, and asparagine—stain weakly. Alanine, glycine, and proline stain more strongly than others. These differences may serve as an internal control for correct pipetting. During later cycles, the intensity of staining diminishes.
6. Peptides of different length: if a peptide library comprises peptides of various length, the shorter peptides can accidentally be elongated at subsequent cycles by transferring excess Fmoc amino acid active esters from other spots during **step 2** from **Subheading 3.5.** However, this effect is very unlikely, and is usually ignored. However, in order to absolutely exclude this peptide elongation, **steps 1** and **2** from **Subheading 3.6.** can be carried out in addition between **steps 2** and **3** from **Subheading 3.5.** Acetylated spots can no longer be stained by bromophenol blue. This procedure also leads to capping of peptides that did not properly react with the last amino acid.
7. If no signals are observed using the protocols described in **Subheadings 3.9–3.11.,** change the protocols as follows (one or more modifications can be adopted simultaneously):
   a. Increase the antibody concentrations of the primary and/or secondary antibody.
   b. Prolong the incubation time with the primary antibody to overnight. This step should be performed at 4°C.
   c. Shorten the washing times. Use washing buffer without Tween 20.
   d. Use a supersensitive chemiluminescence substrate (e.g., SuperSignal® Ultra, Pierce, Rockford, IL).
   e. Perform a simultaneous incubation of primary and secondary antibody.
   f. Check the antibodies and enzymes in an alternative system.
8. Clear spots on dark background can be observed with the chemiluminescence protocol (**Subheading 3.11.**). In this case, the primary and/or secondary antibody concentrations may be too high. A high amount of antibody conjugate on the spots results in all the substrate being used up before the X-ray film can be exposed on the membrane or the imaging system can be started. To avoid this change, follow these protocols:
   a. Wash extensively with T-TBS and re-detect or, if the problem persists, instead of.
   b. Regenerate the membrane and incubate with lower concentrations of antibodies.
9. Results with directly labeled antibodies are often much better, especially for low-affinity binding antibodies, and may even be essential for the mapping of discontinuous epitopes. Conjugation of antibodies with peroxidase is easy to perform, and protocols are available ***(15)***.
10. If the background is too high, change the protocols as follows:
    a. Increase the detergent concentration in the washing buffer.
    b. Increase the washing times and/or the washing volumes.

References

1. Frank, R. (1992) Spot synthesis: an easy technique for the positionally addressable, parallel chemical synthesis on a membrane support. *Tetrahedron* **48,** 9217–9232.

2. Frank, R. (Date of filing: 31. 08. 1990) Verfahren zur parallelen Herstellung von trägergebundenen oder freien Peptiden, trägergegundene Peptide und ihre Verwendung. DE 4027675.
3. Geysen, H. M., Rodda, S. J., Mason, T. J., Tribbick, G., and Schofs, P. G. (1987) Strategies for epitope analysis using peptide synthesis. *J. Immunol. Methods* **102,** 259–274.
4. Geysen, H. M. (Date of filing: 08. 03. 1984) Method for determining antigenically active amino acids sequences. EP 0138855.
5. Pinilla, C., Appel, J. R., and Houghten, R. (1993) Functional importance of amino acid residues making up antigenic determinants. *Mol Immunol.* **30,** 577–585.
6. Molina, F., Laune, D., Gougat, C., Pau, B., and Granier, C. (1996) Improved performances of spot multiple peptide synthesis. *Peptide Res.* **9,** 151–155.
7. Reineke, U., Kramer, A., and Schneider-Mergener, J. (1999) Antigen sequence- and library-based mapping of linear and discontinuous protein-protein-interaction sites by spot synthesis. *Curr. Top. Microbiol. Immunol.* **243,** 23–36.
8. Wenschuh, H., Volkmer-Engert, R., Schmidt, M., Schulz, M., Schneider-Mergener, J., and Reineke, U. (2000) Coherent membrane supports for parallel microsynthesis and screening of bioactive peptides. *Biopolymers (Peptide Science)* **55,** 188–206.
9. Reineke, U., Volkmer-Engert, R., and Schneider-Mergener, J. (2001) Applications of peptide arrays prepared by the SPOT-technology. *Curr. Opin. Biotechnol.* **12,** 59 64.
10. Frank, R. and Schneider-Mergener, J. (2002) SPOT-synthesis: scope and applications, in *Peptide Arrays on Membrane Supports—Synthesis and Applications, Springer Laboratory Manual* (Koch, J., and Mahler, M., eds.), Springer-Verlag, Heidelberg, Germany, pp. 1–22.
11. Reimer, U., Reineke, U., and Schneider-Mergener, J. (2002) Peptide arrays: from macro to micro. *Curr. Opin. Biotechnol.* **13,** 315–320.
12. Ast, T., Heine, N., Germeroth, L., Schneider-Mergener, J., and Wenschuh, H. (1999) Efficient assembly of peptomers on continuous surfaces. *Tetrahedron Lett.* **40,** 4317–4318.
13. Van Regenmortel, M.H.V. (1994) The recognition of proteins and peptides by antibodies, in *Immunochemistry* (van Oss, C. J., and van Regenmortel, M.H.V., eds.), Marcel Dekker, New York, NY, pp. 277–300.
14. Kramer, A., Reineke, U., Dong, L., Hoffmann, B., Hoffmüller, U., Winkler, D., et al. (1999) Spot synthesis: observations and optimizations. *J. Peptide Res.* **54,** 319–327.
15. Wilson, M. B. and Nakane, P. K. (1978) Recent developments in the periodate method of conjugating horseradish peroxidase (HRPO) to antibodies, in *Immunofluorescence Related Staining Techniques* (Knapp, W., Holubar, K., and Wick, G., eds.), Elsevier, Amsterdam, the Netherlands, pp. 215–224.

27

Mapping Antibody:Antigen Interactions by Mass Spectrometry and Bioinformatics

Alexandra Huhalov, Daniel I.R. Spencer, and Kerry A. Chester

1. Introduction

Characterization of the antigen/antibody complex is an important tool in the rational design of antibody-based therapeutics. This can be achieved using several methods. Conventional methods for mapping an antibody-binding region on an antigen rely on synthetic polypeptide libraries, which limit the identification of epitopes to continuous amino acid sequences *(1)*. These techniques preclude the characterization of conformational epitopes, and subsequently, a major category of antibody/antigen interactions.

In order to identify both continuous and conformational epitopes, a more direct approach has been developed to identify binding sites based on antigen proteolysis in which proteins are digested with a range of proteases and the resulting peptides analyzed by polyacrylamide gel electrophoresis (PAGE) *(2)* or high-performance liquid chromatography (HPLC) *(3)*. However, these methods may fail to identify unresolved peptides accurately. Therefore, to identify both continuous and conformational epitopes and to overcome the ambiguity problems that are possible in PAGE and HPLC-based methods, several groups have turned to mass spectrometry (MS) to analyze proteolytic peptide fragments *(3)*.

A new approach to this procedure has been developed using surface-enhanced laser desorption/ionization-affinity mass spectrometry (SELDI-AMS), in which an antibody (or antibody fragment) is covalently coupled to a preactivated ProteinChip® Array and reacted with antigen. The resulting antibody/antigen complex is digested on the chip with proteolytic enzymes to remove all but the antibody-binding site of the antigen, which is protected by the bound antibody, and the remaining antibody-binding peptides are identified by MS. SELDI-AMS has several advantages as compared to exist-

From: *Methods in Molecular Biology, Vol. 248: Antibody Engineering: Methods and Protocols*
Edited by: B. K. C. Lo © Humana Press Inc., Totowa, NJ

ing MS-based epitope-mapping techniques, which can be expensive, time-consuming, and labor-intensive *(4)*. Because the SELDI-AMS technique is based on chemically modified affinity surfaces, the entire process can be done *in situ;* from binding of the complex onto the affinity surface to mass spectrometric analysis. This greatly reduces the experimental time because steps such as removal of non-binding antigen fragments can be done by washing the chip, rather than by the lengthy separation methods required when using non-derivatized mass spectrometry chip platforms in matrix-assisted laser desorption/ionization mass spectrometry (MALDI-MS). Limited sample loss occurs because the remaining antibody-associated peptides are laser-desorped from the chip-bound antibody in the mass spectrometer, and therefore require very little starting material. Because it is a chip-based technology, several samples can be analyzed in parallel, and there are a variety of chip surfaces available, resulting in protocols that are versatile and readily optimized (Ciphergen Biosystems; http://www.ciphergen.com). Recently, an interface has been developed to allow the samples from the ProteinChip Array to be further analyzed using Laser Desorption Ionization-Tandem MS to enable the direct sequencing of peptides identified by SELDI-AMS *(5)*.

This chapter outlines the methods we have developed to routinely characterize protein-protein interactions based on SELDI-AMS. We demonstrate this procedure by presenting two examples. First, to determine the antigen-binding domain of a clinically useful, phage-derived scFv *(6)* and second, to map the polyclonal response of patients to a recombinant scFv fusion protein using scFv libraries to determine immunogenic sites *(7)*.

In the first example, we partially characterized the binding domain of a phage-derived scFv reactive with carcinoembryonic antigen (CEA), MFE-23 *(8)*, which has been used in two clinical trials *(9, 10)*. CEA is a 180-kDa-membrane-associated protein and is made up of seven, heavily glycosylated immunoglobulin-like domains (N-A1-B1-A2-B2-A3-B3) *(11)*. MFE-23 has been previously shown to bind recombinant N-A1 domains, and these domains were tested in our system. The N-A1 domains were bacterially expressed in the presence of MFE-23 containing a hexahistidine tag, which was used to capture, purify, and couple the subsequent N-A1/MFE-23 complex to immobilized metal affinity chromatography (IMAC) chips for SELDI-AMS analysis. The complex was digested with trypsin, and the placement of the resulting antibody-bound peptides was studied in the homology model of CEA *(12)* to predict the MFE-23 binding site *(6)*.

In the second approach, we modeled the polyclonal response in patients to a fusion protein consisting of MFE-23 genetically linked to carboxy-peptidase G2 (CPG2) *(13)*, used in antibody-directed enzyme prodrug therapy (ADEPT) *(14)* in order to neutralize its immunogenicity in patients. A murine scFv phage library was raised against CPG2, and a CPG2-reactive scFv was isolated from this library on the basis of its ability to inhibit CPG2 binding of sera from patients who had an immune response to this antigen *(7)*. This CPG2-reactive scFv was then covalently coupled to a preactivated ProteinChip Array. CPG2 was captured by the bound scFv, and their interaction was mapped by partial digestion with Glu-C. The subsequent peptides were modeled to two nonlinear sites in the crystal structure of CPG2 ***(15)***, and the epitope was determined to be conformational. This was confirmed when subsequent alanine-scanning mutagenesis of one of the predicted sites resulted in a reduction in immunogenicity of the mutant fusion protein when tested against patient sera *(7)*.

The methods used to generate scFv libraries, screen against antigen (*see* Chapters 6 and 8), and produce recombinant scFv (*see* Chapter 14) have been extensively described elsewhere. Here, we present our generally applicable methods to epitope map an antibody/antigen complex by SELDI-AMS on i) preactivated chips and ii) IMAC, as well as iii) a method for the determination of protein molecular weight on normal-phase chips and iv) the use of simple bioinformatics for the prediction of epitopes in protein structures using software available in the public domain (**Fig. 1**).

2. Materials

2.1. SELDI (for Subheadings 3.1.–3.4.)

All chemicals are from Sigma-Aldrich Ltd., and stored at room temperature unless otherwise stated.

1. SELDI ProteinChip Biology System II (Ciphergen Biosystems).
2. NP20, IMAC3, PS10, or PS20 ProteinChip Array (Ciphergen Biosystems).
3. Hydrophobic pen (ImmEdge™ Pen; Vector Laboratories).
4. Dulbecco's phosphate-buffered saline without calcium chloride (PBS; Sigma Cat. #D5773).
5. PBST: 0.1% Triton X-100 in PBS.
6. 1 *M* ethanolamine (BIAcore AB).
7. Acetonitrile.
8. 50 m*M* nickel sulfate.
9. IMAC wash buffer: 300 m*M* sodium chloride in PBST.
10. Imidazole solution: 200 m*M* imidazole in IMAC wash buffer.
11. Energy-absorbing matrices (EAM) (Ciphergen Biosystems):
    a. α-cyano-4-hydroxy cinnamic acid (CHCA): to a microfuge tube of solid CHCA (5 mg) add 500 µL of 100% acetonitrile and 500 µL of 1% trifluoroacetic acid in ddH_2O. Vortex the solution for 10 s. This solution is light-sensitive, and should be freshly prepared each day.
    b. Sinapinic acid (SPA): To a microfuge tube of solid sinapinic acid (5 mg) add 500 µL of 100% acetonitrile and 500 µL of 1% trifluoroacetic acid in ddH_2O.

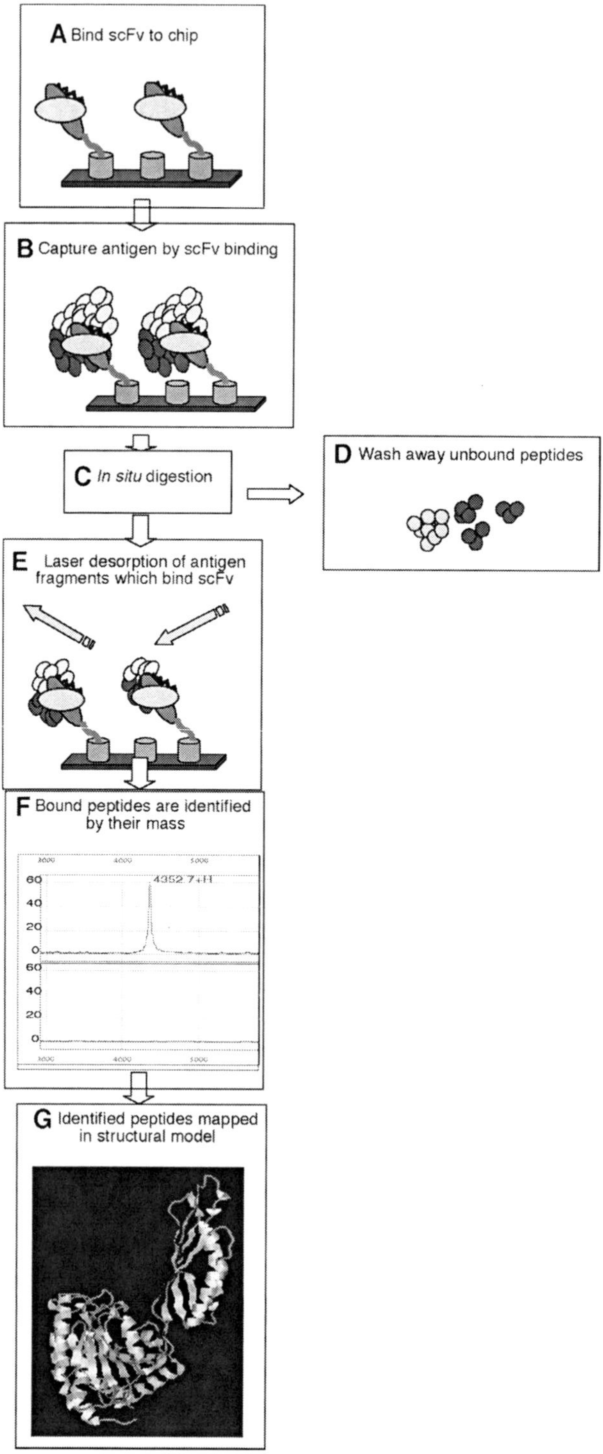
A Bind scFv to chip
B Capture antigen by scFv binding
C In situ digestion
D Wash away unbound peptides
E Laser desorption of antigen fragments which bind scFv
F Bound peptides are identified by their mass
4352.7+H
G Identified peptides mapped in structural model

Fig. 1. SELDI-AMS protocol to epitope-map scFv binding interactions. (**A**) scFv are bound to the chip surface, either covalently or via the hexahistidine tag on the scFv. (**B**) Antigen is captured by scFv (or was present as complex with scFv when bound to chip via histidine tag). (**C**) Complex is digested with proteolytic enzymes. (**D**) Unbound peptides are washed away. (**E**) scFv-bound peptides are laser-desorped and (**F**) identified by their mass. (**G**) Corresponding sequences are then mapped in the structural model.

Vortex the solution for 10 s. The SPA is close to saturation, and will not always completely dissolve. This solution is light-sensitive, and should be freshly prepared each day.

12. Endoproteinase Glu-C (Roche Molecular Biochemicals): Dissolve lyophilized enzyme in ddH_2O to 250 µg/mL. Aliquot and store at 4°C for 1–2 d.
13. Trypsin, modified, sequencing-grade (Roche Molecular Biochemicals): Dissolve lyophilized trypsin in 1 m*M* HCl to 250 µg/mL. Aliquot and store at 4°C for up to 1 wk.
14. Peptide and Protein All-in-One mW standards kit (Ciphergen Biosystems).
15. Humidity chamber (*see* **Note 1**).

2.2. Analytical Software (for Subheadings 3.5.–3.7.)

1. PAWS software available on Protein Information Retrieval Online World Wide Web Lab website: http://prowl.rockefeller.edu/contents/resource.htm.
2. MS-Digest program available on Protein Prospector website: http://prospector.ucsf.edu/.
3. Rasmol 2.7.1 software available on Rasmol homepage: http://www.umass.edu/microbio/rasmol/ *(16)*.
4. Solved protein structures available on the Protein Databank website: http://www.pdb.org/ *(17)*.

3. Methods

Several ProteinChip Array surfaces are available that are suitable for epitope-mapping studies by limited proteolysis and MS using SELDI-AMS. Two of these are outlined here, and include i) preactivated chips (**Subheading 3.1.**), in which scFv are covalently coupled to the chip surface, and ii) IMAC chips (**Subheading 3.2.**) in which scFv are coupled to the chip surface via their hexahistidine tags. Methods for iii) validation of protein mol wt on normal-phase chips (**Subheading 3.3.**), iv) calibration of the instrument and data acquisition (**Subheading 3.4.**), v) the analysis of data using protein software packages (**Subheadings 3.5.** and **3.6.**) and vi) the use of bioinformatics to fit the resulting information into structural models (**Subheading 3.7.**) are also described.

3.1. Preactivated ProteinChip Arrays

The PS10 and PS20 ProteinChip Arrays are used for immunoassays and other protein-protein interactions. They have eight spots that are preactivated with carbonyl diimidazole chemistry that covalently bind to free primary amine groups.

1. Dilute the test scFv, nonspecific scFv (if available) and antigen to 100–300 μg/mL in double-distilled water (ddH_2O) (*see* **Note 2**).
2. Add 1–5 μL of diluted test scFv, nonspecific scFv, or PBS to individual spots and incubate in a humidity chamber for 1 h at room temperature. Do not allow the spots to dry.
3. Using a pipet, remove solution from spot and immediately add 5 μL of PBST. Do not allow the spots to dry. Repeat this procedure for all the spots used on the chip. Using a pipet, pass several streams of PBST over each spot making sure they do not cross over neighboring spots.
4. Submerge the chip into a tube filled with PBST and rotate for 5 min in a carousel shaker at high speed. Then repeat this step two more times with fresh PBST (*see* **Note 3**).
5. Rinse the chip twice in PBS as in **step 4.** Do not allow the spots to dry.
6. Block the spots by adding 5 μL of 1 *M* ethanolamine, pH 8.0 to each spot and incubate in the humidity chamber for 15 min. Further blocking is optional (*see* **Note 4**).
7. Wash the chip as in **steps 3–5.** Do not allow the spots to dry.
8. Add 1–5 μL of antigen diluted in ddH_2O to 1 μL of PBST on spots and incubate in a humidity chamber at room temperature for 1 h. Do not add antigen to one PBS-coated spot as a control.
9. Wash the chip as in **steps 3–5**. Do not allow the spots to dry.
10. Add 2 μL of Glu-C, trypsin, or the protease of choice to each spot diluted in PBST (including the spot containing only antigen for use as a control) added at ratios of between 1:100 to 1:20 of the protein by weight (*see* **Note 5**). Incubate at room temperature in the humidity chamber for 30–90 min. Include one PBS-coated spot containing no scFv or antigen as a control.
11. Wash spots as in **steps 3–5** with a final rinse in ddH_2O.
12. Apply 0.5–1 μL of the appropriate EAM to spots (*see* **Note 6**).
13. Dry the spots by placing chip on top of the ProteinChip Reader (*see* **Note 7**).
14. Load the chip into the ProteinChip Reader.
15. Collect peptide mass data over the desired mass range (*see* **Subheading 3.4**).
16. This will result in a set of mass spectra (**Fig. 2**).

3.2. IMAC ProteinChip Arrays

The IMAC3 ProteinChip Arrays are used for protein capture, profiling, and on-chip purification. They are coated with a nitrilotriacetic acid (NTA) functional group, which binds transitional metals for the capture of polyhistidine affinity-tagged proteins.

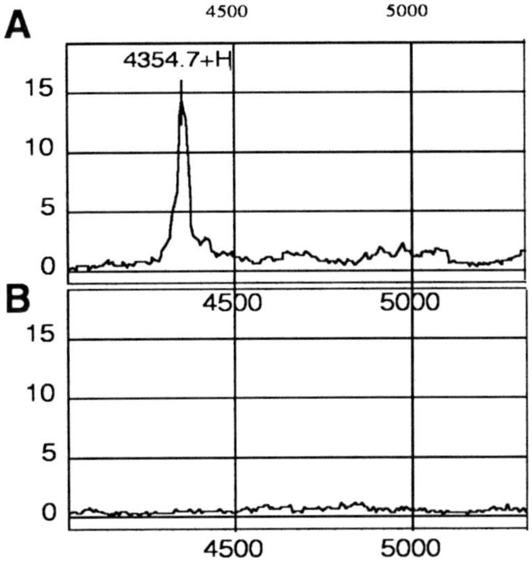

Fig. 2. Mass spectra obtained from the epitope-mapping of scFv interactions using preactivated chips. This method is illustrated by CPG2-reactive scFv, which are covalently bound to a preactivated ProteinChip Array and used to capture CPG2. The subsequent complex is then digested with Glu-C, resulting in an array of peptides. Nonbinding peptides are washed away, and retained peptides are measured by their mass. **(A)** A peak is detected at 4,354.7 Daltons, and **(B)** no peak is detected in the control spot containing a nonspecific scFv.

1. Draw an outline around each spot, using the hydrophobic pen.
2. Add 5 µL of 100% acetonitrile to each spot and incubate at room temperature for 1 min, then wipe off with a tissue.
3. Add 1 µL of 50% acetonitrile diluted in ddH_2O to each spot, and before it dries, add 5 µL of 50 m*M* nickel sulfate and incubate in a humidity chamber at room temperature for 30 min (*see* **Note 8**).
4. Reapply 5 µL of 50 m*M* nickel sulfate and incubate for 30 min in the humidity chamber at room temperature.
5. Submerge the chip in a tube filled with IMAC washing buffer and wash by rotating in a carousel shaker at high speed for 5 min (*see* **Note 9**).
6. Add 5 µL of polyhistidine-tagged scFv alone or in complex with antigen (containing no polyhistidine tag) diluted to 100–200 µg/mL in IMAC washing buffer. As controls, add 5 µL of imidazole solution or IMAC washing buffer to spots and incubate in the humidity chamber for 30–60 min (*see* **Note 10**). Do not allow the spots to dry.
7. Using a pipet, remove solution from spot and immediately add 5 µL of IMAC washing buffer. Be sure to avoid drying the spots. Repeat this procedure for all the spots used on the chip.

8. Using a pipet, pass several streams of IMAC washing buffer over each spot making sure they do not cross over neighboring spots.
9. Submerge the chip into a tube filled with IMAC washing buffer and rotate in a carousel shaker at high speed for 5 min. Then repeat this step two more times. Do not allow the spots to dry.
10. If scFv was added alone, add 5 µL of antigen (containing no polyhistidine tag) diluted to 100–200 µg/mL in IMAC washing buffer or 40 m*M* imidazole in IMAC washing buffer and incubate in the humidity chamber for 1 h. Include one scFv-coated spot with no antigen as a control.
11. Wash the chip as in **steps 7–9.** Do not allow the spots to dry.
12. Add 2 µL of Glu-C, trypsin, or the protease of choice to each spot diluted in PBST added at ratios of between 1:100 to 1:20 of the protein by weight (*see* **Note 5**). Incubate at room temperature in the humidity chamber for 30–60 min. Include one PBS-coated spot containing no scFv or antigen as a control. If scFv and antigen were captured onto the IMAC chip as a complex, include a control digestion done in solution. After the required digestion time, apply the control digestion to scFv and PBS-coated spots to control for nonspecific binding of proteolytic fragments to the nickel-saturated matrix.
13. Wash the chip as in **steps 7–9.**
14. Apply 0.5–1 µL of the appropriate EAM to spots (*see* **Note 6**).
15. Dry the spots by placing chip on top of the ProteinChip Reader (*see* **Note 7**).
16. Load the chip into the ProteinChip Reader.
17. Collect peptide mass data over the desired mass range (*see* **Subheading 3.4.**).
18. This will result in a set of mass spectra (**Fig. 3**).

3.3. Normal-Phase ProteinChip Arrays

These chips are used to confirm protein mol wts and for QC assays. They mimic normal-phase chromatography.

1. Dilute protein into ddH_2O to a concentration of 100–200 µg/mL to attain a final volume of 50 µL (*see* **Note 2**).
2. Add 1 µL of the diluted protein to the first spot.
3. Add 0.5–1 µL of the appropriate EAM (*see* **Note 6**) immediately to the protein on the first spot.
4. Add 1 µL of diluted protein to the three following spots. Do not add the EAM to these spots at this stage.
5. Dry the samples on top of the ProteinChip Reader (*see* **Note 7**).
6. To the three spots containing dried samples, add 5 µL of ddH_2O and leave for 5 s. Using a fresh tip for each spot, pipet in and out 4× and draw up and discard the solution after the fourth mix. Add 0.5–1 µL of the appropriate EAM to the protein on the first of these spots, and 5 µL of ddH_2O to the remaining two spots.
7. Repeat the wash step for the remaining two spots. Add 0.5–1 µL of the appropriate EAM to the protein on the first of these spots and 5 µL of ddH_2O to the remaining spot. Repeat the wash step for the remaining spot. Add 0.5–1 µL of the appropriate EAM to the protein on the remaining spot.

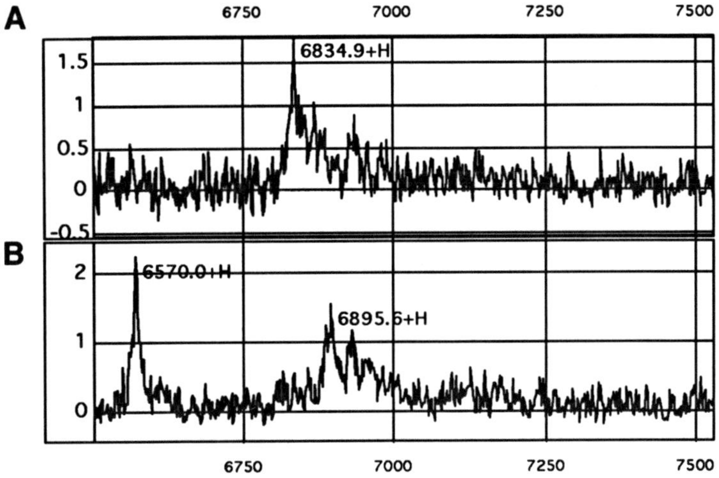

Fig. 3. Mass spectra obtained from the epitope-mapping of scFv interactions using IMAC ProteinChip Arrays. This is illustrated by hexahistidine-tagged MFE-23, in complex with the N-A1 domains of CEA, which are captured by an IMAC surface via immobilized copper ions. The complex is digested with trypsin, resulting in an array of peptides. Nonbinding peptides are washed away, and retained peptides are measured by their mass. **(A)** One peak is detected at 6,834.9 Daltons and **(B)** is not detected in the control spot containing MFE-23 digestion only. The two peaks at 6,570.0 and 6,895.6 Daltons correspond to MFE-23 proteolytic peptides.

8. Dry the spots by placing chip on top of the ProteinChip Reader (*see* **Note 7**).
9. Load the chip into the ProteinChip Reader.
10. Collect peptide mass data over the desired mass range (*see* **Subheading 3.4.**).

3.4. Calibration and Collection of Mass Data

Calibration of the instrument is an important component in converting time-of-flight data to accurate mass measurements. There are two types of calibration for Proteinchip arrays—external and internal. External calibration uses the calibration from one sample to determine the mass on another. In other words, the masses of calibrants from one spot are applied to masses of proteins or peptides on another spot. External calibration results in mass accuracy of approx 0.5% or better. Internal calibration uses known masses in the sample for calibration. This type of calibration results in mass accuracy of approx 0.05% or better (information provided in the Proteinchip® software manual 3.0 Operation Manual).

Mass data may be collected either manually or automatically. Generally, the data collection parameters are optimized manually and then applied to an automated protocol, which collects mass data from all the spots identically. The following protocol is an example of collecting mass data obtained from the All-in-One Protein standard mix. The same process is applied when collecting data from any sample.

1. Apply sample to spot. For example, apply 0.5–1 μL of All-in-One Protein standard to a blank spot for external calibration and to a spot containing sample for internal calibration (*see* **Note 11**).
2. Air-dry the spot and add 0.5–1 μL of the appropriate EAM (*see* **Note 6**). For example, for the Protein standard, apply SPA.
3. Dry the spot by placing chip on top of the ProteinChip Reader (*see* **Note 7**).
4. Load the chip into the ProteinChip Reader.
5. For manual collection, select the "manual protocol" button.
6. Set "high mass." For example, for the All-in-One Protein standard set to 200,000 Daltons.
7. Set "start optimization range." For example, for the All-in-One Protein standard set to 10,000 Daltons.
8. Set "end optimization range." For example, for the All-in-One Protein standard set to 150,000 Daltons.
9. Set "focus" by "mass." For example, for the All-in-One Protein standard set to 66,000 Daltons (*see* **Note 12**).
10. Set "laser intensity." For example, for the All-in-One Protein standard set to 230.
11. Set "initial detector" sensitivity to 9.
12. Click the "fire continuously" button to "on."
13. Adjust the laser intensity and sensitivity settings so that the signal and intensity remain to scale.
14. Turn off the "fire continuously" button.
15. Set up an automatic protocol using the laser and sensitivity settings optimized in the manual protocol.
16. Select "new spot protocol" from the file menu.
17. Enter a name for the protocol when prompted.
18. Select "SELDI quantitation" from the "method type" list.
19. Click on "next" to go to the "auto setup" window.
20. Set "high mass to acquire." For example, for the All-in-One Protein standard set to 200,000 Daltons.
21. Set "start optimization range." For example, for the All-in-One Protein standard set to 10,000 Daltons.
22. Set "end optimization range." For example, for the All-in-One Protein standard set to 150,000 Daltons.
23. Set "focus" by "mass." For example, for the All-in-One Protein standard set to 66,000 Daltons.
24. Set "laser intensity" to the optimized setting determined by the manual protocol.
25. Set "detector sensitivity" to the optimized setting determined by the manual protocol.
26. Click "next" to go to the "SELDI data acquisition settings" window.

27. Set "collect transients per position" to 10.
28. Set the "starting at position" to 20.
29. Set the "ending at position" to 80.
30. Set the "moving positions" to 4.
31. Set the "warming position" to 2 laser shots with the "intensity" set to 250.
32. Deselect "include warming shots in spectra."
33. Click "finish" and save protocol.
34. This spot protocol can now be applied to the entire chip.

3.5. Interpretation of Mass Spectra Using PAWS

PAWS is a protein sequence analysis software package freely distributed over the Internet. It allows the user to paste in the amino acid sequence of a protein and modify it in a number of ways. It is particularly useful for determining proteolytic peptides that arise from the cleavage of a protein.

1. Open the PAWS program, select "new" from the file menu and paste in the protein sequence of interest.
2. Select the appropriate enzyme from the "cleavage" menu. This will provide a list of all possible peptides that may result from the cleavage of the protein. For example, when analyzing the CPG2 sequence by Glu-C cleavage, 26 peptides are revealed (*see* **Note 13**).
3. With this window open, select "a list of peptides" under the "find" menu and type in any peptide masses identified by SELDI-AMS. The program will provide a graphic representation of their placement in the sequence. Four proteolytic fragments were found for Glu-C digestion of CPG2, and were identified according to the CPG2 SWISS-PROT protein database reference (1CG2) (**Fig. 4**). The observed peptide masses were: m/z 4354.7 assigned to CPG2 Ser 379- Lys 415 average mass of 4354.1 Daltons, m/z 2794.9 assigned to CPG2 Tyr391- Lys415 average mass of 2794.4 Daltons, m/z 3539.9 assigned to CPG2 Tyr 159-Glu189 average mass of 3539.8 Daltons, and m/z 2092.2 Daltons assigned to CPG2 Tyr159-Glu176 average mass of 2092.2 Daltons. The peptide fragments identified by SELDI-AMS were located in two distinct and remote regions of the CPG2 sequence.

3.6. Interpretation of Mass Spectra Using MS-Digest

MS-Digest is another protein sequence analysis software package, which is freely distributed over the Internet and provides a comprehensive list of proteolytic peptide fragments resulting from the cleavage of a user-specified protein.

1. Open the MS-Digest page and select "user protein" in the "database" menu.
2. Select the appropriate enzyme in the "digest" menu.
3. Set the maximum number of missed cleavages to between 5 and 10.
4. Set the program to recognize oxidized methionines and pyro-glumatic acid containing peptides in the "considered modifications."

A

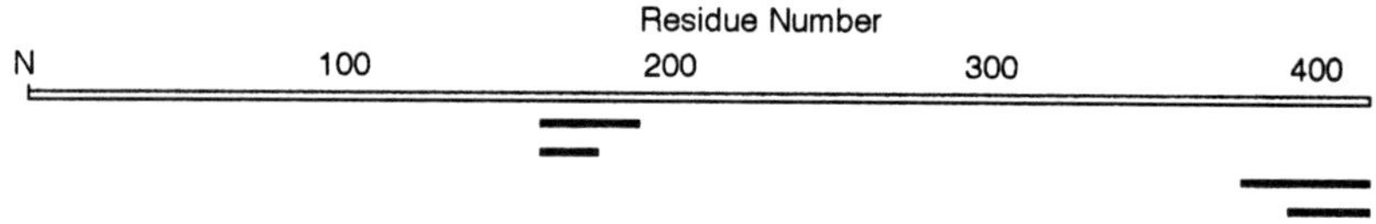

B

#	Mass	Matching sequence
1	3539.900	Y[159-189]E
2	2092.800	Y[159-176]E
3	4354.700	S[376-415]K
4	2794.900	Y[391-415]K

Fig. 4. The use of PAWS to analyze SELDI-AMS-identified proteolytic peptide fragments illustrated using data obtained from the SELDI-AMS analysis of CPG2 with a reactive scFv. The program determines all possible peptides that result from cleavage of a protein sequence by a user-defined protease. Peptide masses obtained by SELDI-AMS are then crosschecked against these and listed together with their placement in the protein sequence, which is graphically represented by lines. **(A)** Graphic representation of CPG2 amino acid sequence with the placement of scFv-binding peptides resulting from Glu-C digestion of CPG2 in complex with a specific scFv bound to pre-activated chips represented by solid lines and **(B)** listed below with their exact placement in the sequence. The peptides correspond to two nonlinear regions along the protein sequence.

5. Set the minimum and maximum fragment mass menus to the appropriate values. For example, when analyzing peptides obtained from the SELDI-AMS of CPG2 using Glu-C, these parameters were set to 1,000 Daltons and 6,000 Daltons, respectively.
6. Paste the entire protein sequence into the "user protein sequence" menu.
7. Select "perform digest."
8. The program will then provide a list of possible peptides. Proteolytic fragment masses obtained from SELDI-AMS can then be cross-referenced against the program-derived list of peptide fragments. All four of the proteolytic fragments derived from the Glu-C digestion of CPG2 were found in this list.

3.7. Prediction of Peptide Fit into Structural Models

Invaluable information on the characterization of protein:protein interactions includes the application of structural models to map the placement of the proteolytic peptides identified by SELDI-AMS. A crystal structure was available for CPG2, enabling the mapping of SELDI-AMS identified peptides resulting from proteolytic digestions. Rasmol 2.7.1 is a freely distributed molecular graphics visualization tool that is available over the Internet. It allows the user to download Protein Data Bank (PDB) Files and study the structure interactively. This simple process enables the mapping of the two regions identified by SELDI-AMS in the structural model of CPG2 and shows that two sequentially remote regions are structurally adjacent, suggesting a conformational epitope.

1. Open the Rasmol program and select open in the "file" menu. Browse downloaded PDB structures to load the protein of choice. For example, for the CPG2 structure, load PDB reference file 1CG2. This will load the tetrameric protein structure.
2. To select the monomeric portion of the structure, enter "select *A" in the Rasmol command line and press return to send the command.
3. Enter "save temp.pdb." This will save the selection onto the desktop.
4. Enter "zap." This will clear the program.
5. Enter "load temp.pdb." This will load the selected monomer.
6. Select "spacefill" under the "display" menu in the main Rasmol window.
7. Select "monochrome" under the "colors" menu in the main Rasmol window.
8. Enter "select 159–189" in the command window. This will select amino acids corresponding to their placement in the structure. These amino acids represent a region identified by SELDI-AMS (sequences identified from mass data corresponding to peaks at 2,092.8 Daltons and 3,539.9 Daltons; *see* **Fig. 4**).
9. Enter "color blue." This will color the selected amino acids blue.
10. Enter "select 376–415."
11. Enter "color red." These amino acids represent a region identified by SELDI-AMS (sequences identified from mass data corresponding to peaks 2,794.9 Daltons and 4,354.7 Daltons) (*see* **Fig. 4**).
12. Maneuver the structure using the mouse left and right buttons.
13. Save the final version by selecting an export file in the "export menu" **(Fig. 5)**.

4. Notes

1. To make a humidity chamber, fill the bottom half of an empty plastic pipet tip box halfway with ddH_2O and place the tip rack back on top to use as a support for the chips. When the lid is replaced firmly, it is sufficiently sealed to prevent drying of the chip for several hours.
2. It is important to use double-distilled water (ddH_2O) when diluting protein for SELDI-AMS, as excess salt can disrupt the laser desorption of peptides.
3. Alternatively, washes can be carried out on a high-frequency shaker.
4. For further blocking, add 5 μL of one of the following in a humidity chamber for 1 h:
   a. Cytochrome C, bovine, 10 pmol/μL;
   b. 5% milk proteins/PBS;
   c. 3% serum albumin, bovine/PBS.
5. Digestion times and ratios are only given as guidelines and must be optimized for individual antibody/antigen interactions. It is advisable to test a range of digestion times, temperatures, and enzyme ratios for optimal results.
6. EAM is an energy-absorbing compound capable of converting laser energy to thermal energy, thus facilitating the desorption/ionization of samples. It is important to

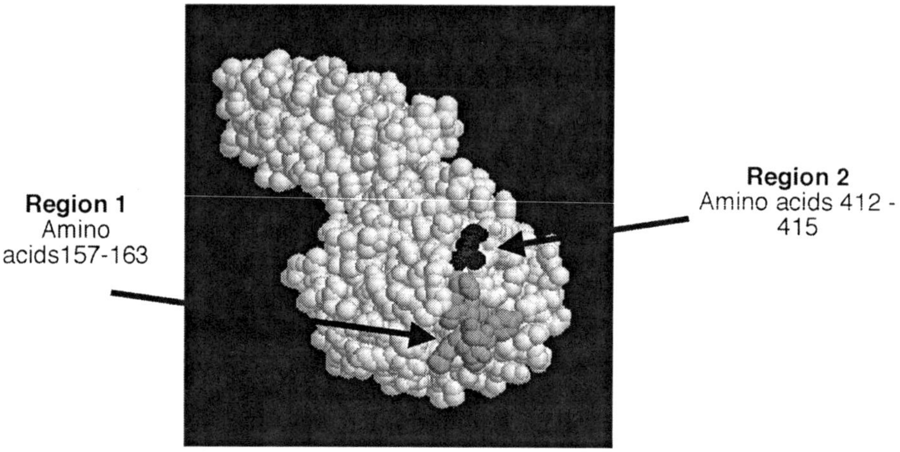

Fig. 5. Representation of the crystal structure of CPG2 showing the surface-exposed conformational epitope determined by SELDI-AMS. Epitope-mapping studies using a specific phage-derived scFv raised against CPG2 revealed reactive peptides corresponding to two nonlinear sites along the CPG2 sequence concentrated at positions Y[159–189]E and S[376–415]K. These peptides were fit into the crystal structure of CPG2, and further characterized to obtain the surface-exposed regions of the epitope as determined by the Define Secondary Structure of Proteins (DSSP) solvent accessibility algorithm ***(18)***, revealing a conformational epitope at Region 1 G[157–163]D *(dark gray)* and Region 2 G[412–415]G *(black)*.

use the appropriate EAM for analysis. For small molecules and peptides between 5 and 15 kDa, use a saturated solution of CHCA. For anything smaller than this, use a 1/5 dilution of the saturated solution. The mol wt of CHCA is 189.2 Daltons. For molecules larger than 15 kDa, use SPA. The mol wt of SPA is 224.2 Daltons.

7. To dry spots more quickly, use a hair dryer set at cool. The dryer is held 10 cm away from the surface of the chip and moved with a slow rotating movement for approx 1 min until the sample(s) has dried.
8. IMAC ProteinChip Arrays work best when coupled to nickel ions. Using Cu(II) and Fe(II) can affect the surface of the chip.
9. Binding and washing buffers benefit from the addition of salt (0.3–1 *M*). To increase selectivity of the polyhistidine-tagged protein, imidazole may be added to the binding buffer (5–10 m*M*) and/or wash buffer (10–100 m*M*).
10. Alternatively, incubations can be done on a high-frequency shaker for 5 min at room temperature, providing that the sample is not allowed to dry out.
11. For a stronger signal, re-apply 0.5 µL of the mol wt standard solution several more times.

12. It is very important to collect sample data at the same focus mass used for calibration. Acquiring sample mass data at different focus masses from those used for calibrants will result in inaccurate masses.
13. When acquiring cleavage peptides by Glu-C in PAWS, select the V8 protease for cleavage of both glutamic acid (E) and aspartic acid (D), as Glu-C is able to cleave at both these positions in phosphate buffer.

References

1. Berzofsky, J. A. (1985) Intrinsic and extrinsic factors in protein antigenic structure. *Science* **229,** 932–940.
2. Sheshberadaran, H. and Payne, L. G. (1988) Protein antigen-monoclonal antibody contact sites investigated by limited proteolysis of monoclonal antibody-bound antigen: protein "footprinting." *Proc. Natl. Acad. Sci. USA* **85,** 1–5.
3. Jemmerson, R. and Paterson, Y. (1986) Mapping epitopes on a protein antigen by the proteolysis of antigen-antibody complexes. *Science* **232,** 1001–1004.
4. Van deWater, W. J., Deininger, S. O., Macht, M., Przybylski, M., and Gershwin, M. E. (1997) Detection of molecular determinants and epitope mapping using MALDI-TOF mass spectrometry. *Clin. Immunol. Immunopathol.* **85,** 229–235.
5. Reid, G., Gan, B. S., She, Y. M., Ens, W., Weinberger, S., and Howard, J. C. (2002) Rapid identification of probiotic lactobacillus biosurfactant proteins by ProteinChip tandem mass spectrometry tryptic peptide sequencing. *Appl. Environ. Microbiol.* **68,** 977–980.
6. Huhalov, A. Spencer, D.I.R., Hawkins, R. E., Perkins, S. J., Begent, R.H.J., and Chester, K. A. (2002) Capture and analysis of interacting tagged-proteins: an application to map the carcinoembryonic antigen binding-site of single chain Fv molecule MFE-23. In preparation
7. Spencer, D. I., Robson, L., Purdy, D., Whitelegg, N. R., Michael, N. P., Bhatia, J., et al. (2002) A strategy for mapping and neutralizing conformational immunogenic sites on protein therapeutics. *Proteomics* **2,** 271–279.
8. Chester, K. A., Begent, R. H., Robson, L., Keep, P., Pedley, R. B., Boden, J. A., et al. (1994) Phage libraries for generation of clinically useful antibodies. *Lancet* **343,** 455–456.
9. Begent, R. H., Verhaar, M. J., Chester, K. A., Casey, J. L., Green, A. J., Napier, M. P., et al. (1996) Clinical evidence of efficient tumor targeting based on single-chain Fv antibody selected from a combinatorial library. *Nat. Med.* **2,** 979–984.
10. Mayer, A., Tsiompanou, E., O'Malley, D., et al. (2000). Radioimmunoguided surgery in colorectal cancer using a genetically engineered anti-CEA single-chain FV antibody. *Clin. Cancer Res.* **6,** 1711–1719.
11. Oikawa, S., Imajo, S., Noguchi, T., Kosaki, G., and Nakazato, H. (1987) The carcinoembryonic antigen (CEA) contains multiple immunoglobulin-like domains. *Biochem. Biophys. Res. Commun.* **144,** 634–642.
12. Boehm, M. K., Mayans, M. O., Thornton, J. D., Begent, R. H., Keep, P. A., and Perkins, S. J. (1996) Extended glycoprotein structure of the seven domains in human carcinoembryonic antigen by X-ray and neutron solution scattering and an

automated curve fitting procedure: implications for cellular adhesion. *J. Mol. Biol.* **259,** 718–736.

13. Michael, N. P., Chester, K. A., Melton, R. G., Robson, L., Nicholas, W., Boden, J. A., et al. (1996) In vitro and in vivo characterisation of a recombinant carboxypeptidase G2: anti-CEA scFv fusion protein. *Immunotechnology* **2,** 47–57.
14. Bagshawe, K. D. (1987) Antibody directed enzymes revive anti-cancer prodrugs concept. *Br. J. Cancer* **56,** 531–532.
15. Rowsell, S., Pauptit, R. A., Tucker, A. D., Melton, R. G., Blow, D. M., and Brick, P. (1997) Crystal structure of carboxypeptidase G2, a bacterial enzyme with applications in cancer therapy. *Structure* **5,** 337–347.
16. Sayle, R. A. and Milner-White, E. J. (1995) RASMOL: biomolecular graphics for all. *Trends Biochem. Sci.* **20,** 374.
17. Berman, H. M., Westbrook, J., Feng, Z., Gilliland, G., Bhat, T. N., Weissig, H., et al. (2000) The Protein Data Bank. *Nucleic Acids Res.* **28,** 235–242.
18. Kabsch, W. and Sander, C. (1983) Dictionary of protein secondary structure: pattern recognition of hydrogen-bonded and geometrical features. *Biopolymers* **22,** 2577–2637.

28

Radiometal Labeling of Antibodies and Antibody Fragments for Imaging and Therapy

Ilse Novak-Hofer, Robert Waibel, Kurt Zimmermann, Roger Schibli, Jürgen Grünberg, Kerry A. Chester, Andrea Murray, Benny K.C. Lo, Alan C. Perkins, and P. August Schubiger

1. Introduction

Monoclonal antibodies (MAbs) are highly selective agents for delivering radiation to tumors both for in vivo imaging (radioimmunoscintigraphy, RIS) and for therapy (radioimmunotherapy, RIT). Antibody-based radiopharmaceuticals can be labeled with iodine nuclides (^{123}I for imaging, ^{131}I for therapy) or with metallic nuclides such as the γ radiation-emitting nuclides ^{111}In and ^{99m}Tc for imaging or the β^- particle-emitting nuclides ^{90}Y, ^{67}Cu, and $^{188/186}$Re for therapy. The labeling of antibodies with three metallic nuclides is described here: ^{67}Cu for both imaging and therapy, ^{99m}Tc for imaging, and $^{188/186}$Re for therapy.

1.1. Labeling of Antibodies with ^{67}Cu

The physical properties of the ^{67}Cu nuclide are well-suited for both RIS and RIT applications. ^{67}Cu is a β^- particle-emitting nuclide with β^- emissions distributed between 577 keV and 395 keV, resulting in a mean energy of 141 keV, which is similar to the mean β^- energy of ^{131}I (180 keV). ^{67}Cu can be produced by bombarding ^{67}Zn with neutrons in high flux reactors, or with cyclotrons by irradiating ^{nat}Zn or enriched ^{68}Zn with protons *(1)*. The therapeutic β^- particles emitted by ^{67}Cu with a mean range of 0.2 mm are very appropriate for the treatment of small tumors in the range of up to 5 mm in diameter, and its half-life of 2.58 d is long enough to permit accumulation of the antibody at the tumor site. ^{67}Cu also emits γ-radiation in the range between 91 keV and 184 keV (49% abundance), an energy range that is particularly suitable for pre-therapy diagnostic imaging with a γ-camera. In contrast to the high level (> 90%) of penetrating γ-radiation associated wih the ^{131}I nuclide, which adds to the whole-body dose of the patient and to the radi-

From: *Methods in Molecular Biology, Vol. 248: Antibody Engineering: Methods and Protocols*
Edited by: B. K. C. Lo © Humana Press Inc., Totowa, NJ

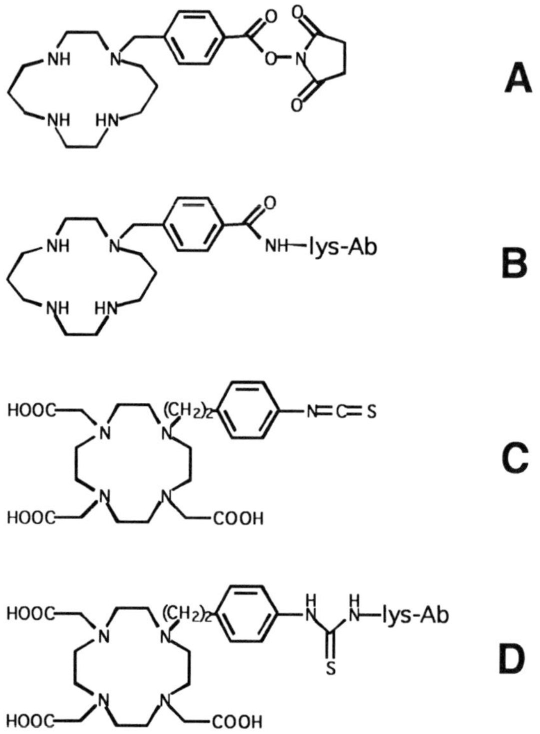

Fig. 1. Structures of the bifunctional chelators CPTA-NHS (**A**), CPTA coupled to a lysine residue of MAb (**B**), DO3A-NCS (**C**), and DO3A coupled to a lysine residue of MAb (**D**).

ation burden of the hospital staff, the lower abundance and energy of the γ-emissions from ^{67}Cu represent a distinct advantage. Preclinical and clinical studies performed with ^{67}Cu-labeled antibodies all indicate the great potential of ^{67}Cu for systemic and local tumor treatment *(2)*. One of the advantages of the copper nuclides compared to other therapeutic nuclides is the availability of excellent tetraaza macrocyclic chelators, which complex copper selectively and rapidly at ambient temperature *(3)*. This property permits convenient postconjugation labeling protocols at temperatures that maintain protein function. In addition to the metal chelating unit, bifunctional copper chelators (BFCs) contain a conjugation group for attachment to the tumor-seeking moiety. Conjugation of chelates to antibodies is performed via electrophilic conjugation groups, activated esters such as the N-hydroxysuccinimide (NHS) group, or the isothiocyanate (NCS) group, attached to a linker, which can be a phenyl group or a short peptide in addition to the phenyl group. In the case of -NCS and -NHS linked chelators, coupling is done directly to the primary amine groups of lysine residues of antibodies or antibody fragments. **Fig. 1** shows the 4-(1,4,8,11-tetraazacyclotetradec-1-yl)-methyl ben-

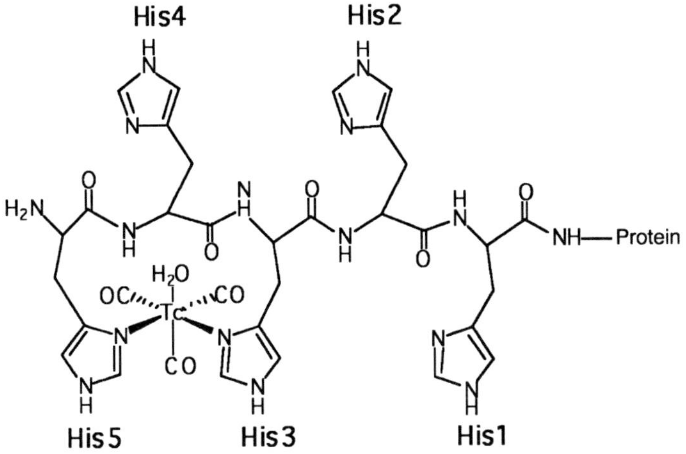

Fig. 2. Structure of a ^{99m}Tc-tricarbonyl-labeled protein. Tc-carbonyl coordinates with two histidine residues in a penta-His-tag.

zoic acid (CPTA) chelators and the 1-(p-nitrobenzyl)-1,4,7,10-tetraazacyclodecane-4,7,10-triacetate (DO3A) chelators attached to MAbs via NHS and NCS coupling groups, respectively. Methods are described to substitute antibodies with the CPTA and DO3A copper chelators and to label the immunoconjugates with ^{67}Cu. The protocols also include methods for the evaluation of the radiochemical purity and immunoreactivity of ^{67}Cu-labeled antibodies.

1.2. Site-Specific Labeling of Antibody Fragments with ^{99m}Tc/^{188}Re

Besides whole MAb molecules, single-chain antibody fragments (scFvs) also have the potential to be good tumor-imaging agents, since they penetrate rapidly into tumor tissue and achieve high tumor-to-background ratios at early time-points. As scFv fragments are becoming increasingly available from large combinatorial libraries, a facile radiolabeling method is needed to exploit these molecules for in vivo tumor imaging. ^{99m}Tc is a γ-radiation emitter with a half-life of 6.02 h, and is the most commonly used radionuclide for radiopharmaceutical applications because of its ready availability and excellent physical properties for imaging.

We recently introduced a site-specific ^{99m}Tc labeling method *(4)* in which ^{99m}Tc-tricarbonyl-trihydrate is stably fused to the polyhistidine tag of recombinant antibody fragments. **Fig. 2** shows the coordination of ^{99m}Tc-carbonyl with His residues in a scFv fragment. This ^{99m}Tc-labeling method avoids the introduction of unpaired cysteines into the scFv (necessary in other ^{99m}Tc-labeling techniques), which would lead to the formation of crosslinked molecules *(5)*. In addition, labeling via the histidine-tag is an elegant approach, because the His-tag is routinely incorporated in most recombinant proteins to allow purification with immobilized ion metal affinity chromatography (IMAC). ^{188}Re is a β^- particle-emitting nuclide with a half-life of 16.7 h, and is available in the form of a generator system. It has similar chemical properties to ^{99m}Tc, and can also be used for radionuclide therapy. Here, we provide protocols for labeling the His-tagged scFv MFE-23, an anti-CEA-antibody fragment *(6)*, with ^{99m}Tc-tricarbonyl and with the ^{188}Re nuclide.

2. Materials

2.1. Labeling of Antibodies with ^{67}Cu

2.1.1. Coupling of MAbs to the Bifunctional Chelators CPTA-NHS and DO3A-NCS

1. 4-(1,4,8,11-tetraazacyclotetradec-1-yl)-methyl benzoic acid (CPTA) (*see* **ref. 7** for chelator synthesis), stored dry at ambient temperature.
2. N-hydroxy succinimide sodium salt (NHS) (Pierce), stored dry at ambient temperature.
3. N-(3-dimethylaminopropyl)-N′-ethylcarbodiimide (EDC) (Pierce), stored dry at –20°C.
4. MAb or MAb $F(ab')_2$ in phosphate-buffered saline (PBS: 100 m*M* NaCl, 50 m*M* sodium phosphate buffer, pH 7.3), stored at 4°C.
5. 100 m*M* sodium phosphate buffer, pH 7.0.
6. 100 m*M* sodium phosphate buffer, pH 8.0.
7. 100 m*M* sodium acetate buffer, pH 5.5.
8. 70 m*M* 1,4,7,10-tetraazacyclodecane-4,7,10-triactetate-L-p-isocyanatophenylalanine (DO3A-NCS) (*see* **ref. 7** for chelator synthesis) stock solution in H_2O, stored at –20°C.
9. MAb or MAb $F(ab')_2$: 6.6 nmol-(1 mg of MAb or 0.7 mg of MAb $F[ab']_2$) in 400 µL of 100 m*M* sodium phosphate buffer, pH 8.0.
10. Na_3PO_4 saturated solution.
11. Microseparation filters (Amicon, Centricon) for concentration by centrifugation dialysis (exclusion limit 30 kDa).

2.1.2. Radiocopper Labeling of Chelate-Substituted MAbs

1. ^{67}Cu stock solution in 0.2 *M* sodium succinate buffer, pH 4.0 (Paul Scherrer Institute, Switzerland). Specific activity of the ^{67}Cu nuclide at the time of labeling is 37.5–75 MBq/µg.
2. PBS.
3. MAb-CPTA and MAb-DO3A (0.5–1.0 mg/mL) in 100 m*M* sodium acetate buffer, pH 5.5, stored at 4°C (from **Subheading 3.1.1.**).
4. 50 m*M* sodium EDTA solution, pH 7.0.
5. Superose 12 gel-filtration chromatography column (Amersham Biosciences) size 15 × 300 mm, equilibrated in PBS.

6. Fast protein liquid chromatography (FPLC) equipment or peristaltic pump.
7. UV/visible spectrophotometer.
8. γ-counter with energy window adjusted to 160–210 keV.
9. Lead shielding.

2.1.3. Assessment of Radiochemical Purity of Radiocopper-Labeled MAbs

1. Alumina-backed silica gel thin-layer chromatography (TLC) foil, 50 × 200 mm (Merck).
2. TLC solvent: a mixture of 1 vol of 10% ammonium acetate in H_2O and 1 vol of methanol.
3. 50 m*M* sodium EDTA solution, pH 7.0.
4. Equipment for electronic autoradiography (e.g., Phosphor Imager or Packard Electronic Imager) or facilities for exposing and developing X-ray film (Amersham Biosciences).
5. Packard γ-spectrometer (Perkin-Elmer).

2.1.4. Assessment of Immunoreactivity of Radiocopper-Labeled MAbs

1. Twenty-four-well microtiter plates coated with target cells, store at –20°C.
2. Blocking buffer: PBS containing 0.5% (w/v) bovine serum albumin (BSA).
3. Radiocopper-labeled-CPTA or -DO3A labeled MAb, and unlabeled MAb.
4. 1 *N* NaOH.
5. Plastic tubes of appropriate size for use in the γ-counter.
6. γ-counter with energy window adjusted to 160–210 keV.

2.2. Site-Specific Labeling of Antibody Fragments with ^{99m}Tc/^{188}Re

2.2.1. Preparation of ^{99m}Tc-Tricarbonyl Precursor

1. $NaBH_4$ puriss p.a., ≥96%, stored dry at ambient temperature.
2. Na_2CO_3 purum, stored dry at ambient temperature.
3. NaK-tartrate MicroSelect, ≥99.5% (Fluka), stored dry at ambient temperature.
4. Carbon monoxide (g), quality 37, 99.9% (Carbagas).
5. Neutralizing solution: 25% mixture of 1 *M* sodium phosphate buffer, pH 7.4 and 75% 1 *M* hydrochloric acid.
6. $^{99}Mo/^{99m}TcO_4^-$ generator (Mallinckrodt-Tyco, Petten, The Netherlands).
7. Sterile 10-mL glass vial or 10-mL serum vial with rubber stopper and aluminum cap.
8. Lead shielding.
9. Well-ventilated hoods.
10. As an alternative to the previous equipment, a lyophilized kit (IsoLink™; Mallinckrodt, Tyco) is also available for the preparation of the Tc-carbonyl core.

2.2.2. Preparation of ^{188}Re-Tricarbonyl Precursor

1. $BH_3{\cdot}NH_3$ pract., ~90%, stored at 0–4°C.

2. Phosphoric acid 98% pure.
3. Carbon monoxide (g), quality 37, 99.9%.
4. $^{188}W/^{188}ReO_4^-$ generator (Oak Ridge National Laboratories).
5. Sterile 10-mL glass vial or a 10-mL serum vial with rubber stopper and aluminum cap.
6. Plexiglas and lead shielding.
7. Well-ventilated hoods.

2.2.3. Assessment of Radiochemical Purity of $^{99m}Tc/^{188}Re$-Tricarbonyl Precursor

1. Glass-backed silica gel TLC plates (Merck $60F_{254}$): 15 × 200 mm.
2. Paper chromatography (PC) strips (Whatman No.1): 15 × 200 mm.
3. TLC solvent: 99.9% methanol, 0.1% concentrated HCl.
4. PC solvent: 99.5% methanol, 0.5% 6 *M* HCl.
5. γ-counter or γ-radiation TLC scanner.
6. HPLC-system equipped with a radiometric γ-detector.
7. C-18 reversed-phase column, h = 150 mm × d = 4.1 mm (Macherey-Nagel, Oensingen, Switzerland).
8. HPLC solvents: 0.05 *M* triethylammonium phosphate (TEAP) buffer, pH 2.25 (solvent A) and methanol (solvent B).

2.2.4. Labeling of His-Tagged ScFv Antibody Fragments

1. scFv antibody fragment: 0.4–2.0 mg/mL in buffer (*see* **Note 1**).
2. $[^{99m}Tc(H_2O)_2(CO)^3]^+$: 3.7 GBq/mL (from **Subheading 3.2.1.**).
3. 1 *M* 2-(N-morpholino) ethanesulfonic acid (MES) buffer hemisodium salt, pH 6.1.
4. Eppendorf tubes.
5. BlueMax 50-mL polypropylene tubes.
6. Bio-Spin 6 (BioRad) columns or HighTrap desalting columns (Amersham Biosciences), equilibrated with PBS.
7. Heating block for polypropylene tubes or incubator (37°C).
8. γ-counter with energy window adjusted to 160–210 keV.

2.2.5. Assessment of Immunoreactivity

1. FPLC system with Superdex75 column (Amersham Biosciences) equilibrated in PBS containing 0.05% (v/v) Tween 20.
2. Carcinoembryonic (CEA) antigen (Sigma).

3. Methods

3.1. Labeling of Antibodies with ^{67}Cu

3.1.1. Coupling of MAbs to CPTA-NHS

1. Dissolve 15.5 mg of CPTA, 8 mg of NHS, and 38.2 mg of EDC in 1.0 mL of 100 m*M* sodium phosphate buffer, pH 7.0 and stir at ambient temperature for 1 h (*see* **Note 2**).

2. Add 20 μL of this ligand/activated ester reaction mixture to 0.5 mL of MAb (1 mg) in PBS. Alternatively, add 25 μL of the reaction mixture to 0.5 mL of $F(ab')_2$ (1 mg) in PBS. Stir the solution at ambient temperature for 1 h.
3. Buffer exchange into 100 m*M* sodium acetate buffer, pH 5.5 by five washes through an Amicon/Centricon centrifugation-dialysis microseparator (exclusion limit 30 kDa). Concentrate the solution to 1 mg/mL (*see* **Note 3**).
4. Determine protein concentration by measuring absorbance at 280 nm (1 OD ≈ 0.7 mg/mL). The ligand-coupled antibody can be stored at 4°C for at least 2 wk.

3.1.2. Coupling of MAbs to DO3A-NCS

1. Add 5 μL of an aqueous solution of DO3A-NCS (70 m*M*) to 200–400 μL of 100 m*M* sodium phosphate buffer, pH 8.0 containing 1 mg of intact MAb or 0.7 mg of $F(ab')_2$ fragments. Adjust to pH 9–10 with a saturated solution of Na_3PO_4 (*see* **Note 2**).
2. Incubate the reaction mixture at 4°C overnight (16 h).
3. Perform buffer exchange as described in **Subheading 3.1.1., step 3.**

3.1.3. Radiocopper Labeling of MAbs Substituted with CPTA- and DO3A-Chelators

Radiocopper labeling is performed in a properly shielded and controlled area (B containment area laboratory). The following protocol describes the labeling of up to 10 MBq (200 μg MAb), amounts that are sufficient for pre-clinical animal studies. It can be scaled up to label diagnostic (37.5 MBq) or therapeutic (1–2 GBq) doses of ^{67}Cu-MAb for clinical studies. In these cases, the procedures should be carried out in a hot cell.

1. Equilibrate 200 μg (50 μL) of the CPTA/DO3A-coupled immunoconjugates (**Subheadings 3.1.1.** and **3.1.2.**) with 100 m*M* sodium acetate buffer, pH 5.5, in a total vol of 500 μL. React this mixture with 18.5 MBq (500 μCi) of neutralized ^{67}Cu solution at ambient temperature for 30 min (*see* **Note 4**).
2. Add 50 m*M* sodium EDTA stock solution to a final concentration of 5 m*M*. Incubate for 5 min to complex free copper. Unchelated copper readily binds to protein and TLC solid phase, and interferes with subsequent quality-control procedures (**Subheading 3.1.4.**).
3. Purify the labeled immunoconjugates by FPLC on a Superose12 gel-filtration column equilibrated with PBS.
4. Collect 0.5-mL elution fractions and monitor A_{280} and radioactivity (cpm) using a γ counter.
5. Pool fractions corresponding to intact MAb (retention time: 24 min) or $F(ab')_2$ fragments (retention time: 26 min). Store at 4 °C prior to use (*see* **Note 5**).

Fig. 3 shows the purification of ^{67}Cu-DO3A-chCE7 by FPLC chromatography on Superose12 (**A**) and a double reciprocal plot of binding to target tumor cells (**B**) for evaluating immunoreactivity.

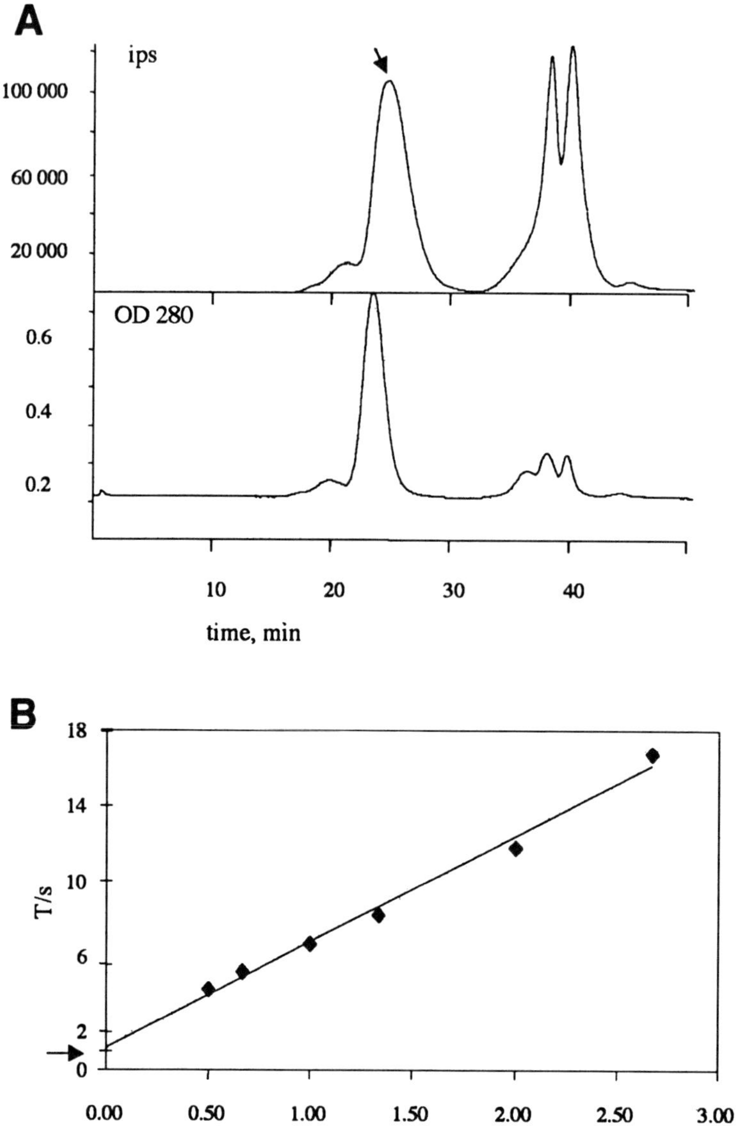

Fig. 3. Purification (**A**) and immunoreactivity (**B**) of ^{67}Cu-DO3A-labeled MAb chCE7. **A:** Fast protein liquid chromatography (FPLC) gel-filtration chromatography on Superose12. Upper trace: Radioactivity (ips). Lower trace: OD 280 nm. The arrow depicts the peak of intact ^{67}Cu-DO3A-MAb chCE7. **B:** Double-inverse immunoreactivity plot for ^{67}Cu-DO3A-chCE7. Y-axis: Total radioactivity (T) divided by specifically bound activity (s). X-axis: Inverse of cell concentration (1/c) (mL/million). The arrow depicts the intercept on the y-axis (1), corresponding to 100% immunoreactivity.

3.1.4. Assessment of Radiochemical Purity and Radionuclide Identity of ^{67}Cu-labeled Antibodies

1. Spot 2 µL of ^{67}Cu-labeled antibody solution (approx 100,000 cpm) on a 50 × 200 mm piece of alumina-backed silica gel foil.
2. Develop the chromatogram in a solvent system consisting of a 1vol:1vol mixture of 10% ammonium acetate in H_2O and methanol. Protein-bound radioactivity remains at the origin, whereas the Cu-EDTA complex migrates to the front.
3. Measure the distribution of radioactivity by electronic autoradiography or with autoradiography using an X-ray film.
4. Determine radionuclide identity by γ-spectroscopy (*see* **Notes 6** and **7**).

3.1.5. Assessment of Immunoreactivity of Radiocopper-Labeled MAbs

1. Coat 24-well microtiter plates with the target tumor cells—e.g., SK-N-AS human neuroblastoma cells in the case of MAb chCE7 *(7)*. Prepare quadruplicate wells with 4 × 10^6, 3 × 10^6, 2 × 10^6, 1.5 × 10^6, 1 × 10^6, 0.75 × 10^6 cells/well, dry the plates at ambient temperature for 48 h, and store dry at –20°C (*see* **Note 8**).
2. Block wells with 500 µL/well of PBS-0.5% BSA (blocking buffer) at 37°C for 1 h.
3. Add to each set of duplicate wells: 50 ng (100,000 cpm) of ^{67}Cu-CPTA or -DO3A-labeled MAb in blocking buffer. In parallel, add 10 µg of the same MAb in unlabeled form in blocking buffer to sets of duplicate wells in order to evaluate nonspecific binding. Incubate antibody solutions at 37°C overnight (16 h) (*see* **Note 9**).
4. Wash wells 3× with 500 µL of ice-cold blocking buffer. Dissolve cells in 500 µL/well of 1 *N* NaOH. Transfer into tubes and count radioactivity in a γ-counter (*see* **Note 10**).

3.2. Site-Specific Labeling of Antibody Fragments with ^{99m}Tc/^{188}Re

3.2.1. Preparation of ^{99m}Tc-Tricarbonyl Precusor

Preparation should take place in a well-ventilated hood (generation of gaseous radioactivity) with appropriate lead shielding.

1. Place 15 mg of NaK-tartrate, 4 mg of Na_2CO_3, and 5.5 mg of $NaBH_4$ in a 10-mL glass vial or a 10-mL serum vial. Seal the vial with an aluminum-capped rubber stopper and flush it with carbon monoxide for 20 min (*see* **Note 11**).
2. Inject 3 mL of the $^{99m}TcO_4^-$ solution ($^{99}Mo/^{99m}TcO_4^-$ generator eluate, 5–8 GBq/mL saline; *see* **Note 12**) into the vial. Avoid injecting any air into the vial during the addition of TcO_4^-. Incubate the reaction mixture at 75°C for 30 min. Balance the pressure from the generated CO gas using a 20-mL syringe.
3. Cool mixture on ice and add 200 µL of the neutralization solution to adjust the pH to 7.0. Typical yield: >98%, determined by TLC and PC (*see* **Note 13**).

3.2.2. Preparation of ^{188}Re-Tricarbonyl Precursor

Preparation should take place in a well-ventilated hood (generation of gaseous radioactivity) with appropriate lead shielding.

1. Place 5 mg of $BH_3{\cdot}NH_3$ in a 10-mL glass vial or a 10-mL serum vial. Seal the vial with an aluminum-capped rubber stopper and flush it with carbon monoxide for 20 min.
2. Mix 1 mL of the $^{188}ReO_4^-$ solution ($^{188}W/^{188}ReO_4^-$ generator eluate; 3–5 GBq/mL) gently with 6 mL of H_3PO_4 (98%). Inject the mixture into the reaction vial carefully without adding any air. Incubate the reaction mixture at 60°C for 15 min. Balance the pressure from the evolving H_2 gas using a 20-mL syringe.
3. Cool the reaction mixture on ice. The final pH of the reaction solution is 7.5. Typical yield: 85 ± 5%, determined by TLC and PC (*see* **Note 13**).

3.2.3. Assessment of Radiochemical Purity of the $^{99m}Tc/^{188}Re$-Tricarbonyl Precursor

3.2.3.1. TLC/PC Analysis

1. Spot 2 µL of the $^{99m}Tc/^{188}Re$-tricarbonyl solution (about 50,000 cpm) on a 15 × 200 mm glass-backed silica gel plate and/or a 15 × 200 mm paper strip.
2. Develop the chromatogram in a mixture of 99% methanol/1% conc. hydrochloric acid (TLC) and/or a mixture of 99.5% methanol/0.5% 6 *M* hydrochloric acid (PC).
3. Analyze the distribution of radioactivity by electronic autoradiography or by scanning with a γ-radiation TLC scanner. On silica gel, the tricarbonyl precursors reveal a retention factor (Rf) of 0.3, and unreacted $^{99m}TcO_4^-$ or $^{188}ReO_4^-$ show a Rf of 0.7. On PC the tricarbonyl precursors reveal a Rf of 0.7, whereas unreacted $^{99m}TcO_4^-$ or $^{188}ReO_4^-$ show a Rf of 0.4. Colloidal TcO_2 or ReO_2 remains in both systems at the origin.

3.2.3.2. HPLC Analysis

1. Load 2–5 mL of the $^{99m}Tc/^{188}Re$-tricarbonyl solution (about 150,000 cpm) into the injection loop.
2. Set up and run the following HPLC gradient, at 1 mL/min:

0–3 min:	100% solvent A
3–9 min:	75% solvent A/25% solvent B
9–20 min:	linear gradient of 66% solvent A/34% solvent B to 100% solvent B
20–22 min:	100% solvent B
22 min onward:	100% solvent A

3. The tricarbonyl precursors show a retention time of 4–5 min, whereas unreacted $^{99m}TcO_4^-$ or $^{188}ReO_4^-$ show a retention time of 9–10 min.

3.2.4. Labeling of His-Tagged ScFvs

1. Mix 2 parts of scFv antibody solution—e.g., 200 µg protein in 200 µL buffer (*see* **Notes 1** and **15**) with 1 part (100 µL) of 1 *M* MES buffer and 1 part (100 µL corresponding to ~370 MBq) of neutralized ^{99m}Tc-tricarbonyl (^{188}Re-tricarbonyl) solution.
2. Incubate at 37°C for 30 min to 1 h.
3. Separate the labeled scFv from free ^{99m}Tc-tricarbonyl and $^{99m}TcO4^-$ by gel-filtration methods such as the HighTrap desalting column (250 µL–1.5 mL volume) or the Bio-spin 6 column (75 µL–100 µL volume) (*see* **Note 15**).

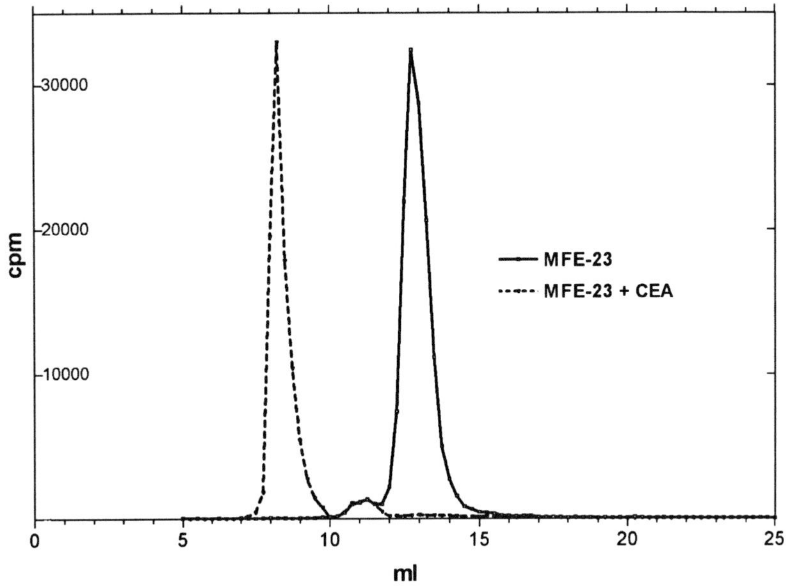

Fig. 4. Purification and immunoreactivity of ^{99m}Tc-tricarbonyl-labeled scFv MFE-23. FPLC gel-filtration chromatography on Superdex75 monitored with online radioactivity detector. In the presence of excess CEA radioactivity is shifted from the scFv peak to a high mol wt CEA-scFv complex

3.2.5. Assessment of Biological Activity

Immunoreactivity after labeling can be evaluated by a cell-binding assay (*see* **Subheading 3.1.5.**) or by a gel-shift assay, as exemplified here for the labeling of scFv MFE-23:

1. After labeling and purification of the scFv antibody by the Bio-Spin desalting column, mix 100–500 ng of the labeled scFv with a 10–100× molar excess of free soluble CEA and incubate at ambient temperature for 30 min.
2. Apply this mixture to a Superdex75 gel-filtration column, equilibrated in PBS containing 0.05% (v/v) Tween 20. Monitor protein separation by A_{280} and a HPLC-radioactivity monitor (Berthold LB-506). Alternatively, collect 0.25-mL fractions and count radioactivity in a γ-counter.
3. Immunoreactive antibody fragments (25 kDa) will bind to CEA (120 kDa), and the radioactive peak will elute with an apparent mol wt of 145 kDa. Non-binding fragments will elute with a mol wt of 25 kDa. The ratio between the area at 145 kDa and 25 kDa can be used as a value for the biological activity of the labeled scFv.

Fig. 4 shows the radioactivity profiles of a gel-filtration separation of ^{99m}Tc-carbonyl-labeled MFE-23 after incubation with excess CEA.

4. Notes

1. The antibody buffer should be free of EDTA, imidazole, azide, citrate, or Tris base, as these substances will interfere with the labeling procedure.
2. The amount of chelator coupled to the antibody influences immunoreactivity and biodistributions. The optimal chelator/antibody ratio should be determined experimentally for every antibody system. For MAb chCE7 *(7)*, which contains 92 lysine residues, we found that derivatizing more than 11 lysine residues leads to impaired immunoreactivity and unfavorable biodistributions. For substitution of MAb chCE7 with CPTA-NHS, our protocol uses a molar ratio of chelator to MAb of 90/1, leading to a mean of 1–2 chelators coupled each MAb molecule. The efficiency of the DO3A-NCS coupling reaction is higher. A molar ratio of chelator to MAb chCE7 of 45/1 will lead to a mean of 8–10 chelators coupled to each MAb.
3. After the coupling reaction, excess chelator can be removed and buffer can be exchanged by centrifugation/dialysis or, alternatively, by FPLC size-exclusion chromatography on a Superose12 column. A longer column is preferred to a short desalting column because, in addition to separating unbound Cu-EDTA, the former is able to remove protein aggregates, which would lead to high levels of unwanted radioactivity in the liver and/or spleen.
4. Some MAbs have a tendency to form aggregates. In these cases, radiocopper labeling can be performed in 0.2 *M* sodium acetate buffer pH 5.5 and subsequently purified by gel-filtration chromatography in threefold concentrated (3X) PBS.
5. After copper labeling and purification of labeled antibody, more than 95% of the radiocopper is associated with antibody. Typical labeling yields are up to 50%, and specific activities are approx 37.5 Mbq (1 mCi)/mg.
6. Radionuclide identity can be determined by γ-spectroscopy. We found that in antibody preparations labeled via the CPTA chelator, the amount of the main contaminants ^{67}Ga ($T_{1/2}$: 78.3 h) and ^{57}Ni ($T_{1/2}$: 36 h) is less than 5%, and there is less than 0.1% ^{65}Zn ($T_{1/2}$: 244.3 d) present in the ^{67}Cu-labeled antibody preparations.
7. ^{64}Cu is present in the ^{67}Cu-labeled antibody preparation, the amount depending on the time that has elapsed after ^{67}Cu production (end of bombardment, EOB). An antibody preparation delivered 72 h after EOB with an actual activity of 37 MBq of ^{67}Cu contains an additional 37 MBq of ^{64}Cu. Because of its shorter half-life, the activity resulting from ^{64}Cu declines more rapidly than ^{67}Cu. Since the photons emitted by ^{64}Cu contribute to the radiation burden of the patient and have an impact on image resolution, it would be advisable to determine the amount of ^{67}Cu and ^{64}Cu that is present at the time of administration by measurement with a γ-spectrometer or a calibrated ionization chamber *(8)*.
8. As an alternative to the protocol described here for MAb chCE7, binding assays can be performed with live target cells, either in suspension or attached to microtiter plates. If the target antigen or its peptide epitope are available, it can be attached to a solid-phase support, and binding is measured at various antigen concentrations *(9)*.
9. The amount of radiocopper-labeled antibody or antibody fragments used for the binding assay should not exceed the concentration necessary for half-maximal

binding (K_d), the amount used to measure nonspecific binding should be up to a 1,000-fold excess.

10. Immuoreactivity is determined by measuring the binding of radiocopper-labeled antibodies to increasing numbers of target cells and the extrapolation of specific binding to infinite cell number according to Lindmo ***(10)***. **Fig. 3B** shows a double reciprocal plot of binding of ^{67}Cu-DO3A-chCE7 to increasing numbers of neuroblastoma cells. Extrapolation to infinite cell-number intercepts on the y-axis to total T/S [total radioactivity (T) divided by specifically bound radioactivity (S)]. The intercept on the y-axis of 1 corresponds to 100% immunoreactivity.
11. A kit formulation of ^{99m}Tc-tricarbonyl is also available from Mallinckrodt-Tyco under the name "Carbonyl Labeling Agent." The preparation of ^{99m}Tc-tricarbonyl with this kit will avoid for need for carbon monoxide handling.
12. Yields of ^{99m}Tc-tricarbonyl drop significantly when the ^{99}Mo/^{99m}Tc-generator is older than 3 wk or after 3–4 d without elution of the generator. Daily elution and use of fresh saline can avoid these problems.
13. Free ^{99m}Tc-tricarbonyl and ^{188}Re-tricarbonyl precursors are stable at neutral and acidic pH for several hours under anaerobic conditions. The main decomposition products are $^{99m}TcO_4^-$ (pertechnetate) or ReO_4^- (perrhenate).
14. Beware of cleavage of the His-tag by prolonged storage of proteins in the presence of imidazole (IMAC eluant) at acidic pH.
15. Labeling yields are 70–95% and depend on protein concentration. At 1 mg/mL final concentration, labeling will be complete after 15 min, and at 0.2 mg/mL complete labeling will take up to 2 h. At low antibody concentrations, the incubation tube should be flushed with with N_2 or CO to reduce the amount of decomposition products (pertechnetate or perrhenate).

References

1. Schwarzbach, R., Zimmermann, K., Blaeuenstein, P., Smith, A., and Schubiger, P. A. (1995) Development of a simple and selective separation of 67-Cu from irradiated zinc for use in antibody labeling: a comparison of methods. *Appl. Radiat. Isot.* **46,** 329–336.
2. Novak-Hofer, I. and Schubiger, P. A. (2002) 67-Cu as a therapeutic nuclide for radioimmunotherapy. *Eur. J. Nucl. Med.* **29,** 821–830.
3. Moi, M., Meares, C. F., Mc Call, M. J., Cole, W. C., and DeNardo, S. J. (1985) Copper chelates as probes of biological systems: stable copper complexes with a macrocyclic bifunctional chelating agent. *Anal. Biochem.* **148,** 249–253..
4. Waibel, R., Alberto, R., Willuda, J., Finnern, R., Schibli, R., Stichelberger, A., et al. (1999) Stable one-step technetium-99m labeling of His-tagged recombinant proteins with a novel Tc(I)-carbonyl complex. *Nat. Biotechnol.* **17,** 897–901.
5. Schmiedl, A., Breitling, F., Winter, C. H., Queitsch, I., and Dubel, S. (2000) Effects of unpaired cysteines on yield, solubility and activity of different recombinant antibody constructs expressed in E. coli. *J. Immunol. Methods* **242,** 101–114.
6. Chester, K. A., Begent, R.H.J., Robson, L., Keep, P., Pedley, R. B., Boden, J. A., et al. (1994) Phage libraries for generation of clinically useful antibodies. *Lancet* **343,** 455–456.

7. Novak-Hofer, I., Zimmermann, K., Maecke, H. R., Amstutz, H., Carrel, F., and Schubiger, P. A. (1997) Tumor uptake and metabolism of copper-67-labeled monoclonal antibody chCE7 in nude mice bearing neuroblastoma xenografts. *J. Nucl. Med.* **38,** 536–544.
8. DeNardo, G. L., Kukis, D. L., Shen, S., and DeNardo, S. J. (1996) Accurate measurment of copper-67 in the presence of copper-64 contaminant using a dose calibrator. *J. Nucl. Med.* **37,** 302–306.
9. Hughes, O. D., Bishop, M. C., Perkins, A. C., Frier, M., Price, M. R., Denton, G., et al. (1997) Preclinical evaluation of copper-67 labeled anti-MUC1 mucin antibody C595 for therapeutic use in bladder cancer. *Eur. J. Nucl. Med.* **24,** 439–443.
10. Lindmo, T. and Bunn Jr, P. A. (1986) Determination of the true immunoreactive fraction of monoclonal antibodies after radiolabeling. *Methods Enzymol.* **121,** 678–691.

29

Production and Characterization of Anti-Cocaine Catalytic Antibodies

Paloma de Prada and Donald W. Landry

1. Introduction

Cocaine abuse has become a major health problem in recent decades, and the problem has been greatly accentuated by the appearance of crack cocaine, the more potent free base of the natural alkaloid. Cocaine's primary molecular target is the dopamine reuptake transporter at the nucleus accumbens. By blocking reuptake, cocaine enhances dopaminergic neurotransmission, which activates the reward pathway. The result is reinforcement of drug-taking behavior in species ranging from mouse to man, and in human subjects, the perception of euphoria. This peculiar mechanism of action has limited the therapeutic approaches to the treatment of cocaine abuse because of the inherent difficulty of "blocking a blocker." One plausible alternative is the neutralization of the drug before it reaches its target in the brain, and antibodies, a class of natural binders, would appear to be well-suited for the task. Several active and passive cocaine vaccines have already been tested with a fair degree of success *(1,2)*, but the elevated concentration of antibody required for binding and neutralizing cocaine blood levels is still a problem to be overcome. This hurdle is less difficult for a catalytic antibody, which not only binds, but also destroys what it binds to.

Jenks *(3)* introduced the concept of catalytic antibodies some 30 years ago, suggesting that by using stable transition-state analogs as haptens during immunization, custom-made catalytic activities could be generated. With the advent of monoclonal antibody (MAb) technology, the field exploded in the 1990s, and today, more than one hundred different activities have been reported *(4)*. We undertook the task of creating a cocaine-hydrolyzing antibody that would target the benzoyl ester group. Cleavage at the benzoyl ester would yield ecgonine methyl ester and benzoic acid—metabolites of cocaine—without

From: *Methods in Molecular Biology, Vol. 248: Antibody Engineering: Methods and Protocols*
Edited by: B. K. C. Lo © Humana Press Inc., Totowa, NJ

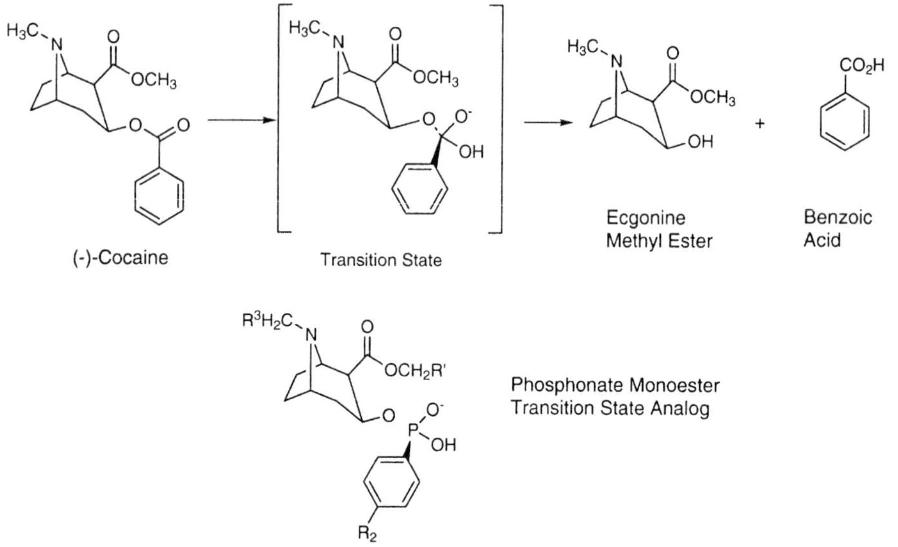

TSA1: $R^1 = (CH_2)_3NHCO(CH_2)_2CONH$-carrier protein; $R^2 = R^3 = H$

Free TSA: $R^1 = R^2 = R^3 = H$

Fig. 1. Cocaine hydrolysis catalyzed by MAb 15A10 and chemical structure of TSA1.

a toxic or reinforcing effect (**Fig. 1**). A transition-state analog resembling the tetrahedral intermediate, and by extension the putative transition state, was obtained by replacing the benzoyl carbonyl with a phosphonate group. The analog was linked to bovine serum albumin (BSA) by a nine-atom tether attached at the carbomethoxyl group. Several MAbs were recovered from mice that possessed cocaine esterase activity *(5)*, and the most active of them, MAb 15A10, has been tested in animal models with very promising results *(6,7)*.

This chapter reviews the generation of anti-cocaine catalytic antibodies, including the processes of immunization, fusion, cloning, screening, and testing of catalytically active antibodies.

2. Materials

2.1. Immunization

1. Six Balb/c female mice per immunization.
2. Complete and incomplete Freund's adjuvant (Sigma).
3. Sterile phosphate-buffered saline (PBS): 1.5 m*M* KH_2PO_4, 8.1 m*M* Na_2HPO_4, 2.7 m*M* KCl, 140 m*M* NaCl, pH 7.2–7.4.

4. Transition-state analog 1 (TSA-1) (**Fig. 1**) conjugated to the carrier protein (BSA), diluted in PBS, and filter-sterilized.

2.2. Cell Fusion and Cloning

1. Equipment needed: laminar flow hood, 5% CO_2 37°C incubator, liquid nitrogen for cell storage, phase-contrast microscope (to monitor cell growth), autoclaved scissors, and forceps (to collect splenocytes).
2. NS-1 myeloma cells (American Type Culture Collection).
3. Growth media: Iscove's Modified Dulbecco's Medium (IMDM) (Gibco-BRL/Invitrogen).
4. HAT-1 media (selection media): 500 mL of IMDM (Gibco-BRL/Invitrogen) containing 10% hybridoma cloning factor (HCF) (Origen), 10% fetal calf serum (FCS) (Hyclone) and 10% HAT-1 (Sigma).
5. Sterilized 50% polyethylene glycol (PEG) 1,500 diluted in growth media.
6. 96-well culture plates (Falcon), culture flasks (NUNC brand), and cryotubes with screw caps (NUNC brand).
7. 70% ethanol.

2.3. ELISA with Anti-Cocaine Catalytic Antibodies

1. 96-well ELISA plates (NUNC brand).
2. PBS, pH 7.4 (**Subheading 2.1**).
3. TSA-1 conjugated to ovalbumin.
4. Blocking solution: 1% BSA diluted in PBS.
5. Washing solution: 0.4% BSA and 0.05% Tween-20 in PBS.
6. Secondary antibody: goat anti-mouse IgG (H + L) antibody conjugated to horseradish-peroxidase, HRP (Zymed), diluted according to the manufacturer's guidelines in washing solution.
7. HRP substrate: 1-Step ABTS (Pierce).
8. ELISA microplate reader (Molecular Devices).
9. Platform shaker.

2.4. Purification of IgG

1. Protein G Sepharose (Amersham Biosciences).
2. Binding buffer: 50 m*M* sodium phosphate buffer, pH 7.4.
3. Elution buffer: 20 m*M* sodium glycine buffer, pH 2.0.
4. Neutralization buffer: 1 *M* Tris-HCl, pH 8.0.
5. Centriplus 30K Centrifugal Filter Devices (Amicon).
6. PD10 Sephadex desalting columns (Amersham Biosciences).
7. PBS, pH 7.4 (**Subheading 2.1., item 3**) to equilibrate the desalting column.

2.5. Catalytic Activity Assay

1. Reaction buffer: 50 m*M* sodium phosphate buffer, pH 7.4.
2. 5 m*M* cocaine stock solution in PBS, pH 7.4.
3. ^{3}H-cocaine, labeled in the phenyl ring 0.05 mCi/mL, 1.25 Ci/mMol (NEN).

4. 3.1 m*M* (1 mg/mL) Free-TSA diluted in PBS.
5. Purified catalytic antibody (from **Subheading 3.4.**).
6. 0.1 *N* HCl.
7. Supelclean LC-SCX SPE anion-exchange mini-columns (Supelco).
8. Methanol.
9. 0.2 *N* NaOH.
10. Ready Protein Scintillation fluid (Beckman Coulter) and 7-mL scintillation vials (Kimble).

3. Methods

3.1. Immunization

1. Mix 100 µg of the TSA 1-BSA conjugate diluted in PBS with an equal volume of Freund's Complete Adjuvant and generate an emulsion thick enough that it does not disperse when a drop is placed in saline solution. Use a total of 200 µL of the emulsion, and inject half of it intraperitoneally and the other half subcutaneously.
2. Perform a second injection 4 wk after the first immunization, preparing the antigen exactly as described in **step 1,** but using Freund's Incomplete as the adjuvant. Collect blood 10 d after the second immunization, and based on the results of the ELISA (*see* **Subheading 3.3.**) with polyclonal sera, select those mice with highest titer to be used for fusion.
3. Boost animals selected for fusion by injecting into the tail vein 50 µg of the antigen in sterile PBS in a volume not to exceed 100 µL. The animals can be sacrificed 5–7 d after the boost injection, or later if further boost injections are planned (*see* **Note 1**).

3.2. Cell Fusion and Cloning

1. Feeder cells are isolated from the spleens of two naive mice sacrificed by CO_2 asphyxiation. Swab the animals with 70% ethanol, carefully remove the spleens, and press them between two frosted microscope glass slides to release the splenocytes. Wash the cells once with growth media, and collect by centrifuging the suspension for 5 min at 400*g*. Discard the supernatant and resuspend the splenocytes in HAT-1 media. Plate the cells on 96-well plates, using 50 µL of the cell suspension per well.
2. Myeloma cells (NS-1) are grown in growth media, harvested, washed twice with growth media, and finally resuspended in IMDM growth media at a cell density of 2.5×10^6 cells/mL.
3. Splenocytes from immunized mice are obtained using the same procedure described for feeder cells, washed 3× with growth media, and resuspended to a cell density of 1×10^8 cells/mL.
4. Pool myeloma cells and splenocytes together and wash twice with growth media.
5. Aspirate the supernatant completely and break up pellet by gently tapping the tube. Add 1 mL of 50% PEG 1,500 dropwise over a 1-min interval while continuously stirring with a pipet tip.
6. Slowly add 9 mL of growth media as follows: 1 mL over a 1-min interval and 3 mL over a 3-min interval, and finally, add the final 5 mL. Incubate at 37°C for 5 min.

Add 40 mL growth media and centrifuge at 400*g* for 7 min. Resuspend pellet in 10 mL of HAT-1 media, add an additional 40 mL of HAT-1 media, and transfer to a 150-cm^2 flask.

7. Incubate overnight for 16–24 h at 37°C.
8. Harvest cells from flask and centrifuge at 400*g* for 10 min. Resuspend in 10 mL of HAT- 1 media and further dilute with an additional 150 mL of HAT-1 before plating 150 µL in each well of the feeder plates (from **step 1**).
9. Incubate at 37°C for 6–8 d and then check for the appearance of colonies. Test the mini-culture supernatant by ELISA (*see* **Subheading 3.3.**) to determine which are producing antibodies of the desired specificity and should be further studied.
10. Cloning is done by limit dilution: cells from growing colonies are diluted in fresh growth media to a cell density of 0.3–0.5 cells per well and grown at 37°C for 7 d. Repeat this process at least two more times to improve cloning efficiency (*see* **Note 2**).

3.3. Screening for Anti-Cocaine Catalytic Antibodies (see Note 3)

1. Coat plates for 2 h at 37°C with a 10 µg/mL solution of the antigen (TSA-1) coupled to ovalbumin in PBS. As an alternative, the plates can be coated overnight at 4°C.
2. Discard the coating solution and block plates with 300 µL of blocking solution for 1 h at 37°C.
3. Discard the blocking solution and wash plates twice with PBS. The plates are now ready for enzyme-linked immunosorbent assay (ELISA).
4. Take 50 µL of supernatant from the fusion plates and incubate in the plates for 1 h at room temperature on a platform shaker.
5. Wash 3× with washing solution before adding 50 µL of the HRP-conjugated goat anti-mouse antibody, and incubate for 1 h at room temperature with constant shaking.
6. Wash again 3× with washing solution and twice with PBS, add 50 µL of HRP substrate, and incubate at room temperature until color develops.
7. Color can be quantitated on an ELISA reader at 405 nm, or simply look for the presence or absence of color.

3.4. Purification of IgG

1. The clones selected for further studies are scaled up to T150 flasks for antibody production. Collect the spent medium by centrifugation and add binding buffer to a final concentration of 0.05 *M*, to adjust the pH of the sample. The sample is now ready to be applied to the column.
2. Equilibrate the Protein-G column (*see* **Note 4**) with running buffer until the pH of the eluant is equal to that of the buffer. Load the sample and wash the column with ten bed-volumes of running buffer, followed by ten bed-volumes of running buffer containing 0.5 *M* NaCl, to release nonspecific binding proteins.
3. Elute the protein with two bed-volumes of elution buffer. Make sure that the eluted fraction is quickly neutralized by adding neutralization buffer to the sample as it drips off the column.

4. Concentrate using Centriplus concentrators and pass the sample through a PD-10 desalting column, previously equilibrated with PBS, pH 7.4, to place the antibody in a buffer that is compatible with the activity assay (*see* **Note 5**).

3.5. Catalytic Activity Assay

1. Two sets of assays will be run, using only catalytic antibody or catalytic antibody pre-incubated with unconjugated transition-state analog (free-TSA), to ensure that catalysis can be inhibited by the original antigen.
2. The reaction mix contains the following: 50 m*M* sodium phosphate, pH 7.4, 1 m*M* cocaine, 5–10 μg of catalytic antibody, a trace amount of ^{3}H-cocaine, and PBS to complete to a final volume of 100 μL.
3. Put the antibody, the phosphate buffer, and the water in microfuge tubes, and label half of them "TSA+" and the other half as "TSA–". Add TSA to half of them and an equal volume of water to the other tubes to compensate for the volume difference. Incubate on ice for 10–20 min, to allow binding of inhibitor to antibody to occur.
4. Next, add 20 μL of a substrate mix containing the 5 m*M* cocaine and ^{3}H-cocaine diluted 100-fold in PBS to the reaction tubes, mix well by vortexing, and briefly spin down the tubes before placing them at 37°C for overnight incubation.
5. Stop the reaction by adding 100 μL of 0.1 *N* HCl, mixing well and spinning briefly in a microfuge.
6. To prepare the anion-exchange mini-columns, wash with 2 mL of methanol followed by 2 mL of water. Place the reaction mix on top of the column and apply vacuum (or open the bottom of the column if it is run by gravity) until it completely enters the resin bed. Be careful not to let the column go dry.
7. Wash the column twice with 0.5 mL of water followed by two washes with 0.75 mL of 0.2 *N* NaOH. Collect all the fractions eluted from the column in four scintillation tubes (two for the water fractions and two for the NaOH ones), add scintillation fluid, and read them on a scintillation counter.
8. The first two fractions will contain the acidic unbound fraction of the reaction mix (benzoic acid), and the fractions from the NaOH wash will contain non-hydrolyzed cocaine. The activity is calculated from the ratio between the eluted benzoic acid and the unreacted labeled cocaine.

4. Notes

1. To achieve higher titer and better quality of antibodies, more boost injections can be done, allowing at least 3 wk between the boosts.
2. Hybridoma cell lines should be cloned 3× at least, to reduce the probability of growth of non-producing variants.
3. Antibody binding is highly variable; try to keep all the assay conditions (incubation times, temperature, buffer pH, composition) as constant as possible from one batch to the other to avoid artifacts.
4. Protein-A Sepharose can also be used for antibody purification of murine antibodies. In our particular case, we obtained better yields and higher purity using Protein G Sepharose.

5. The purification run can be easily followed by checking the eluted fractions on ELISA plates coated with TSA-ovalbumin conjugate.
6. ^{3}H-cocaine is usually diluted in methanol to make stock solution. Since methanol can inhibit the reaction, it is recommended to evaporate it under nitrogen before mixing it with the non-radioactive cocaine to make the substrate mix.
7. Instead of using anion-exchange columns to quantitate the reaction products, the reaction mix can be extracted using an Ether/Hexane (1:1) solvent; the organic phase will contain the hydrolyzed products. It is important to acidify the reaction mix and repeat the extraction 3× to minimize unrecovered products, pooling all the extracted fractions together and counting them in scintillation fluid.

References

1. Carrera, M. R., Ashley, J. A., Wirsching, P., Koob, G. F., and Janda, K. D. (2001) A second-generation vaccine protects against the psychoactive effects of cocaine. *Proc. Natl. Acad. Sci. USA* **98(4),** 1988–1992.
2. Kosten, T. R., Rosen, M., Bond, J., Settles, M., Roberts, J. S., Shields, J., et al. (2002) Human therapeutic cocaine vaccine: safety and immunogenicity. *Vaccine* **20(7–8),** 1196–1204.
3. Jencks, W. P. (1969) *Catalysis in Chemistry and Enzymology,* McGraw Hill, New York, NY.
4. Hilvert, D. (1999) Stereoselective Reactions with Catalytic Antibodies in *Topics in Stereochemistry* (Denmark, S. E., ed.), Wilex New York, NY, p. 135.
5. Yang, G., Chun, J., Arakawa-Uramoto, H., Wang, X., Gawinowicz, M. A., Zhao, K., et al. (1996) Anti-cocaine catalytic antibodies: a synthetic approach to improved antibody diversity. *J. Am. Chem. Soc.* **118,** 5881–5890.
6. Baird, T. J., Deng, S. X., Landry, D. W., Winger, G., and Woods, J. H. (2000) Natural and artificial enzymes against cocaine. I. Monoclonal antibody 15A10 and the reinforcing effects of cocaine in rats. *J. Pharmacol. Exp. Ther.* **295,** 1127–1134.
7. Koetzner, L., Deng, S., Sumpter, T. L., Weisslitz, M., Abner, R. T., Landry, D. W., et al. (2001) Titer-dependent antagonism of cocaine following active immunization in rhesus monkeys. *J. Pharmacol. Exp. Ther.* **296,** 789–796.

30

Recombinant Immunotoxins in the Treatment of Cancer

Ira Pastan, Richard Beers, and Tapan K. Bera

1. Introduction

Recombinant immunotoxins are chimeric proteins composed of the Fv portion of a monoclonal antibody (MAb) fused to a portion of a toxin. The Fv replaces the cell-binding domain of the toxin and directs the toxin to cancer cells that express a target antigen. There are several features that make toxins attractive agents designed to kill cancer cells. They are very potent, and they are able to kill cells that are resistant to standard chemotherapy. Recombinant immunotoxins (RITs) have been produced that kill different types of cancer cells, as well as human immunodeficiency virus (HIV)-infected cells, taking advantage of gp120 on the surface of HIV-infected cells.

We have used Pseudomonas exotoxin A (PE) to produce RITs using the three-dimensional (3D) structure of PE as a guide. The structure shows that PE is made up of three major domains *(1)*. Production of each of these domains in *E. coli* and associated functional studies have shown that domain 1a (a.a. 1-252) is the cell-binding domain, domain 2 (a.a. 253–364) is the translocation domain, and domain 3 (a.a. 396–613) is the adenosine diphosphate (ADP)-ribosylation domain that modifies elongation factor 2, leading to arrest of protein synthesis and programmed cell death *(2)*. Domain 1b is a minor domain with a function that is unknown; it can be removed without affecting cytotoxic activity. A summary of the RITs now in clinical trials is presented in **Table 1** *(3–6)*.

To make a RIT, the Fv portions of the light and heavy chains of a MAb are cloned, starting with RNA extracted from a hybridoma (**Fig. 1**). Then the two chains are assembled into a single-chain Fv (scFv) and fused to a 38-kDa form of PE (PE38) to make a single-chain recombinant immunotoxin (**Fig. 2**). Because scFvs are often unstable, we have developed a method of stabilizing the Fvs by connecting them together with a disulfide bond. In this approach,

From: *Methods in Molecular Biology, Vol. 248: Antibody Engineering: Methods and Protocols*
Edited by: B. K. C. Lo © Humana Press Inc., Totowa, NJ

Table 1
Immunotoxins in Clinical Trials

Clinical Name	Lab Name	Antigen	Tumor Type	Ref.
BL22	RFB4(dsFv)PE38	CD22	B-cell lymphomas and leukemias	*3*
LMB-2	Anti-Tac(Fv)PE38	CD25	T-cell malignancies and some B-Cell	*4*
LMB-9	B3(dsFv)PE38	LE^{Y}	Colon and breast	*5*
SS1P	SS1(dsFv)PE38	Mesothelin	Ovarian and mesothelioma	*6*

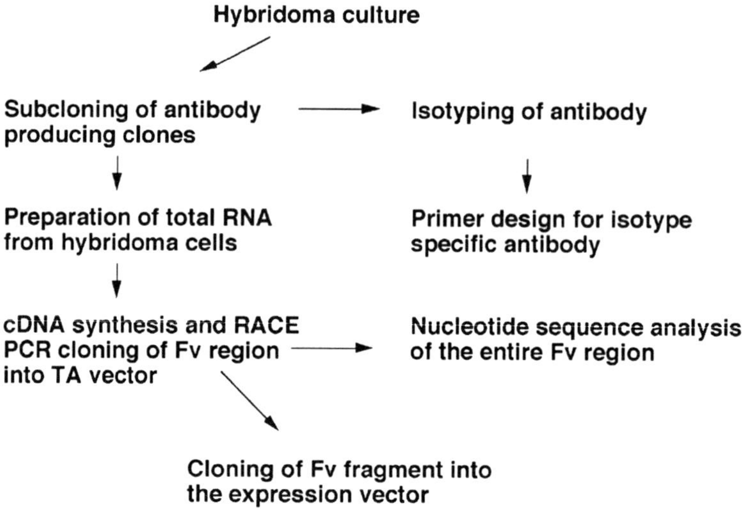

Fig. 1. Flow chart of Fv cloning from hybridoma cell lines.

the light chain and heavy chains of the Fv are first cloned. Then a cysteine residue is inserted into the framework region (FR) of each chain so the two chains can assemble into a disulfide-linked recombinant immunotoxin (dsFv RIT) (**Fig. 2**). As mentioned previously, dsFv RITs are preferred over single chain-containing molecules because of their greater stability *(7)*. Usually, the heavy chain of the Fv is fused to PE38 and the light chain is inserted into a separate expression vector. To express the proteins, we use T7-based vectors inducible with isopropyl-β-D-galactopyranoside (IPTG) as originally described by Studier et al. *(8)*. The two components of the RITs are expressed separately, and inclusion bodies are prepared and dissolved in guanidinium chloride containing a reducing agent, pooled, and renatured, and the RIT is

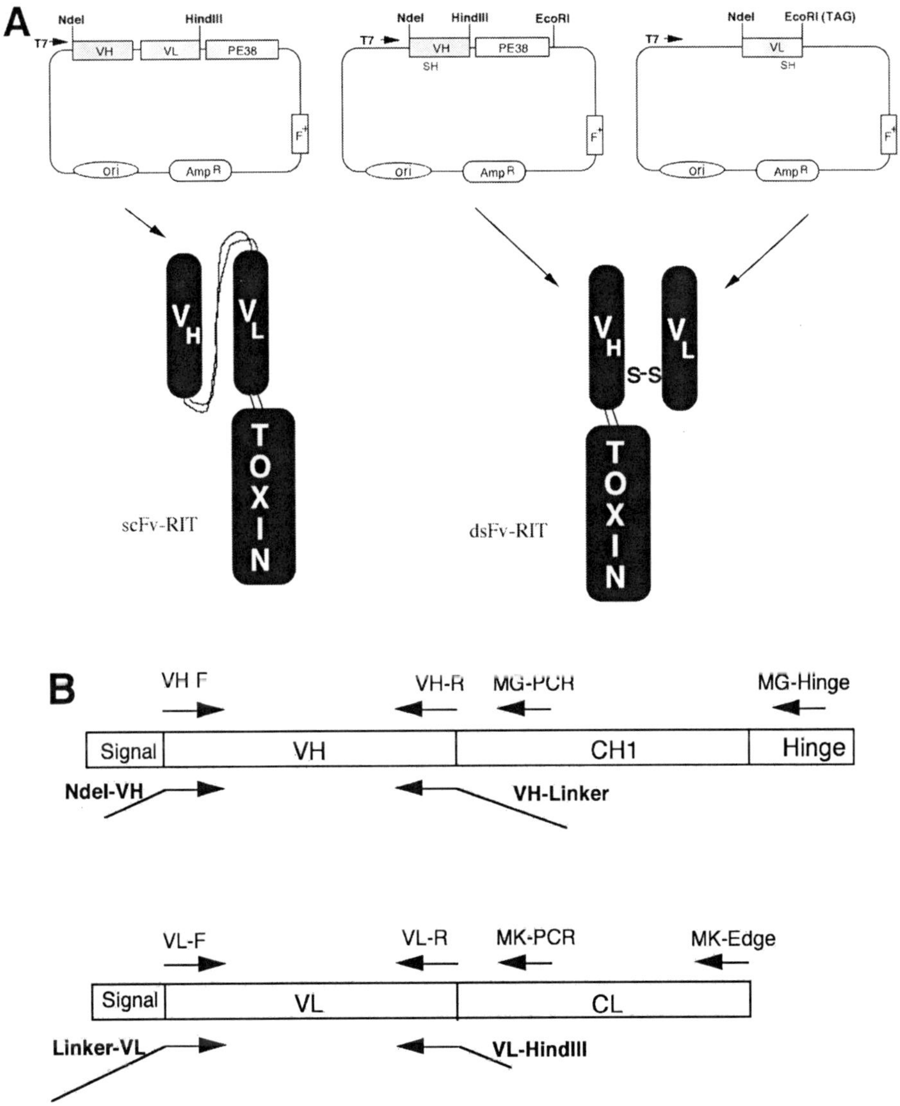

Fig. 2. Primer design and plasmid structures for expression cloning of Fv fragments. (**A**) Plasmids for expression of scFv and dsFv immunotoxin components. Schematic of a scFv and a dsFv immunotoxin is also shown. (**B**) Scheme of the Fv cloning and the relative position of the PCR primers used for the amplification and cloning procedure.

E. coli transformed with expression plasmid for V_H-PE38 or V_L component	**Subheading 3.2.1**
↓	
IPTG induction at OD600 2.0 to 3.0	**Subheading 3.2.2**
↓	
Harvesting of induced cells	**Subheading 3.2.2**
↓	
Cell lysis by lysozyme and TritonX-100	**Subheading 3.2.3**
↓	
Centrifugation of lysate to collect inclusion body protein	**Subheading 3.2.3**
↓	
Inclusion body washing	**Subheading 3.2.3**
↓	
Denaturation and combination of V_H-PE38 and V_L components	**Subheading 3.2.4**
↓	
Refolding of V_H-PE38 and V_L components	**Subheading 3.2.5**
↓	
Dialysis of refolded V_H-PE38 and V_L components	**Subheading 3.2.5**
↓	
Q-Sepharose Chromatography	**Subheading 3.2.6**
↓	
MonoQ Chromatography	**Subheading 3.2.6**
↓	
TSK Chromatography	**Subheading 3.2.6**

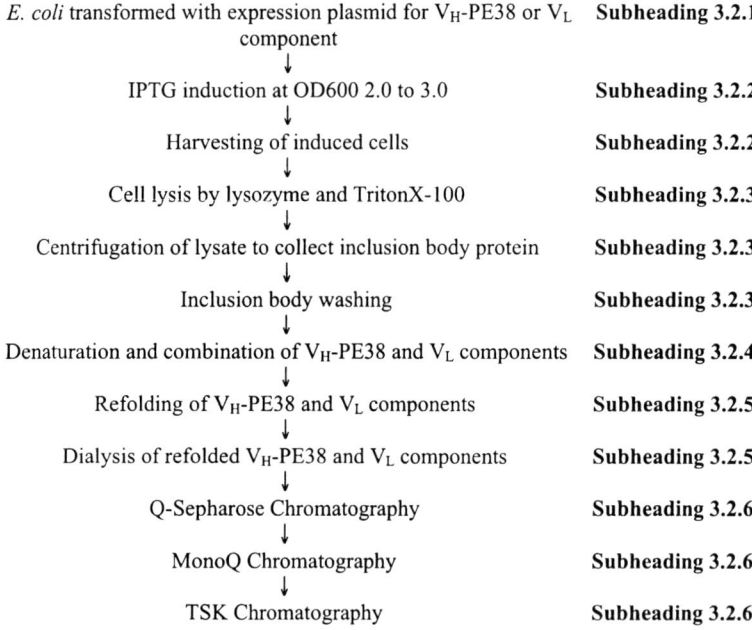

Fig. 3. Expression, refolding, and purification of RITs. Steps are as noted in **Subheading 3.2.**

purified. Typical yields are 10% of the total protein present in inclusion bodies. RITs are stored frozen at –70°C. Their cytotoxic activity is measured on cultured cancer-cell lines expressing the appropriate antigen. A summary of the steps used to make RITs is shown in **Fig. 3.**

2. Materials

2.1. Construction of Plasmids That Encode RITs

2.1.1. Isolation of Total RNA

1. Hybridoma cell line secreting antibody of interest.
2. Trizol reagent (Invitrogen; Cat. #15596).
3. Chloroform.
4. Isopropyl alcohol.
5. Diethylpyrocarbonate (DEPC)-treated water.
6. 75% ethanol in DEPC-treated water.

2.1.2. cDNA synthesis, 5′-Rapid Amplification of cDNA Ends (5′-RACE) and Analysis of V_H and V_L Immunoglobulin Sequence

1. Purified total RNA from hybridoma (**Subheading 3.1.1.**).
2. SMART RACE cDNA amplification kit (Clontech; Cat. #K1811-1).

3. Isotype-specific oligo primers (5′→3′):
   a) V_H: MG1-Hinge: ACC ACA ATC CCT GGG CAC AAT TTT CT; MG1-PCR: AGG GGC CAG TGG ATA GAC AGA TGG GGG TGT; MG2a-Hinge: TCT GGG CTC AAT TTT CTT GTC CAC C; MG2a-PCR: AGG GGC CAG TGG ATA GAC CGA TGG GGC TGT; MG2b-Hinge: GCT GGG CTC AAG TTT TTT GTC CAC C; MG2b-PCR: AGG GGC CAG TGG ATA GAC TGA TGG GGG TGT;
   b) V_L: MK-Edge: CTC ATT CTT GTT GAA GCT CTT GAC AAT; MK-PCR: GGA TGG TGG GAA GAT GGA TAC AGT TGG TGC AGC.

2.1.3. Purification and TA Cloning of 5′-RACE Products

1. 5′-RACE PCR mixture (**Subheading 3.1.2.**).
2. SeaPlaque GTG low-melting-point agarose (FMC Bioproducts; Cat. #50112).
3. Electrophoresis buffer, 1X TAE: 40 m*M* Tris-HCl, 1 m*M* EDTA, 20 m*M* acetic acid, pH 8.0.
4. QIAquick gel extraction kit (Qiagen; Cat. #28704).
5. TOPO TA cloning kit (Invitrogen; Cat. #450640).
6. Max efficiency DH5 α *E. coli* (Invitrogen; Cat. #18258-012).
7. LB/Amp agar plates: Luria-Bertani (LB) agar plates with 100 μg/mL ampicillin.
8. 50 mg/mL X-gal solution (Promega; Cat. #V394A).
9. QIAprep8 miniprep kit (Qiagen; Cat. #27142).
10. 10 mg/mL ethidium bromide solution.
11. EcoRI restriction enzyme and reaction buffer (Roche).
12. Basic local alignment search tool (BLAST) for immunoglobulin sequences, accessible at http://www.ncbi.nlm.nih.gov/igblast.

2.1.4. Cloning of the scFv Fragments into Expression Vector and Conversion of Fv Fragment to dsFv

1. 5′ oligo with V_H framework region 1 (FR1) sequence and NdeI restriction site (NdeI-VH; **Fig. 2B**).
2. 3′ oligo with V_H FR4 sequence and part of linker sequence that will overlap with V_L 5′ primer (VH-linker; **Fig. 2B**).
3. 5′ oligo with 3′ linker sequence and V_L FR1 sequence (Linker-V_L; **Fig. 2B**).
4. 3′ oligo with V_L FR4 and HindIII restriction sequence (VL-HindIII; **Fig. 2B**).
5. Taq DNA polymerase and reaction buffer (Applied Biosystems).
6. Deoxynucleotide 5′ triphosphate (dNTP), 2.5 m*M* each.
7. SeaPlaque GTG low-melting-point agarose.
8. Electrophoresis buffer, 1X TAE (*see* **Subheading 2.1.3.**).
9. QIAquick gel extraction kit.
10. TOPO TA cloning kit with Max Efficiency DH5α *E. coli.*
11. LB/Amp agar plates (*see* **Subheading 2.1.3.**).
12. Oligonucleotide to mutate V_H44 and V_L100 residues to Cys (Kabat numbering).
13. QuikChange site-directed mutagenesis kit (Stratagene; Cat. #200518).
14. Nde1, EcoRI, and HindIII restriction enzymes and reaction buffers (Roche).

15. T4 DNA ligase and reaction buffer (Roche).
16. Immunotoxin expression vector pRB98-Amp obtainable from the corresponding author (IP).

2.1.5. Construction of Immunotoxin Plasmid

1. Cloned scFv or dsFv fragment in TA vector (**Subheading 3.1.4.**).
2. Immunotoxin expression vector pRB98-Amp.
3. NdeI and HindIII restriction enzymes and reaction buffers.
4. SeaPlaque GTG low-melting-point agarose.
5. Electrophoresis buffer, 1X TAE.
6. QIAquick gel extraction kit.
7. T4 DNA ligase and reaction buffer.
8. Max Efficiency DH5α *E. coli.*
9. LB/Amp agar plates.
10. QIAprep8 miniprep kit.

2.2. Preparation of RITs

2.2.1. Transformation of E. coli

1. Competent BL21 (λ DE3) *E. coli* (Invitrogen; Cat. #C6000-03).
2. Falcon 2059 and 2052 tubes.
3. SOC media (Invitrogen; Cat. #15544-034).
4. LB agar plates with selective antibiotic (LB/Amp if the expression plasmid is ampicillin-resistant).

2.2.2. Fermentation

1. Super Broth (Biosource International).
2. Incubator/shaker set to 37°C.
3. 2-L baffled culture flasks.

2.2.3. Inclusion Body Preparation

1. TES buffer: 50 m*M* Tris-HCl, pH 8.0, 20 m*M* EDTA, 100 m*M* NaCl.
2. Lysozyme (Roche; Cat. #837 059).
3. 25% Triton X-100.
4. 250-mL centrifuge bottles.
5. Sorvall RC5B centrifuge with GSA and SS34 rotors.
6. Tissuemizer and Tissuemizer probes, large and small (Janke & Kunkel, Ultra Turrax T25).
7. Sonicator.
8. Pierce Coomassie Plus reagent (Pierce; Cat. #1856210).

2.2.4. Solubilization and Denaturation

1. GTE buffer: 6 *M* guanidine HCl, 100 m*M* Tris-HCl, pH 8.0, 2 m*M* EDTA.
2. Dithioerythritol (DTE).

2.2.5. Refolding

1. Refolding buffer: 100 m*M* Tris-HCl, 1 m*M* EDTA, 0.5 *M* arginine, chill to 10°C, adjust pH to 9.5, add 551 mg/L reduced glutathione just before use.
2. Dialysis buffer: 20 m*M* Tris-HCl, pH 7.4, chill to 4°C then add urea to a concentration of 100 m*M* just before use.
3. 0.45-µm ZapCap Filter units (Scheleicher & Schuell).
4. Dialysis tubing (mW cut-off <50 kDa).
5. 10°C cold room or cold box (VWR Scientific).
6. 50-L dialysis tank.
7. Conductivity meter.

2.2.6. Chromatography

1. Chromatography buffer A: 20 m*M* Tris-HCl, pH 7.4, 1 m*M* EDTA.
2. Chromatography buffer B: 1 *M* NaCl in Chromatography buffer A.
3. Phosphate-buffered saline (PBS): 20 m*M* KH_2PO_4, 50 m*M* Na_2HPO_4, 0.15 *M* NaCl, pH 7.4.
4. Chromatography column HR 10/10 (Amersham Biosciences).
5. Q-Sepharose Fast Flow chromatography media (Amersham Biosciences).
6. Mono Q HR 10/10 column (Amersham Biosciences).
7. Progel TSK G3000SW column (Tosoh Corp. of Japan).
8. Chromatography Pump P-500 (Amersham Biosciences).

2.3. Cytotoxicity Assay

1. 96-well tissue-culture dishes.
2. Appropriate tissue-culture media for the cell line to be assayed.
3. PBS.
4. Human serum albumin (HSA).
5. ^{3}H-Leucine, 37 MBq/mL (1 mCi/mL) (Amersham Biosciences; Cat. #TRK510).
6. Plate harvester (Tomtek).
7. Micro Beta Trilux scintillation counter (Wallac).

3. Methods

3.1. Construction of Plasmids That Encode RITs

3.1.1. Isolation of Total RNA

1. Take 5 × 10^6 cells from a frozen or growing hybridoma culture and add 1 mL of Trizol reagent in a microcentrifuge tube (1.5 mL). Lyse cells by continuous pipetting.
2. Incubate the lysate at room temperature (15–20°C) for 5 min.
3. Add 0.2 mL of chloroform and shake for 15–20 s.
4. Incubate at room temperature for 5 min.
5. Centrifuge the samples at 12,000*g* for 15 min at 4°C.
6. Transfer the colorless upper aqueous phase to a fresh microcentrifuge tube.

7. Add 0.5 mL of isopropyl alcohol, mix, and incubate at room temperature for 10 min.
8. Centrifuge the samples at 12,000*g* for 15 min at 4°C.
9. Remove and discard the supernatant.
10. Wash the RNA pellet with 1 mL of freshly prepared and chilled 75% ethanol in DEPC-treated water.
11. Vortex to resuspend pellet and centrifuge the samples at 12,000*g* for 5 min at 4°C.
12. Discard the ethanol completely without disturbing the pellet and air-dry briefly for 5–10 min.
13. Resuspend the pellet in 20–25 µL of DEPC-treated water and incubate at 55°C for 10 min to ensure total resuspension.
14. Quantify the RNA in a spectrophotometer (concentration, µg/mL) = OD_{260nm} × dilution factor × 40.
15. Store the samples at –70°C until needed.

3.1.2. cDNA Synthesis and 5′-RACE Reaction

1. Use 2.5 µg of total RNA for each reaction along with 10 pmol of isotype-specific V_H hinge primer (heavy chain) or V_L MK-Edge primer (light chain). Set up the reaction as described in the SMART RACE cDNA amplification kit.
2. Incubate the reaction mix at 42°C for 90 min.
3. Heat inactivate the reaction by incubating the tubes at 72°C for 7 min.
4. Set up the 5′-RACE PCR as described in the SMART RACE cDNA amplification kit with 10 pmol of isotype specific V_H- and V_L-PCR primer (**Fig. 2B**). Use 1 µL of the RACE-ready cDNA in 50 µL reaction volume. Heat the mixes to 92°C for 4 min and add 1 U of Taq polymerase per reaction.
5. Perform the following cycles: 94°C for 1 min, 60°C for 1 min and 72°C for 1 min, of total 30 cycles, followed by incubation at 72°C for 5 min.

3.1.3. Purification and TA Cloning of RACE Products

1. Prepare a 1.2% (w/v) low-melting-point agarose gel in 1X TAE buffer with 1 µg/mL of ethidium bromide (preparative gel) and load the entire RACE PCR mixture.
2. Excise the agarose gel fragment containing the expected size (V_H ~850 bp; V_L ~700 bp) PCR fragment using long-wavelength ultraviolet (UV) lamp and purify the DNA using the QIAquick gel extraction kit (*see* **Note 1**).
3. Take 5–10 ng of purified PCR product in 4 µL volume and add 1 µL each of salt solution and TOPO TA vector (both from the TOPO TA cloning kit). Incubate at room temperature for 5 min.
4. Transform 2 µL of the ligation mix into 100 µL of competent Max Efficiency DH5α *E. coli*. After transformation and heat shock at 42°C, add 500 µL of SOC media and incubate shaking at 37°C for 1 h.
5. Plate 100 µL of the culture onto an X-gal overlaid LB/Amp agar plate and incubate at 37°C overnight (X-gal plates are prepared by overlaying 30 µL of a 50-mg/mL X-gal solution on each plate, then air-dry.).

6. Select the isolated white colonies for plasmid DNA isolation using the QIAprep8 miniprep kit, following the instruction described in the manual.
7. Identify clones with insert by digesting the plasmid DNA from individual clone with EcoRI restriction enzyme and analyze by electrophoresis on a 1.0% agarose gel.
8. Deduce the nucleotide sequence and corresponding open reading frames (ORFs) of six independent clones and align them. Ideally, all clones should have the same sequence (*see* **Note 2**).
9. Analyze the sequence by the BLAST program (Blastn) using the Kabat database (http://www.ncbi.nlm.nih.gov/igblast/).

3.1.4. Cloning of the scFv Fragment into TOPO TA Vector

1. After the analysis of the nucleotide sequences of the heavy and light chain, design and synthesize PCR primers for V_H and V_L fragment. As shown in **Fig. 2B,** the 5′ primer for V_H fragment (NdeI-VH) will start with a NdeI restriction site (CAT ATG) followed by the nucleotide sequence that corresponds to the first seven amino acids (21 bp) of FR1. The 3′ primer sequence for V_H (VH-linker) will start with 5′ TCC AGA TCC GCC ACC ACC TGA TCC GCC TCC GCC followed by the anti-sense sequence corresponding to the last six amino acids (18 bp) of FR4. For V_L, the 5′ primer sequence (Linker-VL) will start with 5′ TCA GGT GGT GGC GGA TCT GGA GGT GGC GGA AGC followed by the nucleotide sequence corresponding to the first six amino acids (18 bp) of FR1 and the 3′ primer sequence (VL-HindIII) will start with 5′ GGA AGC TTT (incorporating the HindIII restriction site) followed by the anti-sense sequence that corresponds to the last seven amino acids (21 bp) of FR4 (*see* **Note 3**).
2. Set up a PCR (100 µL) as follows: 10 ng of heavy- or light-chain template (TA-cloned DNA; **Subheading 3.1.3.**), 10 pmol of V_H or V_L 5′ primer, 10 pmol of V_H or V_L 3′ primer, 8 µL of dNTP mix (2.5 m*M* each), 10 µL of 10X Taq polymerase buffer, and water to make up the volume to 99 µL. Heat the mix to 92°C for 4 min, add 1 µL of Taq polymerase per reaction, and perform the PCR with 25 cycles of: 94°C for 1 min, 55°C for 1 min, and 72°C for 1 min, followed by an extension for 7 min at 72°C.
3. Gel-purify the PCR product as described in **Subheading 3.1.3.** and quantify the DNA by measuring the absorbance at OD_{260nm}.
4. Splicing of V_H and V_L fragments: Take 10 ng of each V_H and V_L PCR fragment from **step 3** in a PCR tube and set up a PCR by adding the following components: 10 µL of 10X Taq polymerase buffer, 8 µL of dNTP mix (2.5 m*M* each) 10 pmol of V_H 5′ primer and 10 pmol of V_L 3′ primer, 1 µL of Taq polymerase and water, to adjust the volume to 100 µL. Perform the PCR as described in **step 2** (*see* **Note 4**).
5. Purify the PCR product and TA-clone the purified PCR fragment into the TOPO TA vector, and analyze the correct clones as described in **Subheading 3.1.3.** (*see* **Note 5**).

3.1.5. Construction of Immunotoxin Plasmid

1. Select the right clone from **Subheading 3.1.4., step 5** and digest approx 2 µg of plasmid DNA in 25 µL reaction volume with NdeI and HindIII (use HindIII reac-

tion buffer). At the same time set up a restriction digestion reaction for the immunotoxin expression vector pRB98-Amp (2 µg) with NdeI and HindIII. Purify the digested insert (approx 700 bp) and the plasmid cassette (3.5 kb) as described in **Subheading 3.1.3.**

2. Set up a ligation reaction as follows: take 30 ng of vector DNA and 75 ng of insert in a 0.5-mL Eppendorf tube and adjust the volume to 8 µL with water, place the tube in a 65°C water bath for 7 min, then chill on ice. Add 1 µL of 10X ligase buffer and 1 U (1 µL) of T4 DNA ligase and mix content by tapping. After a brief spin in the microcentrifuge, incubate the mixture at 4°C overnight. Also set up a control ligation reaction without the insert to determine the false-positive colonies from the digested vector only.
3. Take 2.5 µL from each of the ligation mixture and transform 100 µL of Max Efficiency DH5α *E. coli* as described in **Subheading 3.1.3.**
4. Plate 100 µL of the culture mixture onto a LB/Amp agar plate and incubate at 37°C overnight. (*see* **Note 6**).
5. Isolate plasmid DNA from eight different colonies using the QIAprep8 plasmid isolation kit; restriction digest the DNA with NdeI and HindIII, and analyze by electrophoresis on a 1.2% (w/v) agarose gel.
6. Select the clones which have the right insert and sequence at least four clones to confirm the in-frame ligation with the fusion toxin protein.

3.1.6. Conversion of Fv Fragment to dsFv

1. For the generation of the disulfide-stabilized Fv (dsFv) molecule, amino acid residue 44 of V_H and amino acid residue 100 of V_L (Kabat numbering) must be changed to cysteines. Design and synthesize the mutagenic primers spanning residue 44 for V_H and residue 100 for V_L by following the recommendation provided by the QuikChange site-directed mutagenesis kit (*see* also Chapter 18). Use the cDNA clone selected from **Subheading 3.1.4., step 5** as template and follow the instructions provided with site-directed mutagenesis kit.
2. Isolate plasmids from eight independent clones from each group and check for the desired mutation at residue 44 for V_H and residue 100 for V_L by sequencing the clones. Select one clone for mutated V_H and one for mutated V_L and use them as polymerase chain reaction (PCR) templates for the next step.
3. Synthesize the PCR primers for V_H and V_L fragments to clone into the T7 expression vector (pRB98-Amp). Forward primer for both V_H and V_L will start with a NdeI site (CAT ATG) followed by the nucleotide sequence that corresponds to the first six amino acids of FR1. Reverse primer for V_H will start with 5′-GGA AGC TTT-3′ (incorporating a HindIII site) followed by the anti-sense sequence that corresponds to the last six amino acids (21 bp) of FR4. For V_L, the reverse primer sequence starts with 5′-GAA TTC ATT A-3′ (incorporating an EcoRI site) followed by the anti-sense sequence corresponding to the last six amino acids (21 bp) of FR4. PCR-amplify the V_H and V_L fragment by using the previously mentioned primer pair and the template from **Subheading 3.1.6., step 2,** following instructions described in **Subheading 3.1.4.**

4. Gel-purify the PCR product and TA-clone the fragments as described in **Subheading 3.1.3.**
5. After sequencing, select the clone with correct sequence and digest approx 2 μg of plasmid DNA with NdeI/HindIII for the V_H clone and NdeI/EcoRI for the V_L clone. At the same time, digest 1 μg of plasmid pRB98-Amp each with NdeI/HindIII and NdeI/EcoRI. Purify the digested insert and the plasmid cassette as described in **Subheading 3.1.3.**
6. Set up the ligation reaction, transform the competent *E. coli* cells, and screen, and analyze the correct clones as described in **Subheading 3.1.5.**

3.2. Preparation of RITs

Two separate transformations, fermentations, and inclusion body preparations are performed, one for each component of the disulfide-linked immunotoxin (V_H-PE38 and V_L). Flowchart of the procedure is shown in **Fig. 3.**

3.2.1. Transformation of E. coli

1. In pre-chilled Falcon 2059 tubes, add 0.1 μg of plasmid DNA (**Subheading 3.1.5., step 6**) to 100 μL of chemically competent BL21 (λ DE3) cells.
2. Incubate the tubes for 30 min on ice.
3. Heat shock the cells by transferring the tubes to a 42°C water bath for 90 s.
4. Place the tubes on ice for 2 min.
5. Add 900 μL SOC media to each tube.
6. Shake the tubes at 225 RPM at 37°C for 1 h.
7. Plate 100 μL of transformed cells on each of ten LB/Amp agar plates.
8. Incubate the plates at 37°C overnight. After overnight incubation, there should be at least 100 colonies/plate.
9. Pre-warm Super Broth at 37°C overnight.
10. Sterilize two 2-L baffled flasks for each liter of culture to be grown.

3.2.2. Fermentation

1. To each liter of prewarmed Super Broth add 20 mL of 20% glucose, 1.68 mL of 1 *M* MgSO4 and the selective antibiotic (100 μg/mL ampicillin).
2. Pipet 5 mL of Super Broth into each plate and dislodge colonies using a sterile glass rod.
3. Transfer the bacteria from the plates to a sterile tube and mix to homogeneity.
4. Add 10 mL of the cell suspension to 1 L of Super Broth, mix well, and measure the OD_{600nm}.
5. Add an appropriate amount of the cell suspension so that the OD_{600nm} is between 0.15 and 0.20.
6. Transfer the inoculated medium to the sterile 2-L baffled flasks (500 mL per flask).
7. Incubate at 37°C, shaking at 250 RPM. Check the OD_{600nm} at 30–40-min intervals.
8. When the OD_{600nm} is between 2.0–3.0 (to measure OD_{600nm} of dense cultures, dilute the culture 1:10 in Super Broth and multiply the OD_{600} reading by 10), save

500 µL of culture as the pre-induction control sample (keep on ice until prepared) and add 5 mL of 0.1 *M* IPTG to each 500-mL culture (0.1 *M* = 25 mg IPTG/mL of broth or water).

9. Continue incubation at 37°C and shaking for 90 min.
10. Save a 250-µL aliquot of culture as post-induction sample (keep on ice until prepared).
11. Harvest bacteria from culture by centrifugation in four 250-mL bottles at 7,500*g* (at 4°C for 10 min) in a Sorvall RC5B centrifuge.
12. Discard supernatant. Cell pellets may be frozen at –70°C for future workup.
13. Aliquots from **steps 8** and **10:** centrifuge at maximum speed in a microfuge for 2 min. Discard the supernatant. Resuspend the pellets in 1 mL of TES buffer. Sonicate for 20 s. Centrifuge in microfuge at maximum speed for 5 min, and discard the supernatant. Resuspend the pellet in 0.1 mL of TES. Sonicate for 10–20 s to resuspend the pellet. Determine the protein concentration and run equal amounts (10–15 µg) of protein on sodium dodecyl sulfate polyacrylamide gel electrophoresis (SDS-PAGE) to verify induction.

3.2.3. Inclusion Body Preparation

1. Resuspend the pellet(s) in each of the 250-mL centrifuge bottles (totaling 1 L of culture) in TES to a final volume of 160 mL. Pool the suspension to one 250-mL Sorvall centrifuge bottle.
2. Add 6.5 mL of lysozyme (at 5 mg/mL), mix using the Tissuemizer, and incubate at room temperature for 30 min. Shake the bottles by hand every 10 min.
3. Add 20 mL of 25% Triton-X100, mix with the Tissuemizer, and incubate at room temperature for 30 min. Shake by hand every 10 min.
4. Centrifuge at 27,000*g,* at 4°C for 50 min, in a Sorvall GSA rotor. Discard the supernatant.
5. Using the large Tissuemizer probe, resuspend pellet in 160 mL of TES. Add 20 mL of 25% Triton-X100. Mix well and incubate for 5–10 min at room temperature.
6. Centrifuge at 27,000*g,* at 4°C for 50 min, in a Sorvall GSA rotor. Discard the supernatant.
7. Repeat **steps 5** and **6** twice (e.g., total of three washes *with* Triton-X100).
8. Using the large Tissuemizer probe, resuspend the pellet in 180 mL of TES.
9. Centrifuge at 27,000*g,* 4°C for 30 min in a Sorvall GSA rotor. Discard the supernatant.
10. Repeat **steps 8** and **9** twice (e.g., total of three washes *without* Triton-X100). Save 25 µL of the washed inclusion body suspension to run on gels to check the inclusion-body preparation.
11. Using the small Tissuemizer probe, resuspend the pellet in 35 mL of TES and transfer to a 40-mL Oak Ridge centrifuge tube. Centrifuge at 12,000*g* for 10 min at 4°C in a Sorvall SS-34 rotor to pellet the inclusion bodies (IBs).
12. Discard the supernatant. Inclusion bodies can be frozen at –70°C at this step.

3.2.4. Solubilization and Denaturation

1. Using a small Tissuemizer probe, resuspend the IB pellet in 5 mL of GTE buffer.
2. Determine the protein concentration using the Pierce Coomassie Plus reagent.

3. Dilute the protein to 10 mg/mL with GTE.
4. Mix 6.67 mL (66.7 mg) of the V_H-PE38 with 3.33 mL (33.3 mg) of the V_L to a total of 100 mg protein per 10 mL. If preparing a single-chain immunotoxin, use 100 mg per 10 mL of the inclusion body.
5. Add dithioerythritol (DTE) powder to 10 mg/mL and mix well but gently.
6. Incubate at room temperature overnight.
7. Centrifuge the denatured protein solution at 12,000*g*, at 4°C for 10 min in Sorvall SS-34 rotor.
8. Save supernatant and recheck the protein concentration using the Pierce Coomassie Plus reagent.
9. If necessary, adjust the protein concentration to 10 mg/mL with GTE containing 10 mg/mL DTE.

3.2.5. Refolding

1. Prepare refolding buffer: 100-fold more than the volume of the denatured protein solution, chill to 10°C, and adjust the pH to 9.5.
2. Add glutathione to 551 mg/L.
3. With the chilled refolding buffer briskly stirring, and using a pipet, add the denatured supernatant from **Subheading 3.2.4., step 8** quickly over a period of 10–15 s. Mix well for 2–3 min.
4. Stop stirring and let stand at 10°C for 36–48 h.
5. Dialyze at 4°C against 50 L of refolding buffer, until conductivity measures below 3.5 mMHO.
6. Filter dialysate through a 0.45 μm ZapCap. If it is very turbid, first centrifuge, and then filter.

3.2.6. Chromatography

After the refolding step, the correctly folded disulfide-linked immunotoxins must be separated from the impurities. These impurities may include improperly folded immunotoxin, other insoluble bacterial proteins, RNA, and DNA. To do this, we use two ion-exchange steps that separate molecules based on charge, and finally a gel-filtration step that separates molecules based on size. The Q-Sepharose is an inexpensive media used to clean up the dialyzed refolding mixture. The Q-Sepharose column binds positively charged molecules and allows negatively charged molecules to pass through. NaCl (350 m*M*) is used to elute proteins from the Q-Sepharose. This step removes most of the contaminants and concentrates the immunotoxin. Mono-Q ion-exchange chromatography is next used to further purify the protein. In this step, we use a linear NaCl gradient (0–500 m*M*) and observe immunotoxins that elute at a NaCl concentration between 250 and 300 m*M*, aggregated immunotoxin will elute at a higher NaCl concentration. Mono-Q chromatography will concentrate the properly refolded immunotoxin into 2–3 mL fractions. The final chromatography step is the gel-filtration column. The observation of a signal peak eluting from this step ensures that we have purified monomeric dsFv-toxin. This step also serves to exchange the buffer from Tris-HCl/EDTA/NaCl to PBS.

1. Q-Sepharose has a binding capacity of 20 mg protein/mL resin. Only a 5-mL Q-Sepharose column is required (*see* **Note 7**).
2. Pour the column and connect it to the loading pump. Wash with five column volumes of buffer A, then five volumes of buffer B, and finally, five volumes of buffer A.
3. Load the protein onto the column using the loading pump at the maximum rate (499 mL/h using the P-500 chromatography pump).
4. Wash the loaded column with five volumes of buffer A at the maximum rate.
5. Attach the column to the FPLC and elute the protein with a one-step gradient from 10% to 35% buffer B collecting 2-mL fractions using a flow rate of 2 mL/min.
6. Pool protein-containing fractions by reading the OD_{280nm} or by using the Pierce Coomassie Plus reagent.
7. Calculate the total protein content, then dilute with five column volumes of buffer A.
8. Put clean filters into the prefilter unit of the FPLC, and change the filter at the top of the MonoQ column (*see* **Note 8**).
9. Wash the MonoQ column with ten column volumes of buffer A, then ten volumes of buffer B, then ten volumes of buffer A (*see* **Note 9**).
10. Load the MonoQ column at 2–5 mL/min.
11. Wash the loaded column with 20 mL of buffer A at 1 mL/min.
12. Attach the column to the FPLC and elute the protein with a 80-mL linear gradient of 0–50% buffer B, collecting 1-mL fractions in Falcon 2052 tubes at 1 mL/min. Immunotoxins will elute from MonoQ between 25% and 30% buffer B.
13. Check peak fractions on SDS-PAGE under reducing conditions.
14. Pool appropriate fractions. If necessary, MonoQ fractions can be concentrated with a Centricon 30. To cleanup MonoQ column between protein purifications, wash it thoroughly with 20–25 volumes of 1 N NaOH, then ten column volumes of buffer A, ten volumes of buffer B, and ten volumes of buffer A.
15. Equilibrate a TSK column (*see* **Note 10**) on a FPLC system in PBS at 0.5–1.0 mL/min.
16. Load column at the same flow rate.
17. Elute at same flow rate with PBS, collecting 1-mL fractions in Falcon 2052 tubes.
18. Check purity of protein on SDS-PAGE and check activity in cytotoxicity assays (**Subheading 3.3.**).
19. If appropriate, pool fractions, determine protein concentration, aliquot, and freeze at –70°C.

3.3. Cytotoxicity Assay

The purpose of this assay is to determine the potency or cytotoxic activity of the immunotoxin on cell lines that is expressed as IC50. The immunotoxin is diluted over a wide range and dispensed onto cells plated in 96-well plates. After 24 h, tritiated leucine is added to the cells. Uptake of the tritiated leucine corresponds to the number of living cells in the wells. A cell harvester is used to process the dishes. The harvester removes the cells from the 96-well plate and transfers them to glass filter mats. The filter mats are soaked in scintillation fluid and sealed in plastic sleeves, which are then placed into a scintillation counter. The amount of tritiated leucine taken up by cells at each immunotoxin dilution is measured by the scintillation counter as counts per minute (CPM). CPM is plotted vs the immunotoxin dilution (ng/mL). The inhibitory concentration of immunotoxin that produces a 50% reduction in tritiated leucine uptake (IC50) is then determined.

1. Plate cells presenting the antigen of interest in 96-well plates at 1.5×10^4 per well (7.5×10^4/mL, 0.2 mL/well).
2. Dilute immunotoxins in PBS containing 0.2% (w/v) HSA to 10,000, 1,000, 300, 100, 30, 10, and 1 ng/mL.
3. Dispense 20 µL of each dilution into three wells each beginning with PBS/HSA (0 ng/mL) then 0.1, 1, 3, 10, 30, 100, and 1000 ng/mL. In this way, the toxins are diluted 10X while they are dispensed onto the cells.
4. Incubate the plates at 37°C for 20–24 h.
5. Pulse the cells with 20 µL of ^{3}H-leucine at 100 µCi/mL which has been diluted in PBS/HSA.
6. Incubate at 37°C for 2.5 h.
7. Freeze cells on dry ice for 30 min, then thaw at 37°C for 1 h and process in the harvester.
8. Dry the filter mats from the harvester. Then saturate the mats with scintillation cocktail and seal in plastic bags.
9. Measure the incorporation of ^{3}H-leucine using the scintillation counter.
10. The cytotoxic potency of the immunotoxin can be plotted as ^{3}H-leucine incorporation vs ng/mL immunotoxin added.

4. Notes

1. Do not add isopropanol after solubilizing the gel slice as instructed in the manual, because it will reduce the yield of the DNA. Elute the purified DNA with 30 µL of elution buffer.
2. Hybridoma cells sometimes contain unrelated immunoglobulin sequences derived from fusion partner myeloma cells.
3. The 3′ primer for V_H and the 5′ primer for V_L will generate a 15-amino acid linker [$(Gly)_4Ser \times 3$] between the V_H and V_L fragments. Before designing the primers, make sure that there are no NdeI and HindIII restriction sites within the V_H and V_L fragments. If they are present, mutate those sites using alternative codons.
4. The expected size of the PCR product will be approx 700 bp (combining V_H and V_L).
5. It would be useful to confirm the presence of NdeI (CAT ATG) site before the V_H sequence and HindIII (GAA TTC) site after the V_L sequence for subsequent subcloning steps.

6. Plates from the vector-only ligation control should have very few or zero colonies. Plates for the insert ligation should have many colonies.
7. Five milliliters of Q-Sepharose will bind 100 mg of protein, the amount in each liter of refolding solution. Not all the starting protein refolds—some aggregates, and some is removed in the filtration step.
8. MonoQ has a binding capacity of 20 mg protein/mL resin.
9. Small amounts of high salt buffer will cause the bound protein to elute from the column. Thus, be sure to rinse all valves and tubing involved in loading the column and the fast protein liquid chromatography (FPLC) with buffer A.
10. For the sizing column use a TSK G3000SW (0.75 cm × 60 cm = 26.5 mL) with guard column attached (0.75 cm × 7.5 cm = 3.3 mL). The total volume of the TSK G3000SW is 30 mL, the protein capacity is 3% of volume or 0.9–1.0 mg, and the maximum loading volume is 5% of volume = 1.5 mL. If the TSK column is being run for the first time, it should be "seasoned" by running 1 mg of BSA in PBS through it. This will bind to all nonspecific protein-binding sites and protect your protein from binding nonspecifically. Wash the column thoroughly with several column volumes of PBS before injecting your protein.

References

1. Hwang, J., FitzGerald, D.J.P., Adhya, S., and Pastan, I. (1987) Functional domains of pseudomonas exotoxin identified by deletion analysis of the gene expressed in *E. coli. Cell* **48,** 129–136.
2. Keppler-Hafkemeyer, A., Brinkmann, U., and Pastan, I. (1998) Role of caspases in immunotoxin-induced apoptosis of cancer cells. *Biochemistry* **45,** 16,934–16,942.
3. Kreitman, R. J., Wilson, W. H., Bergeron, K., Raggio, M., Stetler-Stevenson, M., FitzGerald, D. J., et al. (2001) Efficacy of the anti-CD22 recombinant immunotoxin BL22 in chemotherapy-resistant Hairy-cell leukemia. *N. Engl. J. Med.* **345,** 241–247.
4. Kreitman, R. J., Wilson, W. H., White, J. D., Stetler-Stevenson, M., Jaffe, E. S., Giardina, S., et al. (2000) Phase I trial of recombinant immunotoxin anti-Tac(Fv)-PE38 (LMB-2) in patients with hematologic malignancies. *J. Clin. Oncol.* **18,** 1622–1636.
5. Benhar, I., Reiter, Y., Pai, L. H., and Pastan, I. (1995) Administration of disulfide-stabilized Fv-immunotoxins B1(dsFv)-PE38 and B3(dsFv)-PE38 by continuous infusion increases their efficacy in curing large tumor xenografts in nude mice. *Int. J. Cancer* **62,** 351–355.
6. Onda, M., Nagata, S., Tsutsumi, Y., Vincent, J. J., Wang, Q-C., Kreitman, R. J., et al. (2001) Lowering the isoelectric point of the Fv portion of recombinant immunotoxins leads to decreased nonspecific animal toxicity without affecting antitumor activity. *Cancer Res.* **61,** 5070–5077.
7. Brinkmann, U., Reiter, Y., Jung, S.-H., Lee, B., and Pastan, I. (1993) A recombinant immunotoxin containing a disulfide-stabilized Fv fragment (dsFv). *Proc. Natl. Acad. Sci. USA* **90,** 7538–7542.
8. Studier, F. W. and Moffatt, B. A. (1986) Use of bacteriophage T7 RNA polymerase to direct selective high-level expression of cloned genes. *J. Mol. Biol.* **189,** 113.

31

Antibodies in Proteomics

Andrew R. M. Bradbury, Nileena Velappan, Vittorio Verzillo, Milan Ovecka, Roberto Marzari, Daniele Sblattero, Leslie Chasteen, Robert Siegel, and Peter Pavlik

1. Introduction

The explosion in genome sequencing, and the DNA array experiments based on those sequences, have provided a wealth of information on gene sequence, organization, and expression. This has led to a desire to carry out similarly broad experiments on all proteins encoded by a genome—the proteome. Although mass spectrometry (MS) is proving to be a powerful tool in the study of proteomes, the information it provides is relatively one-dimensional. The only feasible way to study the rich complexity of protein expression, modification, and interaction on a genomic scale may be to derive well-characterized specific antibodies (or other binding ligands) that recognize each individual protein. Traditional immunization approaches will be unable to generate such banks of antibodies in sufficient quantity, quality, or reproducibility, and as a result, biomolecular-diversity selection methods—such as phage, ribosome, or mRNA display—will probably be used.

Because different genome projects are in different phases of development, it is unlikely that a single method will be relevant to all genomes. When purified proteins are available, physical selection methods (phage or mRNA-displayed libraries) will probably be most effective. When only the sequenced genome is available, genetic methods (two-hybrid or enzyme complementation)—in which interaction between the binding ligand and target confers survival on the cell containing the interacting pair, or the use of synthetic peptides as selectors—are likely to prove more effective, avoiding the need to produce purified protein. Regardless of the methods that are used, validation of the specificity of selected antibodies will be an important yet difficult component of the complete process.

From: *Methods in Molecular Biology, Vol. 248: Antibody Engineering: Methods and Protocols*
Edited by: B. K. C. Lo © Humana Press Inc., Totowa, NJ

This chapter examines the difficulties and challenges in selecting and characterizing such antibodies on a large scale, and discusses their possible applications once derived.

2. Background

Full or draft genome sequences are available for increasing numbers of organisms, including human *(1,2)*, yeast *(3)*, and many others (e.g., see http://www.tigr.org/tigr-scripts/CMR2/CMRHomePage.spl for a list of microbial genome sequences). This has enabled the implementation of experiments using DNA chips in which the expression levels of all genes in a genome are studied in response to different stimuli, and has led to efforts to apply similar genome-wide studies of the corresponding proteomes. The most extensive studies have been carried out in yeast, with individual gene knockouts *(4)*, overexpression and proteome chips *(5)*, intracellular localization by tagging *(6)*, protein-protein interaction studies by phage display *(7)*, yeast two-hybrid *(8,9)*, and widespread mass spectrometric analysis of purified complexes *(10,11)*, providing large amounts of information. One reason that yeast has been used so extensively is the availability of homologous recombination, permitting the replacement of endogenous genes by modified copies. In fact, most of the studies cited here would not have been possible without exploiting this technique, which often involves the genetic fusion of a tag: either a detection peptide such as *myc* *(12)*, or a "tandem affinity purification tag" *(13)*, in the case of MS of complexes. Because this powerful technology is not available for most genomes, the only alternative to the fusion of a general tag (using a single detection reagent) is the derivation of specific binding ligands for all gene products that can be used for techniques in which antibodies have been traditionally used (e.g., Western blotting, fluorescence-activated cell sorting, [FACS], immunoprecipitation, immunofluorescence, immunohistochemistry, and purification), but within a proteomic context, rather than on a single-gene scale. In addition to traditional uses, such binding ligands will also be useful in new techniques that are still in development in a proteomics context (e.g., antibody chips), and potentially in applications such as biosensors. A greater understanding of protein function at a genomic level is likely to come when such banks of binding ligands are derived and made generally available, as has been done virtually for DNA chips by the publication of genomic sequences.

The traditional binding ligand is the antibody, and polyclonal antibodies are usually produced by immunization of rodents or rabbits with proteins, conjugated peptides *(14)*, or recently, with DNA expression vectors *(15)*. With the introduction of hybridoma technology *(16)*, it became possible to avoid polyclonal antibodies—with their problems of reproducibility, crossreactivity, and background—and to produce large amounts of monoclonal antibodies (MAbs) of defined specificity. Although extremely useful, this technology is not easily amenable to high-throughput antibody generation, and cannnot overcome the problems of toxicity or poor immunogenicity found with conserved proteins. It is hoped that the adoption of a new suite of technologies, generically termed the "biomolecular diversity" technologies, will be useful in this regard. In general, these technologies involve the selection of specific binders from large libraries of binding ligands and have a number of common features, which are also applicable to the selection of polypeptides with non-binding activities such as enzymes: i) the generation of diversity at the nucleic acid level to create the "library," usually of binding ligands; ii) the linking of genotype to phenotype (e.g., translated protein with encoding gene; or information to function); iii) the application of selection pressure—e.g., binding to a specific target; and iv) amplification of selected clones after selection. These cycles of selection are usually carried out two or three times before binding ligands can be directly screened, and if the library is of good quality, binding ligands can usually be obtained against all targets. The first examples of biomolecular diversity focused on peptides ***(17–19)***, and used phage display, in which displayed peptides are fused to one of the coat proteins of filamentous bacteriophage, as the platform to identify MAb binding epitopes. Specific binders have also been isolated from large naïve antibody libraries ***(20–26)***, while the display of hormones ***(27)*** and many other proteins (*see* **refs. *28–30*** for reviews) have been carried out with the predominant goal of optimizing binding affinity. In addition to phage display, bacteria ***(31)*** and yeast ***(32)*** have been developed as micro-organismal display methods, while ribosome display ***(33)*** and puromycin display ***(34)***, in which mRNA (or cDNA) is directly coupled to the polypeptides they encode, have been developed as in vitro methods. All these are physical selection methods, and require significant quantities of the selector to perform selection and screening. In addition to these methods, a number of genetic methods for selection of binding ligands have also been developed. Such methods do not require the physical selector during the selection procedure and rely on the in situ synthesis of—and subsequent interaction between—the binding ligand and target, to confer a selectable phenotype. Both the yeast two-hybrid system ***(35)*** and the protein complementation assay ***(36)*** have been adapted to single-chain variable fragment (scFv) selections ***(37–40)*** in model systems, and offer the possibility of selection without the need for antigen synthesis and purification.

As effective as the biomolecular diversity strategies are, they are still not yet very user-friendly, and this represents the biggest obstacle to their widespread adoption. Although difficult to generate in high throughput, once available, MAbs obtained by hybridoma fusion can be produced in relatively large amounts—they are dimeric, very stable, and contain an enormous "tag"—the

Fc domain. In contrast, binding ligands obtained using molecular diversity techniques are usually produced at lower levels—they are monomeric, less stable, and contain a small peptide tag. For these technologies to be widely adopted, it is important that these problems be overcome. The fact that the gene encoding the binding ligand is cloned simultaneously with the selection of the binding ligand in these technologies is the key feature that will lead to the most significant improvements, allowing either straightforward downstream genetic manipulation or the upstream generation of libraries of binding ligands with desirable properties.

This chapter focuses on the processes of selecting and using binding ligands, and not on the different alternative binding ligands themselves, which were recently reviewed ***(41)***. As much of this technology was developed within the context of antibody fragments—scFvs in particular—these are the most commonly described binding ligands. However, most of the technology described is also directly applicable to alternative scaffolds.

3. Physical Selection Methods

In order to be able to select binding ligands against protein targets using physical selection methods, a selector must be available in a purified, synthetic, or recombinant form. Depending upon the method of selection used, 200–500 μg will usually be required to carry out selection and screening, and most of the antigen will be used to identify positive binders. There are two general methods of physical selection. In the first, antigen is fixed to a solid support, such as a polystyrene tube or pin, and incubated with the library of antibodies. Those that recognize the antigen bind and can be eluted after non-binding antibodies are washed away. In the second method, antigen is labeled, usually with biotin or fluorescein, and used to separate antibodies that bind from those that do not. In the case of biotin, this is carried out using streptavidin-coated magnetic beads (MACS), and with fluorescein, FACS is used. In phage, bacterial, and yeast display, living organisms are responsible for amplification, display, and the coupling of phenotype and genotype, and ribosome- and puromycin-based mRNA display systems rely on polymerase chain reaction (PCR) for amplification, and in vitro translation of RNA to produce the binding ligand, which is then attached to the encoding RNA (or cDNA) either covalently (puromycin) or non-covalently (ribosome display). In all methods, more than one round of selection is usually required, although screening thousands of clones can yield a greater diversity of binders after a single round ***(42)***, although with a far lower percentage of positives.

As binders have been isolated from large naïve phage and in vitro display libraries, these are further discussed here. Bacterial ***(43)*** and yeast display ***(44)*** libraries are predominantly used for affinity maturation, although yeast display has also has been used for the improvement of antibody expression and folding *(45)*, and more recently for the selection of antibodies from large naïve libraries with affinities similar to phage display *(145)*.

3.1. Production of Targets

In the use of physical selection methods, the availability of sufficient quantities of antigen for selection and screening is one of the major bottlenecks, and screening is the more antigen-intensive process. There are two general approaches to selection and screening: gene-based and proteome-based. In gene-based selection, the identity of the selector is known in advance, and as a result, the specificity of selected antibodies is also known. Synthetic peptides *(46)*, polypeptide fragments, or recombinant full-length proteins would be examples of suitable selectors. In proteome-based selection, natural sources (e.g., tissue or cell extracts) are used. Although individual known proteins may be purified and used as selectors, this is not really applicable to genomic-scale selections, and the use of either complex extracts or biochemical fractions of such extracts would be preferred. Once selected, the identity of the antigen recognized by such selected antibodies would have to be determined subsequently, a procedure that might be carried out by immunoprecipitation (IP) and MS *(47)*, if this could be implemented in a high-throughput fashion. Selection of phage antibodies on proteins spotted onto polyvinylidene fluoride (PVDF) membranes after separation by two-dimensional (2D) gel electrophoresis has recently been described *(48)*. Although this may be a solution to the use of natural protein sources for selection, screening and identification of appropriate binders in high throughput remains a key issue.

In general, the major advantage of gene-based selection methods is that antigen identity is known in advance, whereas proteome-based selection methods have the advantage that the antigen is in its most natural state, and will include appropriate post-translational modifications. The general differences between these two targets are outlined in **Table 1.** If gene-based selection systems are used, the form of antigen used for selection and screening must be considered. Peptides or polypeptide fragments and full-length recombinant proteins are the most likely choices, and each of these has advantages and disadvantages, as outlined in **Table 2.**

If all things were equal, selectors based on cDNA-derived polypeptides would be preferable to peptide selectors because of the greater antigenicity, and ease of screening, of the former. However, full-length proteins can suffer from problems related to poor expression, folding, or solubility, resulting in difficulties in obtaining sufficient protein for use in selection and screening. It may be possible to eliminate these by using either specific domains or fragments selected for their ability to fold, in a fashion that has been already described for

Table 1
Gene- vs Proteome-Based Selections

Gene-based selection	Proteome-based selection
Sequence-based	No prior knowledge of sequence
Cloning, expression, and purification may be required	Some fractionation (e.g., chromatography, 2D gels) may be required
Antigen specificity known in advance	Antigen specificity must be determined after selection—e.g., by immunoprecipitation and mass spectrometry
Selection against single gene products	Selection may be against more than one gene product
All open reading frames are pre-identified and require systematic selection	Some antigens may not be present in selector
96-well selection possible, using biotinylated peptides/polypeptides or histidine tags	Fractionated samples can be used in 96-well format
No post-translational modifications if bacteria used for production, polypeptide antigen only	Most natural antigen, containing post-translational modifications
Concentration of selector can be predetermined	Selector concentration more difficult to determine
All genes can be represented equally	Some antigens present at very low concentrations or not at all

Table 2
Peptide- vs Polypeptide-Based Selections

Peptide-based selections	cDNA derived polypeptide-based selections
Less antigenic: only linear epitopes	More antigenic: linear/conformational epitopes
Crossreactive antibodies more likely	Crossreactive antibodies less likely
Screening protein binding clones can be difficult	Selection polypeptide also used for screening
No definitive method to identify suitable peptides	Single selector
More than one peptide required	Single selector
Cheaper: less labor-intensive	More expensive: very labor-intensive
More easily automated	More difficult to automate
Relatively simple workup	Long workup per clone
Synthetically produced, no specific purification	Individual purification for each polypeptide
Can use sequence data directly—genomic or expressed sequence tags (EST)	Requires physical cDNA: identification/cloning

random mutations *(49,50)*, although this has not been done on a high-throughput scale. The creation of large banks of soluble proteins is still in its infancy, although some studies *(5,51)*, that emphasize the importance of extensive automation, have been initiated. Some protein-production problems are related to expression in *E. coli,* and may be solved by using other systems such as yeast *(52,53)*, *D. discoideum* *(54)*, filamentous fungi *(55)*, or baculovirus *(56)*—either instead of, or in a hierarchical fashion following failure in, *E. coli.* An alternative approach is to avoid living systems completely and adopt the use of high-efficiency in vitro translation systems *(57–59)*. In model systems, these appear to be very promising, and offer the advantage of many parameters that can be precisely controlled. If such systems can be automated, they may provide powerful methods to produce relatively large amounts (milligrams) of proteins in high throughput, providing enough for both selection and screening.

Peptides have a long history as surrogate immunogens in animals *(14,60)*. There have been many successes with a number of "rules"—e.g., peptide flexibility of hydrophilicity, derived to predict the most immunogenic peptides (reviewed in **refs.** *61–62)*, as well as many more unpublished failures. Some of these rules were brought together in a single "antigenic index" *(63)*, which is intended to identify the most "antigenic" portion of a protein when it is used as a peptide immunogen. However, the antibody response elicited by a whole organism when immunized with an antigen is the result of a complex interaction between T-cells, B-cells, and antigen-presenting cells. Phage antibody libraries can be considered to be in vitro humoral immune systems in which the antibody response is reduced to the bare minimum of antigen recognition. Peptides have been used as selectors with phage antibody libraries *(46,47,64)*, but a systematic examination of their use has not been published. Preliminary experiments in the author's laboratory (submitted) indicate that features predictive of animal antigenicity are not applicable to selection using phage antibody libraries, and that solvent accessibility is a better predictor. Peptides are desirable because they can be produced directly from genomic sequence, and do not require expression or purification. However, their biggest drawback is the possibility of crossreactivity with irrelevant proteins, leading to erroneous results upon use. In one example, some (but not all) antibodies selected against a 15-mer peptide derived from Ku86 (a protein involved in DNA repair) recognized myosin, a crossreaction believed to be the result of the presence of a very similar pentapeptide in both: DIDDL in myosin, DVDDL in Ku86. It is unlikely that peptides lacking such crossreactive epitopes could be identified in advance, indicating the need for careful screening after selection has been carried out. Ideally such screening would entail obtaining concordant results (e.g., Western blots, immunohistochemistry, IP, MS) on natural antigen sources with scFvs selected against two different peptides.

Physical selection methods can also be used to select antibodies against post-translational modifications, using either modified proteins or peptides as selectors in the presence of an excess of the unmodified protein or peptide. Screening is then carried out on both modified and unmodified forms. It is likely that in addition to stable modifications such as phosphorylation, more unstable ones that would never survive an immune response long enough to permit antibody generation, such as sulfation, may also be amenable to selection.

3.2. Phage Display

Most phage antibody libraries have been created by cloning large numbers of different antibody genes upstream of the gene 3 coat protein gene and using phage ***(21)*** or phagemid ***(22–26)*** as the display vehicle. Most of these use scFvs as the antibody format, with one large Fab library ***(24)*** also published. The antibody genes are derived either from natural sources [e.g., human peripheral-blood lymphocytes (PBLs); *see also* Chapters 6 and 8] or created synthetically by introducing diversity using oligonucleotides into frameworks with desirable properties. When an antibody gene is cloned upstream of gene 3, the antibody is displayed as a fusion protein with the gene 3 coat protein. A library of such phage antibodies theoretically consists of as many as 10^{11} different members (the diversity is generally measured by counting the number of independent colonies), with each specificity represented by a relatively small number of phage in a library. In general, diversity is limited by the transformation efficiency of bacteria. However, recombinatorial methods of library creation ***(23,64,65)***, in which V_H and V_L genes are shuffled using *cre* recombinase after an initial cloning step, are capable of creating far larger libraries, and as recombination is associated with amplification, almost unlimited supplies of such libraries.

Although antibodies with sub-nanomolar affinities can be directly selected ***(22,25)***, or affinity-matured, from these libraries ***(66–68)***, this usually requires a considerable degree of effort, which would be difficult to marshall at a genomic scale. As a result, the usual affinity range of antibodies selected directly from such libraries is 10–1,000 n*M*. However, by genetically fusing multimerization domains, as described previously (e.g., jun/fos for dimerization), to the ends of such selected scFvs, the effective affinity can be significantly increased ***(69)***.

Selection of phage antibodies is usually carried out against single antigens. A potentially high-throughput method using antigen immobilized on polystyrene pins in a microtiter format has recently been described ***(70)***, and similar methods may also be developed for biotinylated antigens ***(71)*** using robotic washing and elution systems. However, selection is usually the least difficult part of the procedure, with the identification of different positive clones being far more time- and antigen-intensive (*see* **Subheading 4.**).

Antibodies selected from naïve phage antibody libraries have highly variable expression levels (ranging from 10 µg to 20 mg per L), but can be evolved to be expressed at higher levels ***(72–74)***, although this approach cannot be easily carried out with a large number of different antibodies on a genomic scale. The alternative approach is to create libraries in which most antibodies are well-expressed. A number of promising libraries ***(26,75–77)*** have been constructed with one or more stable scaffolds and synthetically introduced diversity. An alternative approach, based on the fact that the size of a library created by recombination is the product of the number of different V_H and V_L genes, is to select stable, well-expressed scFvs directly from natural naïve libraries, and then to recombine the selected V_H and V_L chains to regenerate diversity ***(65)***. This could be done by selecting for binding to staphylococcal protein A ***(78)*** and/or peptostreptococcus protein L ***(79)***, which recognize conformational epitopes on human V_H3 chains ***(80)*** and $V_\kappa 1$, $V_\kappa 3$, and $V_\kappa 4$ chains ***(81)***, respectively. It has been shown ***(82)***, at least in the case of yeast, that scFv stability, expression, and display are related, indicating that improvement in anyone of these parameters is likely to simultaneously improve the others, and suggesting that the adoption of any of the strategies described here to increase the intracellular stability of scFvs (*see* **Subheading 3.,** below) will also increase expression levels.

3.3. In Vitro Display Systems

In in vitro display systems, genes are coupled to the proteins they encode after translation in an in vitro translation mix. In puromycin display ***(34,83)***, puromycin covalently links the RNA to the encoded protein, while in the case of ribosome display ***(33,84)***, the ribosome itself acts as a noncovalent linker between gene and encoded protein. Selection from such libraries involves binding, washing, elution, and amplification by PCR, and the affinities of scFvs isolated from primary selections are similar to those from phage antibody libraries, although the theoretical library sizes are much larger than most phage libraries (no transformation is required). This may be a result of low folding efficiency in the reducing nature of the in vitro translation mix, as antibodies generally require oxidizing environments to fold correctly. Although positive binders can be selected after two or three rounds using phage display, in vitro display systems usually require many more cycles. One major advantage of in vitro display systems is the possibility of incorporating built-in affinity maturation ***(85,86)*** by using rounds of error-prone PCR, or DNA shuffling ***(87)*** between selection rounds. Although this requires even more selection rounds, antibodies with picomolar affinities have been achieved using this method, and this is more likely to be amenable to automation for high-throughput selection. Once selected, antibodies or other binding ligands selected by in vitro display systems are usually cloned into bacterial expression systems. This represents a bottleneck in the procedure, as not all antibodies selected by in vitro display are subsequently well-expressed in bacteria. It is possible that bacteria may be eliminated altogether if robust picking and amplification of individual genes using plate-based PCR methods to create "polonies" ***(88)*** can be developed and adapted to highly efficient in vitro expression systems ***(57–59)***. Alternatively, pools of polyclonal antibody genes may be used as templates to create in vitro translated polyclonal antibodies.

4. Genetic Selection Methods

Physical selection methods are appropriate for the selection of binding ligands for which a physical selector is available. Although this field is advancing rapidly, with a few groups producing small amounts of proteins on a genomic scale ***(5,51)***, sufficient quantities of selectors are usually not available for most proteome projects. One way around this is to avoid the use of physical selectors altogether and to develop genetic selection methods that use DNA encoding either the whole or part of the gene of interest. The selection of binders from libraries is essentially a protein-protein interaction problem, and the yeast two-hybrid system ***(35)*** is the most widely used genetic selection method to identify such protein-protein interactions. Under ideal circumstances, it would be possible to clone the gene of interest as the bait and to transform an antibody library as the prey. However, most current antibody libraries ***(22–25)*** contain more than five billion clones, which far exceed the transfection capability of yeast. In order to overcome this problem, Visintin and colleagues ***(37,89)*** carried out a single round of selection on a protein of interest, and cloned the output of this selection into a yeast two-hybrid vector. This reduced the diversity of the library to 10^{5-6}, which is amenable to yeast transfection, and permitted the selection of a number of different scFvs. This technology was initially developed with the goal to select scFvs that are functional intracellularly ***(90)***, although it clearly has the potential to be used as a general genetic selection method. However, there are two important issues related to the use of the yeast two-hybrid system in such an approach. The first is the need to carry out a physical selection prior to performing the genetic selection, which eliminates some of the advantages of a genetic approach, and the second is related to the fact that antibodies are secreted proteins in which the disulfide bonds account for a large degree of stability, and as a result, scFvs are not usually functional under the intracellular conditions used in this genetic approach.

The need for a physical selection prior to genetic selection is the result of the low transfection efficiency of yeast as compared to *E. coli,* and may be overcome, through the use of bacterial genetic systems rather than yeast, although recent experiments with the yeast two-hybrid system suggest that it may be possible to select antibodies directly without preselection ***(40)***. Bacterial two-hybrid systems ***(91–94)*** have also been developed, and one of these ***(93,94)*** has recently been commercialized, although relatively few publications have described their use. In addition to interaction systems relying on transcriptional activation, so-called protein complementation assays (PCA) have also been described ***(36,95,96)***, in which an enzyme required for cell survival (e.g., dihydrofolate reductase or β-lactamase) is divided into two parts in such a way that the enzyme activity can be reconstituted by bringing the parts together, but only if those two parts are attached to two interacting species, such as coiled coils or antigen-antibody. Preliminary experiments ***(39)*** using model antibodies and DHFR showed that specific antigen-antibody pairs conferred survival more than 10^7 times more effectively than nonspecific pairs. This, coupled with the high transformation efficiency of *E. coli,* may make this method very useful for proteomic-scale antibody selections.

Most of these bacterial systems also require that antibodies must be stable in the cytoplasm, since with the exception of β-lactamase—they all involve cytoplasmic interactions. In fact, antibody stability, rather than affinity, appears to be more important for function under intracellular conditions ***(97–99)***. The stability of any single scFv can be significantly increased by appropriate mutation (reviewed in **ref. *100***), either rationally, by grafting CDRs onto a stable framework ***(101,102)*** or by incorporating stabilizing mutations (*see* **ref. *100***). Antibodies can also be selected to be more stable by a variety of biomolecular diversity techniques, including phage display ***(103,104)***, ribosome display ***(105)***, intracellular expression ***(106)***, and yeast display ***(45)*** in which random mutagenesis is coupled with selection for increased stability or display. However, such an approach cannot be applied to genomic-scale selection of individual binding ligands, and it would be more appropriate to develop libraries of binding ligands, in which a high proportion of binders are functional under intracellular conditions, rather than to optimize individual scFvs. Although libraries of scFvs with greater stability are not yet available, a greater understanding of the factors underlying stability will allow the future creation of such libraries. An analysis of the sequences of antibodies that are functional intracellularly ***(89)*** has revealed their similarity to Kabat consensus sequences ***(107)***, a result that is consistent with analyses carried out using statistical mechanical methods and confirmed experimentally ***(99,108–110)***. Antigen-independent selection for antibody stability using a modified version of the yeast two-hybrid system ***(38)*** has shown that only 0.1–0.2% (D. Escher, personal communication) of antibodies are stable intracellularly. By using such antibodies or consensus frameworks as scaffolds with synthetic oligonucleotides that provide diversity, one would expect to be able to select stable scFvs without the need for further manipulation. Such libraries of antibodies would also be extremely useful as intracellular antibodies ***(90)*** in downstream target investigation, as well as in genetic selection methods.

The selection for antibody/antigen interactions using β-lactamase complementation ***(96)***, in a fashion similar to that used for DHFR, would avoid the need for libraries of antibodies that were stable intracellularly. However, it does require that antigens are secreted into the periplasm, and it is expected that cytoplasmic antigens may not be stable under these conditions. This caveat also applies to selection by avidity capture ***(111)***, an approach that can best be described as a combined genetic/physical method, which relies on co-expression of antigen and phage antibody in the periplasm. After antigen and antibody have interacted in the periplasm, positive clones are screened using a filter-based approach ***(42)***, which is capable of identifying many different specific antibodies after one round of traditional phage-display selection.

5. Screening

Screening is required to identify the antibodies that bind specifically to the target and to eliminate those that bind nonspecifically to irrelevant targets. At present, this is usually carried out using an ELISA format ***(112)***, with selected phage antibodies tested on both the selecting antigen and an irrelevant antigen, although membrane-based screening ***(42)*** has also been described. In general, 96 antibodies (a single microtiter plate) are tested against the antigen used for selection. After two rounds of phage antibody selection, 10–100% of clones are usually positive, depending upon the library, antigen, and selection conditions, and as many as 20 different antibodies can be identified. Antibodies that show good specific binding and low levels of nonspecific binding can then be carried forward for further analysis.

Screening is relatively straightforward when a polypeptide antigen is used for selection, but becomes conceptually more complex when peptides or genetic selection is used. In the case of peptides, although binding to the peptide used for selection can be confirmed, this does not guarantee that the antibodies bind to the polypeptide from which the peptide was derived. Although screening on such polypeptides can be carried out, the reason that peptides are usually used is precisely because such polypeptides are not available. Similarly, genetic methods are likely to be used to select antibodies when polypeptides are not available. This indicates the need for general screening protocols that will allow the determination of binding specificity. There are two general approaches to this problem. The first involves the use of natural antigen sources (e.g., tissue sections or cell or tissue extracts), and requires that different antibodies selected using surrogate methods provide identical reactivity patterns using such natural antigen sources (e.g., cellular distribution or Western blot pattern). In the case of peptides, antibodies selected against two different peptides would also be required to provide identical reactivity patterns. The second involves the use of in vitro translation, recombinant bacteria, or cell lines that express the gene of interest, even if at very low levels, and requires appropriate recognition (e.g., ELISA or Western blot) of extracts containing the gene of interest, but not those containing other genes. The latter is likely to be appropriate for genetic selection methods, in which the gene product is expressed as an essential part of the selection process, and could also be used for peptide selections if clones are available.

Although it is relatively straightforward to screen 96 clones by ELISA for a single selection manually, this becomes more difficult if alternative screening methods, such as immunohistochemistry, are used. Similarly, screening 96 clones for each of 48 or 96 antigens simultaneously can only be considered if the procedure is automated, which may be possible using modifications of published membrane screening ***(42)*** or array ***(113–115)*** methods.

Once positive clones have been identified, the number of different clones must be determined. Ideally, this is best done by sequencing, although fingerprinting using PCR and restriction enzymes provide similar results. At a high-throughput level, it is likely that sequencing with appropriate programs as a method to identify identical antibodies and gene families is likely to be the most efficient procedure.

6. Using Selected Antibodies

6.1. Improving the System

Antibodies selected from phage display or other libraries can be used in all applications in which MAbs are traditionally used (Western blotting, immunohistochemistry, immunofluorescence, fluorescence-activated cell sorting (FACS), inhibition studies. However, they require one extra step to reveal them, usually comprised by the binding of a MAb recognizing the antibody binding tag (e.g., *myc* [***12***] or SV5 [***116***]). Just as monoclonals are variable in their utility in different applications, the same is also true of antibodies derived by phage display: some are effective in immunoprecipitation, and others are not; some are functional in Western blotting and others are not. Natural antibodies are dimeric, and this gives them greater avidity and greater effective affinities when used for research purposes. In contrast, antibodies selected from phage antibody libraries are monomeric. However, they can be made dimeric or multimeric by using coiled coils or tetramerization domains ***(69,117)***, so theoretically providing the same advantages. In addition, it has been shown that "chelating antibodies," in which two scFvs that recognize different epitopes on the same target are cloned head to tail, provide enormous increases in affinity ***(118)***. These strategies, or modifications of them, represent a relatively easy way to increase the effective functional affinities of selected scFvs, and will be important in the high-throughput use of scFvs. Alkaline phosphatase is a dimeric enzyme that is frequently used in antibody applications, and scFv-alkaline phosphatase fusions *(119–122)* have been used in a number of different applications, combining the advantages of dimerization with a detection domain.

Some scFvs selected from phage antibody libraries have been shown to have low yields and stability. Although these can be increased, using methods described here, this represents an important issue for the use of these binding ligands in high throughput. The stability issue can be overcome by using Fabs *(24)*, although the expression levels, in molar terms, are usually lower, and would require recloning, unless the library is in the Fab format initially. One method to overcome the problem of low stability and yield is to create libraries in which high percentages of the scFvs are stable and well-expressed, as described here. The development of such libraries will be essential to the adoption of this technology in proteomics.

The vectors used for antibody selection are not usually the best for subsequent use. As a result, some groups have worked toward the creation of integrated systems in which selected scFv genes can be excised from selection vectors and recloned into downstream vectors with different useful properties. A non-comprehensive list of examples includes vectors for enhanced expression *(123)*, miniantibodies and dimerization domains *(75,123)*, Fabs *(75)*, alkaline phosphatase fusion proteins *(121,123)*, targeting vectors for intracellular immunization *(124)*, or vectors that allow the transfer of V regions from scFvs to full-length immunoglobulins for mammalian expression *(75,125)*. These all rely on the use of restriction sites, which are rare in V genes, to be present at corresponding sites in the different vectors. An alternative approach is the use of recombination to transfer selected antibodies from the selection vector to a downstream expression vector. Until now, this problem has been faced mainly by researchers who clone large collections of individual genes with specific primers (e.g. *C. elegans [126]*), and has been tackled using lambda-based recombination *(127)* (the Gateway™ system), in which genes are cloned into a vector containing mutated attL sites. Once in the attL (entry) vector, inserts can be easily transferred to destination vectors containing compatible attR sites, with recombination resulting in the gene of interest, flanked by attB sites, in the destination vector. Although this method is very effective for individual genes, the asymmetrical nature of the recombination sites renders it inappropriate for library-based selection methods. In particular, recombination sites that flank binding ligands in libraries should be short and translatable into short peptides that have little effect on the function of the binding ligand, with either the same short peptides, or equally innocuous peptides, being present after recombination. Although attB sites are short (24 bases), attP, L, and R sites are all greater than 100 bases. As a result, although a binding ligand library flanked by attB sites can be made, the transfer of selected clones to a useful secondary vector would have to involve two rounds of recombination, first into an attL-containing vector and then out to another attB-containing destination vector. *Cre* recombinase-mediated recombination of lox sites is conceptually far simpler than lambda-based recombination. Each lox site is 34-bp long, and homologous lox sites recombine at high efficiency to reproduce the same lox site. Heterologous sites recombine with one another at very low efficiencies ***(128–130)***. The 511 lox site has been used as a linker between V_L and V_H in one phage-display library ***(23)***, with display levels as good as or better than those of the standard Gly-Ser linker. Providing the addition of translated lox sites at the N and C termini of scFvs has no effect on expression or display levels, and is not toxic to bacteria, a phage antibody library in which scFvs were flanked by such sites would allow the shuttling of selected scFvs from a selection vector to a downstream vector, completely avoiding the need for any manipulation of DNA whatsoever.

6.2. Using Antibodies in Proteomics Applications

Antibodies are traditionally used individually, or in small numbers, to determine characteristics of a gene product such as tissue distribution, cellular localization, post-translational modifications, expression levels, and complexes formed, using such techniques as immunofluorescence, immunohistochemistry, Western blots, and immunoprecipitation. The application of such traditional antibody techniques on a proteomic scale using antibodies selected from phage antibody libraries will require significant modification of current protocols to allow the reliable high-throughput accumulation of data that is potentially possible. However, once developed, such protocols will be able to provide immense amounts of useful information that has the potential to be stored in easily accessible databases, analogous to that which has been created by the tagging of single genes in yeast.

6.2.1. Antibody Chips

The real promise of antibodies selected from phage display and other libraries on a genomic scale lies in global proteomic experiments analogous to those that have been carried out in the RNA sphere using DNA chips, in which all proteins can be examined simultaneously for increases or decreases in either levels or specific post-translational modifications following experimental manipulations. Initial experiments with protein or antibody chips ***(114,115,131–133)*** are promising, and small changes in protein levels can be detected. Many of these have been carried out with antibodies purchased from commercial suppliers, and as a result have been applied to the detection of well-characterized targets. There are three standard formats for these experiments (*see* **Fig. 1**): i) *Antibody arrays with labeled extracts* ***(114,115)***: capture antibodies are fixed to a solid support (chip) in a spatially addressable manner, cell extracts labeled with fluorescent dyes are incubated with the chip, and after washing, those antibodies with bound components present in the labeled cell extract can be identified. This can be carried out on single extracts, or by the use of two different dyes with non-overlapping excitation and emission spectra, protein levels between two different conditions that are analogous to similar mRNA experiments, can be compared. ii) *Antibody arrays with labeled detection antibodies* ***(134)***: capture antibodies are similarly fixed to a solid surface in an addressable fashion. This is incubated with nonlabeled extracts, and labeled detection antibodies, which recognize different epitopes than those recognized by the capture antibodies, are subsequently added to identify the capture antibodies with bound specific components in the added extract. iii) *Reverse phase protein arrays* ***(133)***: cell extracts are fixed to a solid surface, and antibodies are arrayed in an addressable fashion directly on the immobilized cell extracts. After washing, antibodies that have bound antigens recognized in the extract are revealed either enzymatically or by fluorescence by the addition of appropriate secondary labeled antibodies. The main problem with the first approach is the need to label extracts. Not all proteins are equally labeled, some labeling may destroy antibody binding sites, and if an antibody recognizes one component of a protein complex, erroneously high labeling levels will be obtained when the whole labeled complex binds. However, this approach does allow the easy comparison between different protein extracts. The second approach is the most specific—as each analyte must be recognized by two antibodies to provide a signal—but suffers from the problem that pairs of antibodies must be identified for each target, and one of these must be specifically labeled. This platform has already been commercialized for a limited number of cytokines and interleukins using rolling-circle amplification as an extremely sensitive detection method ***(134)***, but it is difficult to imagine being applied on a genomic scale. The last approach appears to be extremely sensitive when enzymatic amplification is used, and is perhaps the simplest conceptually, but the stability of antigens recognized when deposited in solid phase is a concern. Each of these approaches could be used with antibodies derived from biomolecular diversity libraries, if such antibodies have the appropriate stability to be able to remain functional when fixed to solid surfaces (in the first two cases). This field is very much in flux, and all aspects, including binding ligands, surfaces, chemistry, dyes, and arraying and scanning tools, are still under investigation. The major advantage that molecular diversity libraries will bring to this field is the possibility of supplying large numbers of defined monoclonal recombinant antibodies against different targets.

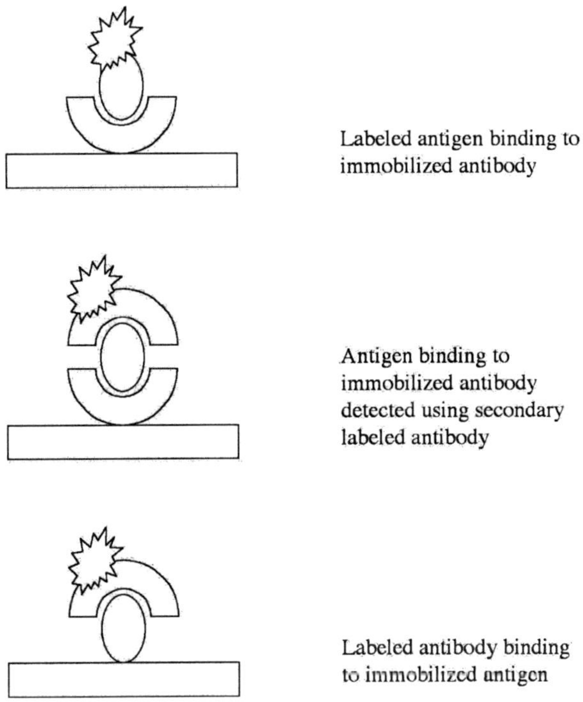

Fig. 1. Three possible detection modalities on antibody chips.

6.2.2. Intracellular Antibodies

In general, the technologies described here provide descriptive information on gene products. Information on function can be obtained if antibodies are expressed intracellularly ***(90)***, and either inhibit the function of a gene of interest ***(135–141)*** or redirect it away from its usual site of action, so that it is no longer able to function ***(142)***. In general, the term "intracellular antibodies" is applied somewhat indiscriminately, but different cellular compartments have very different properties: the cytoplasm and nucleus are reducing, and the endoplasmic reticulum is oxidizing. As a result, the expression challenges vary considerably. It is generally not too difficult to express antibodies or their fragments within the endoplasmic reticulum—this is their normal route of production, and many reports of successful intracellular antibody function are within this compartment ***(140,143,144)***. Expression within the cytoplasm or nucleus is far more difficult, and stability within the cytoplasm, rather than affinity, appears to be the most important predictor of intracellular functionality ***(97–99)***, indicating the importance of selecting antibodies from libraries of stable antibodies. Studies on the identification of stable antibody scaffolds ***(89,99,108–110)*** will probably lead to libraries of antibodies in which a high proportion are stable under intracellular conditions. By transferring antibodies selected from such libraries directly into intracellular expression vectors ***(124)***, and transfecting constructs into cell lines, it should be possible to adopt this technology in high throughput. Consideration must be given to the assay used to evaluate the function of antibodies expressed intracellularly, and in general, this technology will be most powerful when specific pathways are examined for the effect of the inhibition of individual components of that pathway. Such inhibition could be specifically directed toward putative components of such pathways (e.g., caspase proteins in apoptosis), or could be carried out with no preconceptions, by selecting intracellular antibodies that are able to inhibit specific pathways and then identifying the targets recognized. Pathways with defined selectable end points such as apoptotic pathways or drug sensitivities, are particularly appropriate for such broad studies. For other uses of intracellular antibodies, consideration must be given to the phenotype under examination.

7. Conclusion

The study of proteomes will be considerably enhanced by the availability of wide panels of well-characterized binding ligands, with the potential to provide information on distribution, post-translational modifications, and complex formation and function. Antibodies are the best-characterized binding ligands, and in vitro methods to select and modify them have been extensively developed. As a result, antibodies are likely to be the most important binding ligands of this type, although other binding ligands are under development and may offer advantages. The selection of such ligands will almost certainly require multiple approaches, depending upon the state of progress of genome characterization, and their use will require efficient production methods. Consideration should be made to the storage, access, and dissemination of the information acquired, and web-based databases are likely to be important (*see also* Chapter 1).

References

1. Venter, J. C., Adams, M. D., Myers, E. W., Li, P. W., Mural, R. J., Sutton, G. G., et al. (2001) The sequence of the human genome. *Science* **291,** 1304–1351.
2. Lander, E. S., Linton, L. M., Birren, B., Nusbaum, C., Zody, M. C., Baldwin, J., et al. (2001) Initial sequencing and analysis of the human genome. *Nature* **409,** 860–921.
3. Goffeau, A., Barrell, B. G., Bussey, H., Davis, R. W., Dujon, B., Feldmann, H., et al. (1996) Life with 6000 genes. *Science* **274, 546,** 563–567.
4. Ross-Macdonald, P., Coelho, P. S., Roemer, T., Agarwal, S., Kumar, A., Jansen, R., et al. (1999) Large-scale analysis of the yeast genome by transposon tagging and gene disruption. *Nature* **402,** 413–418.
5. Zhu, H., Bilgin, M., Bangham, R., Hall, D., Casamayor, A., Bertone, P., et al. (2001) Global analysis of protein activities using proteome chips. *Science* **293,** 2101–2105.

6. Kumar, A., Agarwal, S., Heyman, J. A., Matson, S., Heidtman, M., Piccirillo, S., et al. (2002) Subcellular localization of the yeast proteome. *Genes Dev.* **16,** 707–719.
7. Tong, A. H., Drees, B., Nardelli, G., Bader, G. D., Brannetti, B., Castagnoli, L., et al. (2002) A combined experimental and computational strategy to define protein interaction networks for peptide recognition modules. *Science* **295,** 321–324.
8. Uetz, P., Giot, L., Cagney, G., Mansfield, T. A., Judson, R. S., Knight, J. R., et al. (2000) A comprehensive analysis of protein-protein interactions in Saccharomyces cerevisiae. *Nature* **403,** 623–627.
9. Schwikowski, B., Uetz, P., and Fields, S. (2000) A network of protein-protein interactions in yeast. *Nat. Biotechnol.* **18,** 1257–1261.
10. Ho, Y., Gruhler, A., Heilbut, A., Bader, G. D., Moore, L., Adams, S. L., et al. (2002) Systematic identification of protein complexes in Saccharomyces cerevisiae by mass spectrometry. *Nature* **415,** 180–183.
11. Gavin, A. C., Bosche, M., Krause, R., Grandi, P., Marzioch, M., Bauer, A., et al. (2002) Functional organization of the yeast proteome by systematic analysis of protein complexes. *Nature* **415,** 141–147.
12. Evan, G. I., Lewis, G. K., Ramsay, G., and Bishop, J. M. (1985) Isolation of monoclonal antibodies specific for human c-myc proto-oncogene product. *Mol. Cell Biol.* **5,** 3610–3616.
13. Rigaut, G., Shevchenko, A., Rutz, B., Wilm, M., Mann, M., and Seraphin, B. (1999) A generic protein purification method for protein complex characterization and proteome exploration. *Nat. Biotechnol.* **17,** 1030–1032.
14. Anderer, F. A. and Schlumberger, H. D. (1966) Cross-reactions of antisera against the terminal amino acid and dipeptide of tobacco mosaic virus. *Biochim. Biophys. Acta* **115,** 222–224.
15. Tang, D.-C., DeVit, M., and Johnston, S. A. (1992) Genetic immunization is a simple method for eliciting an immune respnse. *Nature* **356,** 152–154.
16. Kohler, G. and Milstein, C. (1975) Continuous cultures of fused cells secreting antibody of predefined specificity. *Nature* **256,** 495–497.
17. Cwirla, S. E., Peters, E. A., Barrett, R. W., and Dower, W. J. (1990) Peptides on phage: a vast library of peptides for identifying ligands. *Proc. Natl. Acad. Sci. USA* **87,** 6378–6382.
18. Scott, J. K. and Smith, G. P. (1990) Searching for peptide ligands with an epitope library. *Science* **249,** 386–390.
19. Kay, B. K., Adey, N. B., He, Y. S., Manfredi, J. P., Mataragnon, A. H., and Fowlkes, D. M. (1993) An M13 phage library displaying random 38-amino-acid peptides as a source of novel sequences with affinity to selected targets. *Gene* **128,** 59–65.
20. McCafferty, J., Griffiths, A. D., Winter, G., and Chiswell, D. J. (1990) Phage antibodies: filamentous phage displaying antibody variable domains. *Nature* **348,** 552–554.
21. Huie, M. A., Cheung, M. C., Muench, M. O., Becerril, B., Kan, Y. W., and Marks, J. D. (2001) Antibodies to human fetal erythroid cells from a nonimmune phage antibody library. *Proc. Natl. Acad. Sci. USA* **98,** 2682–2687.

22. Vaughan, T. J., Williams, A. J., Pritchard, K., Osbourn, J. K., Pope, A. R., Earnshaw, J. C., et al. (1996) Human antibodies with sub-nanomolar affinities isolated from a large non-immunised phage display library. *Nat. Biotechnol.* **14,** 309–314.
23. Sblattero, D. and Bradbury, A. (2000) Exploiting recombination in single bacteria to make large phage antibody libraries. *Nat. Biotechnol.* **18,** 75–80.
24. de Haard, H. J., van Neer, N., Reurs, A., Hufton, S. E., Roovers, R. C., Henderikx, P., et al. (1999) A large non-immunized human Fab fragment phage library that permits rapid isolation and kinetic analysis of high affinity antibodies. *J. Biol. Chem.* **274,** 18,218–18,230.
25. Sheets, M. D., Amersdorfer, P., Finnern, R., Sargent, P., Lindqvist, E., Schier, R., et al. (1998) Efficient construction of a large nonimmune phage antibody library; the production of panels of high affinity human single-chain antibodies to protein antigens. *Proc. Natl. Acad. Sci. USA* **95,** 6157–6162.
26. Knappik, A., Ge, L., Honegger, A., Pack, P., Fischer, M., Wellnhofer, G., et al. (2000) Fully Synthetic Human Combinatorial Antibody Libraries (HuCAL) Based on Modular Consensus Frameworks and CDRs Randomized with Trinucleotides. *J. Mol. Biol.* **296,** 57–86.
27. Lowman, H. B. and Wells, J. A. (1993) Affinity maturation of human growth hormone by monovalent phage display. *J. Mol. Biol.* **234,** 564–578.
28. Lowman, H. B. (1997) Bacteriophage display and discovery of peptide leads for drug development. *Annu. Rev. Biophys. Biomol. Struct.* **26,** 401–424.
29. Hoogenboom, H. R., de Bruine, A. P., Hufton, S. E., Hoet, R. M., Arends, J. W., and Roovers, R. C. (1998) Antibody phage display technology and its applications. *Immunotechnology* **4,** 1–20.
30. Bradbury, A. and Cattaneo, A. (1995) The use of phage display in neurobiology. *Trends Neurosci.* **18,** 243–249.
31. Francisco, J. A., Earhart, C. F., and Georgiou, G. (1992) Transport and anchoring of β-lactamase to the external surfaceof Escherichia coli. *Proc. Natl. Acad. Sci. USA* **89,** 2713–2717.
32. Boder, E. T. and Wittrup, K. D. (1997) Yeast surface dispay for screening combinatorial polypeptide libraries. *Nat. Biotechnol.* **15,** 553–557.
33. Hanes, J. and Plückthun, A. (1997) In vitro selection and evolution of functional proteins by using ribosome display. *Proc. Natl. Acad. Sci. USA* **94,** 4937–4942.
34. Roberts, R. W. and Szostak, J. W. (1997) RNA-peptide fusions for the in vitro selection of peptides and proteins. *Proc. Natl. Acad. Sci. USA* **94,** 12,297–12,302.
35. Fields, S. and Song, O. (1989) A novel genetic system to detect protein-protein interactions. *Nature* **340,** 245–246.
36. Pelletier, J. N., Arndt, K. M., Pluckthun, A., and Michnick, S. W. (1999) An in vivo library-versus-library selection of optimized protein-protein interactions. *Nat. Biotechnol.* **17,** 683–690.
37. Visintin, M., Tse, E., Axelson, H., Rabbitts, T. H., and Cattaneo, A. (1999) Selection of antibodies for intracellular function using a two-hybrid in vivo system. *Proc. Natl. Acad. Sci. USA* **96,** 11,723–11,728.
38. Auf der Maur, A., Escher, D., and Barberis, A. (2001) Antigen-independent selection of stable intracellular single-chain antibodies. *FEBS Lett.* **508,** 407–412.

39. Mossner, E., Koch, H., and Pluckthun, A. (2001) Fast selection of antibodies without antigen purification: adaptation of the protein fragment complementation assay to select antigen-antibody pairs. *J. Mol. Biol.* **308,** 115–122.
40. Auf der Maur, A., Zahnd, C., Fischer, F., Spinelli, S. L., Honeggar, A., Cambillau, C., et al. (2002) Direct in vivo screening of intrabody libraries constructed on a highly stable single-chain framework. *J. Biol. Chem.* in press.
41. Skerra, A. (2000) Engineered protein scaffolds for molecular recognition. *J. Mol. Recognit.* **13,** 167–187.
42. de Wildt, R. M., Mundy, C. R., Gorick, B. D., and Tomlinson, I. M. (2000) Antibody arrays for high-throughput screening of antibody-antigen interactions. *Nat. Biotechnol.* **18,** 989–994.
43. Daugherty, P. S., Chen, G., Olsen, M. J., Iverson, B. L., and Georgiou, G. (1998) Antibody affinity maturation using bacterial surface display. *Protein Eng.* **11,** 825–832.
44. Boder, E. T., Midelfort, K. S., and Wittrup, K. D. (2000) Directed evolution of antibody fragments with monovalent femtomolar antigen-binding affinity. *Proc. Natl. Acad. Sci. USA* **97,** 10,701–10,705.
45. Shusta, E. V., Kieke, M. C., Parke, E., Kranz, D. M., and Wittrup, K. D. (1999) Yeast polypeptide fusion surface display levels predict thermal stability and soluble secretion efficiency. *J. Mol. Biol.* **292,** 949–956.
46. Persic, L., Horn, I. R., Rybak, S., Cattaneo, A., Hoogenboom, H. R., and Bradbury, A. (1999) Single-chain variable fragments selected on the 57–76 p21Ras neutralising epitope from phage antibody libraries recognise the native protein. *FEBS Lett.* **443,** 112–116.
47. Siegel, R. W., Allen, B., Pavlik, P., Marks, J. D., and Bradbury, A. (2000) Mass spectral analysis of a protein complex using single-chain antibodies selected on a peptide target: applications to functional genomics. *J. Mol. Biol.* **302,** 285–293.
48. Liu, B., Huang, L., Sihlbom, C., Burlingame, A., and Marks, J. D. (2002) Towards proteome-wide production of monoclonal antibody by phage display. *J. Mol. Biol.* **315,** 1063–1073.
49. Waldo, G. S., Standish, B. M., Berendzen, J., and Terwilliger, T. C. (1999) Rapid protein-folding assay using green fluorescent protein. *Nat. Biotechnol.* **17,** 691–695.
50. Maxwell, K. L., Mittermaier, A. K., Forman-Kay, J. D., and Davidson, A. R. (1999) A simple in vivo assay for increased protein solubility. *Protein Sci.* **8,** 1908–1911.
51. Lesley, S. A. (2001) High-throughput proteomics: protein expression and purification in the postgenomic world. *Protein Expr. Purif.* **22,** 159–164.
52. Lin Cereghino, G. P., Sunga, A. J., Lin Cereghino, J., and Cregg, J. M. (2001) Expression of foreign genes in the yeast Pichia pastoris. *Genet. Eng. (NY)* **23,** 157–169.
53. Rosenfeld, S. A. (1999) Use of Pichia pastoris for expression of recombinant proteins. *Methods Enzymol.* **306,** 154–169.
54. Jung, E. and Williams, K. L. (1997) The production of recombinant glycoproteins with special reference to simple eukaryotes including Dictyostelium discoideum. *Biotechnol. Appl. Biochem.* **25** (Pt 1), 3–8.

55. Punt, P. J., van Biezen, N., Conesa, A., Albers, A., Mangnus, J., and van den Hondel, C. (2002) Filamentous fungi as cell factories for heterologous protein production. *Trends Biotechnol.* **20,** 200–206.
56. Coleman, T. A., Parmelee, D., Thotakura, N. R., Nguyen, N., Burgin, M., Gentz, S., et al. (1997) Production and purification of novel secreted human proteins. *Gene* **190,** 163–171.
57. Kim, D. M., Kigawa, T., Choi, C. Y., and Yokoyama, S. (1996) A highly efficient cell-free protein synthesis system from Escherichia coli. *Eur. J. Biochem.* **239,** 881–886.
58. Sawasaki, T., Hasegawa, Y., Tsuchimochi, M., Kamura, N., Ogasawara, T., Kuroita, T., et al. (2002) A bilayer cell-free protein synthesis system for high-throughput screening of gene products. *FEBS Lett.* **514,** 102–105.
59. Madin, K., Sawasaki, T., Ogasawara, T., and Endo, Y. (2000) A highly efficient and robust cell-free protein synthesis system prepared from wheat embryos: plants apparently contain a suicide system directed at ribosomes. *Proc. Natl. Acad. Sci. USA* **97,** 559–564.
60. Anderer, F. A. and Schlumberger, H. D. (1965) [Cross reactions of antiserums against heterologous terminal amino acid sequences and two strains of the tobacco mosaic virus]. *Z. Naturforsch. B* **20,** 564–568.
61. Van Regenmortel, M., Briand, J., Muller, S., and Plaué, S. (1988) *Synthetic polypeptides as antigens. Laboratory Techniques in Biochemistry* and *Molecular Biology* (Burdon, R. and van Knippenberg, P., eds.), 19, Elsevier, Amsterdam, the Netherland.
62. Van Regenmortel, M. H. and Pellequer, J. L. (1994) Predicting antigenic determinants in proteins: looking for unidimensional solutions to a three-dimensional problem? *Pept. Res.* **7,** 224–228.
63. Jameson, B. A. and Wolf, H. (1988) The antigenic index: a novel algorithm for predicting antigenic determinants. *Comput. Appl. Biosci.* **4,** 181–186.
64. Griffiths, A. D., Williams, S. C., Hartley, O., Tomlinson, I. M., Waterhouse, P., Crosby, W. L., Kontermann, R. E., Jones, P. T., Low, N. M., Alison, T. J., Prospero, T. D., Hoogenboom, H. R., Nissim, A., Cox, J.P.L., Harrison, J. L., Zaccolo, M., Gherardi, E., et al. (1994) Isolation of high affinity human antibodies directly from large synthetic repertoires. *EMBO J.* **13,** 3245–3260.
65. Sblattero, D., Lou, J., Marzari, R., and Bradbury, A. (2001) In vivo recombination as a tool to generate molecular diversity in phage antibody libraries. *J. Biotechnol.* **74,** 303–315.
66. Yang, W. P., Green, K., Pinz-Sweeney, S., Briones, A. T., Burton, D. R., and Barbas, C. F.R. (1995) CDR walking mutagenesis for the affinity maturation of a potent human anti-HIV-1 antibody into the picomolar range. *J. Mol. Biol.* **254,** 392–403.
67. Schier, R., McCall, A., Adams, G. P., Marshall, K. W., Merritt, H., Yim, M., et al. (1996) Isolation of picomolar affinity anti-c-erB-2 single-chain Fv by molecular evolution of the complementarity determining regions in the center of the antibody binding site. *J. Mol. Biol.* **263,** 551–567.

68. Thompson, J., Pope, T., Tung, J. S., Chan, C., Hollis, G., Mark, G., et al. (1996) Affinity maturation of a high-affinity human monoclonal antibody against the third hypervariable loop of human immunodeficiency virus: use of phage display to improve affinity and broaden strain reactivity. *J. Mol. Biol.* **256,** 77–88.
69. Pack, P. and Pluckthun, A. (1992) Miniantibodies: use of amphipathic helices to produce functional, flexibly linked dimeric FV fragments with high avidity in Escherichia coli. *Biochemistry* **31,** 1579–1584.
70. Lou, J., Marzari, R., Verzillo, V., Ferrero, F., Pak, D., Sheng, M., et al. (2001) Antibodies in haystacks: how selection strategy influences the outcome of selection from molecular diversity libraries. *J. Immunol. Methods* **253,** 233–242.
71. Hawkins, R. E., Russell, S. J., and Winter, G. (1992) Selection of phage antibodies by binding affinity: mimicking affinity maturation. *J. Mol. Biol.* **226,** 889–896.
72. Kipriyanov, S. M., Moldenhauer, G., Martin, A. C., Kupriyanova, O. A., and Little, M. (1997) Two amino acid mutations in an anti-human CD3 single chain Fv antibody fragment that affect the yield on bacterial secretion but not the affinity. *Protein Eng.* **10,** 445–453.
73. Duenas, M., Ayala, M., Vazquez, J., Ohlin, M., Soderlind, E., Borrebaeck, C. A., et al. (1995) A point mutation in a murine immunoglobulin V-region strongly influences the antibody yield in Escherichia coli. *Gene* **158,** 61–66.
74. Coia, G., Ayres, A., Lilley, G. G., Hudson, P. J., and Irving, R. A. (1997) Use of mutator cells as a means for increasing production levels of a recombinant antibody directed against Hepatitis B. *Gene* **201,** 203–209.
75. Krebs, B., Rauchenberger, R., Reiffert, S., Rothe, C., Tesar, M., Thomassen, E., et al. (2001) High-throughput generation and engineering of recombinant human antibodies. *J. Immunol. Methods* **254,** 67–84.
76. Desiderio, A., Franconi, R., Lopez, M., Villani, M. E., Viti, F., Chiaraluce, R., et al. (2001) A semi-synthetic repertoire of intrinsically stable antibody fragments derived from a single-framework scaffold. *J. Mol. Biol.* **310,** 603–615.
77. Holt, L. J., Bussow, K., Walter, G., and Tomlinson, I. M. (2000) By-passing selection: direct screening for antibody-antigen interactions using protein arrays. *Nucleic Acids Res.* **28,** E72.
78. Lofdahl, S., Guss, B., Uhlen, M., Philipson, L., and Lindberg, M. (1983) Gene for staphylococcal protein A. *Proc. Natl. Acad. Sci. USA* **80,** 697–701.
79. Bjorck, L. (1988) Protein L. A novel bacterial cell wall protein with affinity for Ig L chains. *J. Immunol.* **140,** 1194–1197.
80. Hillson, J. L., Karr, N. S., Oppliger, I. R., Mannik, M., and Sasso, E. H. (1993) The structural basis of germline-encoded V_H3 immunoglobulin binding to staphylococcal protein A. *J. Exp. Med.* **178,** 331–336.
81. Nilson, B. H., Solomon, A., Björck, L., and Akerström, B. (1992) Protein L from Peptostreptococcus magnus binds to the kappa light chain variable domain. *J. Biol. Chem.* **267,** 2234–2239.
82. Boder, E. T. and Wittrup, K. D. (2000) Yeast surface display for directed evolution of protein expression, affinity, and stability. *Methods Enzymol.* **328,** 430–444.

83. Nemoto, N., Miyamoto-Sato, E., Husimi, Y., and Yanagawa, H. (1997) In vitro virus: bonding of mRNA bearing puromycin at the 3′-terminal end to the C-terminal end of its encoded protein on the ribosome in vitro. *FEBS Lett.* **414,** 405–408.
84. He, M. and Taussig, M. J. (1997) Antibody-ribosome-mRNA (ARM) complexes as efficient selection particles for in vitro display and evolution of antibody combining sites. *Nucleic Acids Res.* **25,** 5132–5134.
85. Hanes, J., Schaffitzel, C., Knappik, A., and Pluckthun, A. (2000) Picomolar affinity antibodies from a fully synthetic naive library selected and evolved by ribosome display. *Nat. Biotechnol.* **18,** 1287–1293.
86. Hanes, J., Jermutus, L., and Pluckthun, A. (2000) Selecting and evolving functional proteins in vitro by ribosome display. *Methods Enzymol.* **328,** 404–430.
87. Stemmer, W. P. (1994) DNA shuffling by random fragmentation and reassembly: in vitro recombination for molecular evolution. *Proc. Natl. Acad. Sci. USA* **91,** 10,747–10,751.
88. Mitra, R. D. and Church, G. M. (1999) In situ localized amplification and contact replication of many individual DNA molecules. *Nucleic Acids Res.* **27,** e34.
89. Visintin, M., Settanni, G., Maritan, A., Graziosi, S., Marks, J. D., and Cattaneo, A. (2002) The intracellular antibody capture technology (IACT): towards a consensus sequence for intracellular antibodies. *J. Mol. Biol.* **317,** 73–83.
90. Cattaneo, A. and Biocca, S. (1997) *Intracellular antibodies: developments and applications.* Biotechnology Intelligence Unit, Landes Bioscience, distributed by Academic Press, Austin, TX.
91. Joung, J. K., Ramm, E. I., and Pabo, C. O. (2000) A bacterial two-hybrid selection system for studying protein-DNA and protein-protein interactions. *Proc. Natl. Acad. Sci. USA* **97,** 7382–7387.
92. Karimova, G., Pidoux, J., Ullmann, A., and Ladant, D. (1998) A bacterial two-hybrid system based on a reconstituted signal transduction pathway. *Proc. Natl. Acad. Sci. USA* **95,** 5752–5756.
93. Dove, S. L. and Hochschild, A. (1998) Conversion of the omega subunit of Escherichia coli RNA polymerase into a transcriptional activator or an activation target. *Genes Dev.* **12,** 745–754.
94. Dove, S. L., Joung, J. K., and Hochschild, A. (1997) Activation of prokaryotic transcription through arbitrary protein-protein contacts. *Nature* **386,** 627–630.
95. Michnick, S. W. (2001) Exploring protein interactions by interaction-induced folding of proteins from complementary peptide fragments. *Curr. Opin. Struct. Biol.* **11,** 472–477.
96. Galarneau, A., Primeau, M., Trudeau, L. E., and Michnick, S. W. (2002) beta-Lactamase protein fragment complementation assays as in vivo and in vitro sensors of protein protein interactions. *Nat. Biotechnol.* **20,** 619–622.
97. Worn, A., Auf der Maur, A., Escher, D., Honegger, A., Barberis, A., and Pluckthun, A. (2000) Correlation between in vitro stability and in vivo performance of anti-GCN4 intrabodies as cytoplasmic inhibitors. *J. Biol. Chem.* **275,** 2795–2803.
98. Zhu, Q., Zeng, C., Huhalov, A., Yao, J., Turi, T. G., Danley, D., et al. (1999) Extended half-life and elevated steady-state level of a single-chain Fv intrabody

are critical for specific intracellular retargeting of its antigen, caspase-7. *J. Immunol. Methods* **231,** 207–222.
99. Ohage, E. and Steipe, B. (1999) Intrabody construction and expression. I. The critical role of V_L domain stability. *J. Mol. Biol.* **291,** 1119–1128.
100. Worn, A. and Pluckthun, A. (2001) Stability engineering of antibody single-chain Fv fragments. *J. Mol. Biol.* **305,** 989–1010.
101. Willuda, J., Honegger, A., Waibel, R., Schubiger, P. A., Stahel, R., Zangemeister-Wittke, U., and Pluckthun, A. (1999) High thermal stability is essential for tumor targeting of antibody fragments: engineering of a humanized anti-epithelial glycoprotein-2 (epithelial cell adhesion molecule) single-chain Fv fragment. *Cancer Res.* **59,** 5758–5767.
102. Jung, S. and Pluckthun, A. (1997) Improving in vivo folding and stability of a single-chain Fv antibody fragment by loop grafting. *Protein Eng.* **10,** 959–966.
103. Proba, K., Worn, A., Honegger, A., and Pluckthun, A. (1998) Antibody scFv fragments without disulfide bonds made by molecular evolution. *J. Mol. Biol.* **275,** 245–253.
104. Jung, S., Honegger, A., and Pluckthun, A. (1999) Selection for improved protein stability by phage display. *J. Mol. Biol.* **294,** 163–180.
105. Jermutus, L., Honegger, A., Schwesinger, F., Hanes, J., and Pluckthun, A. (2001) Tailoring in vitro evolution for protein affinity or stability. *Proc. Natl. Acad. Sci. USA* **98,** 75–80.
106. Martineau, P., Jones, P., and Winter, G. (1998) Expression of an antibody fragment at high levels in the bacterial cytoplasm. *J. Mol. Biol.* **280,** 117–127.
107. Kabat, E. A., Wu, T. T., Perry, H. M., Gottesman, K. S., and Foeller, C. (1991) *Sequences of Proteins of Immunological Interest,* 5th ed., 3 vols, U.S. Department of Health and Human Services, U.S. Government Printing Office.
108. Ohage, E. C., Wirtz, P., Barnikow, J., and Steipe, B. (1999) Intrabody construction and expression. II. A synthetic catalytic Fv fragment. *J. Mol. Biol.* **291,** 1129–1134.
109. Steipe, B., Schiller, B., Pluckthun, A., and Steinbacher, S. (1994) Sequence statistics reliably predict stabilizing mutations in a protein domain. *J. Mol. Biol.* **240,** 188–192.
110. Wirtz, P. and Steipe, B. (1999) Intrabody construction and expression III: engineering hyperstable V(H) domains. *Protein Sci.* **8,** 2245–2250.
111. de Wildt, R. M., Tomlinson, I. M., Ong, J. L., and Holliger, P. (2002) Isolation of receptor-ligand pairs by capture of long-lived multivalent interaction complexes. *Proc. Natl. Acad. Sci. USA* **99,** 8530–8535.
112. Marks, J. D., Hoogenboom, H. R., Bonnert, T. P., McCafferty, J., Griffiths, A. D., and Winter, G. (1991) By-passing immunization. Human antibodies from V-gene libraries displayed on phage. *J. Mol. Biol.* **222,** 581–597.
113. Robinson, W. H., DiGennaro, C., Hueber, W., Haab, B. B., Kamachi, M., Dean, E. J., et al. (2002) Autoantigen microarrays for multiplex characterization of autoantibody responses. *Nat. Med.* **8,** 295–301.
114. Haab, B. B. (2001) Advances in protein microarray technology for protein expression and interaction profiling. *Curr. Opin. Drug Discov. Devel.* **4,** 116–123.

115. Haab, B. B., Dunham, M. J., and Brown, P. O. (2001) Protein microarrays for highly parallel detection and quantitation of specific proteins and antibodies in complex solutions. *Genome. Biol.* **2,** RESEARCH0004.
116. Hanke, T., Szawlowski, P., and Randall, R. E. (1992) Construction of solid matrix-antibody-antigen complexes containing simian immunodeficiency virus p27 using tag-specific monoclonal antibody and tag-linked antigen. *J. Gen. Virol.* **73,** 653–660.
117. Pluckthun, A. and Pack, P. (1997) New protein engineering approaches to multivalent and bispecific antibody fragments. *Immunotechnology* **3,** 83–105.
118. Neri, D., Momo, M., Prospero, T., and Winter, G. (1995) High-affinity antigen binding by chelating recombinant antibodies (CRAbs) *J. Mol. Biol.* **246,** 367–373.
119. Lindner, P., Bauer, K., Krebber, A., Nieba, L., Kremmer, E., Krebber, C., et al. (1997) Specific detection of his-tagged proteins with recombinant anti-His tag scFv-phosphatase or scFv-phage fusions. *Biotechniques* **22,** 140–149.
120. Muller, B. H., Chevrier, D., Boulain, J. C., and Guesdon, J. L. (1999) Recombinant single-chain Fv antibody fragment-alkaline phosphatase conjugate for one-step immunodetection in molecular hybridization. *J. Immunol. Methods* **227,** 177–185.
121. Griep, R. A., van Twisk, C., Kerschbaumer, R. J., Harper, K., Torrance, L., Himmler, G., et al. (1999) pSKAP/S: An expression vector for the production of single-chain Fv alkaline phosphatase fusion proteins. *Protein Expr. Purif.* **16,** 63–69.
122. Bourin, P., Servat, A., Lataillade, J. J., Goyffon, M., Vaux, D., and Billiald, P. (2000) Immunolabeling of CD3-positive lymphocytes with a recombinant single-chain antibody/alkaline phosphatase conjugate. *Biol. Chem.* **381,** 173–178.
123. Krebber, A., Bornhauser, S., Burmester, J., Honeggar, A., Willuda, J., H. R., B., and Pluckthun, A. (1997) Reliable cloning of functional antibody variable domains from hybridomas and spleen cell repertoires employing a reengineered phage display system. *J. Immunol. Methods* **201,** 35–55.
124. Persic, L., Righi, M., Roberts, A., Hoogenboom, H. R., Cattaneo, A., and Bradbury, A. (1997) Targeting vectors for intracellular immunisation. *Gene* **187,** 1–8.
125. Persic, L., Roberts, A., Wilton, J., Cattaneo, A., Bradbury, A., and Hoogenboom, H. (1997) An integrated vector system for the eukaryotic expression of antibodies or their fragments after selection from phage display libraries. *Gene* **187,** 9–18.
126. Reboul, J., Vaglio, P., Tzellas, N., Thierry-Mieg, N., Moore, T., Jackson, C., et al. (2001) Open-reading-frame sequence tags (OSTs) support the existence of at least 17,300 genes in C. elegans. *Nat. Genet.* **27,** 332–336.
127. Hartley, J. L., Temple, G. F., and Brasch, M. A. (2000) DNA cloning using in vitro site-specific recombination. *Genome Res.* **10,** 1788–1795.
128. Siegel, R. W., Jain, R., and Bradbury, A. (2001) Using an in vivo phagemid system to identify non-compatible loxP sequences. *FEBS Lett.* **505,** 467–473.
129. Lee, G. and Saito, I. (1998) Role of nucleotide sequences of loxP spacer region in Cre-mediated recombination. *Gene* **216,** 55–65.

130. Hoess, R. H., Wierzbicki, A., and Abremski, K. (1986) The role of the loxP spacer region in P1 site-specific recombination. *Nucleic Acids Res.* **14,** 2287–2300.
131. Miller, J. C., Butler, E. B., Teh, B. S., and Haab, B. B. (2001) The application of protein microarrays to serum diagnostics: prostate cancer as a test case. *Dis. Markers* **17,** 225–234.
132. MacBeath, G. and Schreiber, S. L. (2000) Printing proteins as microarrays for high-throughput function determination [see comments]. *Science* **289,** 1760–1763.
133. Paweletz, C. P., Charboneau, L., Bichsel, V. E., Simone, N. L., Chen, T., Gillespie, J. W., et al. (2001) Reverse phase protein microarrays which capture disease progression show activation of pro-survival pathways at the cancer invasion front. *Oncogene* **20,** 1981–1989.
134. Schweitzer, B. and Kingsmore, S. F. (2002) Measuring proteins on microarrays. *Curr. Opin. Biotechnol.* **13,** 14–19.
135. Biocca, S., Pierandrei-Amaldi, P., and Cattaneo, A. (1993) Intracellular expression of anti-p21ras single chain Fv fragments inhibits meiotic maturation of Xenopus oocytes. *B.B.R.C.* **197,** 422–427.
136. Gargano, N., Biocca, S., Bradbury, A., and Cattaneo, A. (1996) Human recombnant antibody fragments neutralising human immunodeficiency virus type 1 reverse transcriptase provide an experimental basis for the structural classification of the DNA polymerase family. *J. Virol.* **70,** 7706–7712.
137. Gargano, N. and Cattaneo, A. (1997) Inhibition of murine leukaemia virus retrotranscription by the intracellular expression of a phage-derived anti-reverse transcriptase antibody fragment. *J. Gen. Virol.* **78,** 2591–2599.
138. Cardinale, A., Lener, M., Messina, S., Cattaneo, A., and Biocca, S. (1998) The mode of action of the Y13-259 scFv fragment when expressed intracellularly in mammalian cells. *FEBS Letts.* (In press.)
139. Mhashilkar, A. M., Bagley, J., Chen, S. Y., Szilvay, A. M., Helland, D. G., and Marasco, W. A. (1995) Inhibition of HIV-1 Tat-mediated LTR transactivation and HIV-1 infection by anti-Tat single chain intrabodies. *EMBO J.* **14,** 1542–1551.
140. Richardson, J. H., Sodroski, J. G., Waldmann, T. A., and Marasco, W. A. (1995) Phenotypic knockout of the high-affinity human interleukin 2 receptor by intracellular single-chain antibodies against the alpha subunit of the receptor. *Proc. Natl. Acad. Sci. USA* **92,** 3137–3141.
141. Werge, T. M., Baldari, C. T., and Telford, J. L. (1994) Intracellular single chain Fv antibody inhibits Ras activity in T-cell antigen receptor stimulated Jurkat cells. *FEBS Lett.* **3517,** 393–396.
142. Lener, M., Horn, I. R., Cardinale, A., Messina, S., Nielsen, U. B., Rybak, S. M., et al. (2000) Diverting a protein from its cellular location by intracellular antibodies. The case of p21Ras. *Eur. J. Biochem.* **267,** 1196–1205.
143. Marasco, W. A., Haseltine, W. A., and Chen, S. Y. (1993) Design, intracellular expression, and activity of a human anti-human immunodeficiency virus type 1 gp120 single-chain antibody. *Proc. Natl. Acad. Sci. USA* **90,** 7889–7893.
144. Mhashilkar, A. M., Doebis, C., Seifert, M., Busch, A., Zani, C., Soo Hoo, J., et al. (2002) Intrabody-mediated phenotypic knockout of major histocompatibility

complex class I expression in human and monkey cell lines and in primary human keratinocytes. *Gene Ther.* **9,** 307–319.

145. Feldhaus, M. J., Siegel, R. W., Opresko, L. K., et al. (2003) Flow-cytometric isolation of human antibodies from a nonimmune *Saccharomyces cerevisiae* surface display library. *Nat. Biotechnol.* **21**(2), 163–170.

32

Targeting of Antibodies Using Aptamers

Sotiris Missailidis

1. Introduction

Antibodies are traditionally viewed as targeting entities that are used to specifically recognize and target other molecules, antigens, or receptors. However, there are occasions in which cognate ligands are generated against the antibodies themselves. The number of useful applications that rely on the recognition and targeting of antibodies extends to various diseases, including cancer, inflammation, and autoimmune disorders. For example, anti-idiotypic antibodies have been generated to target surface immunoglobulins on neoplastic B-lymphocytes and plasma cells for the treatment of lymphomas and leukaemias *(1,2)*. Such idiotypic determinants are tumor-specific, and have been exploited in a number of immunotherapeutic approaches, either in the form of vaccines *(3–11)* or as the target of anti-idiotypic antibodies *(12–24)*. The latter, in particular, have been used successfully both in the unconjugated mono- and bi-specific forms *(13,14,17)*, or as conjugates to other agents such as interleukin-2 *(18–20)*, cytotoxic drugs *(23)*, or radioisotopes *(12,22)*. In addition to haematological malignancies, antibodies are also implicated in autoimmune disorders and transplant rejection, and they could become possible targets in the management of these conditions. On the other hand, antigen-mimics, whether structural, functional or both, can be produced against a target antibody. Such mimics have been used in raising antibodies *(25)* and as anti-inflammatory and anti-tumor agents *(26)*. In addition, RNA aptamers have recently been used as antigen mimics to elude patient autoantibodies from binding to acetylcholine receptors in the control of myasthenia gravis *(27)*. Finally, molecules with binding specificity for antibodies could also be used in the generation of immunoaffinity matrices for the purification of antibodies. To fulfill the promise of antibody targeting, significant interest has emerged in the generation of peptide ligands against antibody targets using phage-display gen-

From: *Methods in Molecular Biology, Vol. 248: Antibody Engineering: Methods and Protocols*
Edited by: B. K. C. Lo © Humana Press Inc., Totowa, NJ

erated peptide libraries *(28)*. However, the structural freedom of peptides and the resulting entropic cost upon target binding limit the use of peptide libraries in which high-affinity and specificity are required *(29, 30)*. Moreover, amino acids are not interactive with each other in the way that nucleotides are, causing most small peptides to be unstructured in solution, whereas structurally stable proteins are large *(31)*.

Aptamers are a novel and particularly interesting targeting modality, with the ability to bind a variety of targets including proteins, peptides, enzymes, antibodies, various cell-surface receptors, and small organic molecules with sub-nanomolar and even picomolar affinities and great specificity *(32)*. Aptamers are single-stranded DNA or RNA oligonucleotides that vary in size between 25 and 50 bases and are derived from combinatorial libraries through an in vitro selection process known as Systematic Evolution of Ligands through Exponential enrichment (SELEX). As compared to other targeting agents, they offer unique benefits because they bind with high affinity and selectivity, are not immunogenic or toxic and have good circulation clearance, are easily and quickly synthesized using in vitro techniques, and robust, stable and consistent, with no batch-to-batch variation *(33)*. These characteristics make them extremely attractive as alternatives to peptides *(29,30)* for use in assays, or as diagnostic *(34)* and therapeutic agents. Considerable success in the use of aptamers as inhibitors of cellular pathways has resulted in clinical trials of aptamers as angiogenic inhibitors for the treatment of cancer *(35)*. Furthermore, current research in the development of aptamers against idiotypic antibodies on the surface of B-lymphocytes may lead to the development of novel reagents for the treatment of B-cell lymphomas. These aptamers are specifically selected against B-cell receptors from individual patients, and thus, they play an important role in the design of patient-specific therapies. Finally, the significantly higher affinity presented by aptamers for their targets (K_ds in the sub-nanomolar to picomolar range), compared to those of peptides (K_ds in the micromolar and sub-micromolar range) as well as their increased stability, make them a preferred modality compared to peptide libraries currently in use. Aptamer recognition has been shown to rely strongly on the exceptional propensity of the nucleic acids to assume secondary and tertiary structural elements. The number of possible thermodynamically stable structural variants available for an oligonucleotide sequence is much higher than the number of variants available for a peptide sequence of the same length *(36)*. This is simply based on the ability of nucleotide bases to interact with each other through canonical Watson-Crick as well as unusual base pairing. The existence of oligonucleotide sequences that could assume a myriad of shapes within a random sequence library is the basis for the remarkable success that has been found in generating aptamers to a wide variety of target molecules. This is done by a technique that involves generating a highly diverse library of oligonucleotide sequences and selecting the molecules that contain the structural features required for binding. The sampling of "shape space" in such libraries has proven to be effective for the isolation of aptamer ligands to targets that are not known to interact with nucleic acids physiologically *(37)*. Moreover, antibodies and proteins usually exhibit extensive surfaces with ridges, grooves, projections, and depressions, all covered with numerous H-bond donors and acceptors, making them excellent targets for aptamer selection. In fact, high-affinity aptamers against IgE antibodies have already been selected with implications in allergic diseases *(38)*.

This chapter presents a protocol for the generation of oligonucleotide aptamers, using a variation of the SELEX methodology, as novel targeting entities against antibodies and an alternative to currently used peptide ligands and antigen-mimics.

2. Materials

1. DNA oligonucleotide library with a 25-base degenerate sequence flanked by primer sites on either side to facilitate amplification. The library sequence used here is the following: **5′-GGGAGACAAGAATAAACGCTCAA-(25N)-TTC-GACAGGAGGCTCACAACAGGC-′3,** but any other primer sequence flanking the 25-base degenerate region (25N) could be used. All four bases (A, T, C, and G) are represented at each of the degenerate positions, and both the libraries and primers are HPLC-purified (*see* **Note 1**).
2. Library amplification primers (for library sequence in **step 1**):
   a. Forward primer: 5′-GGGAGACAAGAATAAACGCTCAA-3′;
   b. Reverse primer: 5′-GCCTGTTGTGAGCCTCCTGTCGAA-3′.
3. Target antibody, store frozen in aliquots at –20°C.
4. Sodium carbonate buffer (0.05 *M*), pH 9.6: dissolve 1.59 g of Na_2CO_3 and 2.93 g of $NaHCO_3$ in distilled water to 1 L final volume.
5. Phosphate-buffered saline (PBS): dissolve one PBS tablet (Sigma; Cat. #P4417) in 200 mL of distilled water to yield 0.01 *M* phosphate buffer, 0.0027 *M* KCl, and 0.137 *M* NaCl, pH 7.4. Sterile-filter and store at room temperature.
6. Blocking buffer: PBS containing 0.1% (w/v) casein.
7. Washing buffer: PBS containing 0.05% (v/v) Tween 20.
8. 10X PCR buffer with 15 m*M* $MgCl_2$, deoxynucleotide 5′ triphosphate (dNTP) mix (25 m*M* of each nucleotide) and 15 m*M* $MgCl_2$ solution (MBI Fermentas).
9. Taq DNA polymerase, native form (Sigma; Cat. #D1806).
10. TOPO TA cloning kit (Invitrogen).

3. Methods

The method presented here is for the selection of aptamers against purified antibodies. The procedure is based on a one-pot experiment in which all selection and amplification rounds take place within a single PCR tube and utilizes the antibody's biophysical properties and characteristics *(39)*. Using the methodological approach that permits the rapid identification of an aptamer(s) for an antibody, in the first instance, the antibody is adsorbed directly to the surface of the PCR tube, prior to the addition of the oligonucleotide library. Unbound aptamers are washed away, PCR reagents are added to the tube, and binding aptamers are amplified using PCR. High temperature during the first PCR cycle unfolds the antibody, thus releasing the selected aptamer and allowing for its amplification in successive PCR cycles. The entire process is then repeated 10× to enrich for high-affinity binders. The use of this one-pot procedure eliminates a number of chromatographic or filter separation steps that would otherwise be necessary. In so doing, it provides a rapid isolation of high affinity and selectivity aptamers against antibody molecules.

3.1. Selection of Aptamers Against Antibodies: The One-Pot Experiment

1. Synthesize the aptamer library at a concentration of 40 µ*M* (40 nmols in 1 mL). This would allow for about 20 copies of each possible sequence to be present in the library.
2. Synthesize the forward primer at 1 m*M* (1 µmol in 1 mL) and the reverse primer at 40 µM (40 nmols in 1 mL).

3.1.1. Amplification of Aptamer Library

1. Prepare the following PCR mix in a thin-walled PCR tube:

Aptamer library	100 µL (*see* **Note 2**)
Forward primer	100 nmol (100 µL)
Reverse primer	4 nmol (100 µL)
10X PCR buffer with 15 m*M* $MgCl_2$	40 µL
dNTP mix	40 µL
15 mM $MgCl_2$	18 µL
Taq DNA polymerase, native form	10 U (2 µL)

2. Perform 99 rounds (instrument limit) of PCR at 94°C (1.5 min), 56°C (0.5 min) and 72°C (1.5 min) in a thermocycler. Include a final extension step at 72°C for 0.5 min and soak at 4°C (*see* **Note 3**). Add 5 U (1 µL) of Taq polymerase in the beginning and the second 5 U of Taq at round 50, to avoid enzyme inactivation.

3.1.2. Immobilization of Antibody for Aptamer Selection

1. Dilute the target antibody in 0.05 *M* sodium carbonate buffer, pH 9.6, to a final concentration of 10 µg/mL (*see* **Note 4**) and add 100 µL of the antibody solution to a 500-µL PCR tube. Incubate at 4°C overnight.
2. Remove the excess unbound antibody using a vacuum pump or a pipet. The use of a pipet allows the recovery of excess antibodies for further use, and also enables the quantification of the level of antibody immobilization.
3. Wash the tube 4× with 100 µL of washing buffer.

4. Add 500 μL of blocking buffer to the tube and incubate at room temperature for 1 h. Ensure that the entire antibody-coated surface is covered by the blocking buffer.
5. Wash the tube once with 500 μL of washing buffer.

3.1.3. Aptamer Selection

1. Add the amplified aptamer library (400 μL, from **Subheading 3.1.1.**) to the antibody-coated and blocked PCR tube and incubate at room temperature (~20°C), for 1 h.
2. Discard the library solution and wash the tube once with 500 μL of washing buffer. All unbound sequences will be washed out and only antibody-binding sequences will remain in the PCR tube.
3. Add PCR reagents to the tube as follows:

Forward primer	400 pmol (40 μL)
Reverse primer	4 pmol (10 μL)
10X PCR buffer with 15 m*M* $MgCl_2$	40 μL
dNTP mix	10 μL
15 m*M* $MgCl_2$	4.5 μL
Taq DNA polymerase, native form	7.5 U (1.5 μL)
Water (18Ω)	24 μL

4. Perform 99 rounds (instrument limit) of PCR at 94°C (1.5 min), 56°C (0.5 min) and 72°C (1.5 min) in a thermocycler include a final extension step at 72°C for 0.5 min and soak at 4°C (*see* **Note 5**).
5. Repeat **steps 2–4** 10×. In each round adsorb the antibody on a new PCR tube, add to it the PCR products of the reaction, incubate, discard the solution, add PCR reagents, and amplify. This allows for affinity maturation through competitive binding of the continuously amplified number of aptamers to the limited number of antibody molecules. In each round, the higher-affinity binding aptamers are enriched and thus displace the nonspecific binders or those that bind with lower affinity. At the end of the selection rounds, only one or two sequences with similar affinities remain.
6. Clone the PCR products from the last round using the TOPO TA cloning kit, according to the manufacturer's instructions. Please note that a standard PCR amplification procedure with equal amounts of both primers should be carried out during cloning in order to obtain double-stranded forms of the PCR products obtained in **steps 1–5.**
7. Isolate plasmid DNA from E. coli colonies harbouring the cloned aptamer inserts, prepare plasmid mini-preps using a standard plasmid purification kit for DNA sequencing. You should receive back a single sequence or more than one with similar affinities. Generally, you will find that you get a single sequence or highly homologous sequences that span different parts of the total aptamer sequence.
8. Synthesize the aptamer sequences for further characterization or application. Modifications of the aptamer may be performed to suit particular applications (*see* **Note 6**).

4. Notes

1. Longer degenerate regions could also be used, and there have been examples of 50-75, or 100-base degenerate regions. However, 25 bases are the minimum length to provide all the possible combinations of secondary structures in the library. Additionally, using this length, it is easy to synthesize a complete library, without any possible sequence missing within the synthesis scale of a standard commercial oligonucleotide preparation.
2. This would allow at least two copies of all possible sequences to be present in the solution.
3. This amount of reagents allows the library to be amplified only once in the direction using the reverse primer, and thus generating one complementary cDNA strand for every sequence. This subsequently gets amplified 99× on the other direction, thus generating 99 single-stranded sequences that are identical to the original DNA.
4. The antibody can be adsorbed on the surface of the PCR tubes, not only in carbonate buffer, but also in PBS. If there is a problem using carbonate buffer, pH 9.6, PBS, pH 7.4 can be used instead. However, the coating efficiency was deemed to be superior with the use of carbonate buffer with a variety of IgG, IgM, and immunoglobulins purified from B-cell lymphomas, especially at lower antibody concentrations (author's unpublished observations).
5. During this process, the following takes place. At the very first denaturation step at 95°C, the coated antibody is unfolded ***(39)***, thus releasing the bound aptamers for amplification in subsequent thermocycling rounds. Since this is a unidirectional PCR, after 99 rounds of amplification (this is the limit in most PCR machines), 99 single-stranded copies of each binding sequence will be generated. In the early rounds, the binding sequences will include nonspecific binding aptamers, which are also amplified during the PCR process. However, these will be competed out by the enrichment of higher binding sequences in subsequent rounds.
6. One useful modification that may be considered is the introduction of 3′-amino modified groups (3′-end-cap). This will allow the coupling of the aptamer to activated Sepharose for the generation of chromatographic materials to use in antibody affinity purification. Moreover, a 3′-amino-modified sugar in the nucleotide will provide a primary amine for the coupling of the aptamer to a chelator carrying a radionuclide to generate a novel targeted radiopharmaceutical. Furthermore, 3′-amino modification also increases the resistance of aptamers to nucleases (the principal serum nuclease is a 3′-exonuclease) and their blood clearance properties, thus making them extremely useful for in vivo targeting applications ***(40)***.

Acknowledgments

Research in the development of the above methodologies for antibody targeting was made possible by the support of The Leverhulme Trust. Professor Mike Price is acknowledged (in memoriam) as the source of inspiration and the initiator of this work.

References

1. Terness, P., Welschof, M., Moldenhauer, G., Jung, M., Moroder, L., Kirchhoff, F., et al. (1997) Idiotypic vaccine for treatment of human B-cell lymphoma. Construction of IgG variable regions from single malignant B-cells *Hum. Immunol.* **56,** 17–27.
2. Wurflein, D., Dechant, M., Stockmeyer, B., Tutt, A. L., Hu, P. S., Repp, R., et al. (1998) Evaluating antibodies for their capacity to induce cell-mediated lysis of malignant B cells *Cancer Res.* **58,** 3051–3058.
3. Bohlen, H., Thielemanns, K., Tesch, H., Engert, A., Wolf, H. J., vanCamp, B., et al. (1996) Idiotype vaccination strategies against a murine B-cell lymphoma: dendritic cells loaded with idiotype and bispecific idiotype X anti-class II antibodies can protect against tumor growth. *Cytokines and Molecular Therapy* **2,** 231–238.
4. Hsu, F. J., Caspar, C. B., Czerwinski, D., Kwak, L. W., Liles, T. M., Syrengelas, A., et al. (1997) Tumor-specific idiotype vaccines in the treatment of patients with B-cell lymphoma—Long-term results of a clinical trial. *Blood* **89,** 3129–3135.
5. Caspar, C. B., Levy, S., and Levy, R. (1997) "Idiotype vaccines for non-Hodgkin's lymphoma induce polyclonal immune responses that cover mutated tumor idiotypes: comparison of different vaccine formulations" *Blood* **90,** 3699–3706.
6. Bianchi, A. and Massaia, M. (1997) Idiotypic vaccination in B-cell malignancies *Mol. Med. Today* **3,** 435–441.
7. Okada, C. Y., Wong, C. P., Denney, D. W., and Levy, R. (1997) TCR vaccines for active immunotherapy of T cell malignancies. *J. Immunol.* **159,** 5516–5527.
8. Schultze, J. L. (1997) Vaccination as immunotherapy for B cell lymphoma. *Hematol. Oncol.* **15,** 129–139.
9. Haimovich, J., Kukulansky, T., Weissman, B., and Hollander, N. (1999) Rejection of tumors of the B cell lineage by idiotype-vaccinated mice. *Cancer Immunol. Immunother.* **47,** 330–336.
10. Reichardt, V. L., Okada, C. Y., Liso, A., Benike, C. J., Stockerl-Goldstein, K. E., Engleman, E. G., et al. (1999) Idiotype vaccination using dendritic cells after autologous peripheral blood stem cell transplantation for multiple myeloma – A feasibility study. *Blood* **93,** 2411–2419.
11. Hefty, P. S. and Kennedy, R. C. (1999) Immunoglobulin variable regions as idiotype vaccines. *Infectuous Disease Clinics of North America* **13,** 27–39.
12. Hansen, H. J., Ong, G. L., Diril, H., Valdez, A., Roche, P. A., Griffiths, G. L., et al. (1996) Internalization and catabolism of radiolabelled antibodies to the MHC class-II invariant chain by B-cell lymphomas. *Biochem. J.* **320,** 293–300.
13. DeJonge, J., Heirman, C., DeVeerman, M., VanMeirvenne, S., Demanet, C., Brissinck, J., et al. (1997) Bispecific antibody treatment of murine B cell lymphoma. *Cancer Immunol. Immunother.* **45,** 162–165.
14. Honeychurch, J., Cruise, A., Tutt, A. L., and Glennie, M. J. (1997) Bispecific Ab therapy of B-cell lymphoma: target cell specificity of antibody derivatives appears critical in determining therapeutic outcome. *Cancer Immunol. Immunother.* **45,** 171–173.

15. DeNardo, S. J., Kroger, L. A., MacKenzie, M. R., Mirick, G. R., Shen, S., and DeNardo, G. L. (1998) Prolonged survival associated with immune response in a patient treated with Lym-1 mouse monoclonal antibody. *Cancer Biother. Radiopharm.* **13,** 1–12.
16. Kohler, S., Prietl, G., Schmolling, J., Grunn, U., Fischer, H. P., Schlebusch, H., et al. (1998) Immunotherapy of ovarian carcinoma with the monoclonal anti-idiotype antibody ACA125—Results of the Phase Ib study. *Geburtsh Frauenheilk.* **58,**180–186.
17. De Jonge, J., Heirman, C., de Veerman, M., Van Meirvenne, S., Moser, M., Leo, O., et al. (1998) In vivo retargeting of T cell effector function by recombinant bispecific single chain Fv (anti-CD3 × anti-idiotype) induces long-term survival in the murine BCL1 lymphoma model. *J. Immunol.* **161,** 1454–1461.
18. Davis, T. A., Maloney, D. G., Czerwinski, D. K., Liles, T. M., and Levy, R. (1998) Anti-idiotype antibodies can induce long-term complete remissions in non-Hodgkin's lymphoma without eradicating the malignant clone. *Blood* **92,** 1184–1190.
19. Penichet, M. L., Harvill, E. T., and Morrison, S. L. (1998) An IgG3-IL-2 fusion protein recognizing a murine B cell lymphoma exhibits effective tumor imaging and antitumor activity. *Journal of Interferon and Cytokine Res.* **18,** 597–607.
20. Liu, S. J., Sher, Y. P., Ting, C. C., Liao, K. W., Yu, C. P., and Tao, M. H. (1998) Treatment of B-cell lymphoma with chimeric IgG and single-chain Fv antibody interleukin-2 fusion proteins. *Blood* **92,** 2103–2112.
21. Maloney, D. G. and Press, O. W. (1998) Newer treatments for non-Hodgkin's lymphoma: Monoclonal antibodies. *Oncology-N .Y.* **12,** 63–76.
22. Link, B. K. and Weiner, G. L. (1998) Monoclonal antibodies in the treatment of human B-cell malignancies. *Leuk. Lymphoma* **31,** 237–249.
23. Tseng, Y. L., Hong, R. L., Tao, M. H., and Chang, F. H. (1999) Sterically stabilized anti-idiotype immunoliposomes improve the therapeutic efficacy of doxorubicin in a murine B-cell lymphoma model. *Int. J. Cancer* **80,** 723–730.
24. Reinartz, S., Boerner, H., Koehler, S., Von Ruecker, A., Schlebusch, H., and Wagner, U. (1999) Evaluation of immunological responses in patients with ovarian cancer treated with the anti-idiotype vaccine ACA125 by determination of intracellular cytokines—a preliminary report. *Hybridoma* **18,** 41–45.
25. Connolly, L., Fodey, T. L., Crooks, S.R.H., Delahaut, P., and Elliott, C. T. (2002) The production and characterisation of dinitrocarbanilide antibodies raised using antigen mimics. *Journal of Immunol. Methods* **264,** 45–51.
26. Thurin, M. and Kieber-Emmons, T. (2002) SA-Le(a) and tumor metastasis: the old prediction and recent findings. *Hybridoma and Hybridomics* **21,** 111–116.
27. Deitiker, P., Ashizawa, T., and Atassi, M. Z. (2000) Antigen mimicry in autoimmune disease. Can immune responses to microbial antigens that mimic acetylcholine receptor act as initial triggers of myasthenia gravis? *Hum. Immunol.* **61,** 255–265.
28. Hwang, B. and Lee, S. W. (2002) Improvement of RNA aptamer activity against myasthenic autoantibodies by extended sequence selection. *Biochem. Biophys. Res. Commun.* **290(2),** 656–662.

29. Smith, G. P. and Petrenko, V. A. (1997) Phage display. *Chem. Rev.* **97(2),** 391–410.
30. Ciesiolka, J., Illangasekare, M., Majerfeld, I., Nickles, T., Welch, M., Yarus, M., et al. (1996) Affinity selection-amplification from randomized ribooligonucleotide pools. *Methods in Enzymology.* **267,** 315–335.
31. Conrad, R. C., Giver, L., Tian, Y., and Ellington, A. D. (1996) In vitro selection of nucleic acid aptamers that bind proteins. *Methods Enzymol.* **267,** 336–367.
32. Bacher, J. M. and Ellington, A. D. (1998) Nucleic Acid Selection as a Tool for Drug Discovery. *Drug Discov. Today* **3(6),** 265–273.
33. Jayasena, S. D. (1999) Aptamers: An emerging class of molecules that rival antibodies in diagnostics. *Clin. Chem.* **45(9),** 1628–1650.
34. Hesselberth, J., Robertson, M. P., Jhaveri, S., and Ellington, A. D. (2000) In vitro selection of nucleic acids for diagnostic applications. *J. Biotechnol.* **74,** 15–25.
35. Bell, C., Lynam, E., Landfair, D. J., Janjic, N., and Wiles, M. E. (1999) Oligonucleotide NX1838 inhibits EGF165-mediated cellular responses in vitro. *In Vitro Cell. Dev. Biol. Anim.* **9,** 533–542.
36. Ohlmeyer, M.H.J., Swanson, R. N., Dillard, L. W., Reader, J. C., Asuline, G., Kobayashi, R., et al. (1993) Complex synthetic chemical libraries indexed with molecular tags. *Proc. Natl. Acad. Sci. USA* **90,** 10,922–10,926.
37. Marshall, K. A., Robertson, M. P., and Ellington, A. D. (1997) A biopolymer by any other name would bind as well: a comparision of the ligand-binding pockets of nucleic acids and proteins. *Curr. Biol.* **5,** 729–734.
38. Macaya, R. F., Waldron, J. A., Beutel, B. A., Gao, H., Joesten, M. E., Yang, M., et al. (1995) Structural and functional characterization of potent antothrombotic oligonucleotides possessing both quadruplex and duplex motifs. *Biochemistry* **34,** 4478–4492.
39. Wiegand, T. W., Williams, P. B., Dreskin, S. C., Jouvin, M. H., Kinet, J. P., and Tasset, D. (1996) High-affinity oligonucleotide ligands to human IgE inhibit binding to fce receptor I. *J. Immunol.* **157,** 221–230.
40. Spencer, D.I.R., Missailidis, S., Denton, G., Murray, A., Brady, K., De Matteis, C. I., et al. (1999) Structure/activity studies of the anti-MUC1 monoclonal antibody C595 and synthetic MUC1 mucin-core-related peptides and glycopeptides. *Biospectroscopy* **5,** 79–91.
41. Dougan, H., Lyster, D. M., Vo, C. V., Stafford, A., Weitz, J. I., and Hobbs, J. B. (2000) Extending the lifetime of Anticoagulant oligodeoxynucleotide Aptamers in Blood. *Nucl. Med. Biol.* **27,** 286–297.

Index

Printed in the United States
145888LV00001B/21/P

9 781588 290922